Practical Phased-Array Antennas Systems

The Artech House Antenna Library

Helmut E. Schrank, *Series Editor*

Blake, L.V., *Antennas*

Brookner, Eli, et al., *Practical Phased-Array Antenna Systems*

Corzine, Robert G., and Joseph A. Mosko, *Four-Armed Spiral Antennas*

Djordjevic, A.R., M.B. Bazdar, G.M. Bazdar, G.M. Vitosevic, T.K. Sarkar, and R.F. Harrington, *Analysis of Wire Antennas and Scatterers: Software and User's Manual*

Evans, Gary E., *Antenna Measurement Techniques*

Gupta, K.C., and A. Benalla, eds., *Microstrip Antenna Design*

Hansen, Robert C., ed., *Moment Methods in Antennas and Scattering*

Hirasawa, K., *Analysis, Design, and Measurement of Small and Low-Profile Antennas*

Hirsch, Herbert L., and Douglas C. Grove, *Practical Simulation of Radar Antennas and Radomes*

Johnson, Richard C., *Designer Notes for Microwave Antennas*

Kitsuregawa, Takashi, *Advanced Technology in Satellite Communication Antennas: Electrical and Mechanical Design*

Kumar, A., *Fixed and Mobile Terminal Antennas*

Kumar, A., and H.D. Hristov, *Microwave Cavity Antennas*

Law, Preston E., Jr., *Shipboard Antennas*

Li, S.T., J.C. Logan, J.W. Rockway, and D.W.S. Tam, *The MININEC System: Microcomputer Analysis of Wire Antennas*

Pozar, David M., *Antenna Design Using Personal Computers*

Scherer, James P., *LAARAN: Linear Antenna Array Analysis Software and User's Manual*

Scott, Craig, *Modern Methods of Reflector Antenna Analysis and Design*

Slater, D., *Near-Field Antenna Measurements*

Sletten, Carlye J., ed., Reflector and Lens Antennas, Version 2.0: Analysis and Design Using Personal Computers

Sletten, Carlyle J., ed., *Reflector and Lens Antennas: Software User's Manual and Example Book*

Weiner, M.M., et al., *Monopole Elements on Circular Ground Planes*

Wolff, Edward A., *Antenna Analysis, Second Edition*

Yamashita, E., ed., *Analysis Methods for Electromagnetic Wave Problems*

Practical Phased-Array Antenna Systems

Eli Brookner
Editor

Artech House
Boston • London

Library of Congress Cataloging-in-Publication Data

Brookner, Eli, 1931-
Practical phased-array antenna systems / Eli Brookner.
p. cm.
Includes bibliographical references.
ISBN 0-89006-563-2
1. Phased array antennas. I. Title.
TK6590.A6B76 1991 91-10512
621.384'135--dc20 CIP

British Library Cataloguing in Publication Data

Brookner, Eli
Practical phased-array antenna systems.
1. Antennas (Radio)
I. Title
621.3824

ISBN 0-89006-563-2

685 Canton Street
Norwood, MA 02062

International Standard Book Number: 0-89006-563-2
Library of Congress Catalog Card Number: 91-10512

10 9 8 7 6 5 4 3 2

To my sons Richard and Larry, who have enriched my life

TABLE OF CONTENTS

PREFACE

This book resulted from a set of notes written for the Microwave Journal Intensive Course "Practical Phased-Array Systems" first given in 1975. These notes for the course have since been used extensively by the engineers in the Antenna and Microwave Department of the Raytheon Company. They have proved invaluable both to the most senior design engineers (some of whom call it their bible) as well as the beginning engineer (a copy is given to each when they start working in the Antenna and Microwave Department). Because of this success, it was decided to publish the notes as a book available to a wider audience. Although the notes were originally written some time ago, they deal with fundamentals that make them essentially as timely now as when they were first written. A short summary of this book's contents is now given.

Lecture 1 introduces different 1- and 2-dimensional electronic scanning techniques (such as phase scanning, frequency scanning, electronic feed switching systems, and time delay steering). The advantages and disadvantages of electronic scanning are discussed. Finally, management of the parameters (beamwidth, frequency, frame-time/data-rate, pulse-repetition-rate/sampling-rate, and energy/waveform) of an electronically scanned array are discussed.

Lectures 2 and 3 give a basic introduction to linear and planar phased-scanned array antennas. In Lecture 2, the principles of phased-array steering for linear arrays are discussed, along with antenna beam pattern formation, the conditions for grating lobe prevention, antenna directivity, antenna efficiency, and different types of antenna weighting (tapering) such as Dolph-Chebyschev and Taylor. The important concept of sine space (T-space and u-v space) is discussed. Array thinning is presented in detail. Many useful design examples are presented, and extensive practical design data and curves are provided.

Lecture 3 discusses the effects of array errors and failures on phased-array sidelobes and gain. This section gives two useful heuristic, physical derivations of the effects errors on the sidelobes and gain of an array. One of these derivations is based on simple paired-echo theory. The approximate results developed are then followed by a presentation of the exact equations together with many useful examples. These results are extended to an antenna having subarrays. Techniques for reducing the degrading affects of phased quantization errors on the antenna sidelobe level and on the antenna angle measurement accuracy are discussed. Again, many useful design examples are presented. Finally, the issues of beam shape loss, beam packing loss, and optimum beam spacing are discussed.

Lectures 4 and 5 cover phased-array radiating elements. Lecture 4 describes various candidate antenna array radiating elements, such as rectangular and circular waveguides, dipoles, polyrods, and notch radiating elements. The conditions for preventing grating lobes from appearing are extended to the case of the two-dimensional array having an arbitrary element lattice arrangement. The antenna reflection coefficient is related to the individual antenna radiating element gain pattern. The antenna blindness phenomenon is presented in simple physical terms.

In Lecture 5, radiating element matching, use of a waveguide to simulate an antenna consisting of an infinite array of elements, small array design, example array elements designs, wideband array elements (notch and spiral), and the issue of designing wideband arrays are all discussed. Extensive measured data obtained with actual antenna designs are presented.

Lecture 6 discusses array feeds. Covered are the issues related to the selection of a feed such as polarization diversity requirements, number of simultaneous beams needed, number of transmitters to be used, efficiency, cost, sidelobe level requirements, and array bandwidth. Also covered is the selection of phase controlled devices--diode *versus* ferrite, low *versus* high power, and discrete (digital) *versus* continuous.

Lecture 7 gives an extremely useful detailed design example of an air defense electronically scanned phased array for detecting and tracking aircraft and missiles. The reader is taken step-by-step through the design tradeoffs, with the reasons for the final parameter selections being given.

Lecture 8 describes the AEGIS (AN/SPY-1) phased-array antenna. Array architecture, array beamforming, array components, antenna performance, and array test facilities and procedures are covered. Finally, extensive detailed measurement performance results are presented.

Lectures 9 and 10 cover the basics of limited scanned phased arrays. Covered are the issues of bandwidth, required number of phase shifters, overlapped subarray antennas, reflector and lens implementations, aperiodic arrays, element lattice selection, as well as techniques for minimizing volume, aperture, cost, and number of phase shifters needed.

Lectures 11 and 12 discuss the computer control for a multifunction radar. Covered are radar resource scheduling, track maintenance, false alarm and redundant detection, batch processing, system saturation, and program architecture.

ACKNOWLEDGMENTS

I would like to thank Howard Ellowitz (former editor of the *Microwave Journal*) and David Barton for inviting me to organize the Microwave Journal Intensive Course "Practical Phased-Array Systems" which led to this book. I am indebted to the individual contributors to the lectures for making this book possible.

I am also indebted to Fritz Steudel of the Raytheon Company for educating me on the simple physical-spatial-frequency and sum-of-random-vectors approach for dealing with array errors which is given in Section 3.1.1. I benefited from discussions on phased arrays with Fernando Beltran, James Howell, and Charles Winter, all of the Raytheon Company. I appreciated Charles Fowler's support for running the Microwave Journal Intensive Course "Practical Phased-Array Systems" at the Raytheon Company. (Fowler was a Vice President and head of the Equipment Division Laboratories of Raytheon at the time.) I am grateful to the Raytheon company management for supporting participation in the lecture series that led to this book.

I am grateful to Mark Walsh and Rebecca Warren of Artech House for seeing this book through its gestation period.

It was a great pleasure to have Natalie Shrader type the original manuscript for Lectures 2 and 3 with accuracy, enthusiasm, professionalism, and always a smile. I would also like to thank my present secretary, Ann Marie Quinn, for her time and dedication in making this book possible. Finally, I would like to thank my wife Ethel for her encouragement during the time that the lecture series was being organized and the notes for the lecture series were being prepared.

BIOGRAPHY OF DR. ELI BROOKNER

Eli Brookner received his BSEE degree from the City College of New York in 1953, and the MS and DSc degrees in Electrical Engineering from Columbia University in 1955 and 1962, respectively.

During the summer of 1952 he was with the Radar Division of the Rome Air Development Center, Rome, New York.

In 1953 he joined the Columbia University Electronics Research Laboratories, where he was engaged in noise studies and the design of long-range radars. In 1957 he was employed as Project Engineer at the Federal Scientific Corporation (now part of Nicholet Instruments, Inc.), working in the analysis and development of coherent radar processors, such as the coherent memory filter.

Since 1962, he has been at Raytheon Company, Wayland, Massachusetts, where he is presently a Consulting Scientist for the Equipment Division. He conceived and helped design the Wake Measurement Radar, the first TWT radar put into space; he has been involved in a number of space-based radar studies; and he has been a technical consultant to Cobra Dan, Seasparrow, SIRCS, Milirad, Hardsite, Cobra Judy, PAVE, PAWS, BMEWS, and marine radar programs. He has also been technical director for a number of space-based radar programs.

Dr. Brookner is a member of Eta Kappa Nu and Tau Beta Pi, an Associate Fellow of the American Institute of Aeronautics and Astronautics, and IEEE Fellow and Member of the International Union of Radar Science (URSI; commissions B and C). He has over 100 papers and talks to his credit. For one of his papers, he received the first Journal of the Franklin Institute Premium Award. He wrote a cover article for *Scientific American* entitled "Phased-Array Radars" (February 1985). He is editor and principal contributor to the books *Radar Technology* and *Aspects of Modern Radar* ,and is a lecturer on radar. He has taught his radar course to over 7000 engineers and scientists around the world.

He has organized and participated in other radar lecture series in the Boston area since 1972, with a registration for these series of approximately 2000. An IEEE video tape was made of the first of these courses. This tape was shown to engineers across the country and on Kwajalain Island.

He has personally organized monthly talks for the Boston IEEE/AESS since 1972. Over 3000 people have attended these meetings. The Central New England Chapter of the AESS won the outstanding AESS chapter of the year for 1977-1978 and 1983-1984 while he was Chairman. He was a 1983-1985 IEEE Antennas and Propagation Society Distinguished Lecturer and a 1987-1990 IEEE Aerospace and Electronics Systems Society (AESS) Distinguished Lecturer. He was a recipient of the IEEE Centennial Medal; the 1986 IEEE Region One Award for continuing educational course development and teaching; and the 1990 IEEE Education Activities Board Meritorious Achievement Award in Continuing Education Activities.

LIST OF SYMBOLS

A = design parameter related to sidelobe level for Taylor weighting; Fig. 2-40.

A = Array area; Fig. 2-52, Eq. (1-1), Sect. 7.2.9.

A,B = defined in diagram of Fig. 7-6; designated as α,β and as θ_{ox}, θ_{oy} in Lessons 2 and 3; designated as $90^\circ-\alpha$ and $90^\circ-\beta$ in Lesson 8.

A' = complement of angle A of Lesson 7, Fig. 7-6; designated as θ'_{ox} in Lesson 2, Fig. 2-32; as α in Lesson 8.

A_c = area of circumscribed hexagon; Lesson 7, Sect. 7.2.9.

A_d = cell area or area of direct lattice in Lessons 4 and 5; designated as A_e in Lessons 2 and 3; as α in Lesson 7.

A_e = area occupied by one array element (= $d_x d_y$ for rectangular lattice, see Fig. 2-34) in Lessons 2 and 3; designated as A_d in Lesson 4 and 5; as α in Lesson 7.

a_h = area of hexagon; Lesson 7, Sect. 7.2.9.

A_I = area of Inscribed hexagon; Lesson 7, Sect. 7.2.9.

A_{mn} = I_{mne} = Actual illumination including phase; Lessons 9 and 10.

A_z = azimuth angle; Lesson 7, Fig. 7.6.

A_1 = πA for Taylor weightings; Fig. 2.A.8.

a = element area in Lesson 7, Sect. 7.2.9; designated as A_e in Lessons 2 and 3, Fig. 2.34; as A_d in Lessons 4 and 5.

a = broadside dimension of waveguide; Fig. 1.5.

a = maximum acceleration of tracked target in Lesson 7, Sect. 7.3.2; a_{max} used in Fig. 7.1.

a_i = incident wave amplitude used with scatter matrix [S]; Lesson 6, Fig. 6.8.

a_{max} = target maximum acceleration; Lesson 7, Fig. 7.1.

B' = complement of angle B of Lesson 7, Fig. 7.6; designated as θ'_{oy} and $90^\circ - \beta$ in Lessons 2 and 3, Fig. 2.32; as β in Lesson 8.

B_o = angle B for zero azimuth angle; Lesson 7, Fig. 7.8.

B_{oL} = angle B_o minus ½ beamwidth in sine space; Lesson 7, Fig. 7.10.

BW = signal bandwidth — Hertz; Eq. (1.3).

b_i = Scattered wave amplitude used with Scatter Matrix [S]; Lesson 6, Fig. 6.8.

C, T = Coupling and transmission coefficients for overlapped array circuit; Lessons 9 and 10.

c = coupling coefficient (voltage); Lesson 6, Fig. 6.18.

c = velocity of light.

D = antenna directivity; Fig. 2.12.

D = antenna diameter

D_a = array directivity; Fig. 2.14.

$D.C._S$ = search duty cycle; Lesson 7, Sect. 7.5.6.

$D.C._{ST}$ = short term duty cycle; Lesson 7, Sect. 7.5.6.

$D.C._T$ = track duty cycle; Lesson 7, Sect. 7.5.6.

D_e = array directivity with errors; Fig. 3.3

D_o = directivity for d = $\lambda/2$; Fig. 2.12 and 2.13.

d = spacing between elements of array.

d = effective element spacing in direction where grating lobe first appears; Lesson 7, Sect. 7.2.9.

d_e = element radiator directivity; Fig. 2.14.

d_x,d_y = distance between row and column elements of planar array having rectangular lattice; Lessons 2, 3, 9, and 10, see Fig. 2.31.

E = Energy for each transmitted pulse; Eq. (1.2) and Sects. 7.2.9 and 7.5.2.

E =elevation angle; Lesson 7, Fig. 7.6.

$\underline{E}$ = electric field vector; Lessons 4 and 5.

E(u,v) = array radiated E field in u, v space.

E(u,v) = $E_x(u)$ $E_y(u)$ when separable.

$E(\theta,\varphi)$ = antenna array pattern; Lessons 2 and 3, see Fig. 2.11.

E_L = elevation angle minus ½ beamwidth in sine space; Lesson 7, Fig. 7.10.

E_a = array factor

$E_a(\theta,\varphi)$ = array factor, Fig. 2.2, Fig. 2.31.

$E_x(\theta,\varphi)$, $E_y(\theta,\varphi)$ = array factors for x and y axis arrays; Fig. 2.31.

$\underline{e}$ = element factor; Lessons 4 and 5; designated as $e(\theta,\phi)$ in Lesson 2; as e(u,v) in Lesson 9.

$e(\theta,\phi)$ = element pattern; Fig. 2.11; designated as e(u,v) in Lesson 9; as $\underline{e}$ in Lessons 4 and 5.

e(u,v) = array element pattern in Lesson 9.

e(u,v) = $e_x(u)$ $e_y(u)$ when separable.

$e_s(n)$ = element pattern for subarray; Lessons 9 and 10.

F = antenna focal length; Lesson 3, Fig. 3.36; designated as f in Lessons 9 and 10.

FAR = false alarm rate; Lesson 7, Fig. 7.3.

F_1 = cosine-squared-on-a-pedestal parameter; Fig. 2.20.

F_1 = preamplifier noise figure; Lesson 7, Sect. 7.2.9.

F_2 = mixer noise figure; Lesson 7, Sect 7.2.9.

f = Focal length, in Lessons 9 and 10; designated as F in Lesson 3.

f = scan rate in Eq. (1.3)

f = transmission frequency in Hz for Eq. (1.3)

f = frequency

f_{BS} = blind speed frequency; Lesson 7, Sect 7.4.

G = Antenna area gain in Lesson 1, Sect. 1.1.1; called directivity D in Lesson 2.

G = main lobe to side lobe ratio in direction of multipath reflected ray; Lesson 7, Sect. 7.3.2.

$G(\theta,\phi)$ = Antenna Gain

G_R = Antenna receive area gain in Lesson 1 and 7; Sect. 1.1.1.

G_T = antenna transmit area gain in Lessons 1 and 7; Sect 1.1.1.

$G_r = G_r(\theta,\varphi)$ = realized array gain; Lessons 2 and 4, see Fig. 2.14.

G_1 = preamplifier gain; Lesson 7, Sect. 7.2.9.

G_2 = mixer conversion gain; Lesson 7, Sect. 7.2.9

g = gravitational constant; Lesson 7, Fig. 7.1.

g_r = realized element gain; Lessons 2 and 4, see Fig. 2.14.

H = pedestal height for cosine-squared-on-a-pedestal and parabola-on-a-pedestal aperture weighting; Figs. 2.20 and 2.39.

H = side of circumscribed hexagon; Lesson 7, Sect. 7.2.9.

h = side of hexagon; Lesson 7, Sect 7.2.9.

I_i = magnitude of current in ith radiating element of linear array; Fig. 2.1.

I_m, I_n = normalized currents for mth row and nth column of separable planar array; Fig. 2.31.

I_{mn} = current in mnth radiator of planar array; Fig. 2.31.

$I(x)$ = Aperture Illumination.

$I_\ell(x)$, $I_c(r)$ = illumination function for linear and circular apertures, respectively; Fig. 2.A.5.

IF = required improved factor; Lesson 7, Fig. 7.24.

IF_{2p} = improvement factor of 2 pulse canceller; Lesson 7, Sect. 7.4.

IF_{3p} = improvement factor of 3 pulse canceller; Lesson 7, Sect. 7.4.

K' = Contains range equation constants; Eq. (1.2); K' related to α of Sect. 7.2.9, specifically: $K'=(4\pi)^2\alpha$ if same units used in Eq. (1.2) as in range equation of Sect. 7.2.9.

K_0 = parameter relating to θ_3 and θ_{30}; Fig. 2.23 and 2.40.

K_{00} = parameter relating to θ_{300} for ideal-Taylor weighting; Fig. 2.40.

k = thinning constant; Fig. 2.46.

k = propagation constant = $2\pi/\lambda$.

L = array effective length = Nd; Lessons 2 and 3, see Fig. 2.3.

L = array or aperture depth; Lessons 9 and 10.

L = loss factor to account for system losses not included in noise factor (1.2) and Sect. 7.2.9.

L_{ANT} = antenna losses; Lesson 7, Sect. 7.2.9.

L_1 = microwave losses before preamplifier; Lesson 7, Sect. 7.2.9.

L_2 = losses between mixer and i.f. stages; Lesson 7, Sect. 7.2.9.

ℓ = length of side of equilateral triangle; Lesson 7, Sect. 7.2.9.

M = total number of scans at which the target can be detected Lesson 7, Fig. 7.14.

M = related to number of elements in array, i.e., N = 2M or 2M + 1; Lessons 2 and 3, see Fig. 2.1.

m = number of independent sidelobes; Fig. 2.52.

m = integer indexing number; Fig. 2.31.

m, n = element location, Lesson 9.

N = number of elements in array; Lessons 2, 3, 9 and 10, see Fig. 2.1.

$N = N_x N_y$ = number of elements in square array; Lessons 2, 3, 9 and 10, see Fig. 2.31.

N = angle between true north and plane perpendicular to face of array; Lesson 7, Fig. 7.6.

N_E = number of elements in thinned array; Fig. 2.49.

N_s = number of elements per subarray for subarranged antenna; Lesson 3, see Fig. 3.21.

N_x, N_y = number of rows and columns, respectively, for rectangular separable array; Lessons 2, 3, 9 and 10, see Fig. 2.31.

NF = noise factor; Eq. (1.2).

$n = s/\lambda g_0$, an integer; Fig. 1.5.

$\bar{n}$ = Tapered-Taylor weighting parameter; specifically specifies that 1st $\bar{n}$ - 1 sidelobes have equal value; Fig. 2.24.

n_s = number of subarrays in subarranged antenna; Lesson 3, see Fig. 3.21.

P = number of bits in radiating element phase shifter; Lesson 3, see Fig. 3.11.

P_{av_s} = average transmitter power for search; Lesson 7, Sect. 7.5.4.

P_{av_T} = average transmitter power for track; Lesson 7, Sect. 7.5.4.

P_c = probability of failure of ith element; Lesson 3, see Fig. 3.2.

P_c, $P_{cumulative}$, P_{CUM} } cumulative probability of detection; Lesson 7, Fig. 7.14.

P_{D_i} = probability of detection on ith scan; Lesson 7, Fig. 7.14.

P_N = false alarm number in Lesson 7, Fig. 7.3; P_n used in Sect. 3.2.1, see Fig. 2.47.

P_n = false alarm number in Sect. 3.2.1, see Fig. 2.47; P_N used in Lesson 7.

P_p = peak transmitter power; Lesson 7, Sects. 7.2.9 and 7.5.6..

p, q = beam locations; Lesson 9.

R = radius of circle inscribed in hexagon; Lesson 7, Sect. 7.2.9.

R = polar coordinate.

R = range; Sect. 1.1.1, in n.m. in radar equation of Sect. 7.2.9.

R = Parameter describing A.C. Schell's antenna; Lesson 9.

$R_{A/C}$ = aircraft maximum detection range; Lesson 7, Figs. 7.2 and 7.15.

R_c = radius of circle of circumsctibed hexagon; Lesson 7, Sect. 7.2.9.

R_I = initial range where process begins for search mode, usually maximum range of target; (1.6), and Fig. 7.14, and Sect. 7.5.2.

R_I = radius of a circle of inscribed hexagon; Lesson 7, Sect 7.2.9.

R_{max} = target maximum range or maximum instrumented range; Lesson7, Fig. 7.1.

R_{min} = minimum radar range; Lesson 7, Sect. 7.2.5.

R_{MSL} = missile maximum detection range; Lesson 7, Figs. 7.2 and 7.17.

R_N = maximum range at N^{th} elevation beam position for search mode; Eq. (1.6),Fig. 7.14 and Sect. 7.5.2.

R_T = range of tracked target; Lesson 7, Sect. 7.5.2.

R_m = maximum range of clutter; Lesson 7, Sect. 7.4.

R_{max} = maximum instrumented range; Sect. 7.2.5.

R_{11}, R_{20}, R_{21} = Signal ratios of higher order mode excitation coefficients used for multi-mode limited scan techniques; Lessons 9 and 10.

r = antenna main lobe to sidelobe voltage ratio; Lessons 2 and 3 see Fig. 2.22.

$\overline{r^2}$ = main lobe to average sidelobe power ratio; Lessons 2 and 3, see Fig. 2.50.

$\overline{r_N^2}$ = main lobe to average sidelobe power ratio for pattern with errors; measurement made at position of null of error free antenna pattern; Lesson 3, see Fig. 3.5.

r_0 = radius of circular aperture.

r_p = main lobe to peak sidelobe voltage level; Lessons 2 and 3, see Fig. 2.52.

r_p(P%) = level r_p that peak voltage sidelobe is below main lobe peak P percent of time; i.e., $P_r[S.L. \leqslant r_p] = P\%$; Lessons 2 and 3, see Fig. 2.53.

S = amplitude coefficient for phase interpolation; Lessons 9 and 10.

S = scattering parameter; Lessons 4 and 5.

S_n = excitation coefficient at subarray input terminal; Lessons 9 and 10.

S/N = R.F. signal-to-noise ratio of single pulse required to obtain detection; Eq. (1.2) and Sect. 7.2.9.

S/N_I = signal-to-noise ratio at maximum range, R_I; Lesson 7, Fig. 7.14.

S/N_N = signal-to-noise ratio at range R_N; Lesson 7, Fig. 7.14.

s = path length between radiator slots for frequency scanning system; Fig. 1.5.

s = period of periodic radiator element phase error; Fig. 3.26, Lesson 3.

s = seconds.

s = parameter used in dielectric filter study; Lesson 10.

T = interpulse period; Lesson 7, Sect. 7.5.6.

T = sample period in Lesson 7, Sect. 7.3.2.

T = tilt angle of array from vertical; Lesson 7, Fig. 7.6.

T = pulse-to-pulse period; dependent on elevation angle for optimum time management; Lesson 7, Sect. 7.4.

T = coupling and transmission coefficients for overlapped array circuit; Lessons 9 and 10.

T_A = antenna temperature °Kelvin; Lesson 7, Sect. 7.2.9.

T_N = Dolph-Chebyshev polynomial of degree N; Fig. 2.29.

T_m = Parameter used in dielectric filter study; Lesson 10.

T_o = ambient temperature in 290° Kelvin; Lesson 7, Sect. 7.2.9.

T_s = system noise temperature in °Kelvin; Lesson 7, Sect. 7.2.9.

u = sin θ; Lesson 2, see Fig. 2.24.

u = sin θ cos φ; a coordinate of u, v space (T-space); Lessons 2, 3, 4, 5, 9, and 10, see Figs. 2.32 and 4.16. Designated as α in Lesson 7; Fig. 7.6.

$\dot{u}$ = normalized sin θ; Lesson 2, see Fig. 2.24 and 2.A.6.

$$\left.\begin{array}{l} u_p = u_o + p/d_x \\ v_q = v_o + q/d_y \end{array}\right\} \text{grating lobe points; Lessons 4 and 9.}$$

u,v space = T-space = direction cosine space = grating lobe space; Lessons 2, 3, 4, 5, 9 and 10. Designated as α, β space in Lesson 7.

$v_q = v_o + q/d_y$ = grating lobe points; Lessons 4 and 9.

V_{BS} = blind speed velocity; Lesson 7, Sect. 7.4.

V_{max} = target maximum velocity; Lesson 7, Fig. 7.1.

$V_{closing}$ = radial closing velocity of target; Lesson 7, Fig. 7.16.

v = sin θ sin ϕ; a coordinate of u, v space (T-space); Lessons 2, 3, 4, 5, 9, and 10, see Figs. 2.32 and 4.16. Designated as β in Lesson 7; Fig. 7.6.

x = x-axis of x, y, z rectangular coordinate system; all Lessons, See Fig. 2.31.

x, y, z = cartesian coordinates in space; all Lessons, see Fig. 2.31.

x'', y'', z'' = cartesian coordinates in plane of array; Lesson 7, Fig. 7.6.

Y = admittance; Lessons 4 and 5.

y = y-axis of x, y, z rectangular coordinate system; all Lessons, see Fig. 2.31.

z = z-axis of x, y, z rectangular coordinate system; all Lessons, see Lessons, See Fig. 2.31.

Lecture 1

Electronic Scanning Radar Systems (ESRS) Design & Architecture

Peter J. Kahrilas

Raytheon Company Bedford, Massachusetts 01730

1.1 Modern Radar Requirements. In the last decade the growing need for radars to provide increased range coverage, target identification, trajectory determination, faster data rates (short reaction times) and accommodate increased target densities; to cope with high speed jet propelled aircraft, guided missiles, ballistic missiles and satellites has imposed new technological and system design requirements.

1.1.1 Increased Range Coverage. Rewriting the energy range equation with

$$G_T\, G_R = G^2 = \left[\frac{4\pi A}{\lambda^2}\right]^2 \qquad (1.1)$$

obtains

$$R^4 = \frac{K'\, E\, \sigma\, A^2}{S/N\; L\; NF\; \lambda^2} \qquad (1.2)$$

that clearly indicates the controllable design parameters affecting range performance.

Design for:

1. Decreased noise power density-

   Use low noise preamplifiers such as: transistors, TDAs (tunnel diode amplifiers), TWTs (traveling wave tubes), uncooled or cooled parametric amplifiers and masers.

   The preamplifier selected in a particular radar application is dependent on performance specifications, "balanced design" based on system trade-off studies, and cost effectiveness.

2. High transmission energy-

   Obtained with high peak power and/or long time duration coded waveforms.

   The maximum peak power obtainable with a single tube is approaching a practical design barrier of approximately 50 megawatts using klystrons.

   The maximum pulse time duration is restrained by transmitter limitations, minimum range performance specifications, and compatibility of time-bandwidth factors with range resolution (signal bandwidth) requirements.

3. Large antenna aperture/high antenna gain-

   Practical construction considerations and manufacturing tolerances limit attainable beamwidths to approximately 0.1°. Corresponding directive antenna gain is approximately 58 dB depending on the illumination taper. At an operating frequency of 1.0 GHz the aperture is in the order of 600 feet diameter.

4. Minimized system losses-

   Locate transmitter and receiver preamplifier as close as feasible to antenna feed to reduce waveguide and cable losses.

   Obtain maximum signal processing efficiency by using matched filter waveform processing for: single tone pulses, frequency coded pulses, pulse bursts, MTI and pulse doppler waveforms. Employ correlator devices like FFT – Fast Fourier Transform or Time Compressed Spectrum Analyzer for good signal processing efficiency.

   Employ pulse-to-pulse frequency diversity to minimize post detection integration losses in applications requiring high probabilities of detection on fluctuating targets.

   Pulse-to-pulse frequency diversity also reduces power-aperture requirements in cases requiring high cumulative probabilities of detection with only 3 or 4 looks on fluctuating targets.

1.1.2 Increased Resolution. High range and angular resolution provide improved accuracy since accuracy is inversely proportional to signal bandwidth and directly proportional to beamwidth. Higher accuracy leads directly to improved trajectory determination for locating launch and impact points of missile targets and satellite orbit determination. High resolution enables target identification by providing length measurement and identification of the major scatterers of various body shapes. Another desirable feature of high resolution is the improved capability of detecting targets in clutter. Less clutter is intercepted by the radar since smaller space cell increments are interrogated on each look. If the resolution is equivalent to signal bandwidths of several hundred megahertz, the radar can "see" in-between the clutter patches and detect targets without MTI or pulse doppler processing.

Design for:

1. Wide instantaneous bandwidth systems in the order of several hundred megahertzs-

   All components, antenna, transmitter, receiver, signal processor, display, data processor and even interconnecting cables and connectors must be capable of wide bandwidth operation. Extreme care is required to avoid or eliminate multiple reflections, differential delays and transients.

2. Narrow pencil beams-

   Narrow antenna beams require low antenna sidelobes for proper radar operation.

   Near-in targets and/or targets with large radar cross sections (50 to 60 dBm^2) located at sidelobe angular positions will provide echo signals that exceed the detection threshold and be reported erroneously to be in the angular position corresponding to the main beam direction unless the one-way sidelobes are significantly lower than 30 dB.

   To obtain very low sidelobes in the order of 40 dB requires precise antenna construction to tight tolerances using a high illumination taper.

1.1.3 Shorter Reaction Time. Very short reaction time or almost instantaneous radar reporting of target presence is required in many modern radar applications. Typical examples are: to allow time to catalog satellites while they are rapidly passing through the radar coverage volume; to discriminate ICBM clouds and take appropriate counteraction against the reentry vehicles prior to impact; and to identify a large number of fast moving targets and to assign and guide missiles to destroy them in complex weapon systems.

Design for:

1. High data rate-

   The radar system must be capable of both gathering and processing large amounts of data very rapidly.

2. Multifunction-

   The radar system must be integrated to perform all the required function of search, multitarget acquisition, multitarget track, multimissile track, multimissile guidance and kill evaluation simultaneouslv or on a time shared basis.

3. Automatic Operation-

   Automatic computer programming of multimode radar operation and processing of data is a necessity in a high data rate system providing very short reaction time.

The logical goal in radar development is to seek a system, design satisfying all the radar requirements in one integrated system.

ESRs have the inherent capability to meet this goal.

1.2 Electronic Scanning Radar Systems (ESRS) Characteristics.

1.2.1 Definition. Electronic scanning is defined as a method of positioning an electromagnetic beam in space by electronic means with the antenna aperture remaining fixed and no mechanical mechanism involved in the scanning process.

For comparison, the definition of mechanical scanning is a method of positioning an electromagnetic beam in space by a mechanical rotation or angular positioning of the radiating aperture of the antenna system.

There are also compromise methods. The most common employs one dimension electronic scanning in elevation and mechanical rotation in azimuth.

1.2.2 Electronic Scanning Techniques. In the basic configuration of the radar (Fig 1-1) the exciter/transmitter generates the signal waveform to be transmitted. The phasing system subdivides the transmitter power into a number, N, of equal (or weighted) packets, and appropriately phase codes each packet to form a wavefront (beam) in the designated direction. The N signals are amplified in the transmit modules (omitted in some designs), and radiated by the antenna elements. The N echo return signals from the target are preamplified in the receive modules (omitted in some designs) and coherently added together in the phasing system to obtain a composite target echo signal that is processed by a conventional receiving system before being displayed. The computer programs the operation of the radar and processes the data.

The essential differences among the various methods of implementing an electronic scanning radar (ESR) occurs in the phasing system.

The selection of a phasing system, or electronic scanning technique, to meet a potential user's performance requirements is one of the major design decisions to be made by the system architect.

The basic electronic scanning techniques are:

Phase,

Real-Time,

Frequency,

Electronic Feed Switching.

Phase. In a phase scanning system, electronic beam position scanning is accomplished by varying the phase of the separate elements of a linear array in a predetermined manner. This operation is illustrated in Fig. 1-2. A number of radiating elements are equally spaced along a straight line. Feeding these elements with in-phase currents of equal or tapered amplitude results in formation of a broadside beam. As indicated in Fig. 1-2, the direction in which the beam is formed is an arc sine function of the interelement differential phase shift. A unique beam position is obtained for each progressive phase shift setting in the network feeding the array.

Continuous phase scanning is accomplished by exciting all radiating elements coherently with a small frequency increment between elements as indicated in Fig. 1-3. The beam is continuously scanned at a rate equal to the frequency differential f.

Real Time. Real time scanning is a special case of phase scanning. Instead of employing phase delay to form a wavefront (beam) in a designated direction, time delay is used.

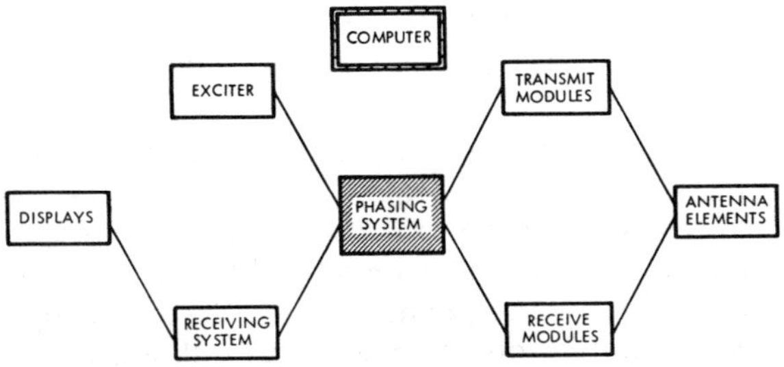

Fig 1-1 Basic elements of an electronic scan radar. ESR radar systems differ primarily in the phasing systems used to steer the antenna beam. The computer connects to all other parts of the system

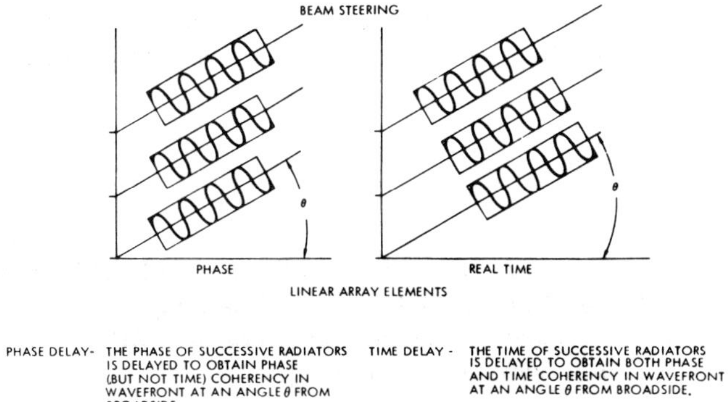

Fig 1-4 Comparison of phase and real time scanning

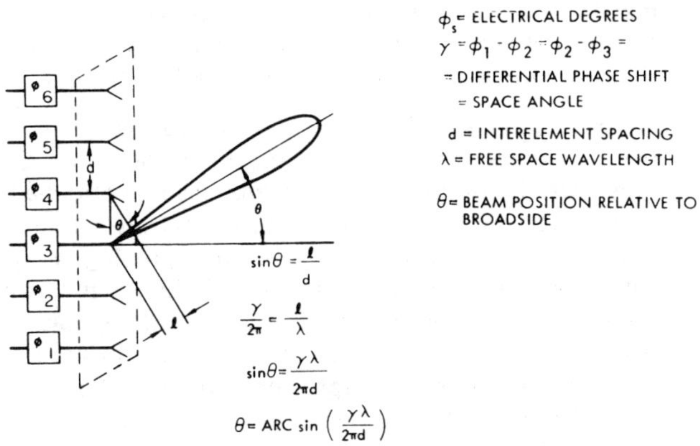

Fig 1-2 Basic operation of a phase scanning system. Note that beam direction depends on phase shift between elements

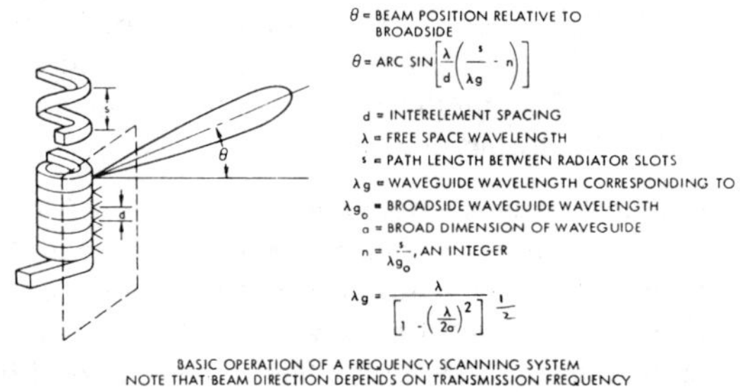

Fig 1-5 Basic operation of a frequency scanning system. Note that beam direction depends on transmission frequency

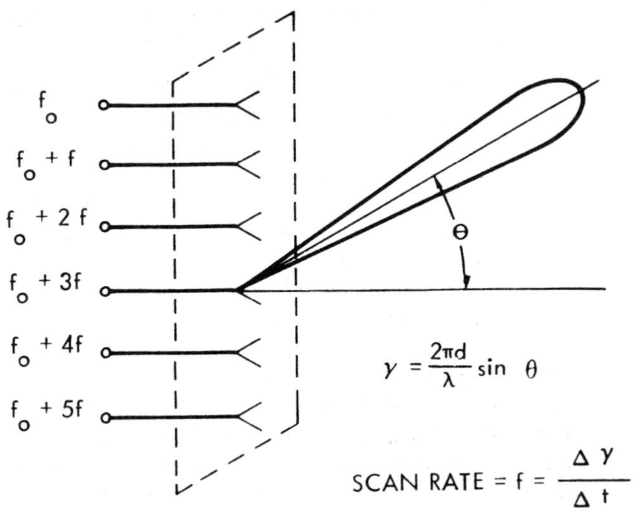

Fig 1-3 Continuous phase scanning

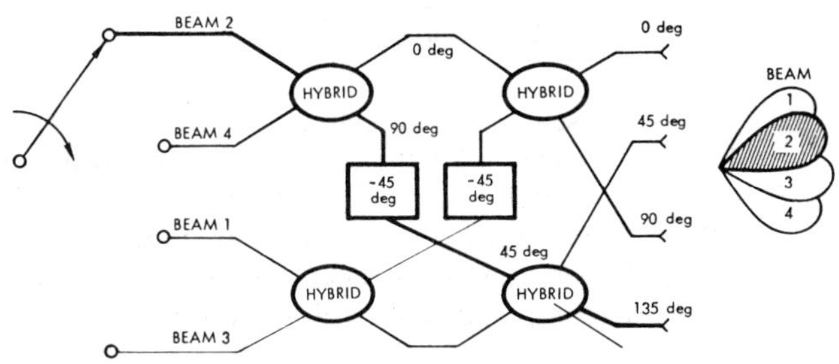

Fig 1-6 Hybrid/Butler matrix
Basic operation of an electronic feed switching system

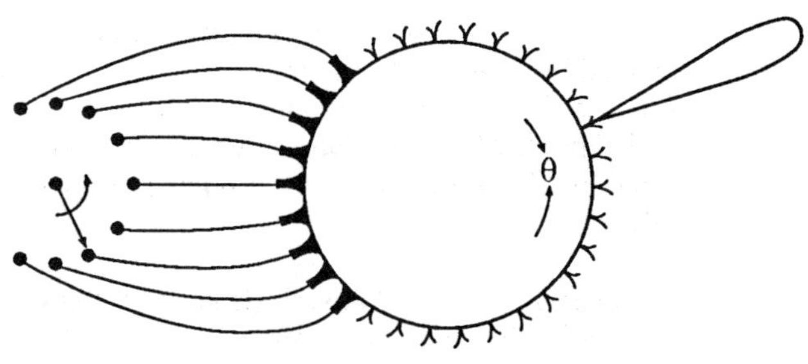

Fig 1-7 Luneburg Lens
Basic operation of an alternative electronic feed switching system

The difference is demonstrated in Fig. 1-4. Time delay is required in ESR systems having wide (instantaneous) signal bandwidths. The relationship among signal bandwidth, beamwidth, transmission frequency and maximum scan angle is derived in Appendix 1-1 and given in equation (1.3).

$$BW \leqslant \frac{\beta f}{128 \sin \theta_m} \tag{1.3}$$

Phase scanning may be used if equation (1.3) is satisfied. For example if $\beta = 1^\circ$, f = 3 GHz and $\theta_m = 60^\circ$, the maximum signal bandwidth that can be used with phase scanning is approximately 27 MHz.

In some applications a compromise design is feasible employing a combination of phase and time delay devices.

Time delay devices consist of different lengths of cable being switched in-and-out in binary increments in a similar manner to the operation of diode phase shifters.

Frequency. In a frequency scanning system, frequency is employed to control the interelement differential phase shift so that each frequency corresponds to a unique beam position. This method is completely passive and reciprocal and is illustrated in Fig. 1-5.

Practical designs employ approximately a 6% bandwidth to scan 90 degrees. Either helical (shown in Fig. 1-5) or serpentine feeds are employed in frequency scanning radars (FSRs). The helical feed can operate over a wider frequency range than the serpentine feed. The relationship among signal bandwidth, beamwidth, frequency scanning bandwidth and maximum scan angle is derived in Appendix 1-2 and given in equation (1.4).

$$BW \leqslant \frac{\beta \Delta f}{256 \sin \theta_m} \tag{1.4}$$

Frequency scanning may be used if equation (1.4) is satisfied. For example, for the same values used in the phase scanning example, i.e., $\beta = 1^\circ$, f = 3 GHz, $\therefore \Delta f$ = 180 MHz (.06 x 3 GHz) and $\theta_m = 60^\circ$, the maximum signal bandwidth that can be used iwth frequency scanning is approximately 0.8 MHz (compared to 27 MHz for phase scanning).

Electronic Feed Switching. Two simple examples of electronic feed switching are illustrated in Fig. 1-6 & 1-7. In both techniques multiple beams can be formed simultaneously. The hybrid/Butler matrix shown in Fig. 1-6 is a special case of phase scanning whereas the Luneburg lens shown in Fig. 1-7 is a real time (wide signal bandwidth) scanning technique.

The basic concept of the hybrid phasing matrix can be explained by referring to the four-element array shown in Fig. 1-6. The system is completely passive and reciprocal, and consists of an interconnection of hybrids, and either fixed phasing elements or differing lengths of strip transmission line.

The beams are formed in accordance with the interferometer equation for a linear array:

$$\Delta\phi = \frac{2\pi d}{\lambda} \sin \theta \tag{1.5}$$

where

$\Delta\phi$ = interelement phase shift
d = interelement spacing
λ = free-space wavelength
θ = beam position relative to broadside

An input connection, transmit or receive, to any of the four input ports results in the formation of the corresponding antenna beam. Formation of beam 2 (in Fig. 1-6) requires a differential phase shift of 45 degrees between elements. It corresponds to excitation of the antenna elements in the sequence zero degrees, 45 degrees, 90 degrees, and 135 degrees. The intrinsic phase difference of 90 degrees between the outputs of a 3 -dB directional coupler (hybrid) provides the proper phase differential for alternate antenna elements. The fixed phase shifter converts the 90 degree difference between the outputs of the input hybrids to a 45 degree differential in order to excite the two groups of alternate antenna elements properly.

The hybrid phasing matrix for a more complex antenna also consists of a structure of hybrids and either fixed phase shifters or differing lengths of strip transmission line. The matrix for an antenna with N elements has N input ports to establish N beams in space; in all instances, it is a repetitive structure of hybrids and fixed phasing elements.

An orthogonal set of (sinx)/x beams is formed crossing at the 4.0 dB point with 13.5 dB maximum sidelobes. Cosine taper with 23 dB maximum sidelobes can be obtained by coherently combining adjacent input ports. Coherent combination of three input ports to the hybrid phasing matrix results in a cosine-squared taper with 32 dB maximum sidelobes. In theory,the coverage sector should approach 180 degrees as the number of radiating elements increases. In practice, the coverage sector is limited by the beamwidth of the individual radiating elements. In most applications, strip transmission line construction can be used.

A geodesic lens may also be used for electronic feed switching. One example is the Luneburg lens shown in Fig. 1-7. A geodesic line is the shortest superficial line between two points on any surface. According to Fermat's principle of least time, this is the path along which energy will propagate between these two points. A geodesic lens is a surface that satisfies these conditions; such a surface is designed by methods of differential geometry. A geodesic lens is inherently a wide bandwidth device.

In both types of electronic feed switching individual transmitters can be added at each aperture in order to provide power amplification and permit low-power electrical switching. In Fig. 1-6 the switch determines which part of the phasing matrix receives or transmits the energy. This selection, in turn, determines the direction of the beam formation. Similarly, the Luneburg lens shown in Fig. 1-7 is employed to obtain the correct phasing to propagate a wave front perpendicular to the axis of the feed being activated.

1.2.3 Implementation. How the basic four techniques are employed, individually or in combination, to implement ESR system designs varies widely. Typical one and two dimensional configurations are described in this section employing line, cylindrical, planar and spherical arrays to form electronically scanned beams. Conformal arrays (where the surface of the array conforms to the shape of, say an aircraft or a missile) are special cases of the above.

One Dimension. A simplified one-dimensional phase scanning radar system is indicated in Fig. 1-8. A linear array is employed to electronically scan the beam in elevation. Three methods are shown to focus the beam in the unscanned azimuth dimension. The parabolic reflector is the first method used. This was replaced by the more efficient planar or billboard array which consists of horizontal rows of elements fed in-phase by strip line corporate feeds. The differential phase shift between elevation elements is obtained in this example by the of 3-bit (180°, 90°, 45°) diode phase shifters. The phase setting determining the elevation beam direction is controlled by the scan programmer. Azimuth coverage is obtained by mechanical rotation of the antenna. The programer settings can be made during the retrace time between transmitted pulses so that no additional time is required to slew the antenna beam to any desired position within the elevation scan sector. The programer is either a digital-analog or completely digital device. Conventional transmitters and receivers are used in this system.

A basic one-dimensional frequency scanning system is shown in Fig. 1-9. Frequency scanning was actually the first technique to be fully implemented in an operational radar. Two methods of forming the pencil beam to obtain one-dimensional frequency scanning are shown in Fig. 1-9. As the transmitted frequency is varied in accordance with the scan program, the interelement phasing will vary and radar beam steerage will result. The radiating elements may be either primary slots or dipole radiators. Aperture amplitude taper (by control of each radiating element) may be employed to reduce sidelobe levels. Azimuth coverage is obtained by mechanical rotation of the antenna.

The receiver system is quite conventional. The exciter-transmitter is usually of a MOPA (Master Oscillator Power Amplifier) design using crystal oscillators to obtain the required frequency stability. The frequency is changed in discrete steps to obtain the various beam positions by appropriate gating, mixing, filtering and frequency multiplication of the energy obtained from a bank of crystal oscillators. The scan programer provides the required gates to accomplish the beam slewing during the retrace time.

A one dimensional electronic feed switching scanning system can be designed in the form of circular cylinder consisting of a stack of Luneburg disks. Vertical corporate feeds supply in-phase energy to the disks. Azimuth electronic scanning is obtained by electronically switching vertical feeds that are sequentially transmitting and receiving the energy. Multiple beams in azimuth can be obtained simultaneously by using multiple transmitters and receivers feeding multiple vertical feeds.

Another form of a one-dimensional electronic feed switching scanning system can be designed using a hybrid phasing matrix in the vertical plane in place of the vertical corporate feed and phase shifters shown in Fig. 1-8a and 1-8b. In this design elevation electronic scanning is obtainable by electronically switching input terminals. As in the previous case, multiple inputs can be fed simultaneously to obtain a stacked beam radar.

Two Dimensions. Various combinations and modifications of the one-dimensional electronic scanning techniques can be made to obtain two-dimensional scanning. In applications where wide bandwidth waveforms are required, real-time scanning techniques are employed such as various types of geodesic lens or replacement of phase delays by time delays. Frequency scanning is not applicable to wide bandwidth applications.

One means of obtaining two-dimensional electronic scanning is illustrated in the functional diagram of Fig. 1-10 where frequency scanning is employed in both dimensions. The elevation frequency determines the elevation beam position indirectly by supplying the differential phase shift, ϕ_E, by means of a delay network. The elevation scanning frequency is eliminated by the mixing process so that only the phase shift appears at the antenna, superimposed on the azimuth scanning frequency, f_A. The antenna consists of a stack of helical or folded (serpentine) waveguides in which the azimuth frequency determines the beam position in the azimuth plane and the differential phase shift between the folded waveguides determines the beam position in the elevation plane. The azimuth and elevation frequencies can be programmed independently to a first order approximation to obtain beam position slewing in either or both directions. The exterior of the antenna has the appearance of a flat billboard. This type of system can scan 90 degrees in elevation and 90 degrees in azimuth without any mechanical motion of the antenna. Four of these systems will provide complete hemispherical coverage with an electronically scanned beam.

Another simple variation employs frequency scanning in elevation and phase scanning in azimuth (Fig. 1-11). The antenna is again a planar array consisting of columns of folded waveguides, each one having an individual power amplifier. The frequency is changed in increments to obtain step scanning in elevation. The azimuth scan is obtained by using a phase shifter at the input of each transmitter to control the differential phase shift between the serpentine columns.

The next logical step is to employ a lens having a cellular structure with a ferrite or diode digital r-f phase shifter in each cell (Fig. 1-12). By combining the phase increment associated with the vertical and horizontal beam positioning in each module, two-dimensional scanning is obtained. This is known as an optical feed or space feed array.

Two-dimensional phase scanning is also obtained by employing a two-dimensional corporate feed input to a cellular structure with a ferrite or diode r-f phase shifter in each cell as shown in Fig. 1-13. This corporate or confined feed array can be transformed to a wide signal bandwidth design by supplementing the phase shifters with real time delay devices.

Two-dimensional electronic feed switching is obtained by cross stacking one-dimensional strip line hybrid phasing matrices. In Fig. 1-14, a 16 element array is illustrated employing eight strip line hybrid phasing matrices. Sixteen orthogonal beams are formed that may be scanned sequentially or interrogated simultaneously. Individual transmitters can be added at each element for power amplification and permit switching at low power.

An alternate two-dimensional electronic feed switching system is obtained by employing a spherical Luneburg lens as a phasing system to switch-scan electronically over a complete hemisphere (Fig. 1-15). It is inherently a wide signal bandwidth device. This device can also form orthogonal beams that may be scanned sequentially or interrogated simultaneously. Individual transmitters can be added at each element for power amplification and permit switch-

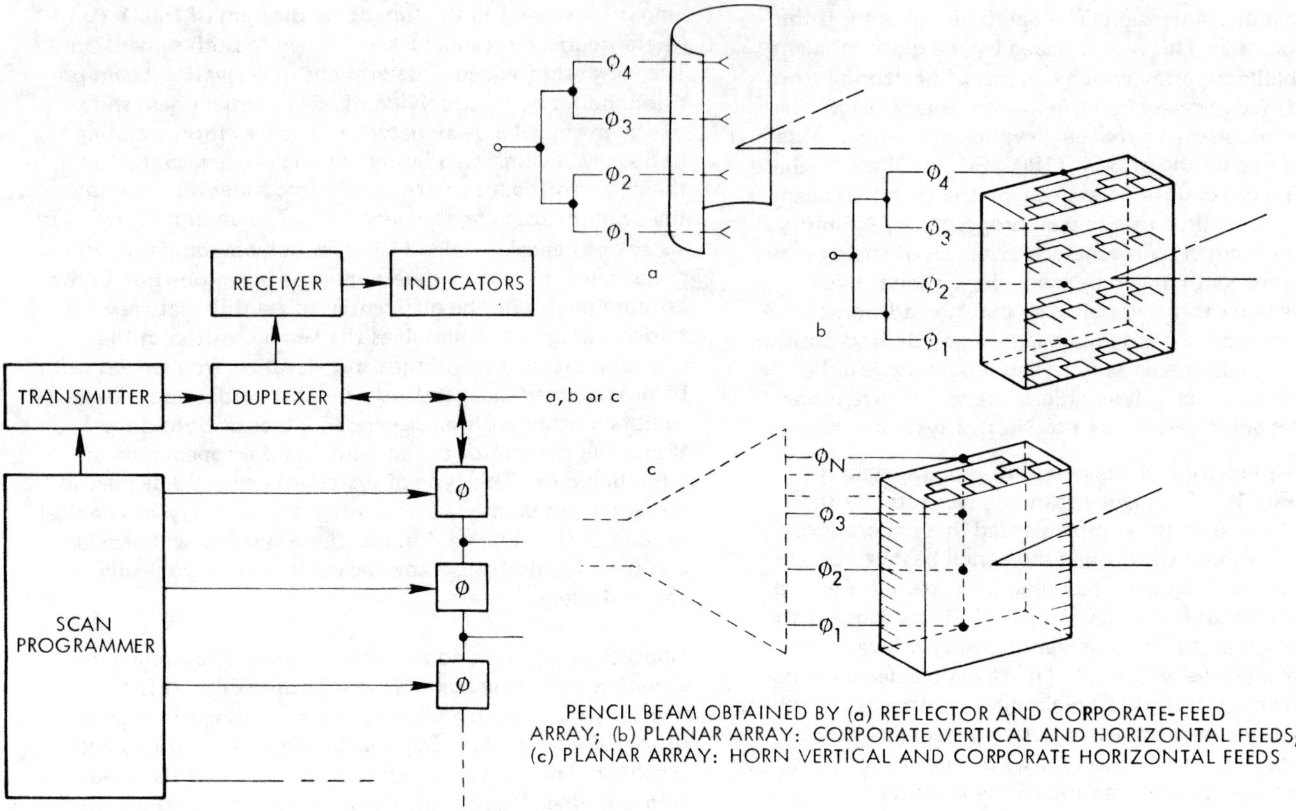

Fig 1-8 One-dimensional phase scanning radar system

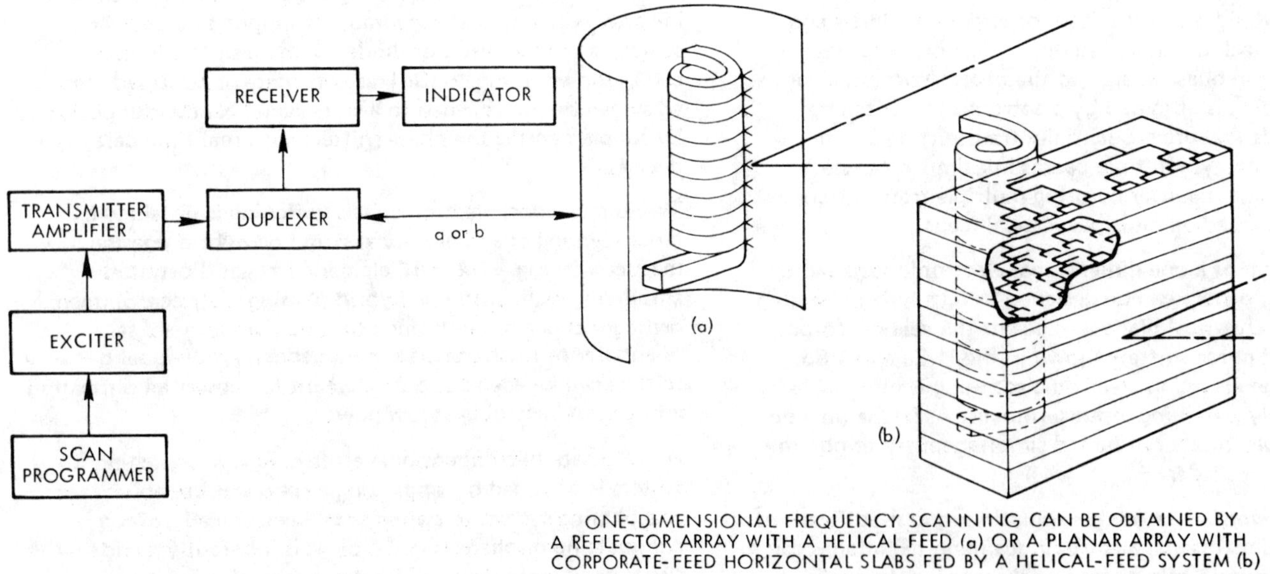

Fig 1-9 One-dimensional frequency scanning radar system

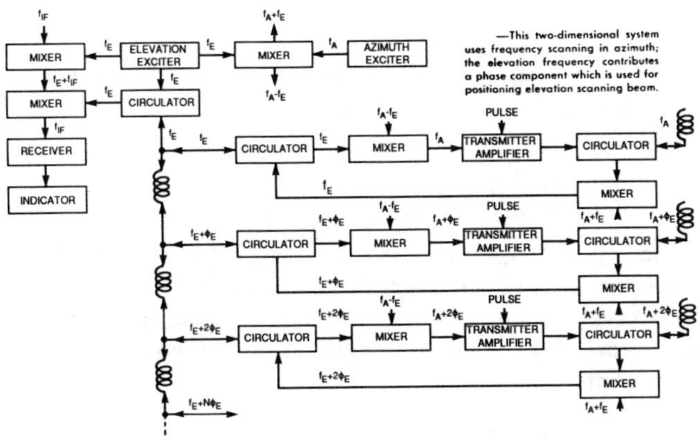

Fig 1-10 Two-dimensional frequency scanning system

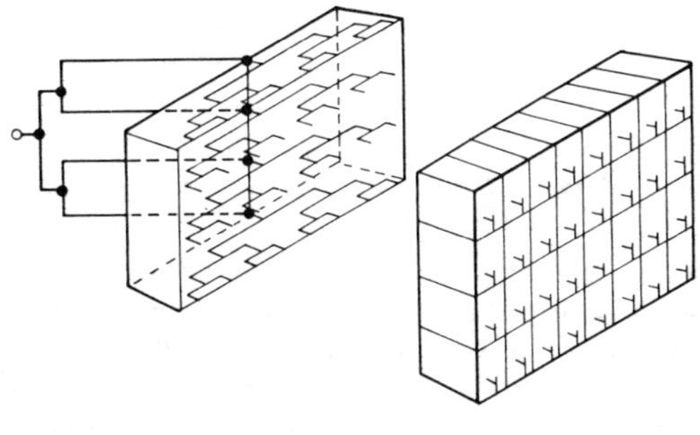

Fig 1-13 Two-dimensional confined feed array

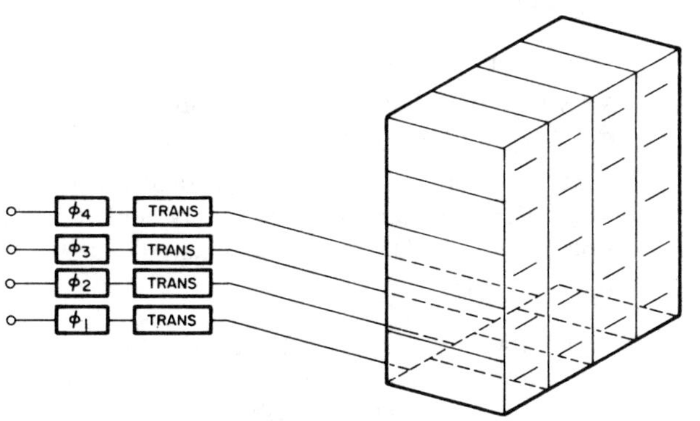

the antenna is a planar array having columns of serpentine waveguides, each one having an individual power amplifier. The frequency is changed in increments to obtain step scanning in elevation. The azimuth scan is obtained by using a phase shifter at the input of each transmitter to control the differential phase shift between the serpentine columns.

Fig 1-11 Two-dimensional frequency/phase scanning system

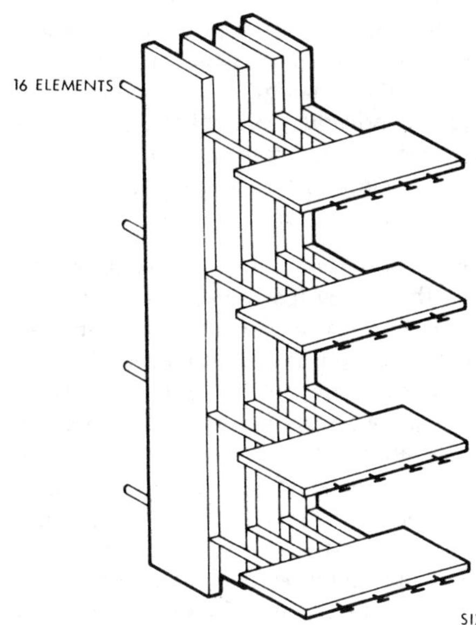

Fig 1-14 Two-dimensional hybrid phasing matrix array

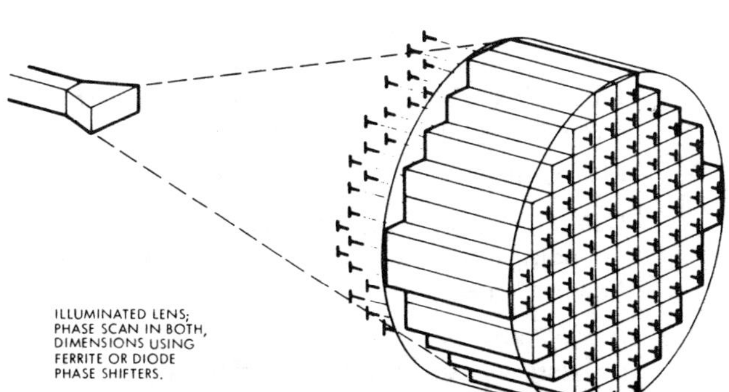

Fig 1-12 Two-dimensional space feed array

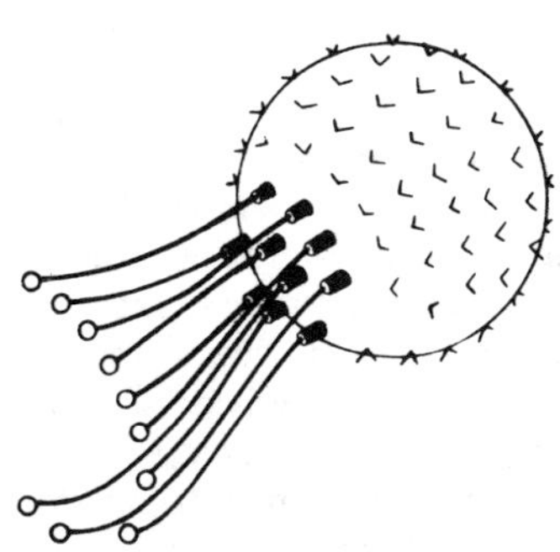

Fig 1-15 Two-dimensional spherical luneburg lens array

ing at low power.

The most popular array for ESR is the planar configuration. In its construction and in ease of electrical design, it has distinct advantages over curved shapes. Lower sidelobes can be achieved because of its superior aperture illumination. Maintenance of a flat structure is simpler than a curved one. Its chief restriction is a limit of 60-75 degrees scan angle off broadside so that 3 or 4 faces are required for full hemispherical coverage. On the other hand, curved arrays, such as a sphere, can use only a portion of the radiating elements for any one beam direction because of the inability of elements on the back side to radiate in the forward direction. In addition, planar arrays make more efficient use of the full aperture.

1.2.4 Illumination Tapering. Phased arrays have made it possible to obtain better tapering of the illumination pattern to minimize sidelobes. In the past designers had only the primary horn pattern to work with and usually settled for a cosine on a pedestal taper providing 23 dB maximum sidelobe levels. Other tapers such as the Taylor with designs providing 40 dB maximum sidelobe levels are applicable to arrays. By individual control of a large number of radiating elements on rigid stationary array structures it is possible to design for and achieve antenna patterns with 40 dB maximum sidelobe levels.

Two basic methods are amplitude taper and space taper. In the amplitude taper, the energy fed to each element is diminished toward the edges of the antenna to approximate the desired pattern. The objection to this method is that more than half the elements are working at low power. Thus cost and weight penalties are incurred for equipment not fully used. Phased arrays, uniquely within the class of radar antennas, permit space taper. In the space tapering method, every active element carries full power, but instead of active elements being located at every radiating point, some are passive. Thus the active elements are less dense toward the outer edge of the antenna. The net effect on the antenna pattern is the same, but for identically sized arrays the space-tapered array has fewer active elements than the amplitude-tapered array. Space tapering is also called density tapering. The basic technique is known as thinning. Approximately 50 percent thinning is employed in thinned array designs where half the antenna elements are active and half are passive. The gain for a thinned array is proportionately less than for a "fully-filled" array. Fifty percent thinning reduces the aperture gain by 3 dB.

In some applications it is advantageous to employ a combination of amplitude and space taper. A space feed array is especially suitable for this combination since the feed horn can be amplitude-tapered and the array can be space-tapered.

Independent taper design is sometimes employed on transmit patterns and receive patterns to obtain maximum energy on target on transmission with relatively poor sidelobe levels and minimum sidelobe levels on reception. It is especially desirable to have independent transmit and receive modules on each radiating element because the transmitters are operated at saturation levels and consequently even illumination or space tapering is employed for transmission. The receive taper is usually very high to obtain very low sidelobe levels to compensate for the poor transmit sidelobe levels. It is preferable to use amplitude taper on reception. It is accomplished with attenuators without loss of sensitivity provided there is adequate preamplification.

In monopulse operation, it is additionally desirable to independently taper the sum and difference illumination in order to minimize the sidelobe level response on both.

1.2.5 Advantages of ESRSs. It was previously indicated that modern radar requirements are:

Increased range coverage
Increased resolution
Shorter reaction time,

and the logical goal in radar development is to seek a system design to satisfy all the radar requirements in one integrated system. Among the advantages of ESRS is the inherent capability to meet this goal and satisfy the above requirements.

1. Simultaneous Multimode Operation-

   Search
   Multitarget Acquisition
   Multitarget Track
   Multimissile Guidance
   Automatic Reacquisition of Lost Targets
   Kill Evaluation
   Passive Detection

2. Large Power-Aperture Products-

   Very high power transmission capability (transmitter per radiating element).

   Large fixed antenna arrays - high angular resolution (large structures can be rigidly constructed since no mechanical motion is required).

3. Fast Reaction Time-

   High data rates are obtainable because of multiple simultaneous beam capability and inertialess slewing of antenna beam position.

4. Automatic Operation-

   ESRSs are particularly suitable for computer control and data processing thus providing rapid, efficient, and accurate radar programming and data processing. (Also contributes to fast reaction time).

5. Electronic Stabilization-

   Space coordinates are obtained referenced to a stable element thus eliminating the requirement for a heavy stable pedestal for moving platform application.

6. Reliability-

   Array component redundancy (active elements, phase shifters, transmit/receive modules) minimizes probability of catastrophic failure and provides graceful degradation.

7. Maintainability-

   Computer controlled self-test diagnostic routines pin-point trouble areas rapidly and minimize downtime for repairs.

   Array component redundancy reduces logistics demands by requiring fewer different types of spare components to be delivered or stored.

8. Growth Potential-

In multiple simultaneous beams ESRS designs - transmitters and receivers can be added at a later date to increase data rate and obtain faster reaction time. Larger computers or additions to computers (Central Processing Units, CPUs, and/or memories) can be added to handle the growth system.

In ESRS designs - larger computers or additions to computers can be added to increase multifunction and/or multitarget track capabilities.

In ESRS thinned array designs - passive elements can be replaced with active elements (and transmit/receive modules).

1.2.6 Disadvantages of ESRSs.

1. Complexity

   The increased capabilities and flexibility of ESRS also leads to more complex designs.

2. Cost

   More complex designs are more expensive.

3. Development of New Components

   Arrays
   Transmitters
   Special Purpose Computers--
   beam steering units
   Phasing Systems --
   frequency scan feeds, phase shifters, real time delay devices, geodesic lens
   Monopulse Receivers--
   with single hit normalization or computer normalization
   Consoles and Displays--
   different and additional controls and displays

The decision of whether to use conventional radars or ESRS is different in each specific application and is dependent on a cost-effective analysis and risk analysis. If both approaches achieve the same performance, which one is least expensive and more likely to be delivered on schedule with minimum risk is the system to be selected. In some applications weight and size are also important considerations.

The more that is required of a radar system (multifunction, high data rate, handle large number of targets) the more likely it is that ESRS designs will be recommended over conventional radars.

1.2.7 Parameter Management.

Definition. Parameter management is a method of obtaining maximum performance for minimum power-aperture and design complexity in an ESRS by judicious manipulation of radar parameters.

The flexibility of a computer controlled ESRS provides the opportunity to obtain considerable economy in energy/power and time with no sacrifice in performance.

Parameter management is accomplished by varying beamwidth, frequency, frame time/data rate, p.r.f./sampling rate, and energy/waveform as functions of operational mode. range and angular beam position.

The results are summarized in a parameter management chart. (Refer to Parameter Management Chart - Case Study).

Beamwidth. Beam generation is one of the most important design considerations in an ESRS. It affects the basic architecture of the entire radar system.

All the options discussed are influenced by system requirements and employ different tradeoffs of power-aperture, energy, time, waveforms, resolution/bandwidth, accuracy, efficiency, design complexity and cost-effectiveness.

Design options are presented in decreasing order of complexity and performance.

1. Transmit/Receive - simultaneous multiple pencil beams.
   Each beam at a different frequency.
   Maximum power - minimum time.
   In limit, the search frame time requires one echo interval and simultaneous multiple target tracks requires a second echo interval.

   On a time-shared basis, the search frame time is equal to two echo intervals and the sampling interval on all tracked targets is also equal to two echo intervals.
   This design requires the highest power, multiple receivers and signal processors (one monopulse set for each search beam), multiple A/D converters, complex phasing networks and displays, and high speed operation and data processing (large capacity computer).
2. Transmit - single broadened (defocused) beam.
   Receive - simultaneous multiple full gain narrow beams to fill volume covered by transmit beam.

   This implementation eliminates complex transmit networks but requires high transmit energy (peak power x pulse duration) and low receive sidelobes.
   Narrow full gain transmit/receive beams are used in track.
3. Transmit - "machine gun" - a group of full gain narrow beams in rapid sucession, possibly employing a different frequency for each beam.
   Receive - simultaneous multiple full gain narrow beams corresponding to transmit beam positions.
   This design reduces the energy per transmit pulse but requires adequate minimum range and a high duty cycle transmitter capability.
4. Transmit/Receive - single broadened (defocused) beam to save search time and narrow (focused) beam in track mode to improve accuracy and conserve energy and power.
   May also vary beamwidth as function of elevation angle and/or as function of echo sweep time (to maintain a constant space cell).

Time. Beamwidth management discussed in the previous section also included certain forms of time management. Additional forms of time management are discussed in this section.

1. Search — In many radar applications, the objects to be detected have a maximum altitude at which they can operate. Time management can be em-

ployed to decrease the interpulse period as the beam is stepped in elevation so that the echo time interval corresponds to the maximum range for each elevation beam position. For a specified maximum altitude limit, the maximum range is determined from a radar refraction chart using the the lower 3 dB angular position of the elevation beam.

Another form of time management is to employ different frame times for different portions of the search volume. One method of implementation is to sweep azimuth fences at low elevation angles twice as often as the upper ones. Possibly alternately employ a long range and short range echo interval (and waveform) consistent with the requirements of detecting near-in pop-up targets.

2. Track — Time management of the sampling interval of tracked targets is a function of target position (R, α, β), track initiation and target dynamics.

Target Position — A set of recursive equations generate a sampled data servo loop to the computer. The equations are identical for range and the two angular coordinates and differ only in the values of the smoothing constants. The sampling rate and bandwidth (smoothing constants) can be decreased continuously or in discrete steps as the range decreases and/or the angular rates increase.

Track Initiation — During track initiation, a larger bandwidth and faster sampling rate are required than under steady state conditions to maintain tenacity of track. The bandwidth (smoothing constants) and sampling rate can be changed in preprogrammed discrete steps as a function of time after target acquisition or changed continuously as a function of the error signal in each coordinate.

Target Dynamics — To improve accuracy and maintain dynamic performance adaptive smoothing rules can be incorporated to vary smoothing constants and sampling rate to match the current target acceleration.

Power/energy and sampling rate requirements can be reduced by the use of higher type or Kalman filter tracking loops in exchange for increased complexity.

Energy. Energy (peak power x pulse duration) management is achieved by adjustment of peak power and/or pulse duration. Since transmitters are mostly peak power limited, peak power control is obtained by turning off the output tube and allowing the driver stage power to pass through the output tube to the antenna. The discrete step power reduction obtained is equal to the gain of the output tube plus the passive losses through the output tube. This method is feasible only when the output tube (cross-field amplifier for example) provides passive coupling when it is turned off. Pulse duration adjustment is employed more extensively because it provides a wider range of control for energy management.

1. Search — As previously mentioned in the discussion of time management, there are many radar applications where the objects to be detected have a maximum altitude operating limit. Energy management can be employed to reduce the transmitted energy as the beam is stepped in elevation consistent with the decrease in maximum range as determined from a radar refraction chart using the lower 3 dB angular position of the elevation beam.

The energy reduction is determined by the following formula:

$$\epsilon_{R_{N_{dB}}} = \left(\frac{R_I}{R_N}\right)^4_{dB} + \Delta_{ATD_{dB}} \tag{1.6}$$

where:

ϵ_{R_N} = energy reduction at Nth elevation beam position

R_I = initial range where process begins

R_N = maximum range at Nth elevation beam position

Δ_{ATD} = Decrease in atmospheric loss between ranges R_I and R_N.

Energy management is normally determined relative to the lowest elevation beam position. An important exception may arise where a specified cumulative probability of detection is required for targets of different cross sections (aircraft and missiles) at different detection ranges. The energy management in these applications may be delayed several elevation beam positions to be consistent with the requirements.

2. Track — Energy management can be employed to reduce the transmitted energy as the range of each tracked target decreases relative to maximum target acquisition range. The formula given for search is still applicable with the following redefinition of R_I and R_N:

R_I = maximum target acquisition range

R_N = next predicted position of tracked target

The Δ_{ATD} term may be dropped for simpler implementation.

3. MTI - Moving Target Indicator — In many applications the specified detection range of the radar is much longer than the maximum clutter range. Sea clutter only extends several miles and distributed land clutter extends to a maximum range of approximately 100 km (54 nm). Mountainous regions may contribute discrete clutter returns at longer ranges. Consequently, energy management is used to reduce appreciably the MTI energy requirements in long range radar system applications. The formula given for search applies for MTI with the following redefinition of R_I and R_N.

R_I= maximum target detection range

R_N = maximum clutter range

4. Beam Scanning — In determining the power/energy requirements of an ESRS, a loss is included in the radar range equation to account for the drop-off in antenna gain as the beam is scanned off broadside. This round trip loss is approximated by $\sec^3 \xi$ where ξ is the angle off broadside. The round trip aperture loss varies as $\sec^2 \xi$. The secant function is raised to the third power to include the round trip effect of antenna element mismatch. For

a scan angle off broadside of 60° this loss is approximately 9 dB. It is therefore highly desirable to employ energy management as a function of scan angle to conserve energy in all modes of operation. The energy can be decreased continuously as ξ decreases or what is more practical it can be controlled in discrete binary increments. A simple computer algorithm computes ξ for every scan angle and determines the energy management for search, track and MTI operational modes.

Waveform. Waveform management is feasible with either digital or analog matched filter implementation. In the most common case of peak power limited transmitters, the pulse duration is varied as a linear function of energy. Energy management is achieved by pulse length management. Both the pulse-length and the time-bandwidth product are adjusted in binary increments to maintain the same compressed pulse width. Keeping the same range resolution (bandwidth) for all operational modes has several advantages. No time is lost nor implementation required to perform a range cell search in going from normal or MTI search to track acquisition. There is less likelihood of multiple target ambiguity in handover from one operational sector to another. The receiver bandwidth is the same for all waveforms and operational modes thus simplifying implementation and threshold level setting. Binary incremental control of the pulse length and time-bandwidth product simplifies the design of the signal processor. In an analog or digital signal processor the various waveforms are obtained by switching in-or-out binary segments of the matched filter network. To avoid excessive design complexity the number of waveforms employed should be limited to approximately ten.

An exception arises to maintaining the same resolution for all waveforms when the energy required for the search function is much higher than required for the track function and the time-bandwidth product is prohibitively large to permit a practical matched-filter processor design. Another exception may arise when a larger bandwidth is required for MTI operation to obtain the required clutter reduction.

Waveform management is employed in search, track and MTI operational modes as a function of range and beam scanning angle in a similar manner to energy management. A lower limit of time-bandwidth product is approximately 25. Below this value it becomes too difficult to achieve acceptable residue levels. In many MTI applications it is possible and desirable to use a single tone waveform with a pulse width equal to the compressed pulse width of the coded waveforms.

Power. In an ESRS design employing parameter management, the average power transmitted in each operational mode is determined by dividing the total energy transmitted by the corresponding total time interval. To assist in this calculation it is recommended a parameter management chart be prepared similar to the one in the Case Study. An appropriate reduction factor is required in the calculation to take into account the energy conserved as a function of beam scanning angle (refer to Case Study).

Time and Power Budgets. The time and power budgets of an ESRS indicate the effect of time sharing of operational modes on time and power allocations. They are usually expressed as fractions or percentages of real time and total average power (refer to Case Study). These budgets are used to size a system and to select suitable subsystem components. Their evaluation can critically affect the system architecture. In the extreme, they can influence the selection between a single beam and multibeam ESR design. For best results, proper management of time and power budgets must be individually tailored for each ESRS application to obtain a balanced design.

Duty Cycle. The ratio peak power to average power or pulse duration to interpulse period is called the duty cycle of the radar. In an ESRS that has multimode operation on a time-shared basis there are different duty cycles for each operational mode. In addition, the short-term duty cycle is different from the long-term duty cycle. Short-term may range from one or two milliseconds (one interpulse period) to approximately 100 milliseconds (computer major cycle time).

Duty cycle management is required to meet the short-term and long-term duty cycle transmitter specifications. It is accomplished by determining all the short-term duty cycles for the various operational modes and then developing an algorithm to obtain an appropriate mixture of search, track and MTI scans per computer major cycle consistent with the transmitter specifications. A second algorithm is developed to continually monitor the long and short term duty cycles of the transmitter and not allow them to exceed acceptable design limits.

Frequency. Pulse-to-pulse frequency agility conserves energy/power in applications where a high probability of detection is desired on non-coherently integrated target returns for Rayleigh and One Dominant plus Rayleigh target models. Swerling cases II and IV involve pulse-to-pulse decorrelation of echo amplitudes (achieved by use of pulse-to-pulse frequency agility) and require less signal-to-noise ratio for a specified n and probability of detection ($\geq$0.5) than Swerling cases I and III that involve scan-to-scan decorrelation but are relatively constant (slow fluctuating) pulse-to-pulse. Frequency agility also economizes on energy/power required for cumulative probability of detection if the individual probabilities of detection are significantly larger than 0.34.

Another use of frequency management, discussed previously, is for beam position identification when multiple receive beams are used.

In addition to the other advantages of frequency management, frequency agility provides an effective electronic counter-countermeasure capability.

1.2.8 ESRS Integration, Test and Evaluation. The following are the major items in the development of an ESRS integration, test and evaluation program.

Program Plan/Schedule

A program plan is required to indicate all the detailed steps beginning with initial design, and concluding with equipment demonstration and evaluation. All tasks, when they occur, the length of time they take, and how they tie in with other tasks should be indicated in this overall plan.

Major Milestones

Successful technical direction of an ESRS development program requires a reliable and timely method of measuring progress. Milestones should be established for each performance objective. They should be selected to represent major technical performance achievements during the progress of the program. The milestones are selected to be measurable by either tests or simulation and provide a time-phased build-up of confidence that the primary objective be achieved.

Hardware/Software Integration and Test Plan

The approach followed to prepare a suitable test plan is to consider first a suitable logical build-up of the operational system and then to modify this system build-up plan on the basis of scheduling and hardware availability constraints. The basis of the test plan is that development and testing of hardware and software proceed independently up to a certain point in order to achieve a high degree of confidence before integration.

The individual subsystems are tested individually and then delivered to a test site for marriage tests until the full system is integrated.

Software Build Concept

A system software "build" is used to add new hardware and software components to an already defined system in order to incorporate a new function within the system. The software in a system build is made of operational modules, some of which have been tested in a previous build and some of which are introduced to the system for the first time.

The following disciplined procedure is employed for successful implementation of the "build concept":

1. Define and freeze system requirements.
2. Write detailed software performance requirements for each build.
3. Develop the building blocks and test software.
4. Checkout the software builds in computer simulated models.
5. Checkout the software builds in hardware simulated models.
6. Checkout the software build in hardware against live targets - fully operational.
7. Store a duplicate tape of each build.

Exercise tight change control. Do not update tape until the changes made on the software build, used for integration, testing and debugging of the system, have been carefully scrutinized by a control committee to assure optimum software implementation and more importantly that the changes have not degraded subsystem and/or system performance.

Performance Demonstration and Evaluation

The performance demonstration begins simply by demonstrating capability of forming suitable antenna patterns and pointing the beam to designated coordinates. Followed by sensitivity measurements and demonstration of other range equation parameters (antenna gain, noise figure, loss budgets) including detection of stationary targets. There should be a gradual demonstration build up to include the track mode and guidance mode (if any) operations. Then multimode operations on one or two targets in a clear environment. Finally culminating in the demonstration of all operating modes with full target density and target dynamics and adverse environment. The various steps must, of course, be consistent with the detailed test plans and milestones. In many cases a full environment of a large number of targets may have to be simulated by the use of a radar target simulator to supplement the limited number of live targets available for test purposes. It is usually desirable to have precision reference radars available at the test site for comparative performance evaluation. In all cases prior agreement should be reached with the customer on what constitutes satisfactory performance demonstration and evaluation - as delineated in the test plans.

1.2.9 ESR System Design Methodology

1. Transform threat model/performance requirements to radar parameters.

   Number and types of targets and expected direction of threat transform to volumetric coverage boundaries and target cross section (and fluctuation model).

   Target density into resolution (beamwidth and waveform).

   Target dynamics and counter weapon reaction time into maximum range and search frame time.

   Environment into MTI or pulse doppler.

2. Performance Analysis

   Search

   Determine coverage pattern in sine space coordinates .

   Solve Range Equation - apply parameter management.

   MTI design.

   Track

   Error analysis, (including tracking loop) - apply parameter management.

3. Select ESR System Design

   1, 2, 3 or 4 antenna arrays

   Single transmit/receive array or separate transmit and separate receive arrays.

   Single time-shared beam or multiple simultaneous beams (cluster or independent)

   Single transmitter or multiple transmitters (tube or solid state)

   Scanning technique - frequency, phase, frequency/phase or real time

   Optical or confined feed

   Beam splitting - amplitude weighting, amplitude monopulse, phase monopulse or amplitude/phase monopulse.

   Illuminating taper - identical taper for transmit and receive antenna patterns

   – different taper for transmit and receive patterns

   or

   – different Δ and Σ taper for both cases

4. Reiterate performance analysis
5. Modify ESR system design--if necessary
6. Tabulate ESRS parameters
7. Tabulate error and loss budgets

8. Tabulate power and time budgets

The specific details of the system design methodology are best understood by studying the illustrative example of the case study in the seventh lecture.

1.3 Problem for Lecture #7.

Case Study - Design of Air Defense ESRS

Threat Model/Performance Requirements. The task is to design a radar to cope with the following threat model:

100 aircraft - primary mission: air superiority

or

50 aircraft - primary mission: air-to-ground attack

plus

50 missiles - aircraft weaponry.

Characteristics:

	cross section	V_{max}	a_{max}	R_{max}
aircraft	1.0 m^2	MACH 2	3g	200 nmi
missiles	0.05 m^2	MACH 1	10g	100 nmi

Tracking coverage: ±45° Azimuth
+60° Elevation

Maximum altitude: 50,000 feet

Minimum range: 10 nmi

Flight formation: 30 meter radial separation

Terrain: Forest (heavily wooded hills)

Reaction time: 8 minutes to send own aircraft off the the ground

2 minutes to fire counter weapons of 70 nmi range

Accuracy: 4 millisines (1σ)
9 meters (1σ)

Based on single hit requirements to determine trajectory for counter weapons fire control against aircraft and missiles.

APPENDIX 1-1

Build-Up Time in a Phase Scan Array.

D = Length of Array Aperture

θ_m = Maximum Scan Angle

x = Space length of travel of field from the last aperture element relative to first element to form wavefront at maximum scan angle θ.

τ = Equivalent Signal Pulse Length

t = Time for signal to travel through distance x

β = Beamwidth in Degrees

f = Transmission Frequency in Hertz

$BW = \frac{1}{\tau}$ = Signal Bandwidth

$$x = D \sin \theta_m$$

$$t = \frac{x}{c} = \frac{D \sin \theta_m}{c}$$

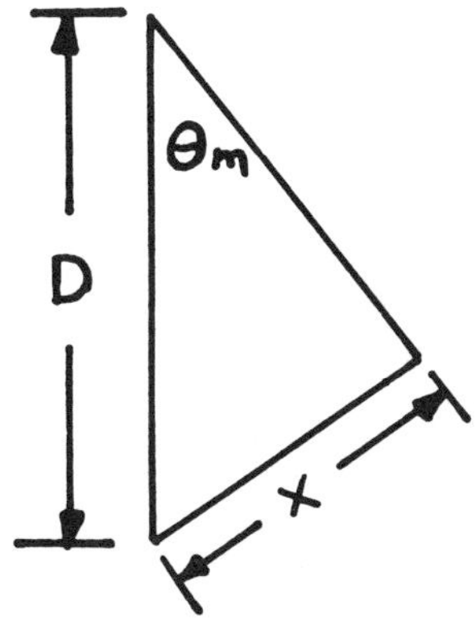

Since

$$c = f\lambda$$

and

$$\beta \approx \frac{64\lambda}{D}$$

$$t = \frac{64 \sin \theta_m}{\beta f}$$

Practical design is to allow ½ pulse width overlap between transmissions from first and last radiating elements.

$$\therefore t \leqslant \frac{\tau}{2}$$

or

$$\tau \geqslant \frac{128 \sin \theta_m}{\beta f}$$

or

$$BW \leqslant \frac{\beta f}{128 \sin \theta_m}$$

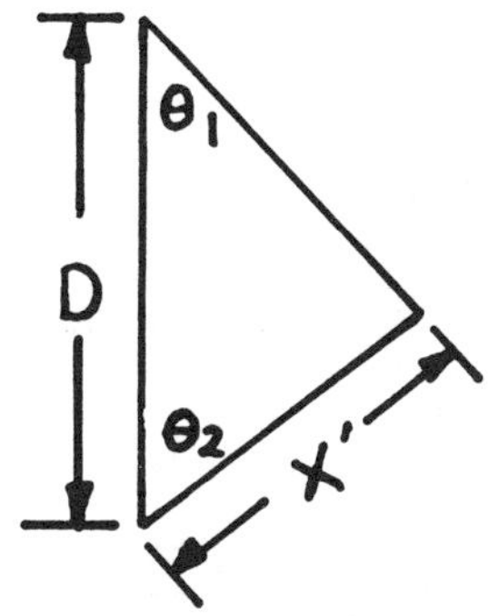

APPENDIX 1.2

D = Length of Array Aperture

θ_1 = Maximum Scan Angle in One Direction

θ_2 = Maximum Scan Angle in Opposite Direction

x' = Space length of travel of field from the last aperture element relative to first element to form wavefront at maximum scan angles.

L = Helical or Serpentine Length

f_1 = Lowest Frequency used for Scanning

f_2 = Highest Frequency used for Scanning

Δf = $f_2 - f_1$ Bandwidth required for Frequency Scanning

τ = Equivalent Signal Pulse Length

x = $x' + L$

t = Time to travel through distance x

β = Beamwidth in Degrees

$BW = \frac{1}{\tau}$ = Signal Bandwidth

c = $f\lambda$

Phase Control for Scanning

$$\theta_1 + \theta_2 = \frac{L}{\Delta\lambda} = \frac{\Delta f L}{c} = \frac{D \sin \theta_1}{\lambda_1} + \frac{D \sin \theta_2}{\lambda_2}$$

or

$$L = \frac{cD}{\Delta f}\left(\frac{\sin \theta_1}{\lambda_1} + \frac{\sin \theta_2}{\lambda_2}\right) = \frac{D}{\Delta f}\,(f_1 \sin \theta_1 + f_2 \sin \theta_2)$$

Total Length of Travel

$$x = L + x' = L + D \sin\theta_1 =$$

$$\frac{D}{\Delta f}\left(f_1 \sin\theta_1 + \Delta f \sin\theta_1 + f_2 \sin\theta_2\right)$$

$$x = \frac{D}{\Delta f}\left[\underbrace{(f_1 + \Delta f)}_{f_2} \sin\theta_1 + f_2 \sin\theta_2\right] =$$

$$\frac{Df_2}{\Delta f}\left(\sin\theta_1 + \sin\theta_2\right)$$

since

$$D = \frac{64\lambda}{\beta} \text{ and } t = \frac{x}{c}, \quad \lambda = \lambda_2$$

$$t = \frac{64}{\beta\Delta f}\left(\sin\theta_1 + \sin\theta_2\right)$$

For

$$t \leqslant \frac{\tau}{2}$$

$$\tau \geqslant \frac{128}{\beta\Delta f}\left(\sin\theta_1 + \sin\theta_2\right)$$

or

$$BW \leqslant \frac{\beta\Delta f}{128\left(\sin\theta_1 + \sin\theta_2\right)} = \frac{\beta\Delta f}{256 \sin\theta_m} \quad \text{for } \theta_m =$$

$$\theta_1 = \theta_2$$

List of Symbols Used

Symbol	Meaning	Section
G_T	Transmit area gain	1.1.1
G_R	Receive area gain	
G	Area gain	
R	Range	
K'	Contains range equations constants	
E	Energy for each transmitted pulse	
σ	radar cross section of target	
A	antenna area	
S/N	R.F. signal-to-noise ratio of single pulse required to obtain detection	
L	loss factor to account for system losses	
NF	noise factor	
λ	transmission wavelength	
ϕ_s	electrical degrees	1.2.2
γ	differential phase shift	
d	interelement spacing	
λ	free space wavelength	
θ	beam position relative to broadside	
f	scan rate	
BW	signal bandwidth - Hertz	
β	beamwidth - degrees	
θ_m	maximum scan angle - degrees	
f	transmission frequency - Hertz	
Δf	bandwidth required for frequency scanning - Hertz	
s	path length between radiator slots	
λg	waveguide wavelength corresponding to λ	
λg_o	broadside waveguide wavelength	
a	broadside dimension of waveguide	
n	$s/\lambda g_o$, an integer	
$\Delta\phi$	interelement phase shift	
d	interelement spacing	
λ	free space wavelength	
θ	beam position relative to broadside	
ϵ_{R_N}	energy reduction at N^{th} elevation beam position	1.2.7
R_I	initial range where process begins	
R_N	maximum range at N^{th} elevation beam position	
Δ_{ATD}	decrease in atmospheric loss between ranges R_I and R_N	

Antenna Array Fundamentals Part 1

Dr. Eli Brookner

Raytheon Company, Wayland, MA

2.1 Linear Arrays. The fundamentals of linear arrays are first developed. The results obtained for linear arrays are then applied to planar arrays.

2.1.1 Uniform Weighting.

A. Antenna Pattern. Fig. 2-1 shows a simple horizontal linear array of N identical radiating elements spaced a distance d apart. Assume that the antenna radiating elements have unity gain in all directions, that is they are isotropic radiators. Consider the phases of the signal along the plane A-A whose normal makes an angle θ_o with respect to the array normal. The signals are assumed to be radiated in phase from all the radiating elements (the phase shifters shown being set for zero phase shift, i.e., $\alpha = 0$). Along plane A-A the phase of the signal from radiator number 1 lags that from radiator number 0 by a phase angle $(2\pi/\lambda)d \sin\theta_o$ because the signal from radiator number 1 has to travel a distance $d \sin\theta_o$ more than from radiator number 0 to reach plane A-A. Similarly, the phase of the signal from the ith radiator lags that from the 0th radiator by an angle $i(2\pi/\lambda)d\sin\theta_o$. Thus, the signals from the various radiators are not in-phase along the plane A-A when they are radiated in phase from the various radiating elements.

Assume a frequency f for the transmitted signals from the radiating elements. Then the signals along the plane A-A due to the radiators will add as

$$E_a(\theta) = \sum_{i=-M}^{M} I_n\, e^{j2\pi f\left(t - \frac{id\sin\theta}{c}\right)} \tag{2-1}$$

$$= \sum_{i=-M}^{M} I_n\, e^{j\left[2\pi ft - \frac{i2\pi fd\sin\theta}{c}\right]}$$

$$= \sum_{i=-M}^{M} I_n\, e^{j\left[2\pi ft - \frac{i2\pi d\sin\theta}{\lambda}\right]} .$$

If the phase shifters were adjusted so that the ith radiating element has a phase advance which equaled the phase lag of the signal from the ith radiator along the plane A-A, then all the signals would be in phase along this plane. Consequently at a point far enough away from the antenna (i.e., in the far field of the antenna) along the line which makes an angle θ_o with respect to the array normal, the signals from the N radiators will add up in phase. The phase advance needed by the ith array element is $i\alpha$ where

$$\alpha = \frac{2\pi}{\lambda} d \sin\theta_o . \tag{2-2}$$

If the currents in all the radiating elements are of equal magnitude, that is, $I_i = 1$ for $i = -M$ to M, corresponding to uniform weighting, then the normalized total received signal for a distant receiver located in the direction θ_o is given by

$$E_a(\theta_o) = \sum_{i=M}^{M} I_i = 2M + 1 = N. \tag{2-3}$$

Fig. 2-2 gives the antenna field $E_a(\theta)$ for angles other than θ_o: Eq. (2-6) of Fig. 2-2*. This antenna pattern obtained for an array of isotropic radiating elements is referred to as the antenna array factor. The antenna array factor gives the antenna pattern (that is, the transfer characteristic of the antenna versus θ) for when operating either as a transmitter or as a receiver.

A plot of the antenna array factor for the case of uniform illumination (that is, $I_i = 1$, for $i = -M$ to M) with $\alpha = 0°$ (so that $\theta_o = 0°$) is given in Fig. 2-3. As expected one has in-phase addition at $\theta_o = 0$. The main lobe 3 dB width is $0.886\,\lambda/L$ radians where L is the effective array length defined as Nd and not as $(N-1)d$; see (2-8). Fig. 2-3 indicates there can be more than one angle for which the in-phase addition of the signals from the radiators occurs. The other angles at which such in-phase addition occurs give rise to what are called grating lobes. Generally one desires only one angle at which in-phase addition occurs. Otherwise, when the antenna is used as a transmitter a large amount of energy is wasted as radiation into these grating lobe angle directions which are not of interest. Furthermore, when operating as a receiver these grating lobes give rise to ambiguities in the angle domain. Also interference can result from the reception of signals from grating lobe directions, which are directions other than where the target of interest is located.

For broadside conditions, i.e., $\theta_o = 0°$, the location of the first grating lobe is at $\theta = \theta_1$ given by

$$\sin\theta_1 = \frac{\lambda}{d} . \tag{2-9}$$

* Equations not given in the text appear in the figures.

LINEAR ARRAYS

1. UNIFORM ARRAYS

2. TAPERED ARRAYS

LINEAR ARRAY

FOR $\alpha = \frac{2\pi}{\lambda} d \sin\theta_0$ AND $I_i = 1$ (2.2)

$$E_a(\theta_0) = \sum_{i=-M}^{M} I_i = 2M+1 = N \quad (2.3)$$

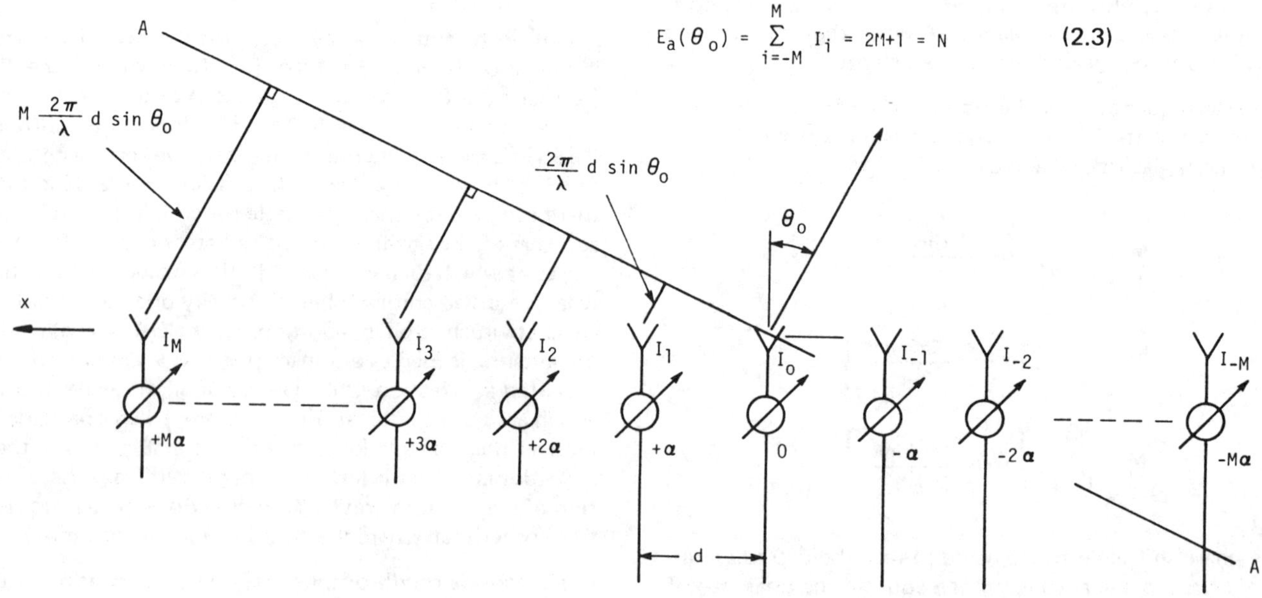

Fig. 2.1 Linear array peak far field signal.

ANTENNA ARRAY FACTOR

ISOTROPIC RADIATOR ELEMENTS HAVING UNIFORM EXCITATION ASSUMED

$$E_a(\theta) = \sum_{n=-M}^{M} I_n \, e^{\, j\left[n\alpha - n \frac{2\pi}{\lambda} d \sin\theta\right]} \quad (2.4)$$

LET $I_n = 1$, $n = -M$ TO M

$$\alpha = \frac{2\pi}{\lambda} d \sin\theta_0 \quad (2.5)$$

THEN

$$E_a(\theta) = \frac{\sin\left[(2M+1)\frac{\pi d}{\lambda}(\sin\theta - \sin\theta_0)\right]}{\sin\left[\frac{\pi d}{\lambda}(\sin\theta - \sin\theta_0)\right]} \quad (2.6)$$

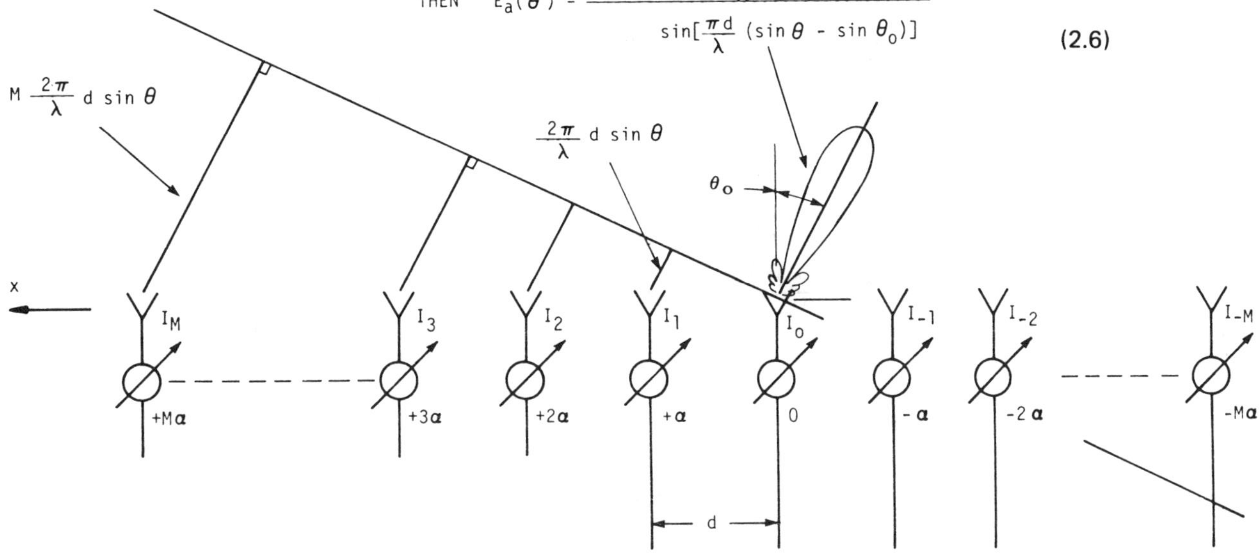

Fig. 2.2 Antenna array factor; isotropic radiator elements having uniform excitation.

ARRAY FACTOR FOR N=10, θ_0=0° (BROADSIDE CASE)

WHERE $N = 2M+1$

$$E_a(\theta) = \frac{\sin\left[N \frac{\pi d}{\lambda}(\sin\theta)\right]}{\left[\sin \frac{\pi d}{\lambda}(\sin\theta)\right]} \quad (2.7)$$

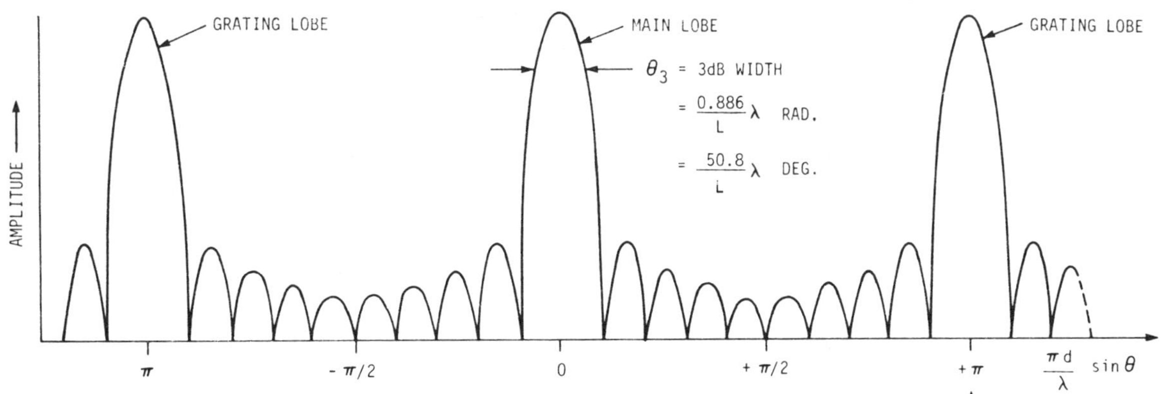

$\sin\theta_1 = \frac{\lambda}{d}$ FOR $\theta_0 = 0°$ (2.9)

L = ARRAY EFFECTIVE LENGTH

= Nd (2.8)

EX. 1 : $d = 2\lambda$

$\sin\theta_1 = 0.5 \Rightarrow \theta_1 = 30°$

EX. 2 : $d = \lambda$

$\sin\theta_1 = 1 \Rightarrow \theta_1 = 90°$

EX. 3 : $d < \lambda \Rightarrow \sin\theta_1 > 1$

$\Rightarrow$ NO GRATING LOBES

Fig. 2.3 Array factor for N=10, θ_0=0° (broadside case).

CRITERION FOR NO GRATING LOBES

$$E_a(\theta) = \frac{\sin\left[N \frac{\pi d}{\lambda}(\sin\theta - \sin\theta_0)\right]}{\sin\left[\frac{\pi d}{\lambda}(\sin\theta - \sin\theta_0)\right]} \qquad (2.10)$$

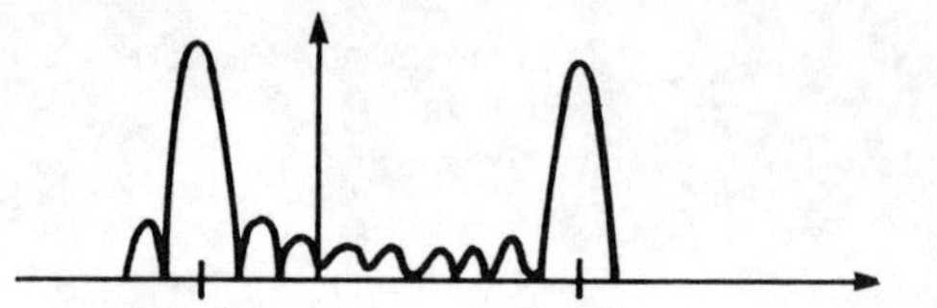

GRATING LOBES OCCUR WHENEVER DENOMINATOR EQUALS ZERO FOR $-90 \leq \theta \leq +90°$, EXCEPT $\theta = \theta_0$ (MAIN LOBE). EQUIVALENTLY WHEN

$$\frac{\pi d}{\lambda}(\sin\theta - \sin\theta_0) = \pm i\pi \quad \text{, FOR i AN INTEGER.}$$

CASE 1 : $\theta_0 < 0$

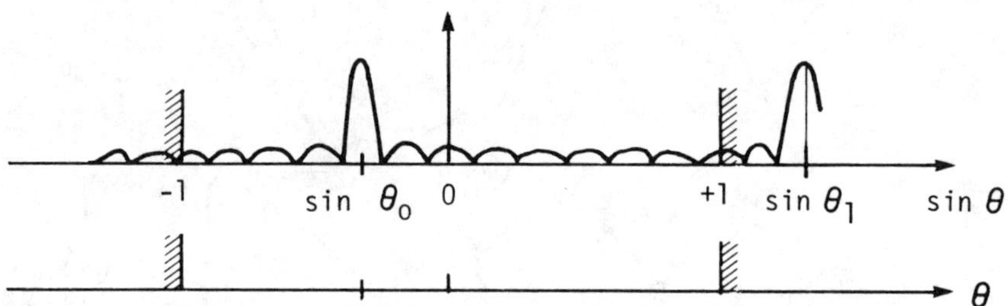

GRATING LOBE OCCURS WHEN : $\frac{\pi d}{\lambda}(1 - \sin\theta_0) = \pi$ OR $\frac{d}{\lambda} = \frac{1}{1 + |\sin\theta_0|}$ (2.11)

NO GRATING LOBE OCCURS IF : $\frac{\pi d}{\lambda}(1 - \sin\theta_0) < \pi$ OR $\frac{d}{\lambda} < \frac{1}{1 + |\sin\theta_0|}$ (2.12)

CASE 2 : $\theta_0 > 0$

NO GRATING LOBE OCCURS IF : $\frac{\pi d}{\lambda}(-1 - \sin\theta_0) > -\pi$ OR $\frac{d}{\lambda} < \frac{1}{1 + |\sin\theta_0|}$ (2.13)

Fig. 2.4 Derivation of criterion for no grating lobes.

SUMMARY OF CRITERION FOR NO GRATING LOBES

$$\frac{d}{\lambda} < \frac{1}{1 + |\sin\theta_0|} \qquad (2.14)$$

EX. 1. $\theta_0 = 0°$

$$\frac{d}{\lambda} < \frac{1}{1} = 1$$

2. $\theta_0 = 60°$

$$\frac{d}{\lambda} < \frac{1}{1.866}$$

3. $\theta_0 = 90°$

$$\frac{d}{\lambda} < \frac{1}{2}$$

Fig. 2.5 Summary of criterion for no grating lobes.

MAXIMUM ALLOWABLE ELEMENT SPACING d/λ VS MAXIMUM SCAN ANGLE $\theta_{0,\,MAX}$

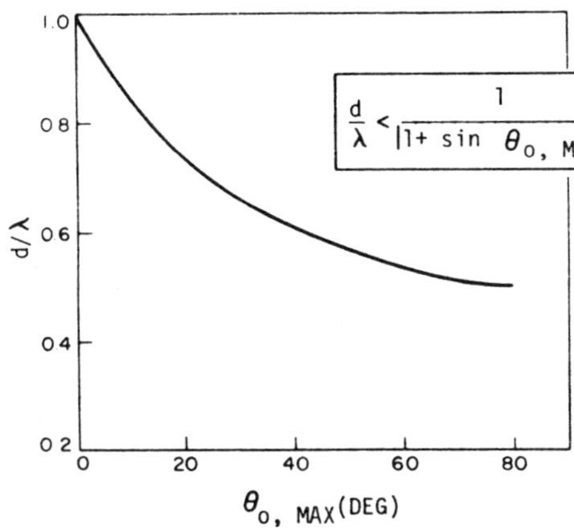

$$\frac{d}{\lambda} < \frac{1}{|1 + \sin \theta_{0,\,MAX}|} \quad (2.15)$$

Fig. 2.6 Maximum allowable element spacing d/λ vs maximum scan angle $\theta_{0,max}$. (After Allen, Lincoln Laboratory[1].)

DIFFERENTIAL PHASE SHIFT α VS BEAM POINTING ANGLE θ_0

$$\alpha = \frac{2\pi}{\lambda} d \sin \theta_0 \quad (2.16)$$

EXS. FOR $d/\lambda = 1/2$ $(\alpha = \pi \sin \theta_0)$

1. $\theta_0 = 0°$ $\alpha = 0$
2. $\theta_0 = \lambda/L$ RAD. $\alpha \doteq \frac{\pi\lambda}{L}$ RAD. FOR $\frac{\lambda}{L} \ll 1$
3. $\theta_0 = 30°$ $\alpha = 90°$
4. $\theta_0 = 60°$ $\alpha = 155.9°$
5. $\theta_0 = 90°$ $\alpha = 180°$

Fig. 2.7 Equation for differential phase shift α vs beam pointing angle θ_0.

DIFFERENTIAL PHASE VS POINTING ANGLE θ_0

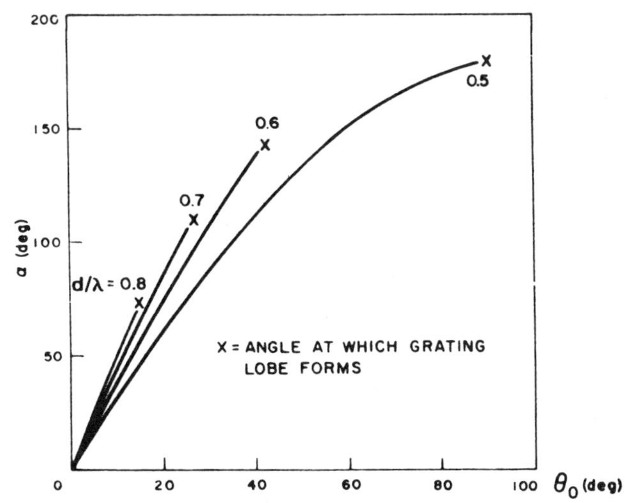

Fig. 2.8 Curve for differential phase vs pointing angle θ_0. (After Allen, Lincoln Laboratory [1].)

1-WAY 3dB BEAMWIDTH VS SCAN ANGLE θ_0

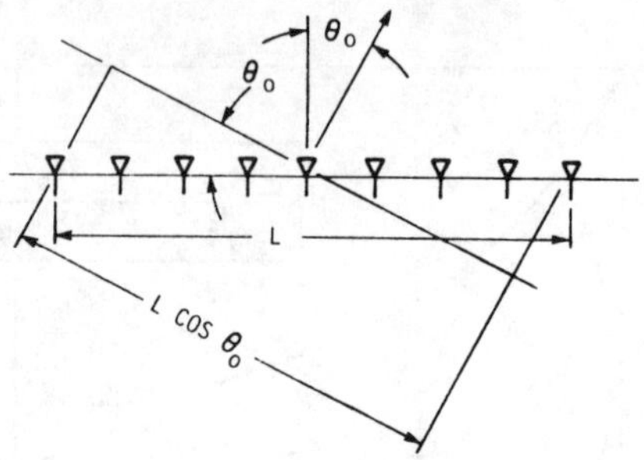

$$\theta_3 \doteq \frac{0.886}{L \cos\theta_0}\lambda \qquad \text{FOR } \theta_3 << 1 \text{ RAD. AND } \theta_0 \text{ NOT AT END FIRE.} \tag{2.17}$$

EX. LET $\theta_3 = \theta_{30}$ FOR $\theta_0 = 0°$

THEN

FOR $\theta_0 = 30°$ $\theta_3 = 1.155\ \theta_{30}$

$\theta_0 = 60°$ $\theta_3 = 2\ \theta_{30}$

Fig. 2.9 1-way 3 dB beamwidth vs scan angle θ_0; uniformly-weighted linear array.

BEAMWIDTH VERSUS LINEAR ARRAY LENGTH AND SCAN ANGLE; UNIFORM ILLUMINATION

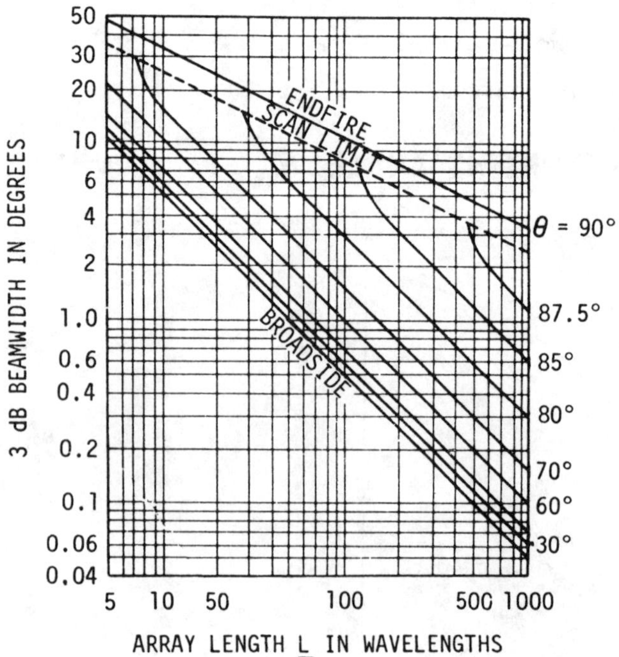

UP TO NEAR ENDFIRE :

$$\theta_3 = \cos^{-1}\left[\sin\theta_0 - 0.443\frac{\lambda}{L}\right] - \cos^{-1}\left[\sin\theta_0 + 0.443\frac{\lambda}{L}\right] \tag{2.18}$$

AT ENDFIRE :

$$\theta_3 = 2\cos^{-1}\left[1 - 0.443\frac{\lambda}{L}\right] \tag{2.19}$$

Fig. 2.10 Beamwidth versus linear array length and scan angle; uniform illumination. (After Elliott, Microwave Journal [2].)

ARRAY PATTERN FOR ARRAY OF NON-ISOTROPIC RADIATORS

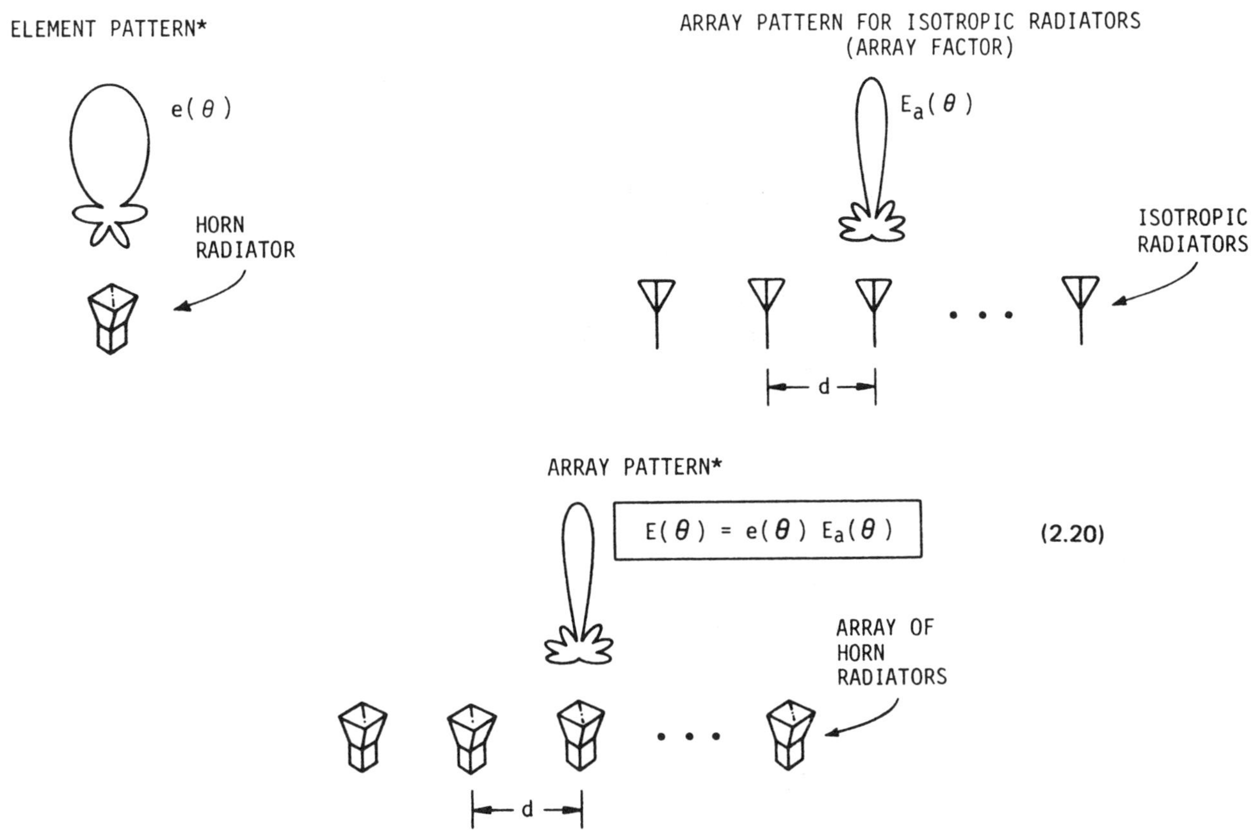

$$E(\theta) = e(\theta)\, E_a(\theta) \qquad (2.20)$$

*NOTE: $e(\theta)$ DETERMINED FOR CENTRAL HORN IN ARRAY OF HORN RADIATORS d APART WITH ALL OTHER RADIATORS TERMINATED.

Fig. 2.11 **Array pattern for array of non-isotropic radiators.**

DIRECTIVITY

$$D = \frac{\text{MAXIMUM POWER RADIATED PER UNIT SOLID ANGLE}}{\text{AVERAGE POWER RADIATED (= TOTAL POWER RADIATED } /4\pi)} \qquad (2.21)$$

HENCE

$$D = \frac{|E(\theta_0)|^2}{|E(\theta)|^2_{AVERAGE}} \qquad (2.22)$$

$|E(\theta)|^2_{AVERAGE}$ = AVERAGE OF $|E(\theta)|^2$ OVER 4π STERADIANS.

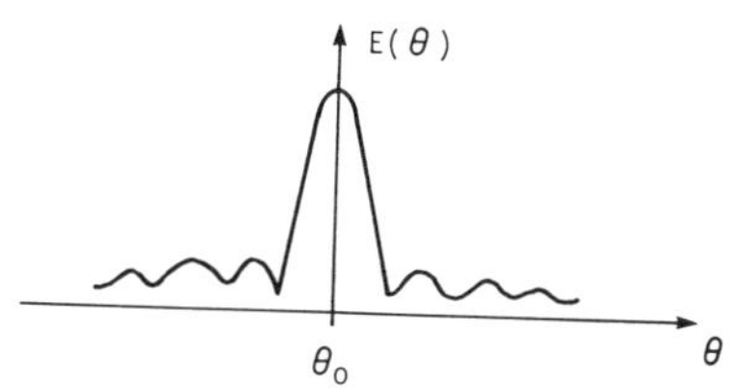

Ex. 1 $d = \lambda/2$, $I_i = 1$ FOR ALL i, ISOTROPIC RADIATORS

$D = N$

EX. 2 $d = \lambda/2$, NONUNIFORMLY WEIGHTED ISOTROPIC RADIATORS

$$D = D_0 = \left(\sum_{i=-M}^{M} I_i\right)^2 \Big/ \sum_{i=-M}^{M} I_i^2 \qquad (2.23)$$

(13b) INDEPENDENT OF α .

Fig. 2.12 **Directivity: definition.**

DIRECTIVITY

ASSUMPTIONS: UNIFORMLY WEIGHTED ARRAY

ISOTROPIC RADIATORS

$\alpha = 0°$, I.E. $\theta_0 = 0°$ (BROADSIDE ARRAY)

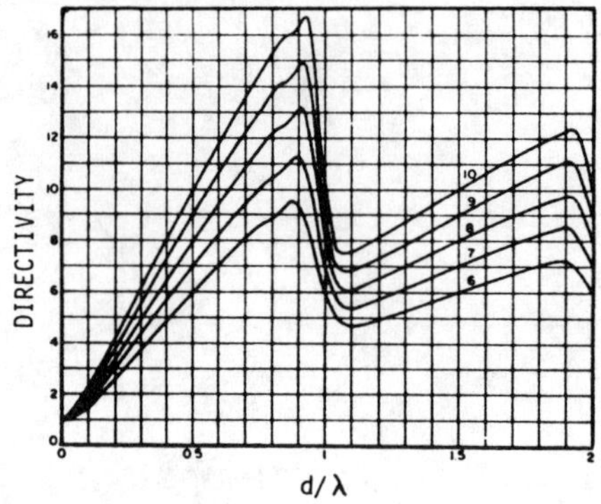

NOTE: FOR $\frac{d}{\lambda} \lesssim 0.9$

$$D \doteq \frac{2d}{\lambda} D_0 \qquad (2.24)$$

WHERE D_0 = DIRECTIVITY FOR $d = \lambda/2$

= N FOR UNIFORM WEIGHTING

EQ. (2.24) HOLDS FOR NONUNIFORMLY WEIGHTED LINEAR ARRAYS AS WELL WITH D_0 NOW GIVEN BY (2.23)

$$D_0 = \frac{\left(\sum_{i=-M}^{M} I_i\right)^2}{\sum_{i=-M}^{M} I_i^2} \qquad (2.23)$$

Fig. 2.13 Directivity of linear array of isotropic radiators versus d/λ. (After Tai, Microwave Journal [5].)

DIRECTIVITY AND REALIZED GAIN FOR LINEAR ARRAY OF NON-ISOTROPIC RADIATORS

FOR DIRECTIVITY, GENERALLY:

$$D \neq D_a d_e \qquad (2.25)$$

$d_e \equiv$ ELEMENT RADIATOR DIRECTIVITY

$D_a \equiv$ ARRAY DIRECTIVITY

INSTEAD FOR REALIZED GAIN OF UNIFORMLY ILLUMINATED ARRAY:

$$G_r = N g_r \qquad (2.26)$$

g_r, G_r = REALIZED RADIATING ELEMENT AND ARRAY GAINS, RESPECTIVELY, (INCLUDES DISSIPATION AND MISMATCH LOSSES.)

Fig. 2.14 Directivity and realized gain for linear array of non-isotropic radiators.

The location of the grating lobes for $d = 2\lambda$ and λ are 30° and 90°, respectively, see Examples 1 and 2 of Fig. 2-3. If $\theta_1 > 90^\circ$, there will be no grating lobes present. Thus, for broadside conditions, when $d < \lambda$, $\sin\theta_1 > 1$ so that no grating lobes exist; Example 3 of Fig. 2.3.

For nonbroadside conditions, i.e., $\theta_o \neq 0$, the conditions for no grating lobes are derived in Fig. 2-4 and summarized concisely by (2-14) of Fig. 2-5. For $\theta_o = 60^\circ$, (2-14) indicates that no grating lobes occur for $d/\lambda < 1/1.866$. A plot of (2-14) is given in Fig. 2-6.

Strictly speaking, (2-14) only gives the condition for the peak of the grating lobe not being present. Although the peak may not be present when (2-14) is satisfied, the rest of the grating lobe may be present. The spacings between the elements must be less than indicated by (2-14) in order for the whole grating lobe not to be present. Alternately, the maximum scan angle $\theta_{o,max}$ must be smaller by ½ the width of the grating lobe. The maximum angle must actually be smaller by ½ the width of the grating lobe measured in terms of the variable $\sin\theta$, called the sine space variable; see Sect. 2.2 and Lecture 4. From a practical point of view, however, an actual antenna will not consist of isotropic elements. As a result (2-14) generally is over restrictive. The element pattern may attenuate (by perhaps 10 to 20 dB) the far out grating lobe in practice so that they can be tolerated in visible space.

Values of the differential phase shift α of Fig. 2-1 needed to point the main beam at angles $\theta_o = 0^\circ$, λ/L*, 30°, 60°, and 90° are given in Fig. 2-7. A plot of α versus θ_o is given in Fig. 2-8.

The shift of the beam by λ/L from broadside in Fig. 2-7 is equivalent to shifting it by about the 3 dB beamwidth.* More precisely shifting a uniformly-weighted-linear-array broadside beam by λ/L is equivalent to shifting the peak of the beam to where the first null of the beam was when pointing broadside. Note that for this shift the differential phase shift across the whole aperture is 360°, since

$$N\alpha = N\frac{\pi\lambda}{L} = N\frac{\pi\lambda}{Nd} = 2\pi$$

when $d = \lambda/2$ as assumed. Note that the differential phase shift across the aperture is defined as $N\alpha$ and not $(N-1)\alpha$ just as L was defined as Nd and not $(N-1)d$; see (2-8). Consequently, $\alpha = 2\pi/N$ in order to have the beam shift λ/L from broadside. This result holds when $d \neq \lambda/2$, i.e., to shift the beam by λ/L the differential phase shift across the aperture must be 360°. Moreover, this result holds for nonbroadside conditions as well except that $\sin\theta$ (i.e., the sine space variable) and not θ is shifted by λ/L. It can be shown that as a result the peak of the beam is still shifted to where the first null was prior to the shift. This corresponds to the condition needed for the orthogonality of adjacent beams. It is the situation that exists between adjacent beams of a multiple beam Butler matrix corporate feed network; see Fig. 1-6.

At broadside ($\theta_o = 0^\circ$), Fig. 2-3 indicates that $\theta_3 = 0.886\ \lambda/L$. For $\theta_o \neq 0^\circ$, θ_3 increases approximately inversely proportionally to $\cos\theta_o$ when $\theta_3 << 1$; see (2-17). Hence, for $\theta_3 = 30^\circ$ and 60°, $\theta_3 = 1.155\theta_{3o}$ and $2\theta_{3o}$, respectively, where θ_{3o} is the 3 dB beamwidth at broadside. Fig. 2-10 plots the exact expressions for θ_3 given by (2-18) and (2-19) of that figure.

---

* $\lambda/L = \theta_3/0.886 \doteq \theta_3$ is the 3 dB main lobe beamwidth.

Up to now we have only given consideration to the determination of the pattern of a linear array consisting of isotropic radiating elements. The antenna pattern for a linear array of identical non-isotropic radiating elements is obtained by multiplying the array pattern of the array when it consists of isotropic radiators by the array pattern of the individual non-isotropic radiating elements; see (2-20).[1],[3] The array pattern for the individual non-isotropic radiating elements is determined for a centrally located element in the array with all of the other elements present but terminated. Eq. (2.20) holds as long as the array is large, typically $N > 50$. Furthermore, it applies for planar as well as linear arrays. (It is possible also to use the isolated element radiation pattern of each of the radiating elements.[3] However, when using this approach for determining the array pattern for the array of isotropic elements, that is, the antenna array factor, it is necessary to take into account the mutual coupling effects between radiating elements which exist when the actual non-isotropic array elements are used in the array. Mutual coupling affects the phasing of the radiating elements, that is, the phases of the radiator currents of each element. The phases of the currents of each element of Fig. 2-1 are no longer just determined by the phases of the phase shifters.)

B. Directivity. In the direction of maximum radiation, the antenna directivity is defined by (2-21) and (2-22) of Fig. 2-12[2],[4] The antenna directivity would equal the antenna gain if there were no ohmic, mismatch, or other losses. The directivity of a uniformly weighted linear array of N isotropic radiators $\lambda/2$ apart equals N independent of the main beam scan angle θ_o; see Example 1 of Fig. 2-12.[2] Example 2 of Fig. 2-12 gives the directivity of a linear array of isotropic radiators $\lambda/2$ apart for which the weightings are not the same for all radiating elements.[2] This second example indicates that the directivity of a nonuniformly weighted aperture is also independent of the main beam pointing angle θ_o when the elements are spaced $\lambda/2$ apart. The independence of the antenna directivity with scan angle is a surprising result. Because of the aperture forshortening (by the factor $\cos\theta_o$) one would expect the directivity to decrease as $\cos\theta_o$. Alternately, because θ_3 increases as $1/\cos\theta_o$ with scan angle, for conservation of energy one might expect the directivity to decrease by $\cos\theta_o$. However, for a linear array of isotropic elements, due to symmetry about the array axis, the axis of the main beam at an angle θ_o forms the surface of a cone of revolution whose axis is along the array axis and whose half angle is $(\pi/2)-\theta_o$. As the beam is scanned more off broadside this cone tends to occupy a smaller solid angle in space. This effect just cancels out the beam broadening.[2]

For spacings between elements of other than $\lambda/2$, the array directivity is given in Fig. 2.13 for linear arrays of N=6, 7, 8, 9, and 10 uniformly weighted isotropic elements at broadside conditions, that is, $\theta_o = 0^\circ$. For $d \lesssim 1$, the directivity is proportional to d/λ and hence it is proportional to the effective array length since L=Nd; see (2-24). For d/λ slightly larger than 1, the two first grating lobes are fully developed. At this point more energy is radiated into the grating lobes than into the main beam. Consequently, there is a sharp drop in the antenna directivity near $d/\lambda = 1$ as indicated in Fig. 2-13.

Recall that radiation patterns multiply; see (2-20). This is not the case for directivity in general. The directivity of a linear array of non-isotropic radiation elements is not equal to the directivity of this array when made up of isotropic elements times the directivity of the individual radiating elements which make up the array; (see Fig. 2-14).[6] There is no simple general expression relating the directivity of the whole array to the directivity of the radiating elements of the array. Instead the basic definition for directivity given by (2-22) has to be used with the array antenna pattern determined from (2-20). Examples worked out using this procedure shall be presented shortly.

Although there is no simple formula for determining the directivity of an array non-isotropic elements, there is a very useful equation for determining the realized gain of an array in terms of the realized gain of the radiating elements for the case of a uniformly weighted large array; see (2-26).[7] Realized gain represents antenna directivity with the inclusion of two factors: 1) antenna ohmic dissipation losses and 2) the variations of the mismatch of the antenna to free space with scan angle[7]; see also Lesson 4. As was the case for (2-20) the realized gain of the individual radiating elements must be determined for a centrally located element in the array with all the other elements terminated. Also, for (2-26) to apply it is necessary that the array be large ($N > 100$). Finally, as was the case for (2-20), (2-26) applies for large planar arrays as well as large linear arrays.

Although (2-26) does not hold in general for directivity[8] it does hold for the case where the radiating elements are non-dissipative and matched to free space for all scan angles. Though this is a specialized case it is found to be useful for the case of planar arrays and shall be presented when discussing them in Sect.2.2.1.

The directivity as a function of d/λ for several broadside arrays consisting of identical non-isotropic radiating elements is given in Fig. 2-15. The isolated power patterns for these non-isotropic radiating elements (consisting of short dipole elements) are given in Fig. 2-16.[9],[10] The directivity results given in Fig. 2-15 were derived using the basic defining equation for directivity given by (2-22). However, the isolated antenna patterns for the non-isotropic radiating elements were utilized without taking into account mutual coupling effects. Hence, the results given are only approximate.

As an interesting sidelight it is pointed out that the directivity of a linear array of isotropic elements is not maximum when the radiating elements are equally weighted, that is, uniformly weighted. Fig. 2-17 compares the array directivity for optimum weighting with that obtained with uniform weighting for the case of broadside radiation. Only in the region of low d/λ is there a substantial difference between the directivity obtained with uniform weighting and that obtained with optimum directivity. The improvement in this region is referred to as supergaining. Generally supergaining is difficult to achieve and not worth the trouble. For $d = \lambda/2$, uniform weighting does give optimum directivity, i.e., $D = N = D_{opt}$.[11]

Most of the results in this section on directivity were presented more to provide further insight into the theory and performance characteristics of arrays than for their practical utility. Also, the material presented should help the reader understand the literature on arrays. Of the results presented, that of greater practical use is (2-26). Various other useful forms of this equation are given later in Fig. 2-34. Eq. (2-23) is employed in the useful concept of aperture taper efficiency discussed in the next section.

2.1.2 Weighted Apertures.

So far uniformly weighted apertures have been considered primarily; that is, $I_i = 1$ for all i. Generally, a weighted illumination is used in order to obtain lower antenna sidelobe levels. This lower sidelobe level arising from the use of a weighted illumination is obtained at the expense of a loss in directivity.

The loss in directivity is measured relative to the directivity obtained with uniform weighting and is referred to as the aperture taper efficiency η_t[1] (it is also known as illumination efficiency[1] and aperture efficiency)[3],[4],[12] (see Fig. 2-18). This efficiency is not to be confused with the many other antenna efficiencies which relate to the other losses that can be present in an array. These other efficiencies are listed and defined in Fig. 2-19. As indicated in Fig. 2-19, the product of these antenna efficiencies form the over-all antenna efficiency η. Not all of the efficiencies listed in the Fig. 2-19 are present in every array antenna. For example, the spillover efficiency would only be present if a reflectarray antenna was used, e.g., the Raytheon reflectarray antenna[12]. Similarly, the polarization loss is of concern for a circularly polarized radar used to penetrate rain but not necessarily for other radars. The tolerance losses will be covered in Lecture 3, Sect. 3.1.1. Taper efficiency is covered in this lecture. The other efficiencies will not be discussed. They are listed primarily to make one aware of the potential losses one may have to take into consideration. Some of these losses are discussed in Ref. [13].

If all the antenna efficiencies of Fig. 2-19 where equal to unity with the exception of the taper efficiency, then the antenna gain G would equal the antenna directivity. The antenna gain is otherwise equal to the antenna directivity times all the efficiencies except for the taper efficiency. The antenna gain is also given by

$$G = \eta D_U \qquad (2\text{-}29)$$

where D_U is defined as the directivity of the antenna for uniform illumination.

Taper efficiencies for two types of antenna weightings are given in Fig. 2-20. Specifically, they are given for the cosine-squared-on-a-pedestal weighting and the Tapered-Taylor weighting which is to be discussed shortly. Also given are the 3 dB beamwidths for the Dolph-Chebyshev weighting which is also to be discussed shortly. Finally, Fig. 2-20a gives taper coefficient and pedestal height versus sidelobe level for the cosine-squared-on-a-pedestal weighting.

The Dolph-Chebyshev weighting is an optimum weighting for linear arrays in the sense that it minimizes the main lobe null-to-null width for a specified maximum allowable sidelobe level.[14] This optimization process yields an antenna pattern for which all the sidelobes have equal amplitude;

(see Fig. 2-21). Equations for determining the antenna element weightings for specified array sidelobe levels are given in the Appendix to this lesson.

As the length of a Dolph-Chebyshev array increases the number of sidelobes increases. Thus, as the array length increases a larger fraction of the radiated power goes into the sidelobes. Consequently, the directivity of a Dolph-Chevyshev array does not increase linearly as the array length is increased when the design sidelobe level is held constant.

COMPARISON OF THE DIRECTIVITIES OF SEVERAL DIFFERENT TYPES OF UNIFORM BROADSIDE ARRAYS

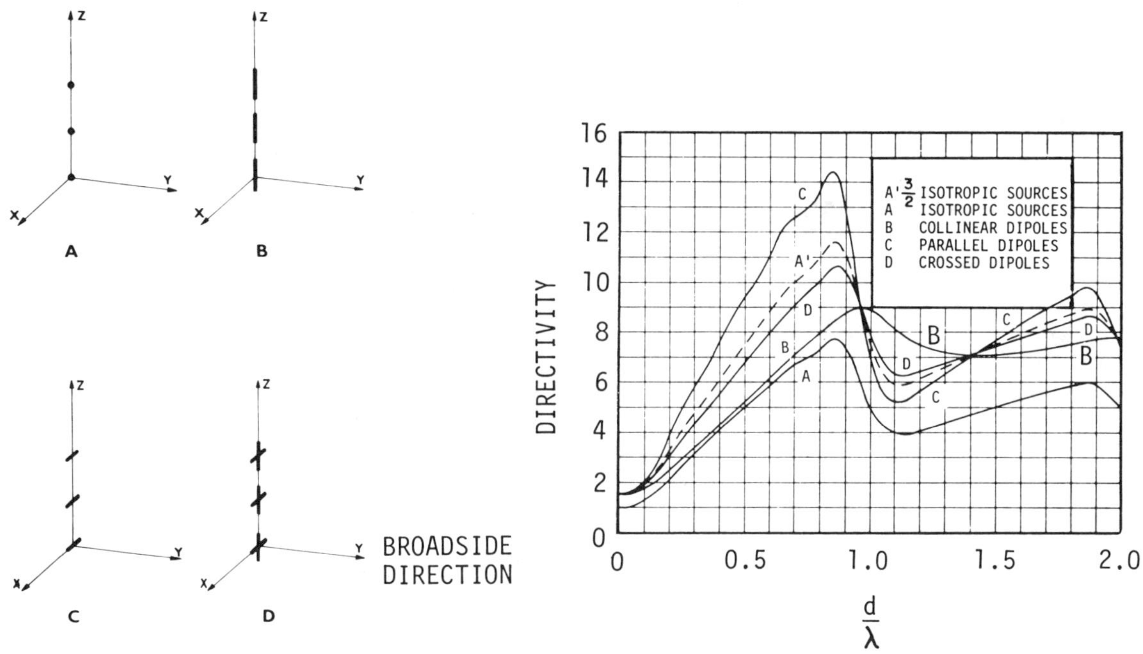

A: Array of isotropic elements.
B: Array of collinear dipoles.
C: Array of parallel dipoles.
D: Broadside array of dipole crosses. *

* DIPOLES OF EACH CROSS EXCITED 90° OUT OF PHASE.

Fig. 2.15 Comparison of the directivities of several linear arrays of non-isotropic elements: uniform weighting;broadside conditions, i.e. $\theta_0=0^{\circ}$. (After Tai [5] and Bach [8], Microwave Journal.)

ISOLATED BEAM PATTERNS FOR TYPICAL SHORT DIPOLE RADIATING ELEMENTS

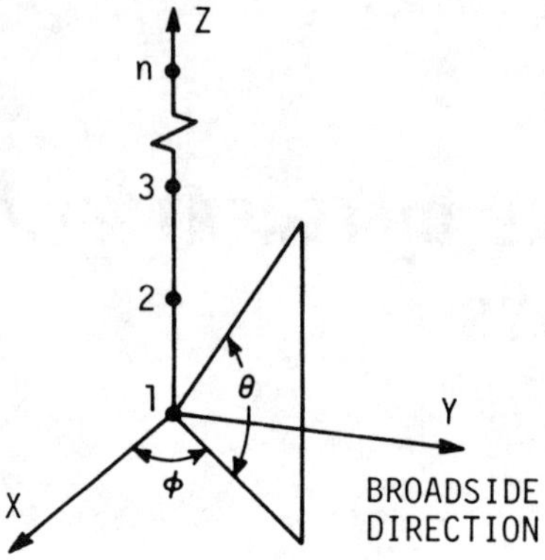

TYPE ELEMENT	POWER PATTERN $\lvert e(\theta,\phi)\rvert^2$	DIRECTIVITY (dB)
ISOTROPIC ELEMENTS	1	0
COLLINEAR DIPOLES	$\cos^2\theta$	>1.76
PARALLEL DIPOLES	$1 - \cos^2\theta\cos^2\phi$	>1.76
BROADSIDE CROSSES	$\frac{1}{2}(1 + \cos^2\theta\,\sin^2\phi)$	

COLLINEAR DIPOLE ELEMENT PATTERN

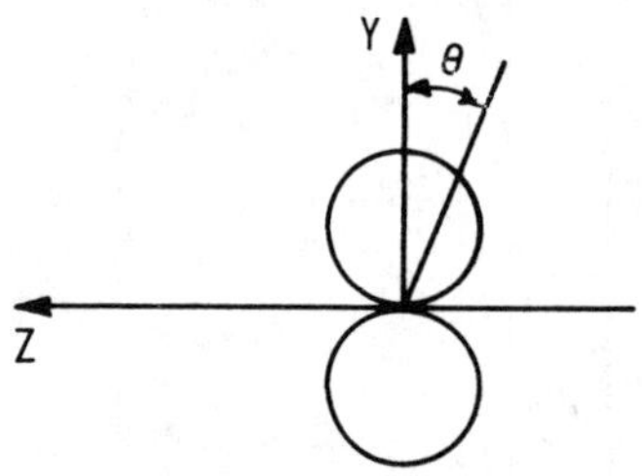

BROADSIDE CROSS ELEMENT PATTERN

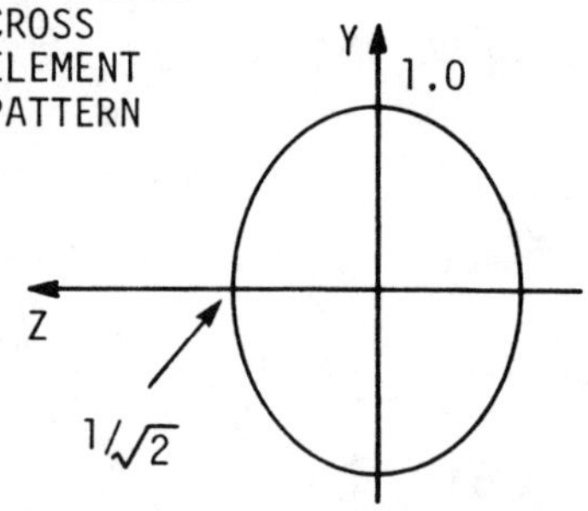

(NOTE THAT A NONSTANDARD θ, φ SPHERICAL COORDINATE SYSTEM IS USED.)

Fig. 2.16 Isolated beam patterns for typical short dipole radiating elements.

OPTIMUM DIRECTIVITY

ISOTROPIC RADIATORS
BROADSIDE ARRAY

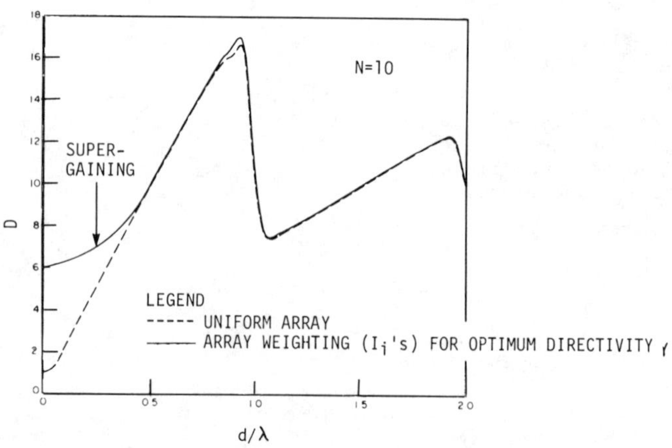

Fig. 2.17 Optimum directivity: broadside linear array; isotropic radiators. (After Tai, IEEE PGAP [10].

TAPER EFFICIENCY*

$$\eta_t = \frac{D \text{ FOR } d = \lambda/2}{D \text{ FOR UNIFORM ARRAY WITH } d = \frac{\lambda}{2}} = \frac{D_o}{N} \tag{2.27}$$

WHERE FROM (2.23)

$$D_o = \frac{\left(\sum_{i=-M}^{M} I_i\right)^2}{\sum_{i=M}^{M} I_i^2} \tag{2.23}$$

* ALSO CALLED ILLUMINATION EFFCIENCY AND APERTURE EFFICIENCY

Fig. 2.18 Taper efficiency (also called illumination efficiency and aperture efficiency).

ANTENNA EFFICIENCIES

$$\eta = \eta_t \eta_\ell \eta_s \eta_b \eta_f \eta_M \eta_\Delta \eta_\delta \eta_p \eta_m \quad (2.28)$$

η = overall antenna efficiency.

η_t = taper efficiency; see Fig. 2.18

η_ℓ = corporate feed, line, and radiating element ohmic losses.

η_s = spillover loss (applies for reflectarray).

η_b = blockage loss (blockage of aperture by feed system and its supports; applies for space fed arrays).

η_f = feed horn ohmic losses (applies for reflectarray).

η_M = mechanical tolerance errors in radiating element locations (see Lecture 3, Sect. 3.1.1).

η_Δ = aperture illumination amplitude errors (see Lecture 3, Sect. 3.1.1).

η_δ = aperture illumination phase errors (see Lecture 3, Sect. 3.1.1).

η_p = polarization loss (far field energy not all radiated in desired principal polarization, some in cross polarization).

η_m = mismatch losses.

Fig. 2.19 Antenna efficiencies.

ANTENNA TAPER EFFICIENCY, PEAK SIDELOBE LEVEL, AND 3dB BROADSIDE BEAMWIDTH

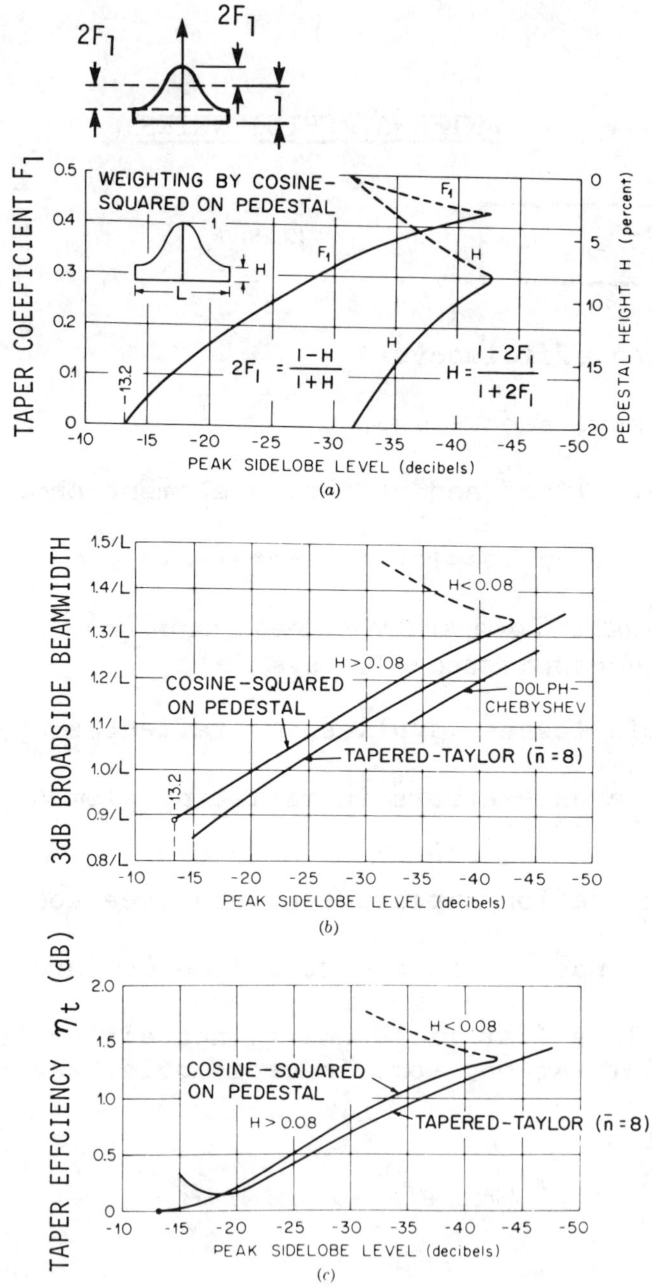

(a) TAPER COEFFICIENT AND PEDESTAL HEIGHT VS. PEAK SIDELOBE LEVEL; (b) 3dB BEAMWIDTH VS. PEAK SIDELOBE LEVEL; (c) TAPER EFFICIENCY, η_t VS. PEAK SIDELOBE LEVEL.

Fig. 2.20 Antenna taper efficiency, peak sidelobe level, and 3 dB broadside beamwidth. (After Farnett, Howard, and Stevens in Skolnik, McGraw-Hill Comp. [11].)

DOLPH-CHEBYSHEV LINEAR ARRAY FACTOR

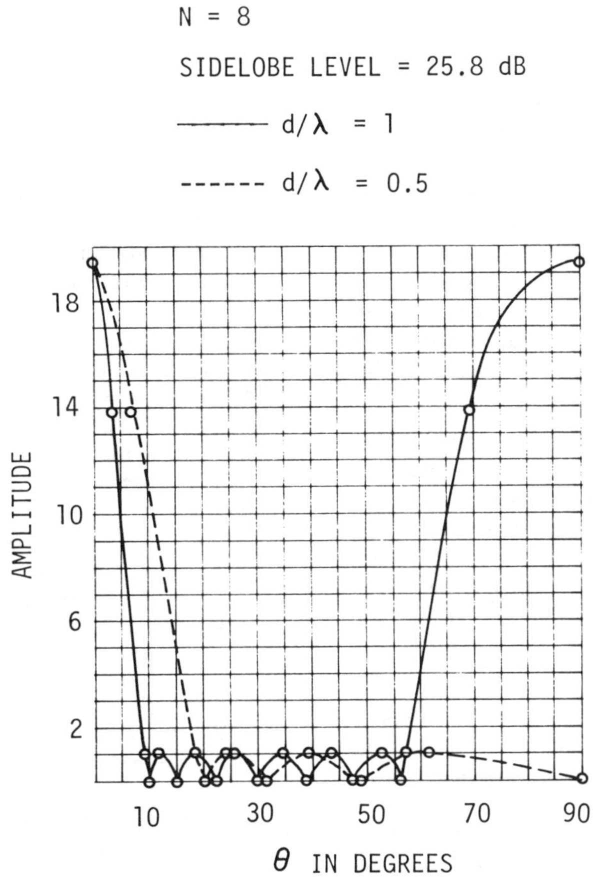

Fig. 2.21 Dolph-Chebyshev linear array factor. (After Dolph, Proc. IRE [12].)

DIRECTIVITY OF DOLPH-CHEBYSHEV LINEAR ARRAYS

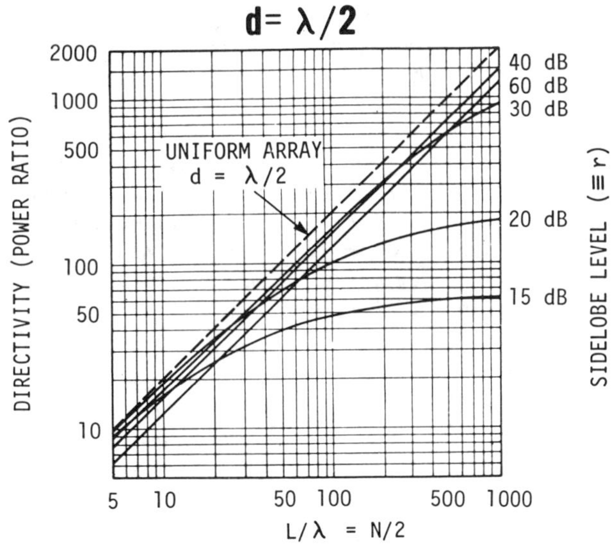

$$\lim_{N \to \infty} D = D_{max} = 2r^2 \quad (23)$$

WHERE r = VOLTAGE SIDELOBE RATIO.
EX. r = 10 = 20 dB D_{max} = 200 = 23 dB

Fig. 2.22 Directivity of Dolph-Chebyshev linear arrays: d = λ/2. (After Elliott, Microwave Journal [2].)

This is illustrated in Fig. 2-22 for $\lambda/2$ element spacing. Given a specified design sidelobe level, a procedure for picking a good array length would be the selection of the point on the curve of Fig. 2-22 where it just begins to bend over.

Curves of the taper efficiency versus sidelobe level for two Dolph-Chebyshev arrays are shown in Fig. 2-23a. These curves indicate that for a given number of array elements there exists a sidelobe level which maximizes the array taper efficiency and hence array directivity; this can also be seen from Fig. 2-22, although not as easily. Results are also given in Fig. 23a for an aperture illumination known as the one-parameter Taylor weighting (to be discussed shortly), as well as for the cosine-squared-on-a-pedestal weighting. The 3 dB beamwidth coefficient is given for all three weightings in Fig. 2-23b.

The Tapered-Taylor weighting distribution is an approximation to the Dolph-Chebyshev weighting.[15] It was developed for continuous linear apertures. However, a discrete spatial sampling of the continuous aperture weighting of the Tapered-Taylor illumination function provides an approximate Tapered-Taylor pattern for large linear arrays. For this spatial sampling the continuous illumination function is sampled at the points corresponding to the locations of the radiating elements of the array.

The Tapered-Taylor weighting developed for linear continuous apertures has an antenna pattern consisting of approximately equal sidelobes out to the $\bar{n}$ - 1st sidelobe to either side of the main lobe. Beyond these lobes the sidelobes decrease as they would for a uniformly weighted aperture, that is, as $\sin \dot{u}/\dot{u}$; (see Fig. 2-24). However, these $\sin \dot{u}/\dot{u}$ lobes are displaced downward by a fixed amount of dB from where they would be for a uniformly weighted aperture; (see Fig. 2-A-8) of the Appendix to this lesson. (For the one-parameter Taylor weighting the sidelobes fall off as $\sin \dot{u}/\dot{u}$ starting with the first sidelobe. However, as with the Tapered-Taylor weighting, these sidelobes are all displaced downward some fixed amount of dB, (see Fig. 2-A-8). The amount they are displaced downard is traded off against the taper efficiency and beam broadening that can be tolerated; (see Fig. 2-A-3).

The spatially sampled approximate Taper-Taylor weighting resulting for linear arrays will have sidelobes that fall down slightly slower than $\sin \dot{u}/\dot{u}$. The same situation arises with respect to the fall off of the sidelobes of a uniformly weighted linear array relative to its continuous aperture counterpart, the former falling off at the slower rate of $\sin \dot{u}/N\ [\sin(\dot{u}/N)]$ while the latter falls off as $\sin \dot{u}/\dot{u}$; (see Fig. 2-3). Moreover, the sidelobes will start to rise if the element-to-element spacing is such as to have a grating lobe just outside of visible space, i.e., at $\sin\ \theta_1$ slightly greater than 1 in Fig. 2-4, visible space (real space) ending for $\sin \theta = 1$, or equivalently, $\theta = 90°$ for a linear array.

The tapered-Taylor weighting permits a more flexible trade-off between the sidelobe level, beamwidth, and aperture efficiency than does the Dolph-Chebyshev weighting. The Tapered-Taylor antenna pattern also has the advantage of being easier to design to for a large linear array, that is, the weightings are easier to obtain for the individual elements than they are for a Dolph-Chebyshev array pattern.

When $\bar{n}$ approaches infinity for the Tapered-Taylor weighting it becomes what is referred to as the ideal-Taylor pattern; see Appendix.[15] As the number of elements of a Dolph-Chebyshev array become infinite for a fixed length array, the antenna pattern also becomes that of the ideal-Taylor.[16] The ideal-Taylor is impractical as it requires an infinite current (delta functions) at the aperture ends and in addition it provides poor taper efficiency; hence the development of the Tapered-Taylor approximation to the ideal-Taylor.

For a given maximum sidelobe level there exists for the Tapered-Taylor weighting an $\bar{n}$ which gives maximum directivity. This is apparent from Fig. 2-25 which gives curves of η_t versus $\bar{n}$ for different sidelobe levels. It is easy to explain physically why there exists an $\bar{n}$ whiich gives maximum antenna directivity. For a given design sidelobe level, as $\bar{n}$ increases more energy is lost in the sidelobes. Hence, as $\bar{n}$ becomes larger and larger eventually the antenna directivity decreases. On the other hand, for small enough $\bar{n}$, as $\bar{n}$ decreases the main lobe width increases and as a consequence more energy is radiated in the main beam but not at the peak of the beam with a resultant loss in directivity at the beam peak. In (2-34) of Fig. 2-25 $2L/\lambda$ represents the directivity of a long ($L/\lambda \geqslant 5$) uniformly-weighted continuous aperture. Fig. 2-26 tabulates the optimum $\bar{n}$ versus sidelobe level.[16],[17] Also given in Fig. 2-26 is the 3 dB beamwidth coefficient for the Tapered-Taylor weighting.[16]

Fig. 2-27 compares the antenna patterns of a 40 dB Tapered-Taylor weighting obtained for $\bar{n} = 6$ with that of a uniform weighting and a special case of the cosine-squared-on-a-pedestal weighting, known as the Hamming weighting. Typical plots of Tapered-Taylor aperture weightings are given in Fig. 2-28. A table of Tapered-Taylor aperture weightings is given in the Appendix. Also given in the Appendix are the equations which can be used for calculating the weightings for Tapered-Taylor illuminations.

The Dolph-Chebyshev patterns given above result in an optimum one-way antenna pattern in the sense discussed. It does not result in the optimum two-way pattern, that is, it is not optimum when the antenna is used simultaneously for the transmission and reception of the signal as is done in monostatic radar systems. The two-way antenna pattern for the above Dolph-Chebyshev array is the square of the one-way pattern. Hence, it is the square of a Chebyshev polynomial and thus not optimum in the sense of providing the minimum two-way main lobe width for a specified design two-way maximum sidelobe level (the unsquared Chebyshev polynomial antenna pattern giving the optimum pattern).

What is needed is a design for which the two-way pattern is a Chebyshev polynomial. The design of such a pattern requires the use of the relationship (2-35).[18] This expression indicates that a Chebyshev polynomial pattern of degree 2N is obtained by transmitting with an antenna pattern consisting of the sum of: 1) a Chebyshev polynomial pattern of degree N and 2) an isotropic pattern; and receiving with an antenna pattern consisting of the difference between these two patterns; (see Fig. 2-29).

To arrive at this optimum two-way antenna pattern for an array of N elements a conventional Dolph-Chebyshev antenna pattern is first designed. Then on transmit the gain of one of the radiating elements is raised sufficiently to achieve the small additional isotropic pattern component needed. On reception the gain of this radiating element is reduced by the same amount (as it is raised on transmit) from that

TAPER EFFICIENCY AND BEAMWIDTH COEFFICIENT VS HIGHEST SIDELOBE LEVEL

LINEAR ARRAYS

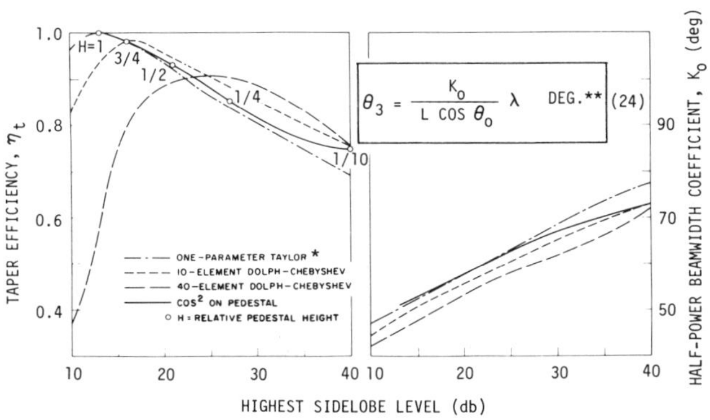

* SEE APPENDIX.

** K_o = 50.8° FOR UNIFORM ILLUMINATION

Fig. 2.23 Taper efficiency and beamwidth coefficient vs highest sidelobe level; linear arrays. (After Allen, Lincoln Laboratory [1].)

APPROXIMATE SKETCH OF TAPERED TAYLOR ANTENNA PATTERN; LINEAR APERTURE

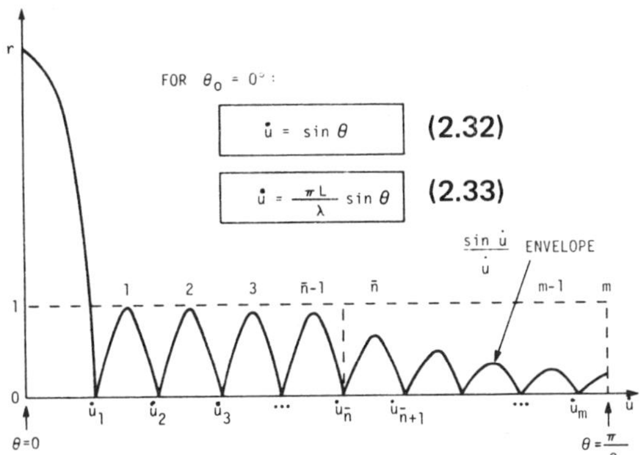

Fig. 2.24 Approximate sketch of Tapered-Taylor antenna pattern; linear aperture. (After Hansen, IEEE PGAP [14].)

ILLUMINATION EFFICIENCY VS $\bar{n}$ FOR TAPERED TAYLOR LONG LINEAR APERTURES

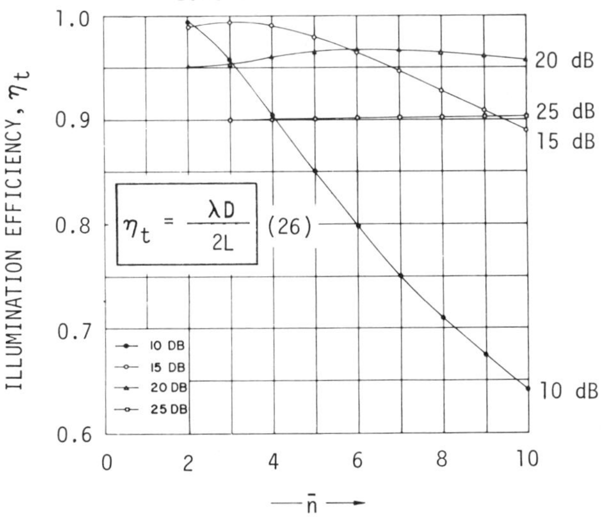

Fig. 2.25 Illumination efficiency vs $\bar{n}$ for Tapered-Taylor long linear apertures, i.e. $L/\lambda \geq 5$; also require that $\bar{n} \leq L/\lambda$. After Taylor, IRE PGAP [13].)

TAPERED TAYLOR $\bar{n}$ FOR MAXIMUM ILLUMINATION EFFICIENCY

LARGE LINEAR APERTURES

MAXIMUM SIDELOBE LEVEL dB	$\bar{n}$	ILLUMINATION EFFICIENCY * η_t	HALF-POWER BEAMWIDTH COEFFICIENT K_o (deg)
15	3	0.99	
20	6	0.97	53.2
25	12≠	0.90	58.2
30	24#≠	0.75	63.1
35	60#≠	0.67	67.5

$$* \eta_t = \frac{\lambda D}{2L} \quad (2.34)$$

\# ESTIMATE

≠ HAVE VERY BROAD MAXIMA

Fig. 2.26 Tapered-Taylor $\bar{n}$ for maximum illumination efficiency; long linear apertures.

COMPARISON OF ANTENNA PATTERNS FOR DIFFERENT LINEAR APERTURE WEIGHTING FUNCTIONS

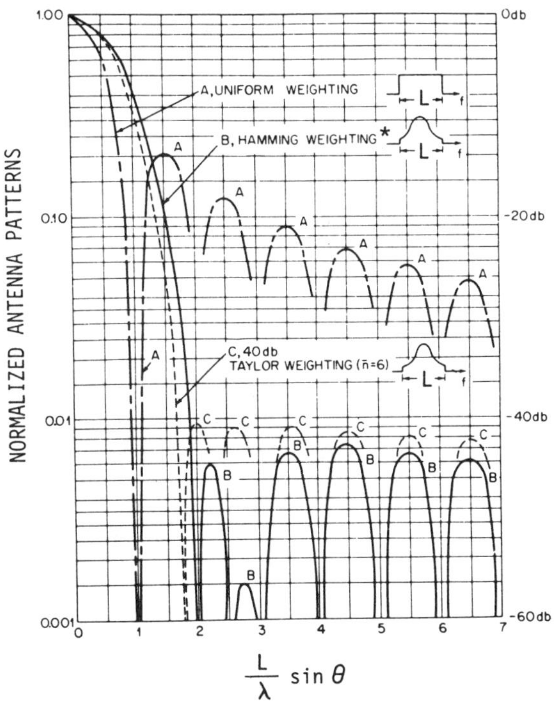

* COSINE-SQUARED-ON-PEDESTAL WITH H=.0800(F_1 = 0.426)

Fig. 2.27 Comparison of antenna patterns for different linear aperture weighting functions. (After Farnett, Howard, and Stevens in Skolnik, McGraw-Hill Book Company [11].)

TYPICAL TAPERED TAYLOR APERTURE WEIGHTING FUNCTIONS

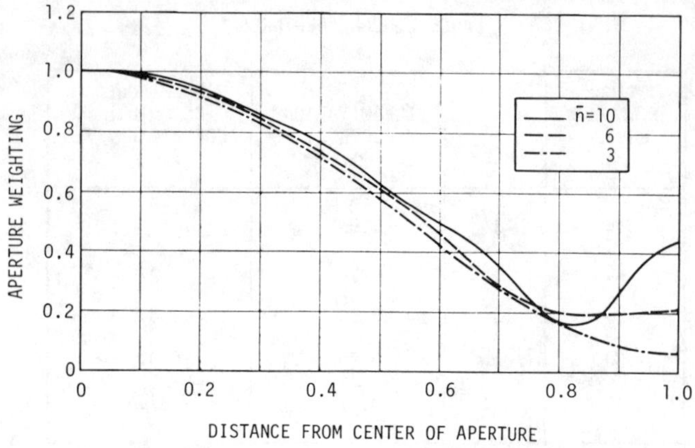

APERTURE WEIGHTING FUNCTION FOR 25-db SIDELOBE LEVEL FOR SEVERAL n̄.

Fig. 2.28 Typical Tapered-Taylor aperture weighting functions for 25 dB sidelobe level for several $\bar{n}$. (After Hansen, IEEE PGAP [14].)

TWO-WAY DOLPH-CHEBYSHEV LINEAR ARRAY FACTOR

$E_a(\theta)_T$ = TRANSMIT ARRAY FACTOR

$E_a(\theta)_R$ = RECEIVE ARRAY FACTOR

$E_a(\theta)_{2-W}$ = 2-WAY ARRAY FACTOR $= E_a(\theta)_T \, E_a(\theta)_R$

DESIRE THAT :

$E_a(\theta)_{2-W}$ = DOLPH - CHEBYSHEV ARRAY FACTOR

SOLUTION :

$$E_a(\theta)_{2-W} = T_{2N} = 2(T_N)^2 - 1 = 2\left(T_N + \frac{1}{\sqrt{2}}\right)\left(T_N - \frac{1}{\sqrt{2}}\right) \quad (2.35)$$

$$= [E(\theta) + C][E(\theta) - C] \quad (2.36)$$

Fig. 2.29 Two-way Dolph-Chebyshev linear array factor.

IMPLEMENTATION OF TWO-WAY DOLPH-CHEBYSHEV LINEAR ARRAY FACTOR

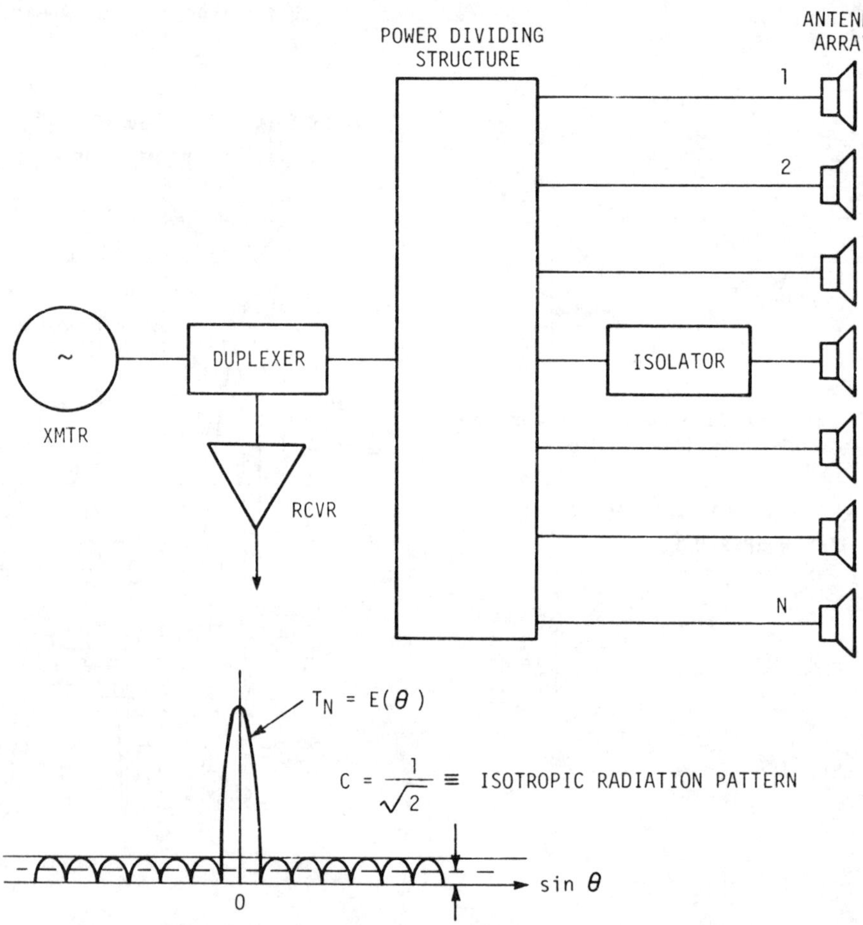

ADVANTAGES: EITHER 1) 8 TO 10 BETTER 2-WAY SIDELOBE LEVEL FOR SAME NULL-TO-NULL WIDTH

OR 2) 10% SMALLER NULL-TO-NULL WIDTH FOR SAME SIDELOBE LEVEL.

} RELATIVE TO CONVENTIONAL DESIGN.

Fig. 2.30 Implementation of two-way Dolph-Chebyshev linear array factor. (After Mattingly, Proc. IRE [16].)

of the conventional design in order to effect the subtraction of the isotropic pattern from the Chebyshev pattern as required. Placing a non-reciprocal element in the path of one of the radiating elements, as shown in Fig. 2-30, can be used to automatically vary the gain the required amount from transmit to receive.[18]

The advantage of this two-way Chebyshev design over the conventional design is that it either provides an 8 to 10 dB better two-way sidelobe level for the same mainlobe two-way width or else it provides a 10% smaller main lobe width for the same two-way sidelobe level. [18]

2.2 Planar Arrays.

First rectangular apertures and then circular apertures will be treated.

2.2.1 Rectangular Aperture. Fig. 2-31 shows the geometry for a rectangular planar array consisting of isotropic elements equally spaced in the x and y coordinate directions. The array has $N_x = 2M_x + 1$ rows and $N_y = 2M_y+1$ columns. Assume that for this rectangular aperture the shape of the set of weightings for any given row is the same as it is for any of the other row, i.e., that the weightings are identical except for a relative gain factor; see (2-38). Similarly, assume that for any given column the set of weightings is the same as it is for any other column except for a gain factor; see (2-39). Fig. 2-31 shows that when these conditions are satisfied, the rectangular aperture array pattern is given by the product of the array pattern for a linear array of row elements times the array pattern for a linear array of column elements; see (2-40). (Without loss of generality the linear array along the x axis shall be chosen for the row array antenna pattern and the array along the y axis shall be chosen for the column array pattern for future discussions.) The row times column array pattern product property of (2-40) is called array separability.*

Fig. 2-32 shows how the main beam for a rectangular array is formed by the intersection of the conical beam patterns formed from the row and column linear arrays. Also given in Fig. 2-32 are Eqs. (2-47) and (2-48) for the differential phase shifts α_x and α_y between successive row elements and column elements, respectively, needed to point the beam at an arbitrary angle θ_o, ϕ_o. These equations are equivalent to (2-16) for a linear array. The equations give the beam pointing angle in terms of the x and y coordinate linear array off-broadside pointing angles θ_{ox} and θ_{oy}, respectively. The complements of the linear array off-broadside angles θ_{ox} and θ_{oy} (designated as θ'_{ox} and θ'_{oy}) are the half angles of the conical surfaces generated by the axes of the main lobes formed by the x and y axes' arrays; (see Fig. 2-32). The two off-broadside angles θ_{ox} and θ_{oy} are used extensively in the literature (where they are designated often ** α and β; see Fig. 3.36a#) to define the main beam direction. Moreover, the direction of the main beam is often given in terms of $\sin\theta_{ox}$ and $\sin\theta_{oy}$ (usually designated as u_o and v_o, respectively) rather than θ_{ox} and θ_{oy}. Furthermore, the shape of the main beam is usually plotted in a rectangular coordinate system given by $u=\sin\alpha$ and $v=\sin\beta$, referred to as sine space, T-space,[20] u, v space, or direction-cosine space (when defined in terms of the direction cosines, $\cos\theta'_{ox}$ and $\cos\theta'_{oy}$). This space shall be discussed more extensively in Lecture No. 4. Sine space has the advantage that the shape of the beam when plotted in this space is invariant to its scan position. Visible space (i.e., the upper half hemisphere, above the planar antenna) in sine space is the region where $\sin^2\alpha + \sin^2\beta \leqslant 1$, that is, the region inside the unit circle drawn about the origin.

When array separability applies, the results obtained for linear arrays can be applied to rectangular planar arrays. The 3 dB beamwidths of a rectangular aperture possessing the separability property are given in Fig. 2-33 in terms of the 3 dB beamwidths of the x and y axes linear arrays.

The directivity of a separable rectangular array is given by the first half of (2-55) in terms of the directivities of the x and y axis linear arrays.[2] An alternate form in terms of N and η_{tx} and η_{ty}, the x and y axes linear array taper efficiencies, respectively, is given by the second half of (2-55). [1] This latter expression for directivity is the equivalent equation to (2-26) for the directivity of a planar array in terms of the directivity of the individual radiating elements mentioned previously. It shall be discussed more fully after its more general form is presented.

A more general expression, that applies for rectangular arrays for which the array separability property need not hold and which applies as well as to circular apertures, is given by (2-56) in terms of the antenna 3 dB beamwidth at broadside.[21] This expression gives the interesting result that the directivity is essentially independent of the antenna taper (to within ± 0.2 dB) as long as the taper provides a first sidelobe level which is $\leqslant -20$ dB and a far out sidelobe level of about -40 dB. The generalized form of the second half of (2-55) is given by (2-57).[1] Expression (2-57) also does not depend on separability holding and applies for circular as well as rectangular apertures. Furthermore, there are no constraints on the sidelobe level as was the case for (2-56), except that no grating lobes exist. For (2-57) and the second half of (2-55) to apply it is also necessary for N to be large, typically $N > 100$. A simple easy to remember approximate form for (2-56) is given by (2-59).[12],[19] The approximate expression on the right hand side of (2-59) gives results 0.3 dB larger than (2-56).

As mentioned before the second half of (2-55) and its more general form given by (2-57) are the equivalent equations to (2-26) for the directivity of a planar array in terms of the directivity of the radiating elements. For the special case where the individual radiating elements are ideally matched to free space and nondissipative, their directivity as a function of θ is given in terms of the physical area alotted to the radiating element in the array (as opposed to the actual size of the radiating element, which is usually smaller) times $\cos\theta$ as*

$$d_e = \frac{4\pi A_e}{\lambda^2}\cos\theta \quad . \tag{2-60}$$

* Also called the principle of pattern multiplication.[2],[3],[19]

** α, β however have been used also in the literature to represent the complements of α and β as defined in this lesson. This is done i Lesson 8; see Fig. 8.45. Also it has been used to represent u and v of this lesson as well. This is done in Lesson 7; see Fig. 7.6.

\# Strictly speaking α and β should be designated as α_o and β_o if they represent θ_{ox} and θ_{oy}, the main beam off-broadside angles. Without the subscript "o" α and β designate an arbitrary direction.

* Note that at broadside d_e equals the gain of an ideal uniformly illuminated broadside antenna whose area equals the physical area allotted to the radiating element, i.e., it is equal to $4\pi A_e/\lambda^2$.

PLANAR ARRAYS

1. RECTANGULAR APERTURES

2. CIRCULAR APERTURES

RECTANGULAR PLANAR ARRAY

$$E_a(\theta,\phi) = \sum_{m=-M_x}^{M_x} \sum_{n=-M_y}^{M_y} \frac{I_{mn}}{I_{oo}} e^{jk \sin\theta [md_x \cos\phi + nd_y \sin\phi]} \quad (2.37)$$

ASSUME:

$$\frac{I_{mn}}{I_{on}} = \frac{I_{mo}}{I_{oo}} = I_m \quad (2.38)$$

$$\frac{I_{mn}}{I_{mo}} = \frac{I_{on}}{I_{oo}} = I_n \quad (2.39)$$

I.E. ALL ROWS HAVE SAME CURRENT DISTRIBUTION GIVEN BY I_n, $n = -M_x$ TO M_x.

ALL COLUMNS HAVE SAME CURRENT DISTRIBUTION GIVEN BY I_m, $m = -M_y$ TO M_y.

THEN

$$\boxed{E_a(\theta,\phi) = E_x(\theta,\phi)\, E_y(\theta,\phi)} \quad \text{"ARRAY SEPARABILITY"} \quad (2.40)$$

WHERE

$E_x(\theta,\phi)$ = ARRAY FACTOR FOR ARRAY HAVING CURRENT DISTRIBUTION I_m, $m = -M_x$ TO M_x

$$= \sum_{-M_x}^{M_x} I_m e^{jm \frac{2\pi}{\lambda} d_x \sin\theta \cos\phi} \quad (2.41)$$

$$= \sum_{-M_x}^{M_x} I_m e^{jm \frac{2\pi}{\lambda} d_x \sin\theta} \quad \text{FOR } \phi = 0 \quad (2.42)$$

$E_y(\theta,\phi)$ = ARRAY FACTOR FOR ARRAY HAVING CURRENT DISTRIBUTION I_n, $n = -M_y$ TO M_y

$$= \sum_{n=-M_y}^{M_y} I_n e^{jn \frac{2\pi}{\lambda} d_y \sin\theta \sin\phi} \quad (2.43)$$

$$= \sum_{n=-M_y}^{M_y} I_n e^{jn \frac{2\pi}{\lambda} d_y \sin\theta} \quad \text{FOR } \phi = 90° \quad (2.44)$$

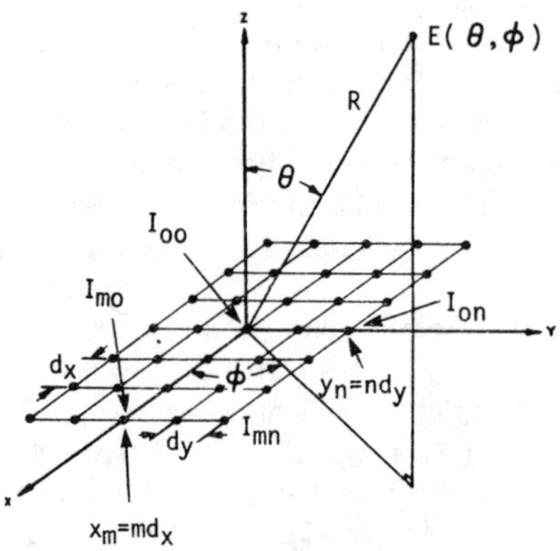

RECTANGULAR PLANAR ARRAY GEOMETRY

$$N_x = 2M_x+1 \quad (2.45)$$

$$N_y = 2M_y+1 \quad (2.46)$$

Fig. 2.31 Rectangular planar array. (After Elliott, Microwave Jour. [2].)

DIFFERENTIAL PHASE SHIFT VS BEAM POINTING ANGLE

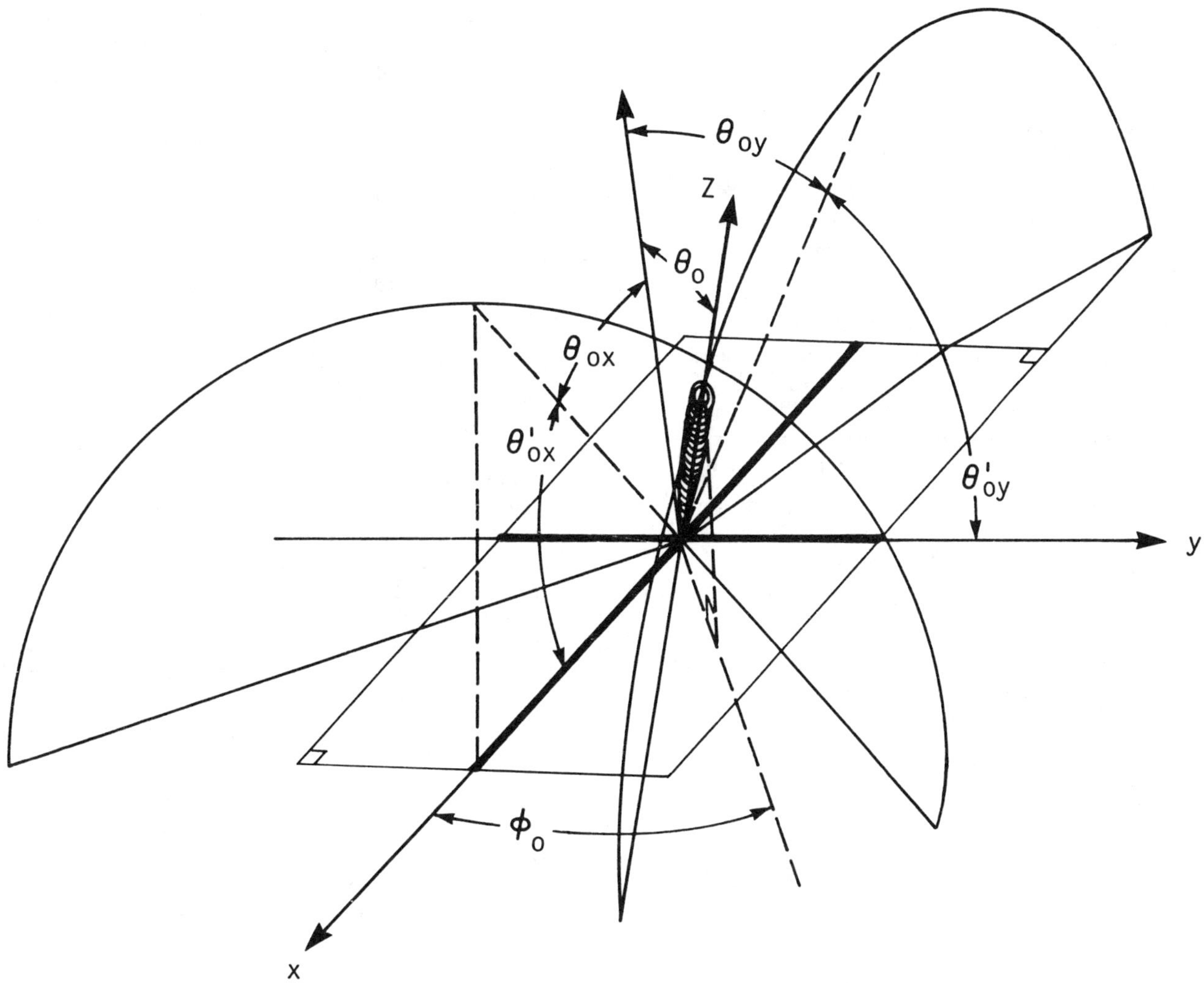

$$\alpha_x = \frac{2\pi}{\lambda} d_x \sin\theta_{ox} = \frac{2\pi}{\lambda} d_x \sin\theta_o \cos\phi_o = \frac{2\pi}{\lambda} d_x u_o \quad (2.47)$$

$$\alpha_y = \frac{2\pi}{\lambda} d_y \sin\theta_{oy} = \frac{2\pi}{\lambda} d_y \sin\theta_o \sin\phi_o = \frac{2\pi}{\lambda} d_y v_o \quad (2.48)$$

WHERE $u_o = \sin\theta_o \cos\phi_o = \sin\theta_{ox}$ (2.49)

$v_o = \sin\theta_o \sin\phi_o = \sin\theta_{oy}$ (2.50)

Fig. 2.32 Differential phase shift as a function of beam pointing angle.

The directivity of the whole array for the case of uniform weighting is then given by the equivalent expression to (2.26), which is

$$D = Nd_e \ . \tag{2-61}$$

Eq. (2-57) is identical to (2-61) for the case of uniform weighting, i.e., for $\eta_t = 1$. As discussed in relation to (2-26), (2-61) applies for the special case where two conditions are met: 1) the antenna radiating elements are nondissipative and 2) they are ideally matched to free space for all scan angles. Although these are somewhat idealized conditions, (2-57) is still a very useful equation. First of all, it provides a bound on the best performance one can aim for. In principle it is possible to nearly obtain an ideal match over all scan angles,[8],[22],[23],[24] (see also Lecture 5, Sect. 5.1.2) although to the authors knowledge there is no operational radar for which this has been done. Secondly, (2-57) can be modified to include the antenna dissipation and mismatch. For example, if due to antenna mismatch with scan angle the antenna gain goes down as $\cos^{3/2}\theta_o$ instead of $\cos\theta_o$ then (2-57) would be appropriately modified by replacing the $\cos\theta_o$ term with a $\cos^{3/2}\theta_o$ term. Including an efficiency factor to account for dissipation would give an expression for the realized antenna gain. Incorporating the other losses of Fig. 2-19 would give the overall antenna gain.

Intuitively, one expects the antenna directivity to ideally go down as $\cos\theta_o$ with scan angle since the projected area of the array goes down as $\cos\theta_o$. As a result the $\cos\theta_o$ term is sometimes referred to as the aperture foreshortening term.*

The $\cos\theta_o$ fall off can also be thought of as arising from the element radiating pattern. That is, the ideally matched array element has a gain pattern given by

$$e(\theta,\phi) = \sqrt{\cos\theta} \ . \tag{2-62}$$

It is interesting to now digress for a moment and discuss a paradox that arises in antenna theory. If the element gain pattern is given by (2-62) then it follows from the basic definition of directivity, given by (2-22), that the directivity of the element radiator having the gain pattern given by (2-62) is 4 (6 dB) at broadside. But it was indicated that (2-60) gives the directivity of the ideally matched radiating element. Assume $\lambda/2$ spacing between array elements arranged in a rectangular lattice. Then (2-60) indicates that the directivity is π (4.97 dB) instead of 4. According to Hannan[8] this discrepancy arises from losses due to mutual coupling and passive reflection not accounted for when using (2-62) and (2-22). As a result the radiated power is $\pi/4$ times the available power for the element having the pattern of (2-62) when $d_x = d_y = \lambda/2$. The reader is referred to Hannan's interesting paper for a discussion of other element spacings and for further details.

In practical arrays one generally matches ideally in one or two directions only. In spite of this, in some instances the fall off can be designed to be close to $\cos\theta$ in a preferred scan plane at the expense of a worse fall off in other planes. This is done by the appropriate choice of antenna polarization, element lattice structure, radiating element size and type, and direction angle for match to free space. A tradeoff thus exists as to what direction to match the antenna for best system performance.

Fig. 2-35 gives more accurate results for the directivity of various rectangular and circular apertures.

The conventional method to design a Dolph-Chebyshev one-way antenna type pattern for a rectangular array is to specify Dolph-Chebyshev weighting apertures for the column and row linear arrays independently. This procedure yields an array for which the one-way antenna array factor is given by the product of the row and column Dolph-Chebyshev array patterns; see (2.40). The Dolph-Chebyshev arrays of Fig. 2.35b were designed in this way.

Assume a 20 dB Dolph-Chebyshev design for the x and y axes linear array patterns of Fig. 2-31. Furthermore, assume that the main beam is pointing at broadside, that is, $\theta_o = 0^\circ$ in Fig. 2-31. Then the sidelobes in the principal planes (that is the planes defined by $\phi = 0^\circ$ and $\phi = 90^\circ$ in Fig. 2-31) will be 20 dB down. (Also, the antenna patterns in these principal planes will be Chebyshev polynomials.) However, outside of the principal planes the sidelobe level will be twice as far down (40 dB down) because the sidelobe pattern is produced by the product of the sidelobes of the x and y axes linear arrays. A rectangular aperture Dolph-Chebyshev array designed in this manner does not represent an optimum array in the sense that the main lobe width is minimized in all directions for a given design maximum sidelobe level. Tseng and Cheng have provided the weightings for a rectangular aperture which give such an optimum antenna pattern.[27] Fig. 2-36 and 2-37 compare the results obtained for the optimum Dolph-Chebyshev array design with those obtained with the more conventional design using Dolph-Chebyshev patterns for the row and column linear arrays. The optimum Dolph-Chebyshev design does not possess the array separability property whereas the conventional array does.

Fig. 2-36a and 2-36b compare the optimum and conventional array antenna patterns. Fig. 2-36c indicates that for a 40 x 40 element square array designed for a 20 dB sidelobe level, the optimum design provides a 10% and 15% decrease in the main lobe 20 dB width for maximum scan angles of 45° and 60°, respectively. As with the linear Dolph-Chebyshev array, for the optimum rectangular aperture Dolph-Chebyshev array there is a sidelobe level which gives maximum directivity. Fig. 2-37e and 2-37f gives the directivity of a square array as a function of N_x, the number of elements along each side of the array, for 30 and 40 dB sidelobe levels. Fig. 2-37f shows that for a 40 dB sidelobe level design the optimum Dolph-Chebyshev array has about the same directivity as the conventional array design. This is because though the conventional design exhibits a lower loss due to radiated energy into the sidelobes, this loss is balanced off a larger loss than the optimum design in the main lobe, the conventional array having a larger main lobe width.

2.2.2 Circular Apertures. The geometry for a circular aperture is shown in Fig. 2-38. The illumination functions, that is the weightings, for the circular apertures to be considered all assumed to be circularly symmetric, that is, they only depend on r. The illumination efficiencies and 3 dB beamwidth coefficients for Tapered-Taylor and $h+(1-h)\times\left[1-\left(\frac{r}{r_o}\right)^2\right]^m$ weightings are given in Fig. 2-39.[1],[28] Fig.

* Also referred to as the obliquity factor.[12],[25]

3dB BEAMWIDTHS OF SQUARE AND RECTANGULAR SEPARABLE PLANAR APERTURES

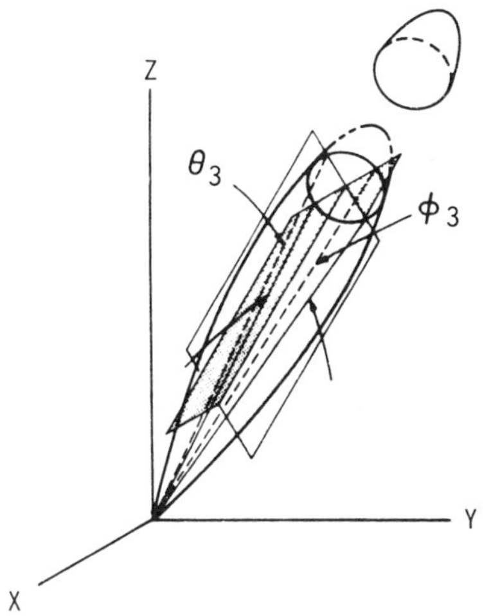

SQUARE ARRAY :

$$\theta_3 = \theta_{3x}/\cos\theta_0 \qquad (2.51)$$

$$\phi_3 = \theta_{3x} \qquad (2.52)$$

Eqs. (2.51) and (2.52) apply if square separable array has identical tapers for rows and columns or, more generally, if $\theta_{3x} = \theta_{3y}$.

RECTANGULAR ARRAY :

$$\theta_3^{-2} = \cos^2\theta_0[\theta_{3x}^{-2}\cos^2\phi_0 + \theta_{3y}^{-2}\sin^2\phi_0] \qquad (2.53)$$

$$\phi_3^{-2} = \theta_{3x}^{-2}\sin^2\phi_0 + \theta_{3y}^{-2}\cos^2\phi_0 \qquad (2.54)$$

WHERE

θ_{3x} = 3db BEAMWIDTH OF x - AXIS LINEAR ARRAY AT BROADSIDE.

θ_{3y} = 3db BEAMWIDTH OF y - AXIS LINEAR ARRAY AT BROADSIDE.

Fig. 2.33 3 dB beamwidths of square and rectangular separable planar apertures. (After Elliott, Microwave Jour. [2].)

DIRECTIVITY OF PLANAR RECTANGULAR AND CIRCULAR APERTURES

SEPARABLE RECTANGULAR ARRAYS* :

$$D \doteq \pi D_x D_y \cos\theta_0 \doteq \frac{4\pi(d_x d_y)}{\lambda^2}\eta_t N\cos\theta_0 \qquad (2.55)$$

WHERE D_x AND D_y ARE THE ROW AND COLUMN ARRAY DIRECTIVITIES, η_t = ARRAY TAPER EFFICIENCY $= \eta_{tx}\eta_{ty}$ FOR SEPARABLE RECTANGULAR ARRAY, AND N = NUMBER OF ARRAY ELEMENTS.

RECTANGULAR AND CIRCULAR ARRAYS*# :

$$D \doteq \frac{38{,}400}{\theta_{3x}\theta_{3y}}\cos\theta_0 \qquad \text{(ERROR} \doteq \pm 0.2\text{dB)} \qquad (2.56)$$

WHEN 1) 1st SIDELOBE $\leq$ -20 dB AND 2) FAR OUT SIDELOBE $\leq$ -40 dB. θ_{3x} AND θ_{3y} ARE THE 3 dB BEAMWIDTHS FOR $\theta_0 = 0°$, I.E., AT BROADSIDE, IN DEGREES.

ALTERNATE EXPRESSION FOR D OF RECTANGULAR AND CIRCULAR ARRAYS* :

$$D = \frac{4\pi A_e}{\lambda^2}\eta_t N\cos\theta_0 \qquad (2.57)$$

WHERE A_e = AREA OCCUPIED BY ONE ARRAY ELEMENT $= d_x d_y$ FOR ARRAY ELEMENT ARRANGEMENT OF Fig. 2.31

η_t = APERTURE TAPER EFFICIENCY

$$= \frac{\text{D FOR APERTURE WITH AMPLITUDE WEIGHTING}}{\text{D FOR UNIFORM WEIGHTING}} = \frac{(\sum_m\sum_n I_{mn})^2/(\sum_m\sum_n I_{mn}^2)}{N} \qquad (2.58)$$

* APPLY AS LONG AS NO GRATING LOBES EXIST.

\# WHEN θ_{3x} AND θ_{3y} ARE IN RADIANS (2.56) BECOMES

$$D = \frac{11.70}{\theta_{3x}\theta_{3y}}\cos\theta_0 \doteq \frac{4\pi}{\theta_{3x}\theta_{3y}}\cos\theta_0 \qquad (2.59)$$

Fig. 2.34 Directivity expressions for planar rectangular and circular apertures.

DIRECTIVITY OF RECTANGULAR SEPARABLE ARRAYS AND ELLIPTICAL APERTURES:

MORE ACCURATE VALUES :

$$D \doteq \frac{C}{\theta_{3x}\,\theta_{3y}} \cos\theta_o \qquad (2.63)$$

WHERE $C = D\,\theta_{3x}\,\theta_{3y}$ = DIRECTIVITY-BEAMWITH PRODUCT FOR $\theta_o = 0°$ IS GIVEN BY TABLES BELOW FOR θ_{3x} AND θ_{3y} IN DEGREES.

(a) [26]

APERTURE ILLUMINATION	*C*	$10 \log_{10} C$
Uniform circular	33,700	45.28
Uniform rectangular	32,300	45.10
Rectangular, weighted:		
$\cos(\pi x)\cos(\pi y)$	37,400	45.73
$\cos^2(\pi x)\cos^2(\pi y)$	37,300	45.72
$\cos^3(\pi x)\cos^3(\pi y)$	36,800	45.66
Tapered Taylor, 20-db sidelobes	~36,000	45.57
" " *30-db* "	~36,900	45.67
" " *40-db* "	~37,000	45.68

(b) [21]

GAIN-BEAMWIDTH PRODUCT FOR DOPH-CHEBYSHEV RECTANGULAR ARRAYS**

Number of Elements	*Sidelobe Level (db)*	*Half-Power Beamwidth (degrees)*	*C*
6	14	17	33,000
10	17	10.5	33,200
20	21	5.4	33,400
24	22	4.6	33,500
28	23	4.0	34,800
32	36	4.2	38,400
32	40	4.5	40,100
40	27	3.0	37,200
40	30	3.1	38,000
40	34	3.3	37,600
40	39	3.5	38,500

** $d_x/\lambda = d_y/\lambda = 1/2$.

(c) [21]

GAIN-BEAMWIDTH PRODUCT OF RECTANGULAR ARRAYS WITH ONE-PARAMETER TAYLOR WEIGHTING

Sidelobe Ratio (db)	η_t	K_o *(deg)*	*C*
20	0.933	58.6	37,700
25	0.863	63.9	38,200
30	0.801	68.8	38,200
35	0.751	73.2	38,000

(d) [21]

GAIN-BEAMWIDTH PRODUCT FOR A DISTRIBUTION $(1-r^2)^m$ OVER A CIRCULAR APERTURE

m	η_t	K_o *(Radians)*	*Sidelobe Level (db)*	*C*
0	1.00	1.02	17.6	33,700
1	0.75	1.27	24.6	39,200
2	0.56	1.47	30.6	39,200
3	0.44	1.65		38,800
4	0.36	1.81		38,200

* APPLIES AS LONG AS NO GRATING LOBES EXIST.

Fig. 2.35 Directivity of rectangular separable apertures and elliptical apertures: more accurate values. (Tables b, c, and d after Stegen, IEEE PGAP [19]; Table a after Barton and Ward, Prentice-Hall [20].)

SCANNABLE DOLPH-CHEBYSHEV PLANAR SQUARE ARRAY WITH UNIFORM SIDELOBES OVER ALL DIRECTIONS

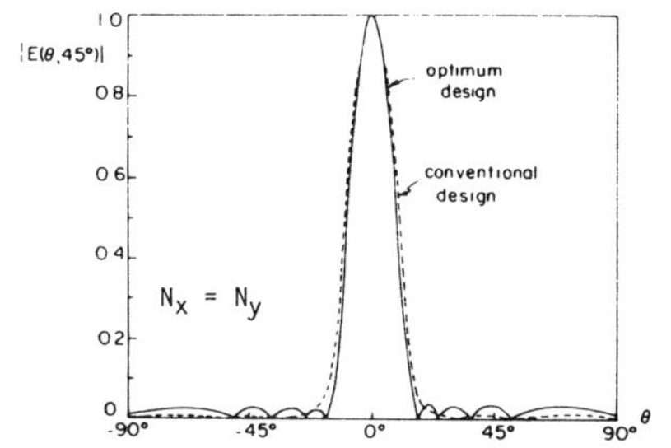

a) ARRAY FACTOR IN ϕ=45° PLANE: N_x=11, d_x=d_y= $\lambda/2$, θ_o= 0°
(DESIGNED SIDELOBE LEVEL=30 dB)

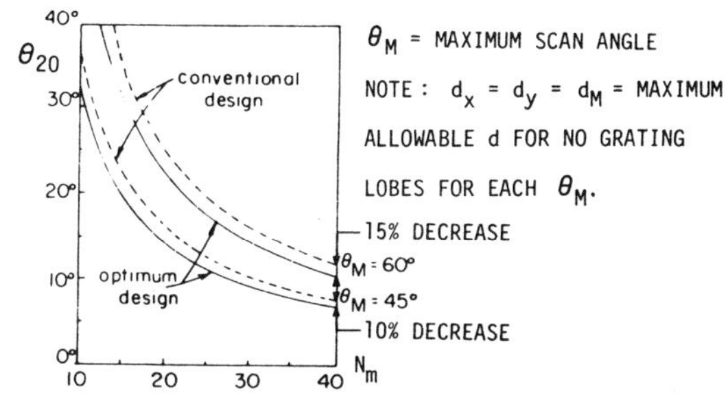

c) 20 dB 1-WAY BEAMWIDTH θ_{20} VERSUS MINIMUM NUMBER OF ELEMENTS N_m PER SIDE
(20 dB SIDELOBES)

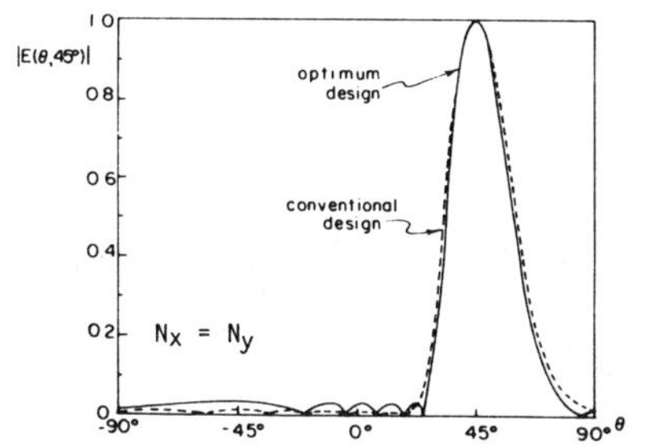

(b) ARRAY FACTOR IN ϕ=45° PLANE: N_x=11, d_x=d_y= $\lambda/2$, θ_o=45°
ϕ_o=45°. (DESIGNED SIDELOBE LEVEL=30 dB.)

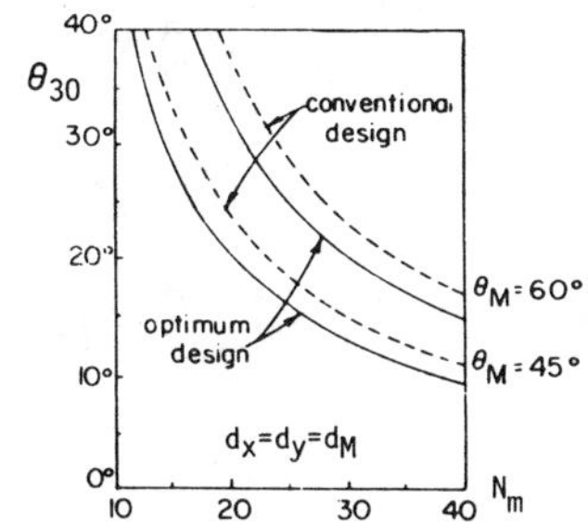

d) 30 dB 1-WAY BEAMWIDTH θ_{30} VERSUS MINIMUM NUMBER OF ELEMENTS N_m PER SIDE
(30 dB SIDELOBES)

Fig. 2.36 Scannable Dolph-Chebyshev planar square array with uniform sidelobes over all directions: array factor; 20 and 30 dB beamwidths. (After Tseng and Cheng, IEEE PGAP [21] .)

SCANNABLE DOLPH-CHEBYSHEV PLANAR SQUARE ARRAY HAVING UNIFORM SIDELOBES OVER ALL DIRECTIONS

3 dB BEAMWIDTH

MAXIMUM ALLOWABLE ELEMENT SPACING

DIRECTIVITY

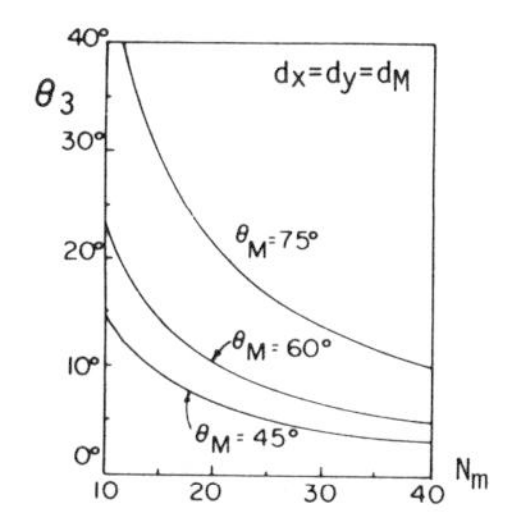

a) 3 dB 1-WAY BEAMWIDTH θ_3 VERSUS N_m
(20-dB SIDELOBES)

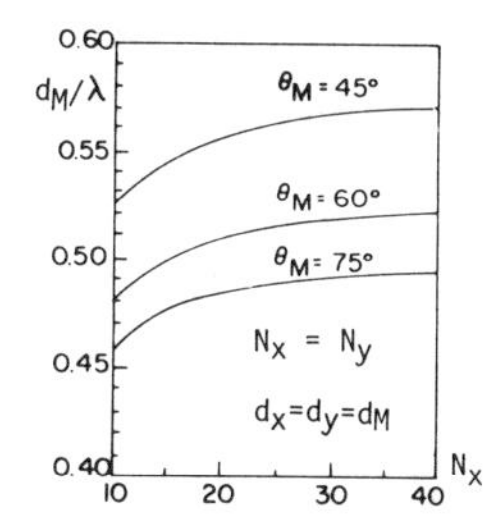

c) d_M/λ VERSUS L FOR NONAPPEARANCE OF GRATING LOBES
(20-dB SIDELOBES)

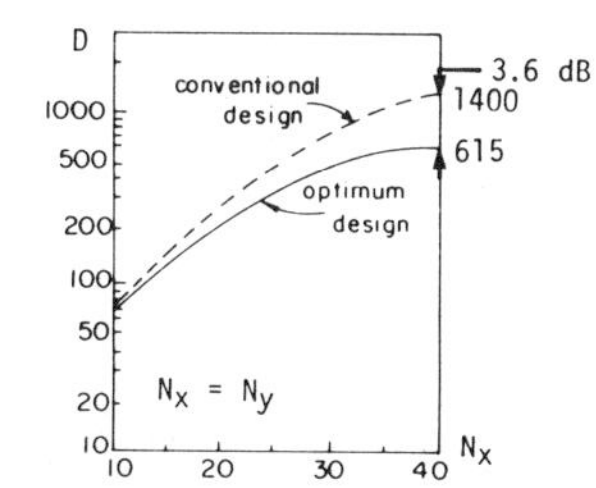

e) DIRECTIVE GAIN VERSUS N_x
(d_x=d_y= $\lambda/2$, θ_o=0°,30 dB SIDELOBES).

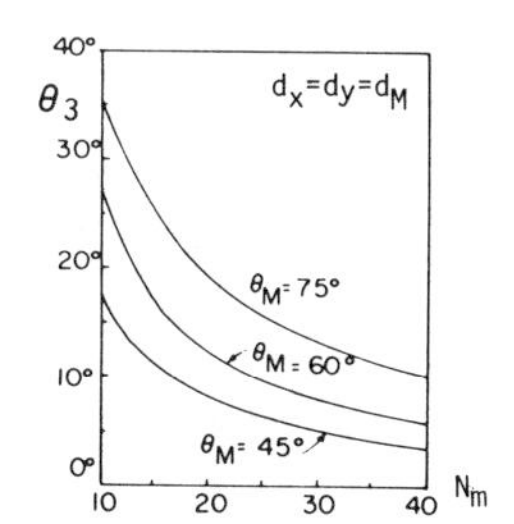

b) 3 dB 1-WAY BEAMWIDTH θ_3 VERSUS N_m
(30-dB SIDELOBES).

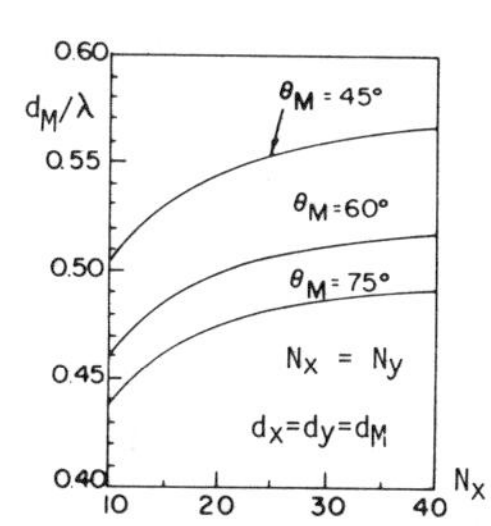

d) d_M/λ VERSUS L FOR NONAPPEARANCE OF GRATING LOBES
(30-dB SIDELOBES)

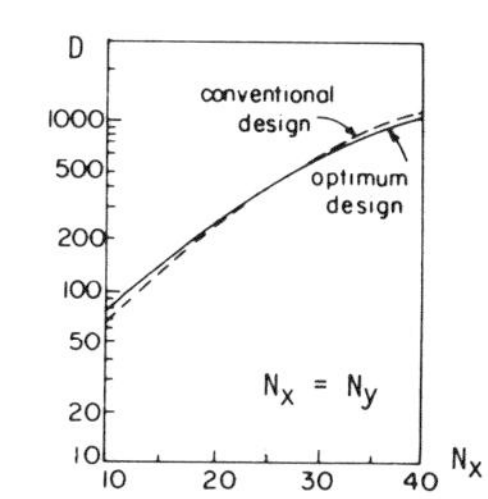

f) DIRECTIVE GAIN VERSUS N_x
(d_x=d_y= $\frac{\lambda}{2}$, θ_o=0°, 40 dB SIDELOBES).

Fig. 2.37 Scannable Dolph-Chebyshev planar square array having uniform sidelobes over all directions: 3 dB beamwidths; maximum allowable element spacings; directivity.(After Tseng and Cheng, IEEE PGAP [21].)

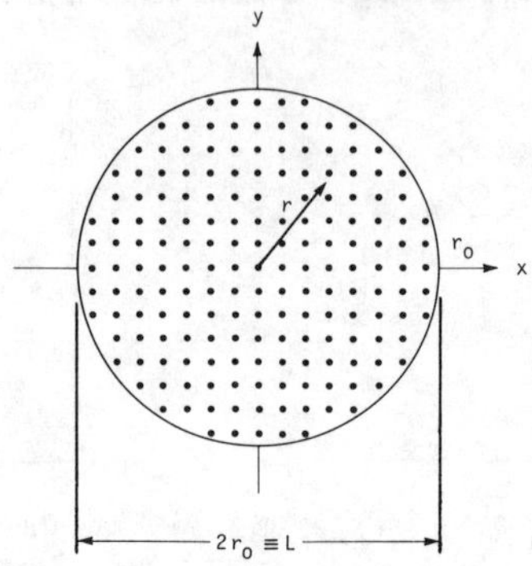

Fig. 2.38 Circular apertures. (After Allen, Lincoln Laboratory [1].)

CIRCULAR APERTURE ILLUMINATION EFFICIENCY AND 3dB 1-WAY BEAMWIDTH FOR SEVERAL TAPERS

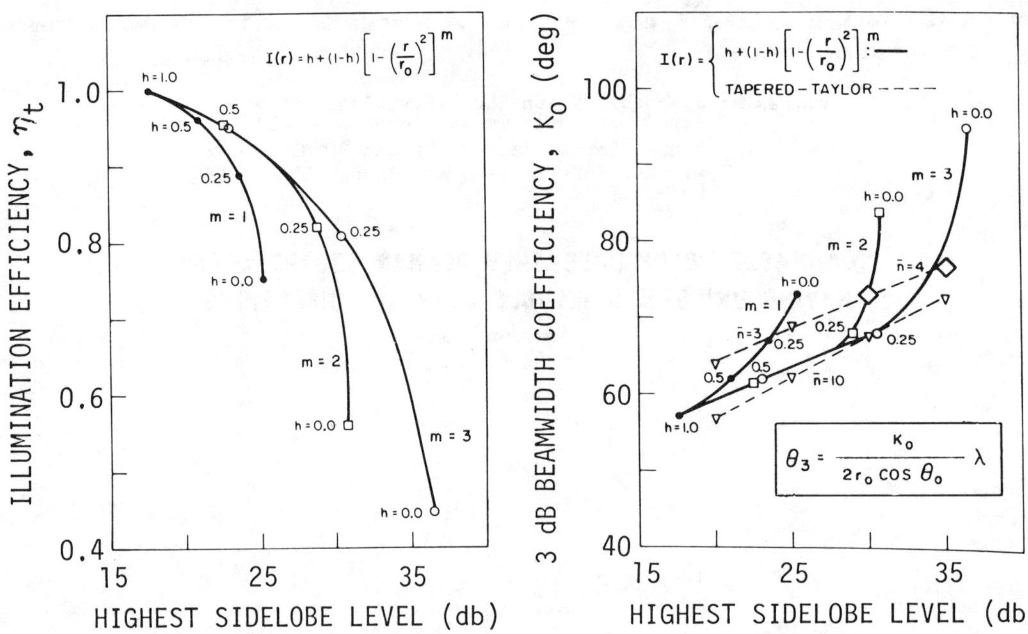

(a) ILLUMINATION EFFICIENCY (b) 3 dB 1-WAY BEAMWIDTH COEFFICIENT.

Tapered-Taylor			$\left[1 - \left(\frac{r}{r_o}\right)^2\right]^m$		
Sidelobe level, db	$\bar{n}$	K_o(rad.)	Sidelobe level, db	m	K_o(rad.)
25	4	1.13	24.6	1	1.27
30	4	1.20	30.6	2	1.47
35	5	1.25	36.0	3	1.65
40	5	1.31	40.9	4	1.81

Fig. 2.39 Circular aperture illumination efficiency and 3 dB 1-way beamwidth for two tapers. (Figs. after Allen, Lincoln Laboratory [1]; Table after Hansen, IRE PGAP [22].)

2.40a gives the beam broadening factor as a function of the parameter $\bar{n}$ and the design sidelobe level for circular aperture Tapered-Taylor weightings. [28]

Fig. 2.40b gives the shapes of the Tapered-Taylor weighting as a function of sidelobe level and $\bar{n}$. This data is of importance in the design of optically fed arrays because generally one can most easily realize the monotonically decreasing tapered weightings for this type of feed system.

As was the case for Tapered-Taylor weighted linear apertures, for circular apertures having a Tapered-Taylor weighting there exists an $\bar{n}$ which gives maximum directivity, or equivalently, maximum η_t, for a specified sidelobe level; (see Fig. 2-41a and 2-41b).[17] Also shown for comparison are the results for a conventional separable rectangular array having the optimum linear aperture Tapered-Taylor weighting for the row and column arrays as specified by Fig. 2-26. Fig. 2-41c shows the rapid fall off in the illumination efficiencies that result if the ideal-Taylor weighting for circular apertures is used instead of the Tapered-Taylor weighting.[17] The fall off is not as drastic for the case of linear arrays; see for example the 30 and 40 dB sidelobe curves for directivity given in Fig. 2.22 for the linear Dolph-Chebyshev arrays.

The illumination function for several Tapered-Taylor circular aperture weightings are tabulated in Fig. 2.42. [28] Fig. 2.43 gives plots of some of these weights. [28] A sample Tapered-Taylor design for a circular aperture is worked out in Fig. 2.44.

2.3 Equivalence Between Circular and Linear Aperture Antenna Patterns.

An equivalence exists between the antenna patterns for circular and linear apertures.[29] ,[30] Specifically, there exists a broadside linear aperture whose antenna pattern is the same as the antenna pattern of a circular aperture having circular symmetry for its illumination function. This equivalence holds as long as the projection of the aperture weighting for the circular aperture onto the x axis is the same as the aperture weighting for the equivalent linear array; see Fig. 2-45 for a mathematical statement of equivalence with some examples. Additional examples are given in the Appendix together with other tapered aperture results.

It is worth pointing out that the linear-aperture Tapered-Taylor and circular aperture Tapered-Taylor patterns are not identical, that is, equivalence does not exist between these antenna patterns. While the far out sidelobes for the linear-aperture Tapered-Taylor fall off as $\sin \dot{u}/\dot{u}$; (see Fig. 2-24), those for the circular-aperture Tapered-Taylor fall off as $\sin \dot{u}/\dot{u}^{3/2}$. However, the ideal-Taylor antenna patterns of continuous linear apertures are equivalent to the ideal-Taylor antenna patterns of continuous circular apertures. As a consequence the 3 dB beamwidths for these two cases are identical for the same sidelobe level; see Figs. 2-40 and 2-A-2 , and their footnotes.

2.4 Monopulse Difference Patterns.

Up to now consideration was only given to the sum patterns of antenna apertures. The sum pattern is principally designed to have maximum detection sensitivity within the main lobe while minimizing possible sidelobe interference. For the difference pattern one designs for maximum angle measurement accuracy while minimizing sidelobe interference. To maximize angle accuracy the slope of the difference pattern is maximized. However, the greater the difference pattern slope the higher the sidelobes are.

Bayliss has developed difference patterns for linear and circular apertures which are the equivalent to the Taylor sum beam patterns.[31] There is an ideal-Bayliss pattern similar to the ideal-Taylor pattern and a Tapered-Bayliss similar to the Tapered-Taylor. The ideal-Bayliss has constant sidelobe levels out to infinity while the Tapered-Bayliss approximation to the ideal-Bayliss has constant sidelobes out to the $\bar{n}$-1 st sidelobe after which they go down as $\dot{u}^{-3/2}$ for the case of a circular aperture. The ideal-Bayliss pattern has the maximum difference slope pattern for a specified sidelobe level. Plots and tables of the Bayliss difference patterns are given in Ref. [31].

That equal sidelobes about the boresight angle may not be optimum is addressed in Ref. [32]. There consideration is given to obtaining difference patterns which have minimum sidelobes or nulls in the region where interference is expected while at the same time maximizing angular accuracy sensitivity.

2.5 Array Thinning.

Array thinning involves the removal of radiating elements from a linear or planar array. Thinning is used to reduce the array cost. Thinning has been effectively implemented on the ESAR,[33] HAPDAR,[34] and the operational FPS-85[35] (on receive) radars. It is to be used on the Cobra Dane radar presently under construction.[36]

The technique allows for increased angle accuracy with minimum system cost increase. Building a thinned array also allows for future growth by way of increased system sensitivity while still maintaining the same system beamwidth. These advantages are achieved at a price in reduced antenna gain, higher far out sidelobes, and longer search times. These trade-offs are discussed further later. First, a technique for array thinning called statistical density tapering is described.

2.5.1 Statistical Density Tapering[33].

It is easiest to illustrate the technique for a linear array. Consider a linear array which would have N=2M+1 radiating elements if it were filled. It is assumed that these elements would be equally spaced, d apart, as illustrated in Fig. 2-46. Assume that it is desired to have an array factor for the thinned array corresponding to that produced if the array were filled and weighted with weightings I_i, i = −M to M. (This full, weighted array whose pattern the thinned array is to match is called the reference design array.) Then the thinned array design is arrived at by either permanently installing in the array or permanently deleting from the array the radiating elements of the full array according to the rules given in Fig. 2-46. The resulting thinned array is illuminated with a uniform weighting, i.e., I_i = 1 for the thinned array. Following this procedure for the statistical design of thinned arrays yields an antenna pattern which on the average (the average being over all possible such random designs) is identical to that of the reference design weighted full array; see (2-69).

The design procedure given in Fig. 2-46 indicates that the ith radiating element will be installed with probability kI_i where the parameter k is a thinning parameter between 0 and 1. The lower k is the greater the amount of thinning. For k = 1, one has natural thinning. The rules of Fig. 2-46 result in a random installation of the elements with the greatest number of elements being installed in the region where I_i is largest. This is illustrated in the example thinned circular array of Fig. 2-47 which was designed for

TAPERED TAYLOR FOR CIRCULAR APERTURE

(a) Ideal Beamwidth and Beam Broadening Factor

Design Side-lobe Level, Decibels	r (Sidelobe Voltage Ratio)	Ideal Taylor K_{oo} * (Degrees)	A^2	Beam Broadening Factor σ							
				$\bar{n}=3$	$\bar{n}=4$	$\bar{n}=5$	$\bar{n}=6$	$\bar{n}=7$	$\bar{n}=8$	$\bar{n}=9$	$\bar{n}=10$
15	5.62341	45.93	0.58950	1.2382	1.1836	1.1485	1.1244	1.1069	1.0937	1.0833	1.0751
16	6.30957	47.01	0.64798	1.2330	1.1809	1.1469	1.1233	1.1061	1.0931	1.0829	1.0747
17	7.07946	48.07	0.70901	1.2276	1.1781	1.1452	1.1222	1.1053	1.0925	1.0825	1.0743
18	7.94328	49.12	0.77266	1.2220	1.1752	1.1434	1.1210	1.1045	1.0919	1.0820	1.0740
19	8.91251	50.15	0.83891	1.2163	1.1723	1.1416	1.1198	1.1037	1.0913	1.0815	1.0736
20	10.00000	51.17	0.90777	1.2104	1.1692	1.1398	1.1186	1.1028	1.0906	1.0810	1.0732
21	11.2202	52.17	0.97927	1.2044	1.1660	1.1379	1.1173	1.1019	1.0899	1.0805	1.0728
22	12.5893	53.16	1.05341	1.1983	1.1628	1.1359	1.1160	1.1009	1.0892	1.0799	1.0723
23	14.1254	54.13	1.13020	1.1920	1.1594	1.1338	1.1146	1.1000	1.0885	1.0793	1.0719
24	15.8489	55.09	1.20965	1.1857	1.1560	1.1317	1.1132	1.0990	1.0878	1.0788	1.0714
25	17.7828	56.04	1.29177	1.1792	1.1525	1.1296	1.1118	1.0979	1.0870	1.0782	1.0708
26	19.9526	56.97	1.37654	1.1726	1.1489	1.1274	1.1103	1.0969	1.0862	1.0775	1.0704
27	22.3872	57.88	1.46395	1.1660	1.1452	1.1251	1.1087	1.0958	1.0854	1.0769	1.0699
28	25.1189	58.78	1.55406	1.1592	1.1415	1.1228	1.1072	1.0946	1.0845	1.0762	1.0694
29	28.1838	59.67	1.64683	1.1524	1.1377	1.1204	1.1056	1.0935	1.0836	1.0756	1.0688
30	31.6228	60.55	1.74229	1.1455	1.1338	1.1180	1.1039	1.0923	1.0827	1.0749	1.0683
31	35.4813	61.42	1.84044	1.1385	1.1298	1.1155	1.1022	1.0911	1.0818	1.0742	1.0677
32	39.8107	62.28	1.94126	1.1315	1.1258	1.1129	1.1005	1.0898	1.0809	1.0734	1.0671
33	44.6684	63.12	2.04473	1.1244	1.1217	1.1103	1.0987	1.0885	1.0799	1.0727	1.0665
34	50.1187	63.96	2.15092		1.1176	1.1077	1.0969	1.0872	1.0790	1.0719	1.0659
35	56.2341	64.78	2.25976		1.1134	1.1050	1.0951	1.0859	1.0779	1.0711	1.0653
36	63.0957	65.60	2.37129		1.1091	1.1023	1.0932	1.0846	1.0769	1.0703	1.0647
37	70.7946	66.40	2.48551		1.1048	1.0995	1.0913	1.0832	1.0759	1.0695	1.0640
38	79.4328	67.19	2.60241		1.1005	1.0967	1.0894	1.0818	1.0748	1.0687	1.0633
39	89.1251	67.98	2.72201		1.0961	1.0938	1.0874	1.0803	1.0737	1.0678	1.0627
40	100.0000	68.76	2.84428		1.0916	1.0910	1.0854	1.0789	1.0726	1.0670	1.0620

LET θ_{300} = 3 dB BROADSIDE BEAMWIDTH FOR IDEAL TAYLOR TAPER.

THEN $$\theta_{300} \doteq \frac{K_{oo}}{2r_o}\lambda \quad \text{FOR } \theta_{300} \leq 10°.^{*} \qquad (42a)$$

AND θ_{30} = 3 dB BROADSIDE BEAMWIDTH FOR TAPERED TAYLOR.

$$\doteq \sigma \frac{K_{oo}}{2r_o}\lambda = \frac{K_o}{2r_o}\lambda \quad \text{FOR } \theta_{30} \leq 10°^{**} \qquad (42b)$$

WHERE $K_o = \sigma K_{oo}$

A = PARAMETER OF TAPERED TAYLOR

$$= (\text{arc cosh } r)/\pi \qquad (42c)$$

DISTRIBUTION SHAPES

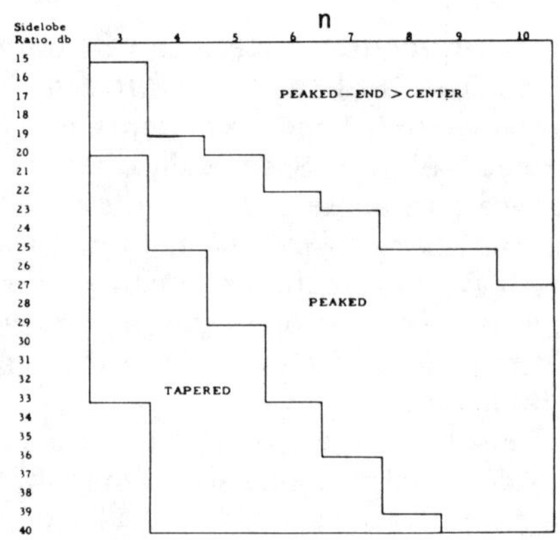

(b)

$$* \quad \theta_{300} = 2 \text{ arc sin } \frac{K_{oo}}{4r_o}\lambda \qquad (42d)$$

$$** \quad \theta_{30} = 2 \text{ arc sin } \frac{K_o}{4r_o}\lambda \qquad (42e)$$

NOTE: VALUES FOR K_{oo} AND IN TURN θ_{300} OBTAINED FOR SPECIFIED r FOR CIRCULAR APERTURE HAVING IDEAL-TAYLOR WEIGHTING ALSO APPLY FOR LINEAR APERTURE HAVING IDEAL-TAYLOR WEIGHTING. SAME IS TRUE FOR VALUES OF A. MOREOVER, (42c) APPLIES FOR CIRCULAR AND LINEAR APERTURES HAVING TAPERED-TAYLOR WEIGHTING. SEE FIG. A-3.

Fig. 2.40 Tapered-Taylor for circular aperture. (After Hansen, IRE PGAP [22].)

OPTIMIZATION OF ILLUMINATION EFFICIENCY FOR TAPERED TAYLOR CIRCULAR APERTURE

(a) ILLUMINATION EFFICIENCY η_t VS $\bar{n}$ AND SIDELOBE LEVEL

Sidelobe Level, db	$\bar{n}$ = 3	4	5	6	8	10
15	**0.9497**	0.8404	0.7166	0.6023	0.4254	0.3083
18	**0.9893**	0.9489	0.8778	0.7938	0.6284	0.4922
20	**0.9820**	0.9723	0.9356	0.8808	0.7506	0.6238
22	0.9601	**0.9684**	0.9565	0.9279	0.8412	0.7398
25	0.9151	0.9324	**0.9404**	0.9379	0.9064	0.8526
30	0.8377	0.8482	0.8623	0.8735	**0.8838**	0.8804
35	—	0.7708	0.7779	0.7880	0.8048	0.8153
40	—	0.7056	0.7063	0.7119	0.7252	**0.7365**

NOTE: VALUES OF η_t IN BOXES GIVEN MAXIMUM TABULATED η_t FOR GIVEN SIDELOBE LEVEL.

(b) $\bar{n}$ FOR MAXIMUM η_t

	TYPE APERTURE			
	Circular		Rectangular #	
Sidelobe Level, db	Optimum Gain	$\bar{n}$	Optimum Gain	$\bar{n}$
20	0.9820	3	0.941	6
25	0.9404	5	0.81	12
30	0.8838	8 to 9	0.5625	24•

• ESTIMATED (VERY BROAD MAXIMUM.)
CONVENTIONAL SEPARABLE WEIGHTING DESIGN.

(c)

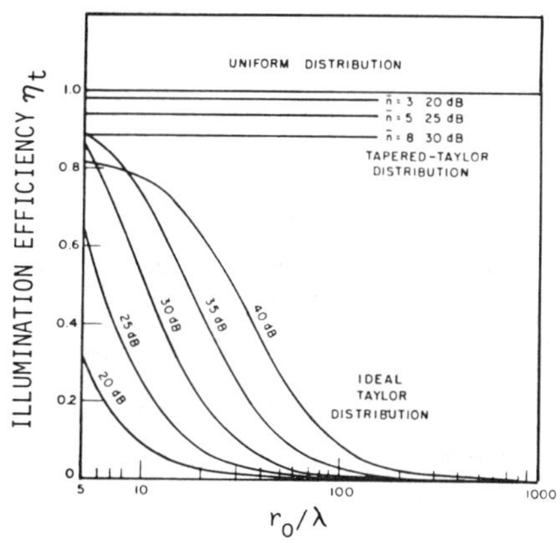

Fig. 2.41 **Optimization of illumination efficiency for Tapered-Taylor circular aperture. (After Rudduck and Wu, Radio Science [15].)**

TABLE OF SAMPLE AMPLITUDE WEIGHTINGS FOR TAPERED TAYLOR PATTERNS FOR CIRCULAR APERTURE

$$I(r) = I\left(\frac{kr_0}{20}\right)$$ for 25 db sidelobe level

k	$\bar{n}=3$	$\bar{n}=4$	$\bar{n}=5$	$\bar{n}=6$	k	$\bar{n}=7$	$\bar{n}=8$	$\bar{n}=9$	$\bar{n}=10$
0	0.36063	0.33057	0.33889	0.30780	0	0.32293	0.29429	0.31251	0.28546
1	0.35908	0.33046	0.33615	0.30936	1	0.31857	0.29778	0.30630	0.29114
2	0.35448	0.32990	0.32868	0.31243	2	0.30844	0.30347	0.29483	0.29786
3	0.34699	0.32819	0.31847	0.31329	3	0.29874	0.30270	0.28924	0.29244
4	0.33686	0.32441	0.30786	0.30862	4	0.29329	0.29297	0.28900	0.28061
5	0.32441	0.31761	0.29845	0.29777	5	0.29021	0.28045	0.28399	0.27480
6	0.31004	0.30716	0.29043	0.28328	6	0.28401	0.27201	0.27059	0.27154
7	0.29420	0.29287	0.28245	0.26914	7	0.27113	0.26663	0.25695	0.25973
8	0.27735	0.27519	0.27223	0.25796	8	0.25370	0.25698	0.24939	0.24310
9	0.25998	0.25514	0.25766	0.24895	9	0.23775	0.23950	0.24153	0.23334
10	0.24258	0.23419	0.23796	0.23816	10	0.22708	0.22022	0.22447	0.22620
11	0.22559	0.21398	0.21432	0.22111	11	0.21856	0.20772	0.20305	0.20819
12	0.20943	0.19607	0.18984	0.19633	12	0.20383	0.20070	0.19047	0.18528
13	0.19448	0.18162	0.16871	0.16752	13	0.17714	0.18611	0.18518	0.17583
14	0.18104	0.17124	0.15478	0.14286	14	0.14308	0.15375	0.16591	0.17068
15	0.16938	0.16488	0.15027	0.13135	15	0.11694	0.11358	0.12166	0.13628
16	0.15968	0.16195	0.15485	0.13805	16	0.11588	0.09509	0.08151	0.07862
17	0.15209	0.16141	0.16570	0.16088	17	0.14629	0.12389	0.09701	0.06963
18	0.14666	0.16210	0.17835	0.19077	18	0.19682	0.19498	0.18478	0.16623
19	0.14344	0.16298	0.18816	0.21541	19	0.24302	0.26935	0.29361	0.31510
20	0.14238	0.16335	0.19180	0.22481	20	0.26143	0.30063	0.34238	0.38688

for 30 db sidelobe level

k	$\bar{n}=3$	$\bar{n}=4$	$\bar{n}=5$	$\bar{n}=6$	k	$\bar{n}=7$	$\bar{n}=8$	$\bar{n}=9$	$\bar{n}=10$
0	0.41139	0.39967	0.39892	0.37942	0	0.38363	0.36543	0.37280	0.35611
1	0.40974	0.39847	0.39661	0.37926	1	0.38048	0.36644	0.36862	0.35833
2	0.40483	0.39480	0.39006	0.37798	2	0.37254	0.36695	0.35995	0.35949
3	0.39672	0.38849	0.38025	0.37378	3	0.36307	0.36235	0.35280	0.35256
4	0.38552	0.37932	0.36834	0.36508	4	0.35412	0.35135	0.34701	0.34066
5	0.37139	0.36707	0.35524	0.35165	5	0.34489	0.33734	0.33726	0.33059
6	0.35458	0.35169	0.34123	0.33489	6	0.33269	0.32416	0.32170	0.32064
7	0.33538	0.33328	0.32595	0.31697	7	0.31584	0.31151	0.30482	0.30499
8	0.31419	0.31221	0.30856	0.29946	8	0.29557	0.29568	0.29024	0.28576
9	0.29146	0.28909	0.28833	0.28226	9	0.27519	0.27488	0.27480	0.26945
10	0.26773	0.26471	0.26507	0.26369	10	0.25691	0.25241	0.25389	0.25402
11	0.24361	0.24001	0.23947	0.24175	11	0.23949	0.23310	0.23018	0.23244
12	0.21974	0.21595	0.21315	0.21581	12	0.21892	0.21671	0.21095	0.20796
13	0.19680	0.19342	0.18824	0.18777	13	0.19238	0.19669	0.19583	0.19067
14	0.17546	0.17316	0.16688	0.16176	14	0.16218	0.16773	0.17390	0.17621
15	0.15633	0.15572	0.15056	0.14251	15	0.13611	0.13481	0.13925	0.14695
16	0.13999	0.14143	0.13973	0.13304	16	0.12316	0.11326	0.10685	0.10573
17	0.12689	0.13044	0.13378	0.13307	17	0.12726	0.11690	0.10397	0.09040
18	0.11738	0.12273	0.13130	0.13899	18	0.14366	0.14426	0.14032	0.13180
19	0.11166	0.11821	0.13066	0.14554	19	0.16106	0.17646	0.19066	0.20350
20	0.10977	0.11674	0.13061	0.14827	20	0.16831	0.19045	0.21382	0.23888

for 35 db sidelobe level

k	$\bar{n}=4$	$\bar{n}=5$	$\bar{n}=6$	$\bar{n}=7$	k	$\bar{n}=8$	$\bar{n}=9$	$\bar{n}=10$
0	0.46024	0.45750	0.44562	0.44456	0	0.43248	0.43396	0.42295
1	0.45821	0.45511	0.44421	0.44174	1	0.43181	0.43058	0.42301
2	0.45212	0.44813	0.43968	0.43406	2	0.42852	0.42253	0.42029
3	0.44211	0.43704	0.43127	0.42323	3	0.42034	0.41310	0.41111
4	0.42832	0.42249	0.41840	0.41043	4	0.40671	0.40237	0.39726
5	0.41100	0.40508	0.40116	0.39548	5	0.38956	0.38768	0.38251
6	0.39046	0.38524	0.38043	0.37728	6	0.37109	0.36821	0.36622
7	0.36706	0.36316	0.35751	0.35525	7	0.35145	0.34658	0.34551
8	0.34124	0.33888	0.33348	0.33026	8	0.32904	0.32504	0.32175
9	0.31353	0.31245	0.30869	0.30426	9	0.30319	0.30227	0.29867
10	0.28452	0.28413	0.28275	0.27874	10	0.27588	0.27609	0.27558
11	0.25488	0.25450	0.25507	0.25350	11	0.24991	0.24810	0.24900
12	0.22537	0.22451	0.22562	0.22691	12	0.22559	0.22246	0.22085
13	0.19679	0.19536	0.19547	0.19780	13	0.19995	0.19946	0.19679
14	0.16998	0.16836	0.16674	0.16739	14	0.17043	0.17375	0.17500
15	0.14576	0.14463	0.14195	0.13963	15	0.13956	0.14223	0.14645
16	0.12491	0.12501	0.12304	0.11924	16	0.11515	0.11251	0.11243
17	0.10808	0.10991	0.11057	0.10883	17	0.10466	0.09886	0.09260
18	0.09580	0.09938	0.10369	0.10699	18	0.10833	0.10728	0.10378
19	0.08837	0.09326	0.10067	0.10915	19	0.11786	0.12620	0.13380
20	0.08592	0.09128	0.09988	0.11052	20	0.12258	0.13565	0.14959

Fig. 2.42 Table of sample amplitude weightings for Tapered-Taylor patterns for circular aperture. (After Hansen, IRE PGAP [22].)

CURVES OF SAMPLE AMPLITUDE WEIGHTINGS FOR TAPERED TAYLOR PATTERNS FOR CIRCULAR APERTURE

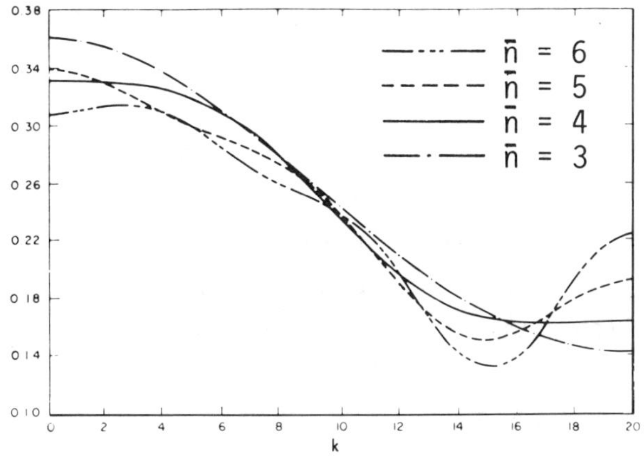

APERTURE WEIGHTING FOR 25-dB SIDELOBES.

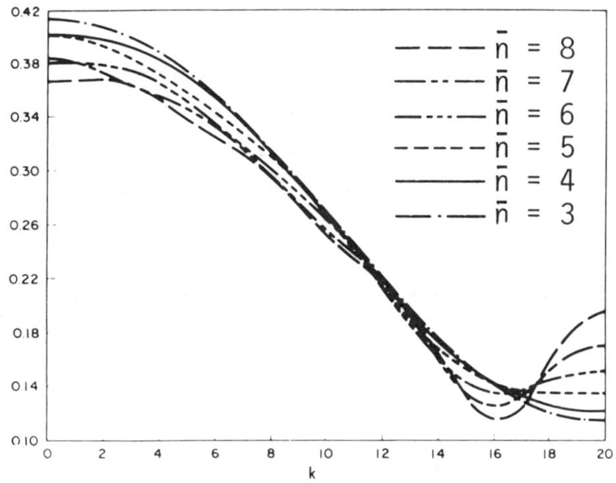

APERTURE WEIGHTING FOR 30-dB SIDELOBES.

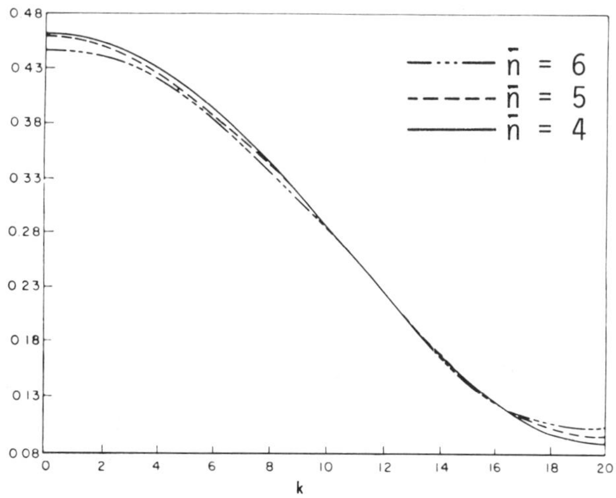

APERTURE WEIGHTING FOR 35-dB SIDELOBES.

Fig. 2.43 Curves of sample amplitude weightings for Tapered-Taylor patterns for circular aperture. (After Hansen, IRE PGAP [22].)

EXAMPLE:
TAPERED TAYLOR FOR CIRCULAR APERTURE

DESIRE : $\theta_{30}=3°$ AND SIDELOBE LEVEL = 30 dB

FIND : $\bar{n}$, $\frac{2r_o}{\lambda}$, η_t.

FROM Fig. 2.41b $\bar{n}$ FOR OPTIMUM η_t IS

$$\boxed{\bar{n}=8}$$

FOR WHICH : $\boxed{\eta_t=0.8838}$

FROM Fig. 2.40b

$$\theta_{30}=\sigma \frac{K_{oo}}{2r_o}\lambda$$

$$3° = 1.0827\ \frac{60.55}{2r_o}\lambda$$

$$\boxed{\frac{2r_o}{\lambda} = 21.85}$$

Fig. 2.44 Example: Tapered-Taylor for circular aperture.

EQUIVALENCE BETWEEN LINEAR AND CIRCULAR APERTURES

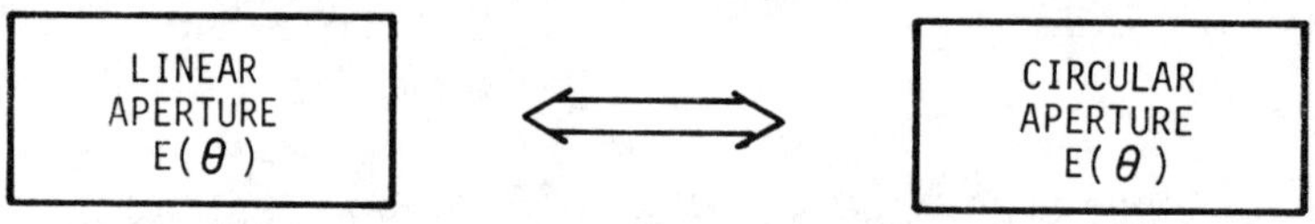

CONDITIONS FOR ABOVE EQUIVALENCE TO BE TRUE:

1. $I_\ell(x)$ OF LINEAR APERTURE GIVEN BY:

$$I_\ell(x) = \int_{-\sqrt{1-x^2}}^{\sqrt{1-x^2}} I_c(x,y)\,dy$$

2. $I_c(x,y)$ FOR CIRCULAR APERTURE IS ONLY FUNCTION OF r, I.E.,

$$I_c(x,y) = I_c(r).$$

EX. 1. $I_c(r) = 1 \Longleftrightarrow I_\ell(x) = \sqrt{1-x^2}$

2. $I_c(r) = (1-r^2)^m \Longleftrightarrow I_\ell(x) = (1-x^2)^{m+\frac{1}{2}}$

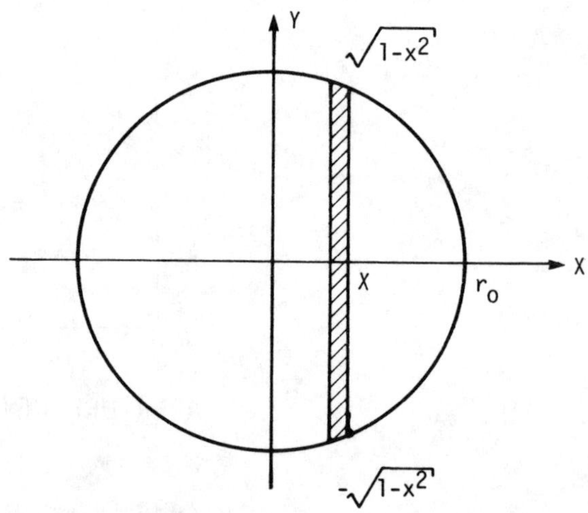

Fig. 2.45 Equivalence between linear and circular apertures.

THINNING BY STATISTICAL DENSITY TAPER

PROBLEM: TO DESIGN THINNED ARRAY HAVING ARRAY FACTOR $E_a(\theta)$ OF FILLED ARRAY WITH WEIGHTS I_i, i = -M TO M, WHERE MAXIMUM $I_i = 1$.

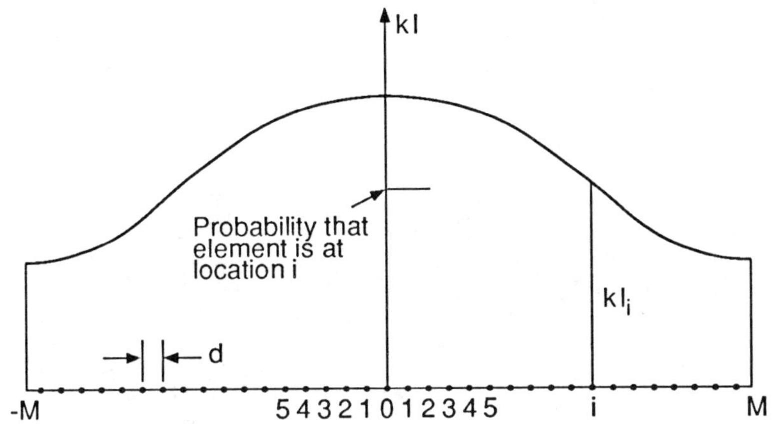

DESIGN PROCEDURE

1. CHOOSE SUCH THAT $0 < k \geq 1$. (κ ΔΕΤΕΡΜΙΝΕΣ THE ΔΕΓΡΕΕ ΟΦ THINNINΓ.)
2. ΛΕΤ χ_i, i = –M TO M, BE INDEPENDENT RANDOM VARIABLES UNIFORMLY DISTRIBUTED BETWEEN 0 AND 1.
3. IF $c_i > k\,I_i$ REMOVE ith RADIATING ELEMENT.
4. IF $c_i \leq \kappa\,I_i$ LEAVE ith ELEMENT IN WITH WEIGHING UNITY

RESULTING ARRAY FACTOR:

LET: $E_a(\theta)_{THIN} \equiv$ ARRAY FACTOR FOR RESULTING THINNED ARRAY.

THEN:

$$\text{AVERAGE OF } E_a(\theta)_{THIN} = \overline{E_a(\theta)_{THIN}} = E_a(\theta) \quad (1)$$

Fig. 2.46 Thinning by statistical density taper. (After Skolnik in Collin and Zucker, McGraw-Hill Book Comp. [1].)

*Note: I_i is the voltage weighting.

THINNED TAPERED TAYLOR CIRCULAR ARRAY

ASSUMPTIONS:		
	WEIGHTING FOR FILLED ARRAY:	TAPERED TAYLOR; $\bar{n}$ = 3
	DESIGN SIDELOBE LEVEL :	30 dB
	NUMBER OF ELEMENTS N IN FILLED ARRAY:	7800
	THINNING CONSTANT k:	UNITY (NATURAL THINNING)
	NUMBER OF ELEMENTS N_E IN THINNED ARRAY:	3773
	% THINNING:	51.6%
	ARRAY DIAMETER:	50 λ
	ELEMENT SPACING (ALONG X AND Y DIRECTIONS):	$\lambda/2$

Fig. 2.47 Thinned Tapered-Taylor circular array. (After Skolnik, in Collin and Zuker, McGraw-Hill Book Comp. [1]; also Skolnik, Sherman, and Ogg, IEEE PGAP [5].)

COMPARISON OF ANTENNA ARRAY PATTERNS FOR THINNED AND FILLED ARRAYS

ASSUMPTIONS:	SAME AS FOR Fig. 2.47 TO WIT:	
	WEIGHTING FOR FILLED ARRAY:	TAPERED TAYLOR; $\bar{n} = 3$
	DESIGN SIDELOBE LEVEL:	30 dB
	NUMBER OF ELEMENTS N IN FILLED ARRAY:	7800
	THINNING CONSTANT k:	UNITY (NATURAL THINNING)
	NUMBER OF ELEMENTS N_E IN THINNED ARRAY:	3773
	% THINNING:	**51.6%**
	ARRAY DIAMETER:	50 λ
	ELEMENT SPACING (ALONG X AND Y DIRECTIONS):	$\lambda/2$

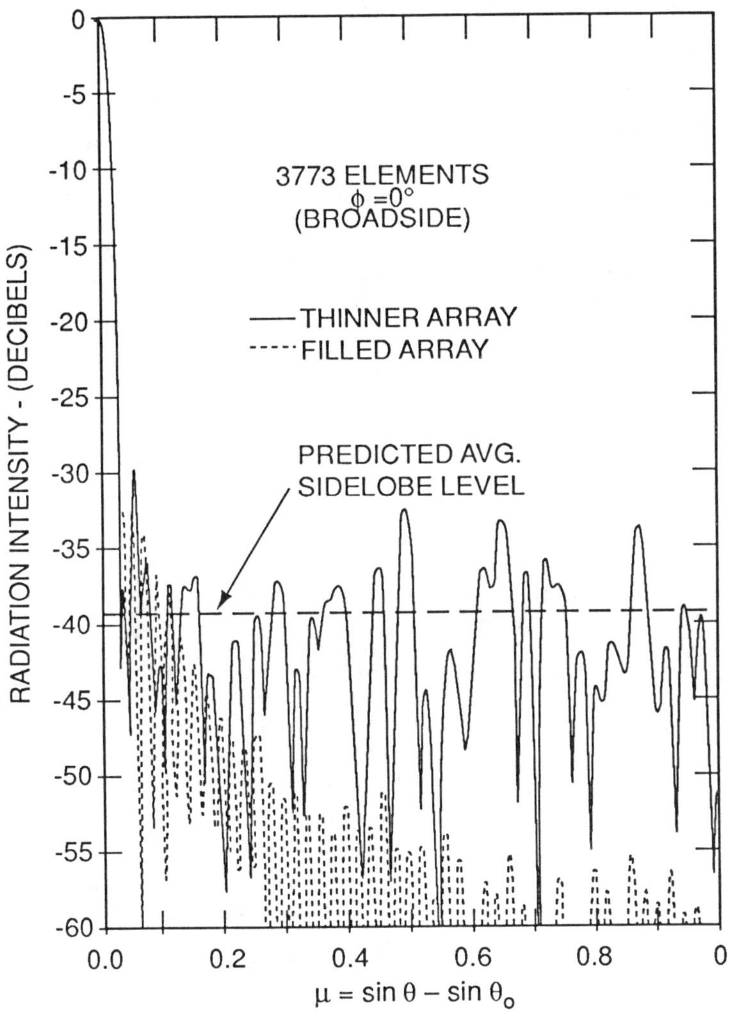

Fig. 2.48 Comparison of antenna array patterns for thinned and filled arrays. (After Skolnik in Collin and Zucker, McGraw-Hill Book Comp. [1]; also Skolnik, Sherman, and Ogg, IEEE PGAP [5].)

an $\bar{n} = 3$, 30 dB Tapered-Taylor weighting. The aperture illumination function for this array is given in Fig. 2-43b.

The resulting thinned array is, as indicated before, illuminated with uniform weighting, that is, $I_i = 1$ for all the remaining radiating elements. Using this uniform illumination, the thinned array gives a far field antenna array pattern whose main beam and first few sidelobes are essentially the same as that which the amplitude tapered full array would have. This is illustrated in Fig. 2-48, which shows the array factor for the thinned array of Fig. 2-47 as well as the array pattern for the reference design full array having a Tapered-Taylor amplitude weighting. The main lobe and first sidelobes are very closely matched. The second sidelobe is closely approximated. The other sidelobes for the thinned array are much higher than those for the Tapered-Taylor full array.

Note that while for the full array the antenna array pattern is obtained using amplitude weighting, for the thinned array the same antenna pattern is achieved using "space tapering".[38]

The above method of thinning is most effective for large arrays, arrays having over 100 elements. An important characteristic of the type of thinning described above is that it only places the elements in positions specified for full array of equally spaced elements. For the example of Fig. 2-47, a rectangular lattice of possible location points was used. A triangular lattice could also have been used. Other space tapering techniques are available which specify completely random locations for the array elements, that is, locations which are not specified by a regular lattice structure as given in Figs. 2-47, 2-31, and 3-36a.

The use of a regular lattice of radiating elements locations has two important advantages over the designs using completely random array element locations. First, it permits the use of lower cost row/column steering. Second, it allows for better control of the array antenna pattern design as shall now be explained. At the locations where the array radiating elements are to be left out in the thinned array, one just does not leave a void. Instead, dummy elements are placed there. The presence of these dummy elements causes the active elements to have a surrounding the same as that which they would have in a full array. Consequently, all the active radiating elements in the thinned array will have the same radiation pattern as they would have in a full array. If the dummy elements were not present, each active radiating element would have a different surrounding (depending on how many nearby elements were removed) and hence a different radiation pattern. As a result, it would be difficult to predict the array pattern. Similarly, if the active radiating elements had completely random locations (instead of being placed on a regular lattice), the surroundings for each radiating element would be different from any other radiating element and the radiation patterns for each radiating element would be different.

The design tradeoffs that were discussed for full apertures having a Tapered-Taylor weighting (see Figs. 2-26 and 2-41 for example) do not apply to thinned arrays, that is, tradeoffs with respect to antenna directivity, n, sidelobe level, and beamwidth. This is because the Tapered-Taylor thinned array only matches the main lobe and first two or three sidelobes of the Tapered-Taylor pattern. Small values of $\bar{n}$ (3 and 4) have been used for thinned arrays in the past. Fig. 2-49 shows the percent thinning obtained for several Tapered-Taylor designs which use natural thinning (k = 1) as did the example in Figs. 2-47 and 2-48. Fig. 2-49 also indicates how to obtain any specified lower percent thinning with non-natural thinning (k < 1).

Example 1 of Fig. 2-49 calculates the number of elements N_E that the naturally thinned array of the example of Figs. 2-47 and 2-48 require. The result obtained is $N_E = 3978$ instead of the value 3773 indicated in Figs. 2-47 and 2-48. This discrepancy results from the statistical nature of the design procedure which makes N_E a random variable. The value $N_E = 3978$ calculated in Fig. 2-49 represents the average expected value for N_E obtained when the design procedure is repeated a large number of times.

2.5.2 Sidelobe Level of Thinned Arrays.

A. Simple Approximate Expression. Using straightforward physical arguments, a simple expression for the average sidelobe level of a thinned array is derived. Assume that the far field pattern of each of the radiating elements of a thinned array is a unit vector. There are assumed to be N_E active radiating elements in the thinned array. In the direction corresponding to the peak of the main lobe of the thinned array, the N_E unit vectors will add up in phase to produce an electric field of strength N_E; (see Fig. 2-50a). In the direction of the sidelobes, these N_E unit vectors will not add up in phase. Because one is looking in a direction other than for which in phase additions occur and because of the random selection of the location of the radiating elements, the N_E unit vectors will add up like random vectors whose phases are uniformly distributed between 0 and 2π; (see Fig. 2-50b). Consequently, the magnitude of the rms value of the electric field in the direction of the sidelobes will have a value of $\sqrt{N_E}$. Thus, the ratio of the peak of the main lobe to the rms level of the sidelobes is given by

$$\text{voltage sidelobe level} \doteq \frac{\sqrt{N_E}}{N_E} = \frac{1}{\sqrt{N_E}} \quad \text{if } E_a(\theta) << \frac{1}{\sqrt{N_E}} \qquad (2\text{-}70)$$

where $E_a(\theta)$ is the normalized electric field pattern of the full array at the sidelobe direction in question, normalized such that $E_a(0) = 1$. The inequality specifies that (2-70) only holds as long as the design sidelobes of the full array are lower than those predicted by (2-70). When the inequality is reversed, the sidelobe level is given by $E_a(\theta)$. The average power sidelobe level is given by the square of (2-70) which results in (2-71). For the example of Figs. 2-47 and 2-48, $N_E = 3{,}773$ so that the average power sidelobe level is predicted to be −35.8 dB by (2-70) and (2-71).

B. More Precise Determination of Average and Peak Sidelobe Levels of a Thinned Array.

i. Average Sidelobe Level — Eq. (2.72) gives a more precise relationship for the determination of the average power sidelobe level of a thinned array.[37] For a high degree of thinning, that is, for $N_E/N << 1$, (2.72) becomes the approximate expression (2-71); (see Fig. 2-51). In Example 1 of Fig. 2.51, the parameters of the example 51.6% thinned array of Figs. 2-47 and 2-48 are applied to (2-72) to yield a predicted average sidelobe level of −39.5 dB, lower than the approximate expression gave, but about the value the actual array pattern indicates in Fig. 2-48. If the thinning were 90% for the array of Figs. 2-47 and 2-48 while all other parameters were unchanged, then the average sidelobe level becomes −36.3 dB which is within 0.5 dB of the result obtained with the approximate Eq. (2-71).

ii. Peak Sidelobe Level — Now consideration is given to the peak sidelobe levels. Define the peak sidelobe level as the level which is exceeded only some small percentage, e.g., 10%, of the time, or equivalently, the level which the sidelobes are below some large percentage, e.g., 90%, of the time. When there are a large enough number of radiating elements in the array, the central limit theorem can be used to determine the distribution of the sidelobes. It indicates that the sidelobe voltage level has a Rayleigh distribution (the magnitude of two Gaussian variates x_o and y_o that are in quadrature and which have mean zero and identical variances, i.e., $r = |x_o + jy_o|$.[39] Thus the sidelobe voltage ratio r has a probability density function given by

$$p(r) = \frac{r}{\sigma^2} \exp - \frac{r^2}{2\sigma^2} \qquad (2\text{-}73)$$

where σ^2 is given in terms of $\overline{r^2}$ by

$$\overline{r^2} = 2\sigma^2 . \qquad (2\text{-}74)$$

(Physically σ^2 is the variance of each of the quadrature Gaussian processes that make up r.) Hence,

$$p(r) = \frac{2r}{\overline{r^2}} \exp - \left[\frac{r^2}{\overline{r^2}}\right] . \qquad (2\text{-}75)$$

From (2-75) the probability that $r \leqslant r_p$, or equivalently, $r^2 \leqslant r_p^{\,2}$, is given by (2.76) of Fig. 2.52.

Eq. (2-76) gives the probability that a single sidelobe is below a specified level r_p. Often one is more interested in the probability that none of the sidelobes exceed a specified level r_p. Let m be the number of independent sidelobes of the thinned array antenna. Then (2-77) gives the probability that all m sidelobes are less than r_p. Approximate expressions for m for linear and planar arrays are given by (2-78) to (2-80).[40] Applying the parameters of the thinned array example of Figs. 2-47 and 2-48 to (2-77) and (2-79) yields that the 90% peak sidelobe level $r_p(90\%)$, that is, the level r_p below which all the sidelobe levels are 90% of the time, is −29.3 dB. This 90% peak sidelobe level is 10.5 dB above the average power sidelobe level r^2 for this example; see Fig. 2-53 for details. The 99% peak sidelobe level r_p (99%) is 0.9 dB higher; (see Fig. 2-53).

Fig. 2-54 gives a convenient plot of Eq. (2-77).[41],[42]

As a final comment, it is interesting to point out that the peak sidelobe level of a linear thinned array of N_E elements is about 3 dB lower than the peak sidelobe level of a planar array having the same number of elements and about the same linear cross dimensions.[40] This is because m for the planar array case is equal to about the square of the m for the linear array; see (2-78) and (2-79).

2.5.3 Directivity of a Thinned Array. Eq. (2-83) gives the directivity of a thinned array.[43],[44] From (2-83), the directivity of the example array of Figs. 2-47 and 2-48 is found to be 40.8 dB which is 3.15 dB below the directivity the array would have if it were full and uniformly weighted; (see Fig. 2-55 for details.) However, a more proper comparison is the directivity of the thinned array to that of the reference design array, i.e., this same array if it were full and used an amplitude weighting such as to yield the same main beam shape and first few sidelobe level values. Making this comparison indicates that using spatial thinning rather than amplitude weighting yields a directivity which is 2.4 dB lower. The loss is 0.8 dB less than obtained for the case of a uniformly weighted full array because the amplitude tapering efficiency loss η_t is present for the full, amplitude-weighted array (and not present for the thinned array); (see Fig. 2-55).

2.5.4 System Aspects of Thinned Arrays. The cost of a phased array is roughly proportional to the number of elements used. The question that arises is: Should one build a full array of N_E elements or should one build a thinned array of N_E elements? What are the trade-offs? A radar using a thinned array of N_E elements has about the same sensitivity as a radar using a full array of the same number of elements N_E if the transmitted power is the same for both cases. However, the thinned array has the advantage of a narrower beamwidth, narrower by a factor $\sqrt{N_E}/N$; where N is the number of elements the thinned array is capable of having. This narrow beamwidth also gives rise to a disadvantage. More time is needed with the narrower beamwidth of the thinned array to search out a given volume of space, about N/N_E more time. Thus, thinning has an advantage of giving better tracking performance while at the same time providing poorer search performance. Hence, if tracking is the driving factor of a radar system, one may want to build a thinned array rather than a full array.

Thinning also has the advantage that it avoids the problem of having to provide amplitude tapering with a corporate fed antenna aperture to achieve a desired near-in sidelobe level and main lobe shape on transmit. In some instances, a thinned array design has the advantage that it provides potential growth in systems sensitivity while keeping the beamwidth fixed. Growth is achieved by filling in the dummy elements of the array with active elements. Thinning has the disadvantage of increasing the sidelobe levels. In some instances, this higher sidelobe level will raise the system noise temperature (because of the larger ground or ocean noise seen through the sidelobes) and reduce system sensitivity. However, if a large enough array is being built, the sidelobe levels achieved can be satisfactory.

The first column of Fig. 2-56 summarizes some of the above advantages to using a thinned array of N_E elements rather than a full array of the same number of elements.

If the choice is between a full array of N elements and the same array thinned out to N_E elements, then the factors given in the second column of Fig. 2-56 must be taken into account. Going from a full array of N elements to a thin array of N_E elements of the same size reduces the received power signal-to-noise ratio per pulse by a factor $(N_E/N)^2$ approximately. Hence, the dwell time per beam position must be increased by one over this factor, i.e., by $(N/N_E)^2$, to compensate for the decreased signal-to-noise ratio per pulse. The degradation in the variance of angle tracking accuracy $\sigma_\theta^{\,2}$ assumes that the same track rate is used per target for the array both with and without thinning.

PERCENT THINNING VS SIDELOBE LEVEL FOR TAPERED-TAYLOR SPACE WEIGHTING

SIDELOBE LEVEL (dB)	$\bar{n}$	PERCENT NATURAL THINNING, (k=1) $\left(\frac{N-N_E}{N}\right)$ IN %	% ACTIVE ELEMENTS REMAINING FOR NATURAL THINNING,* (N_E/N) IN %
25	3	42	58
30	3	49	51
35	3#	55	45
40	4#	60	40

*RESULTS FOR k < 1 OBTAINED SIMPLY BY MULTIPLYING k = 1 VALUES BY k.

APPROXIMATELY SAME RESULTS OBTAINED FOR $\bar{n}$ = 3 AND 4 FOR THESE CASES.

EX. 1: GIVEN: 30 dB TAPERED TAYLOR WEIGHTING WITH $\bar{n}$ = 3. N = 7800

PROBLEM: N_E = ? FOR NATURAL THINNING (k = 1).

SOLUTION: FROM TABLE N_E/N = 51% HENCE N_E = 3978.

EX. 2: GIVEN: SAME AS EX. 1.

PROBLEM: k = ? FOR 90% THINNING.

SOLUTION: FROM TABLE: % THINNING = 49% FOR k = 1.

THUS k(51%) = 10% OR k = 0.1961

*Note: Applies for circular aperture

Fig. 2.49 Percent natural thinning vs sidelobe level for Tapered-Taylor space weighting [5].

APPROXIMATE EXPRESSION FOR AVERAGE SIDELOBE LEVEL OF THINNED ARRAY

LET N_E = NUMBER OF ELEMENTS IN THINNED ARRAY

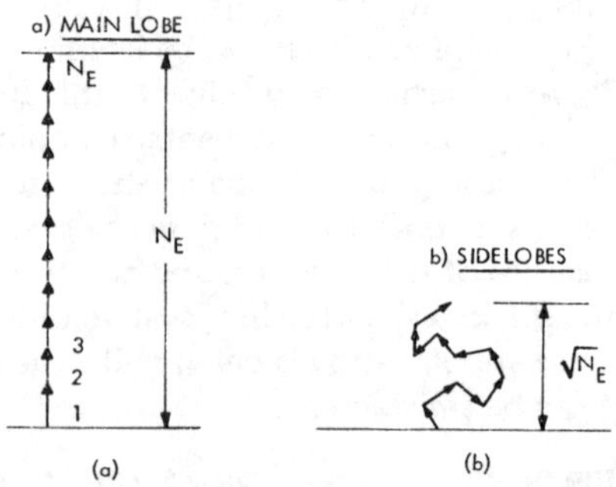

RMS VOLTAGE SIDELOBE LEVEL $\doteq \frac{\sqrt{N_E}}{N_E} = \frac{1}{\sqrt{N_E}}$ IF $E_a(\theta) << \frac{1}{\sqrt{N_E}}$ **(2.70)**

$\overline{r^2}$ = AVERAGE POWER SIDELOBE LEVEL $\doteq \frac{1}{N_E}$ IF $E_a(\theta) << \frac{1}{\sqrt{N_E}}$

$$\boxed{\overline{r^2} = \text{S.L.} = \frac{1}{N_E}} \quad \text{IF } E_a(\theta) << \frac{1}{\sqrt{N_E}} \qquad (2.71)$$

EX: SAME AS **Figs. 2.47 and 2.48:** N_E = 3773 ELEMENTS

AV. S.L. = $\overline{r^2} = \frac{1}{3773}$ = -35.8 dB.

Fig. 2.50 Approximate expression for average sidelobe level of thinned array.

MORE ACCURATE EXPRESSION FOR AVERAGE SIDELOBE LEVEL OF THINNED ARRAY

$$\boxed{\overline{r^2} = \frac{1 - \frac{N_E}{\eta_t N}}{N_E}} \quad \text{WHEN } |E_a(\theta)|^2 << r^2 \qquad (2.72)$$

WHERE N = NUMBER OF ELEMENTS IN FILLED ARRAY.

η_t = TAPER EFFICIENCY OF FILLED ARRAY.

NOTE THAT $\overline{r^2} \doteq \frac{1}{N_E}$, I.E. **(2.71)** FOR $N_E/\eta_t N << 1$, i.e., WHEN SUFFICIENT THINNING IS USED.

EX. 1: SAME AS **Figs. 2.47 and 2.48:**

GIVEN: CIRCULAR ARRAY

N = 7800

N_E = 3773

WEIGHTING FOR FILLED ARRAY: 30 dB TAPERED TAYLOR WITH $\bar{n}$ = 3.

THUS: η_t = 0.838 SEE **Fig. 2.41**

AND FROM (5):

$\boxed{\overline{r^2} = -39.5\text{ dB}}$ AS INDICATED IN **Fig. 2.48**

Fig. 2.51 More accurate expression for average sidelobe level of thinned array [5].

MORE ACCURATE EXPRESSION FOR AVERAGE SIDELOBE LEVEL OF THINNED ARRAY (CONT.)

NOTE THAT APPROXIMATE EXPRESSION **(2.71)** GIVES 3.7 dB TOO HIGH AN $\overline{r^2}$; SEE EXAMPLE OF **Fig. 2.49**

EX. 2: GIVEN: SAME AS FOR EXAMPLE 1 EXCEPT THAT N = 37730

NOW $\boxed{\overline{r^2} = -36.3\text{ dB}}$ AND RESULT OBTAINED WITH **(2.71)** NOW ONLY 0.5 dB TOO HIGH.

Fig. 2.51 (con't.) More accurate expression for average sidelobe level of thinned array. (Cont.)

ACCURATE PROCEDURE FOR CALCULATING PEAK SIDELOBE LEVEL OF THINNED ARRAY

NORMALIZE ANTENNA ARRAY FACTOR SO THAT:

$E_a(0)_{THIN} = 1$

LET r = SIDELOBE LEVEL VOLTAGE

THEN r IS RAYLEIGH DISTRIBUTED WITH $\overline{r^2}$ GIVEN BY **(2.72)**

THUS $$\boxed{P(r^2 \le r_p^2) = 1 - \exp - r_p^2/\overline{r^2}} \quad \text{WHEN } |E_a(\theta)|^2 << r^2 \qquad (2.76)$$

(2.76) GIVES PROBABILITY THAT PEAK OF ONE SIDELOBE $\le r_p$.

THE PROBABILITY THAT ALL SIDELOBES ARE $\le r_p$ IS GIVEN BY

$$\boxed{P\,[\text{ALL SL.} \le r_p] = P(r^2 \le r_p^2;\ \text{ALL SIDELOBES}) \doteq [1 - \exp - r_p^2/\overline{r^2}]^m \doteq 1 - me^{-r_p^2/\overline{r^2}}} \qquad (2.77)$$

FOR $me^{-r_p^2/\overline{r^2}} << 1$.

WHERE $m \equiv$ NUMBER OF INDEPENDENT SIDELOBES

$$\boxed{m \doteq \frac{L}{\lambda}} \quad \text{FOR LINEAR ARRAY} \qquad (2.78)$$

$$\boxed{m \doteq \frac{2A}{\lambda^2}} \quad \text{FOR CIRCULAR AND RECTANGULAR ARRAYS} \qquad (2.79)$$

FOR ~ λ/2 ELEMENT SPACING

$$\boxed{m \doteq N/2} \quad \text{FOR LINEAR AND PLANAR ARRAYS} \qquad (2.80)$$

Fig. 2.52 Accurate procedure for calculating peak sidelobe level of thinned array.

EXAMPLES USING MORE ACCURATE PROCEDURE FOR CALCULATING PEAK SIDELOBE LEVEL OF THINNED ARRAY

EX. 1: SAME AS Figs. 2.47 and 2.48

GIVEN: CIRCULAR APERTURE OF DIAMETER 50 λ

30 dB TAPERED TAYLOR WEIGHTING

N_E = 3773, N = 7800

FOR WHAT r_p IS P [ALL SL. $\le r_p$] = 90% ?

SOLUTION: FROM (2.77)

$$P\,[\text{ALL SL.} \le r_p] \doteq 1 - m e^{-r_p^2/\overline{r^2}} = 0.9 \qquad (2.77)$$

FROM EX. 1 OF Fig. 2.51

$\overline{r^2}$ = -39.5 dB = 1.122X10^{-4}

FROM (2.79)

m = 3927

SUBSTITUTING INTO (2.77) YIELDS THAT 90% PEAK POWER SIDELOBE LEVEL, r_p^2 (90%), IS

r_p^2 (90%) = -29.3 dB

WHICH IS 10.5 dB HIGHER THAN AVERAGE SIDELOBE LEVEL $\overline{r^2}$.

EX. 2: SAME AS EX. 1 ABOVE EXCEPT DESIRE r_p^2 (99%).

r_p^2 (99%) = -28.4 dB

$\therefore$ r_p^2 (99%) > r_p^2 (90%) BY 0.9 dB FOR THIS EXAMPLE.

Fig. 2.53 Examples using more accurate procedure for calculating peak sidelobe level of thinned array.

NINETY PERCENTILE (90%) PEAK SIDELOBE LEVEL FOR THINNED ARRAYS

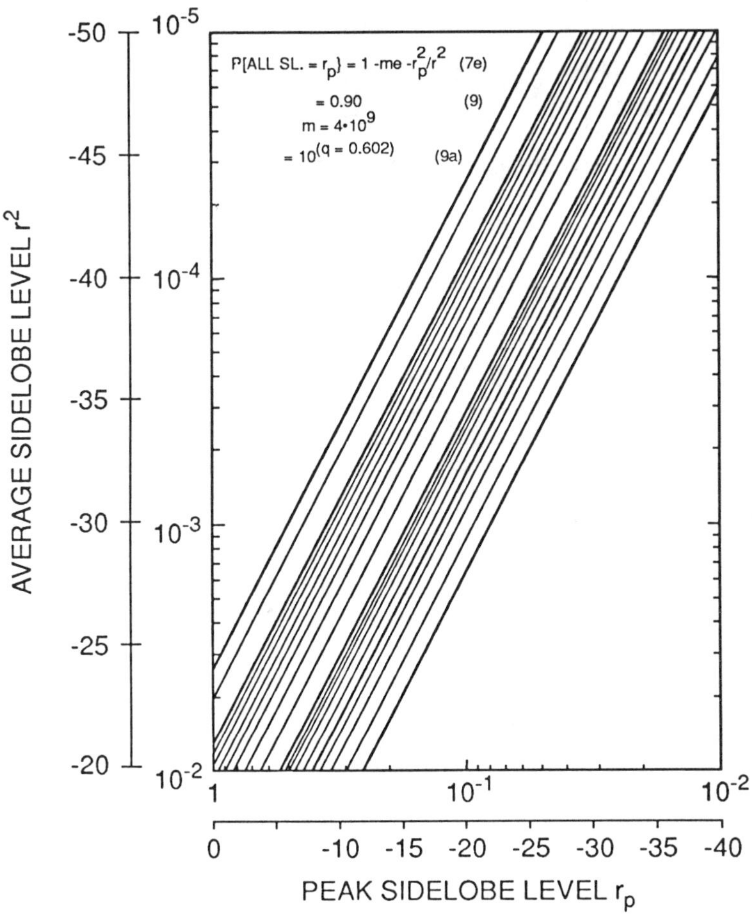

Fig. 2.54 Ninety percentile (90%) peak sidelobe level for thinned arrays. (After Lo, IEEE PGAP [9].)

DIRECTIVITY OF THINNED ARRAY

$$D_{THIN} = (D_{UN.\ FULL}) \frac{N_E}{N} \qquad (2.83)$$

WHERE: D_{THIN} = DIRECTIVITY OF THINNED ARRAY

$D_{UN.\ FULL}$ = DIRECTIVITY OF FULL ARRAY WITH UNIFORM WEIGHTING

$$D_{UN.\ FULL} = \frac{C}{\theta_{3x}\,\theta_{3y}} \cos\theta_o \qquad (2.63)$$

C = 33,700 FOR UNIFORM CIRCULAR ARRAY

= 32,300 FOR UNIFORM RECTANGULAR ARRAY

θ_{3x}, θ_{3y} ARE 3 dB BROADSIDE BEAMWIDTHS IN DEGREES.

ALTERNATELY FROM (2.57)

$$D_{UN.\ FULL} = \frac{4\pi}{\lambda^2} A_e N \cos\theta_o \qquad (2.57a)$$

FOR BOTH CIRCULAR AND RECTANGULAR ARRAYS WHERE A_e = AREA OCCUPIED BY ONE ARRAY ELEMENT = $d_x d_y$ FOR RECTANGULAR ARRAY ELEMENT ARRANGEMENT OF FIG. 2.31

[NOTE: 2.63 OBTAINED FROM 2.57a BY APPLYING THE FACT THAT θ_{30} = 0.886 λ/L AND 1.02 $\lambda/2r_o$ RADIANS FOR LINEAR AND CIRCULAR ARRAYS, RESPECTIVELY.]

Fig. 2.55 Directivity of thinned array.

DIRECTIVITY OF THINNED ARRAY (CONT.)

EX. SAME AS FIGURES 2.47 AND 2.48

GIVEN: CIRCULAR APERTURE OF DIAMETER 50 λ

$d_x = d_y = \lambda/2$

N_E = 3773, N = 7800

FROM FIG. 2.35 $\theta_{3x} = \theta_{3y} = \frac{K_o \lambda}{2r_o} = \frac{1.02}{50}$ = 0.0204 RAD. = 1.169°

HENCE FROM 2.63

$D_{UN.\ FULL}$ = 43.92 dB AT BROADSIDE

AND FROM (10)

D_{THIN} = 43.92 dB - 3.15 dB = 40.8 dB

FROM FIG. 2.41a η_t = 0.8377 = -0.77 dB FOR FULL TAPERED TAYLOR ARRAY.

HENCE THINNED ARRAY DIRECTIVITY 2.4 dB BELOW THAT OF EQUIVALENT FULL WEIGHTED ARRAY.

Fig. 2.55 (con't.) Directivity of thinned array.

SYSTEM ASPECTS OF THINNING

FUNCTION OR PARAMETER EFFECTED	EFFECT OF GOING FROM FULL ARRAY OF N_E ELEMENTS TO THIN ARRAY OF N_E ELEMENTS. THIN ARRAY CAPABLE OF HAVING N ELEMENTS IF FULL. AVERAGE POWER AND TARGET DETECTION OR MEASUREMENT RANGE HELD FIXED	EFFECT OF REMOVING ELEMENTS FROM FULL ARRAY OF N ELEMENTS LEAVING N_E ELEMENTS. AVERAGE POWER AND TARGET DETECTION OR MEASUREMENT RANGE HELD FIXED
SEARCH TIME	INCREASED BY FACTOR: N/N_E	INCREASED BY FACTOR: $(N/N_E)^2$
TRACK	IMPROVES VARIANCE OF ANGLE MEASUREMENT, σ_θ^2, BY FACTOR: N_E/N	DEGRADES σ_θ^2 BY FACTOR: $(N/N_E)^2$
POWER SNR/PULSE	UNCHANGED	DEGRADES BY FACTOR: $(N_E/N)^2$

Fig. 2.56 System aspect of thinning.

APPENDIX A
ADDITIONAL
LINEAR AND PLANAR APERTURE
DESIGN DATA

$\cos^N(\frac{\pi X}{2})$, $1-|X|$, AND $1-(1-\Delta)X^2$

LINEAR APERTURE ILLUMINATION FUNCTION ANTENNA PARAMETERS

$\theta_o = 0°$

ILLUMINATION	3 dB BEAMWIDTH (RAD)	PEAK SIDELOBE, dB	η_t	FIRST NULL POSITION (RAD)
$I_\ell(x) = \cos^N(\pi x/2)$; $\lvert x\rvert < 1$				
$N = 0$	$0.88\lambda/l$	13.2	1.00	$1.0\lambda/l$
$N = 1$	$1.20\lambda/l$	23	0.81	$1.5\lambda/l$
$N = 2$	$1.45\lambda/l$	32	0.667	$2.0\lambda/l$
$N = 3$	$1.66\lambda/l$	40	0.575	$2.5\lambda/l$
$N = 4$	$1.94\lambda/l$	48	0.515	$3.0\lambda/l$
$I_\ell(x) = 1 - \lvert x\rvert$; $\lvert x\rvert \leq 1$				
	$1.28\lambda/l$	26.4	0.75	$2.0\lambda/l$
$I_\ell(x) = 1 - (1-\Delta)x^2$; $\lvert x\rvert < 1$				
$\Delta = 1.0$	$0.88\lambda/l$	13.2	1.0	$1.00\lambda/l$
$\Delta = 0.8$	$0.92\lambda/l$	15.8	0.99	$1.06\lambda/l$
$\Delta = 0.5$	$0.97\lambda/l$	17.1	0.97	$1.14\lambda/l$
$\Delta = 0$	$1.15\lambda/l$	20.6	0.83	$1.43\lambda/l$

Fig. 2.A.1 $\cos^N(\frac{\pi X}{2})$, $1-|x|$, and $1-(1-\Delta)x^2$ Linear aperture illumination function antenna parameters. (After Sherman in Skolnik, McGraw-Hill Book Comp. [23].)

TAPERED TAYLOR LINEAR APERTURE BEAM PATTERN PARAMETERS AND SAMPLE AMPLITUDE WEIGHTINGS

$\theta_o = 0°$

SIDELOBE LEVEL, dB	πA	K_o RAD	σ		
			$\bar{n}=3$	$\bar{n}=5$	$\bar{n}=10$
20	2.993	0.893	1.121	1.087	
25	3.571	0.978	1.092	1.077	
30	4.146	1.057	.....	1.066	1.043
35	4.722	1.131	.....	1.054	1.040
40	5.298	1.200	.....	.....	1.036

X	$I_\ell(X)$ FOR L = 2			
	$\bar{n}=3$, 20 dB	$\bar{n}=5$, 25 dB	$\bar{n}=5$, 30 dB	$\bar{n}=5$, 35 dB
0	1.317	1.429	1.555	1.665
0.1	1.301	1.408	1.530	1.635
0.2	1.254	1.351	1.458	1.546
0.3	1.182	1.268	1.347	1.407
0.4	1.093	1.159	1.203	1.230
0.5	0.996	1.021	1.032	1.029
0.6	0.900	0.864	0.848	0.818
0.7	0.815	0.722	0.669	0.614
0.8	0.749	0.626	0.520	0.437
0.9	0.706	0.581	0.422	0.315
1.0	0.692	0.569	0.388	0.271

NOTE: LINEAR APERTURE 3 dB BROADSIDE BEAMWIDTHS θ_{300} AND θ_{30} FOR IDEAL-TAYLOR AND TAPERED-TAYLOR WEIGHTINGS, RESPECTIVELY, ARE OBTAINED FROM VALUES IN ABOVE TABLES USING EQUATIONS OF FIG 2.40 FOR CIRCULAR APERTURE. MOREOVER VALUES FOR K_{oo} AND IN TURN θ_{300} (SEE FIG 2.40) OBTAINED ABOVE FOR SPECIFIED VOLTAGE SIDELOBE LEVEL r APPLY FOR IDEAL-TAYLOR WEIGHTED CIRCULAR APERTURE AS WELL FOR IDEAL-TAYLOR LINEAR APERTURE. SAME IS TRUE FOR A.

Fig. 2.A.2 Tapered-Taylor linear aperture beam pattern parameters and sample amplitude weightings. (After Sherman in Skolnik, McGraw-Hill Book Comp. [23].)

ONE PARAMETER TAYLOR*
ANTENNA PATTERN PARAMETERS

$$E(\dot{u}) = \frac{\sin\sqrt{\dot{u}^2-A^2}}{(\dot{u}^2-A^2)^{1/2}} \qquad (2.A.1)$$

$$= \Lambda_{1/2}\left(\sqrt{\dot{u}^2-A^2}\right) \qquad (2.A.2)$$

$$\theta_o = 0°$$

Sidelobe level, db	A	K_o# rad	η_t **
13.2	0	0.885	1.000
15	1.118	0.923	0.993
20	2.320	1.024	0.933
25	3.214	1.116	0.863
30	4.009	1.200	0.801
35	4.755	1.278	0.751
40	5.471	1.351	0.709

* SPECIAL CASE OF BICKMORE - SPELLMIRE TWO-PARAMETER FAMILY:

$$E(\dot{u}) = \Lambda_\nu\left(\sqrt{\dot{u}^2-A^2}\right) = \Gamma(\nu+1)\left(\frac{2}{\sqrt{\dot{u}^2-A^2}}\right)^\nu J_\nu\left(\sqrt{\dot{u}^2-A^2}\right) \qquad (2.A.3)$$

\# TO CALCULATE θ_{30} USE (2.68) OR (2.65)

** GIVEN FOR LINEAR SOURCE. K_0 AND A VALUES APPLY FOR LINEAR AND CIRCULAR APERTURE.

Fig. 2.A.3 One parameter Taylor antenna pattern parameters. (After Sherman in Skolnik, McGraw-Hill Book Comp. [23].)

LAMBDA FUNCTIONS

AN IMPORTANT CLASS OF ANTENNA PATTERNS

$$\Lambda_{-1/2}(x) = \cos x$$

$$\Lambda_0(x) = J_0(x)$$

$$\Lambda_{1/2}(x) = \frac{\sin x}{x}$$

$$\Lambda_1(x) = \frac{2J_1(x)}{x}$$

$$\Lambda_{3/2}(x) = \frac{3}{x^2}\left(\frac{\sin x}{x} - \cos x\right)$$

$$\Lambda_2(x) = \frac{8J_2(x)}{x^2}$$

$$\Lambda_{5/2}(x) = \frac{15}{x^3}\left(\frac{3\sin x}{x^2} - \frac{3\cos x}{x} - \sin x\right)$$

$$\Lambda_3(x) = \frac{48J_3(x)}{x^3}$$

$$\Lambda_\nu(x) - \Lambda_{\nu-1}(x) = \left(\frac{x}{2}\right)^2 \frac{\Lambda_{\nu+1}(x)}{\nu(\nu+1)}$$

$$\frac{\Lambda_{\nu+1}(x)}{\nu+1} = -\frac{2}{x}\Lambda_\nu'(x)$$

$$\Lambda_{\nu-1}(x) - \Lambda_\nu(x) = \frac{x}{2\nu}\Lambda_\nu'(x)$$

$$\Lambda_\nu'(x) = \frac{d}{dx}\Lambda_\nu(x)$$

$$\Lambda_\nu(x) = \Gamma(\nu+1)\left(\frac{2}{x}\right)^\nu J_\nu(x)$$

Fig. 2.A.4 Lambda functions. (After Sherman in Skolnik, McGraw-Hill Book Comp. [23].)

DATA FOR EQUIVALENT LINEAR AND CIRCULAR APERTURE ILLUMINATION FUNCTIONS

LINEAR APERTURES : $I_{\ell}(x) = (1-x^2)^{m+\frac{1}{2}}$

CIRCULAR APERTURES : $I_c(r) = (1-r^2)^m$

$\theta_0 = 0°$

m	$E(\dot{u})$ *	First Sidelobe Level, db	3-db # Beamwidth θ_3 (rad.)	Amplitude Illumination: Linear Aperture $I_l(x)$		η_t	Circular Aperture $I_c(r)$		n_t
$-\frac{3}{2}$†	$\Lambda_{-\frac{1}{2}}(\dot{u})$	....	....	$\delta(x-1)+\delta(x+1)$					
-1†	$\pi\Lambda_0(\dot{u})$	....	....	$\frac{1}{\sqrt{1-x^2}}$		.....	$\frac{1}{2}\delta(r)$		
$-\frac{1}{2}$	$2\pi\Lambda_{\frac{1}{2}}(\dot{u})$	13.2	$0.88\frac{\lambda}{l}$	π	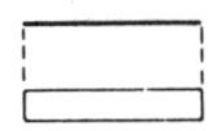	1.00	$\frac{1}{\sqrt{1-r^2}}$		
0	$\frac{2\pi}{2}\Lambda_1(\dot{u})$	17.6	$1.02\frac{\lambda}{l}$	$2\sqrt{1-x^2}$	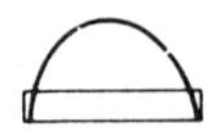	0.865	1		1.00
$\frac{1}{2}$	$\frac{2\pi}{3}\Lambda_{\frac{3}{2}}(\dot{u})$	20.6	$1.15\frac{\lambda}{l}$	$\frac{\pi}{2}(1-x^2)$		0.833	$\sqrt{1-r^2}$		0.75
1	$\frac{2\pi}{4}\Lambda_2(\dot{u})$	24.6	$1.27\frac{\lambda}{l}$	$\frac{4}{3}(1-x^2)^{\frac{3}{2}}$		0.75	$(1-r^2)$		0.64
$\frac{3}{2}$	$\frac{2\pi}{5}\Lambda_{\frac{5}{2}}(\dot{u})$	28.6	$1.36\frac{\lambda}{l}$	$\frac{2\pi}{5}(1-x^2)^2$	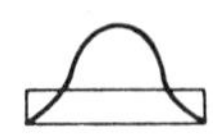	0.68	$(1-r^2)^{\frac{3}{2}}$		
2	$\frac{2\pi}{6}\Lambda_3(\dot{u})$	30.6	$1.47\frac{\lambda}{l}$	$\frac{3}{8}(1-x^2)^{\frac{5}{2}}$	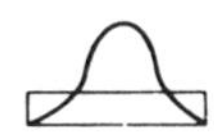	.....	$(1-r^2)^2$		0.55

* $\dot{u} = (\pi L \sin\theta)/\lambda$ FOR LINEAR APERTURES AND $\dot{u} = (\pi 2r_0 \sin\theta)/\lambda$ FOR CIRCULAR APERTURES FOR $\theta_0 = 0°$.

\# FOR CIRCULAR APERTURE ℓ REPLACED BY $2r_0$.

Fig. 2.A.5 Data for equivalent linear and circular aperture illumination functions. (After Sherman in Skolnik, McGraw-Hill Book Comp. [23].)

NORMALIZED ANTENNA PATTERNS FOR EQUIVALENT LINEAR AND CIRCULAR APERTURE ILLUMINATION FUNCTIONS

LINEAR APERTURES : $I_\ell(x) = (1-x^2)^{m+\frac{1}{2}}$

CIRCULAR APERTURES : $I_c(x) = (1-r^2)^m$

$\theta_0 = 0°$

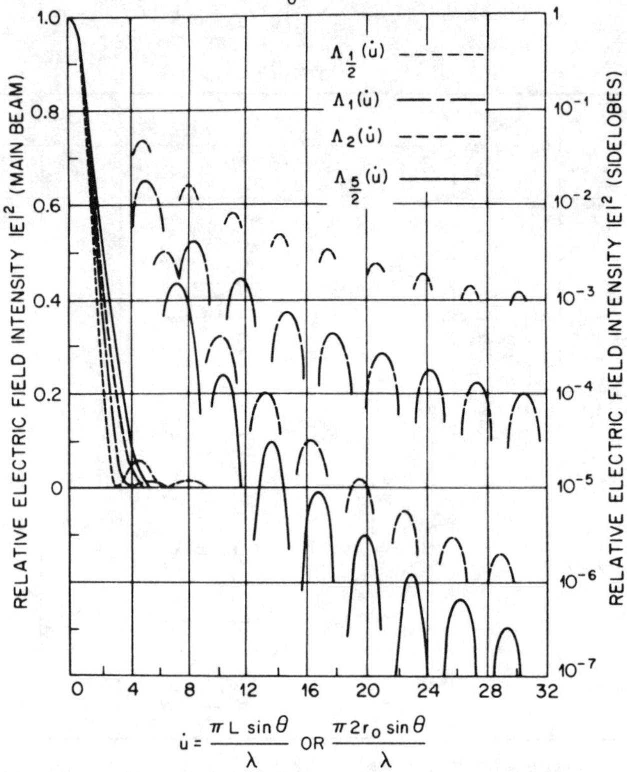

Fig. 2.A.6 Normalized antenna patterns for equivalent linear and circular aperture illumination functions. (After Sherman in Skolnik, McGraw Hill Book Comp. [23].)

BICKMORE-SPELLMIRE PATTERN
K_0 VS SIDELOBE LEVEL

$$E(u) = \frac{J_\nu(\sqrt{\dot{u}^2-A^2})}{(\dot{u}^2-A^2)^{\nu/2}} \Longleftrightarrow \Lambda_\nu(\sqrt{\dot{u}^2-A^2}) \qquad (2.A.4)$$

$$\dot{u} = \frac{\pi L \sin\theta}{\lambda} \text{ OR } \frac{\pi 2r_0 \sin\theta}{\lambda} \quad \text{FOR } \theta_0 = 0° \qquad (2.A.5)$$

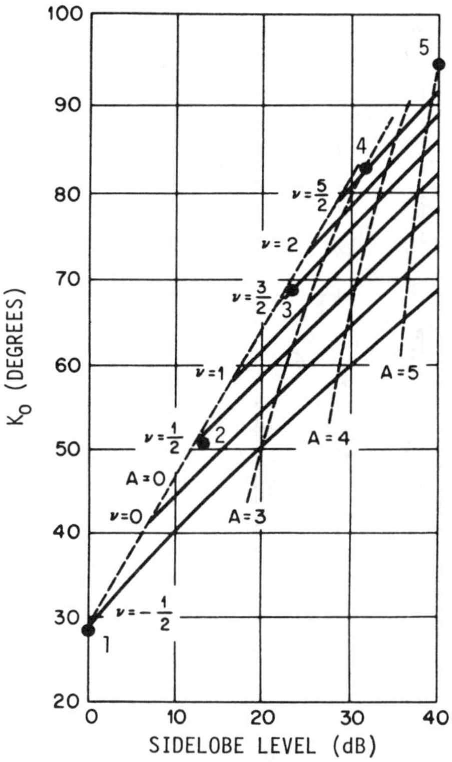

LEGEND FOR POINTS 1-5:

1) TWO-ELEMENT INTERFEROMETER $f(\dot{u})=\cos \dot{u}$.
2) UNIFORM ILLUMINATION $f(\dot{u})=\sin \dot{u}/\dot{u}$.
3) COSINE ILLUMINATION

$$f(\dot{u}) = \frac{\cos \dot{u}}{1 - (2\dot{u}/\pi)^2} \qquad (2.A.6)$$

4) COSINE-SQUARED ILLUMINATION

$$f(\dot{u}) = \frac{\sin \dot{u}}{u\left\{1 - (\dot{u}/\pi)^2\right\}} \qquad (2.A.7)$$

5) COSINE-CUBED ILLUMINATION

$$f(u) = \frac{\cos \dot{u}}{\left\{1 - (2\dot{u}/\pi)^2\right\}\left\{1 - (2\dot{u}/3\pi)^2\right\}} \qquad (2.A.8)$$

NOTE 1. EQUIVALENCE USED IN (2.A.4) SINCE EXPRESSIONS DIFFER BY CONSTANT; SEE (2.A.3)

NOTE 2. FOR GNu = -½, THAT IS, TAPERED TAYLOR WEIGHTING, A REPLACED BY $A_1 = \pi A$

Fig. 2.A.7 Bickmore-Spellmire pattern K_0 vs sidelobe level. (After Sherman in Skolnik, McGraw-Hill Book Comp. [23]; see also [25].)

COMPARISON OF SEVERAL ILLUMINATIONS THAT HAVE 25 dB PEAK SIDELOBE LEVELS. $\Lambda_{1/2}(\dot{u})$ IS SHOWN FOR COMPARISON

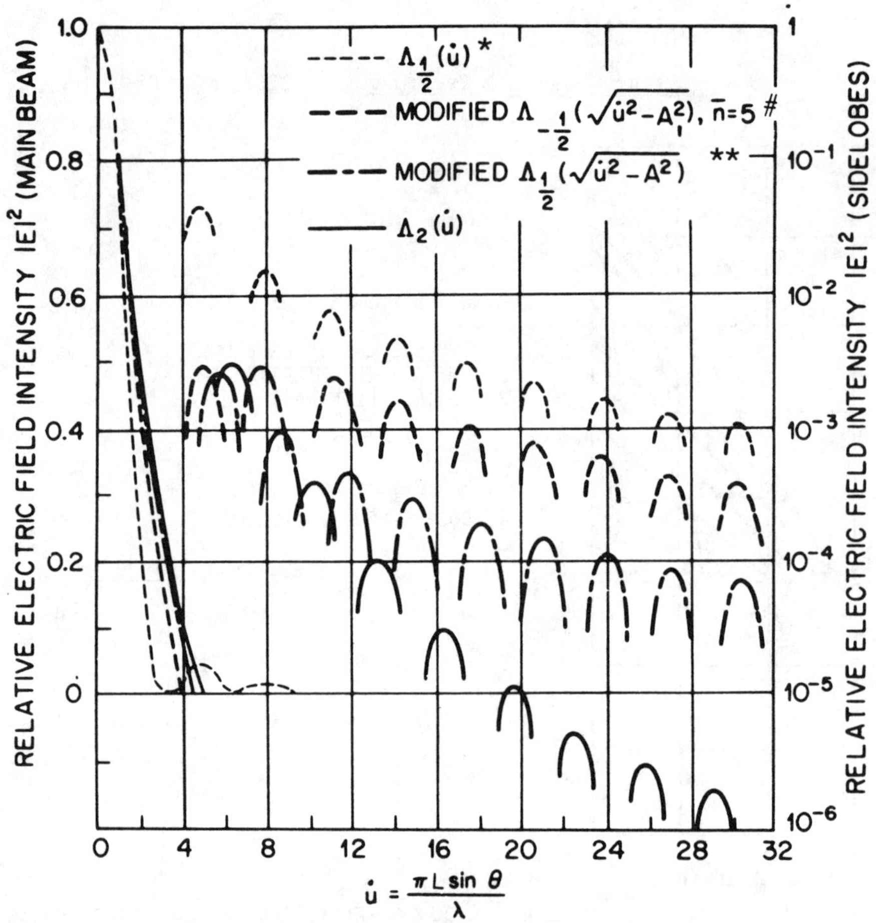

* CORRESPONDS TO UNIFORM ILLUMINATION FOR LINEAR APERTURE.

\# CORRESPONDS TO TAPERED TAYLOR WEIGHTING FOR $\bar{n}=5$ FOR LINEAR APERTURE; $A_1 = \pi A$.

** ONE-PARAMETER TAYLOR. FOR LINEAR APERTURE.

Fig. 2.A.8 Comparison of several illuminations that have 25 dB peak sidelobe levels. $\Lambda_{\frac{1}{2}}(u)$ is shown for comparison. (After Sherman in Skolnik, McGraw-Hill Book Comp. [23].)

BICKMORE-SPELLMIRE APERTURE ILLUMINATIONS FOR LINEAR AND CIRCULAR APERTURES

$E(\dot{u}) = \int_{-1}^{1} I_l(x)\Lambda_{-1/2}(\dot{u}x)\,dx \qquad E(\dot{u}) = 2\int_{0}^{1} I_c(r)\Lambda_0(\dot{u}r)r\,dr \qquad I_l(x) = \frac{1}{2\pi}\int_{-\infty}^{\infty} E(\dot{u})\Lambda_{-1/2}(\dot{u}x)\,d\dot{u} \qquad I_c(r) = \frac{1}{2}\int_{0}^{\infty} E(\dot{u})\Lambda_0(\dot{u}r)\dot{u}\,d\dot{u}$

ν	$E(u)$	$I_l(x)$	$I_c(r)$
$-1/2$	$\Lambda_{-1/2}(\sqrt{\dot{u}^2 - A_1^2})$	$\frac{\delta(x-1)+\delta(x+1)}{2} + \frac{A_1^2}{2}\Lambda_{-1}(jA_1\sqrt{1-x^2})$	
0	$\Lambda_0(\sqrt{\dot{u}^2 - A^2})$	$\frac{1}{\pi\sqrt{1-x^2}}\Lambda_{-1/2}(jA\sqrt{1-x^2})$	$\frac{1}{2}\delta(r-1) + \frac{A^2}{2}\Lambda_{-1}(jA\sqrt{1-r^2})$
$1/2$	$\Lambda_{1/2}(\sqrt{\dot{u}^2 - A^2})$	$\frac{1}{2}\Lambda_0(jA\sqrt{1-x^2})$	$\frac{1}{2\sqrt{1-r^2}}\Lambda_{-1/2}(jA\sqrt{1-r^2})$
1	$\Lambda_1(\sqrt{\dot{u}^2 - A^2})$	$\frac{2}{\pi}\sqrt{1-x^2}\Lambda_{1/2}(jA\sqrt{1-x^2})$	$\Lambda_0(jA\sqrt{1-r^2})$
$3/2$	$\Lambda_{3/2}(\sqrt{\dot{u}^2 - A^2})$	$\frac{3}{4}(1-x^2)\Lambda_1(jA\sqrt{1-x^2})$	$\frac{3}{2}\sqrt{1-r^2}\Lambda_{1/2}(jA\sqrt{1-r^2})$
2	$\Lambda_2(\sqrt{\dot{u}^2 - A^2})$	$\frac{8}{3\pi}(1-x^2)^{3/2}\Lambda_{3/2}(jA\sqrt{1-x^2})$	$2(1-r^2)\Lambda_1(jA\sqrt{1-r^2})$
$5/2$	$\Lambda_{5/2}(\sqrt{\dot{u}^2 - A^2})$	$\frac{15}{16}(1-x^2)^2\Lambda_2(jA\sqrt{1-x^2})$	$\frac{5}{2}(1-r^2)^{3/2}\Lambda_{3/2}(jA\sqrt{1-r^2})$
ν	$\Lambda_\nu(\sqrt{\dot{u}^2 - A^2})$	$\frac{\Gamma(\nu+1)(1-x^2)^{\nu-1/2}}{\Gamma(1/2)\Gamma(\nu+1/2)}\Lambda_{\nu-1/2}(jA\sqrt{1-x^2})$	$\nu(1-r^2)^{\nu-1}\Lambda_{\nu-1}(jA\sqrt{1-r^2})$

NOTE: FOR IDEAL-TAYLOR WEIGHTING $\nu = -\frac{1}{2}$ AND $A_1 = \pi A$.

Fig. 2.A.9 Bickmore-Spellmire aperture illuminations for linear and circular apertures. (After Sherman in Skolnik, McGraw-Hill Book Comp. [23]; also after W. D. White, Airborne Instruments Laboratory.)

FAR-FIELD PATTERN PARAMETERS FOR $b+(1-r^2)^m$ CIRCULAR-APERTURE ILLUMINATIONS

m	SIDELOBE LEVEL (dB)			3 dB BEAMWIDTH COEFFICIENT K_0 (RAD.)*		
	$b = 0$	$b = 1/4$	$b = 1/2$	$b = 0$	$b = 1/4$	$b = 1/2$
0	17.6	....	....	1.02		
1	24.6	23.7	22.0	1.27	1.17	1.13
2	30.7	32.3	26.5	1.47	1.23	1.16
3	36.1	32.3	30.8	1.65	1.25	1.16

m	η_t			FIRST NULL POSITION $\lambda/2r_0$ COEFFICIENT		
	$b = 0$	$b = 1/4$	$b = 1/2$	$b = 0$	$b = 1/4$	$b = 1/2$
0	1.00	....	....	1.22		
1	0.75	0.87	0.92	1.61	1.49	1.33
2	0.65	0.81	0.88	2.03	1.68	1.51
3	0.45	0.79	0.87	2.42	1.97	1.56

* TO CALCULATE θ_{30} USE (2.68) or (2.65).

Fig. 2.A.10 Far-field pattern parameters for $b+(1-r^2)^m$ circular-aperture illuminations. (After Sherman in Skolnik, McGraw-Hill Book Comp. [23].)

DOLPH-CHEBYSHEV LINEAR ARRAY ILLUMINATION FUNCTION

$$I_K = \frac{N-1}{N-K} \sum_{s=0} \binom{K-2}{s} \binom{N-K}{s+1} \alpha^{s+1} \tag{2.A.9}$$

for $K \neq 1$ and $K \neq N$,

$I_K = 1$ for $K = 1$ and $K = N$,

where

$$\alpha = \tanh^2 \left[\frac{\ln(r + (r^2 - 1)^{\frac{1}{2}})}{N-1} \right]. \tag{2.A.9a}$$

The series ends for $s = K-2$, if $K \leq (N+1)/2$, and for $s = N-K-1$, if $K \geq (N+1)/2$.

r = VOLTAGE SIDELOBE RATIO

FOR N LARGE AND α SMALL

$$\alpha \approx \frac{\ln^2(r + (r^2 - 1)^{\frac{1}{2}})}{(N-1)^2}. \tag{2.A.9b}$$

Fig. 2.A.11 Dolph-Chebyshev linear array illumination function. (After Van der Maas, J. Appl. Phys. [26]; see also [28].)

TAPERED-TAYLOR APERTURE ILLUMINATION FUNCTIONS

LINEAR ARRAY ILLUMINATION FUNCTION $I_\ell(x')$:

$$I_\ell(x') = \frac{1 + 2 \sum_{m=1}^{\bar{n}-1} F_m \cos(2\pi m x')}{1 + 2 \sum_{m=1}^{\bar{n}-1} F_m} \tag{2.A.10}$$

$$F_m = \frac{(-1)^{m+1} \prod_{n=1}^{\bar{n}-1} [1 - m^2/\sigma^2[A^2 + (n - \frac{1}{2})^2]]}{2 \prod_{n=1}^{\bar{n}-1} [1 - (m^2/n^2)]_{m \neq n}} \tag{2.A.10a}$$

$$\sigma^2 = \frac{\bar{n}^2}{A^2 + (\bar{n} - \frac{1}{2})^2} \tag{2.A.10b}$$

$$SLL = -20 \log[\cosh(\pi A)] \tag{2.A.10c}$$

Providing that: $\bar{n} \geq 2A^2 + \frac{1}{2}$.

$$x' = x/L \tag{2.A.10d}$$

CIRCULAR ARRAY ILLUMINATION FUNCTION $I_c(r')$:

$$I_c(r') = \frac{\sum_{m=0}^{\bar{n}-1} \frac{J_0(2\pi\mu_m r') F(\mu_m, A, \bar{n})}{[J_0(\pi\mu_m)]^2} + 1}{\sum_{m=0}^{\bar{n}-1} \frac{F(\mu_m, A, \bar{n})}{[J_0(\pi\mu_m)]^2} + 1} \tag{2.A.10e}$$

$$F(\mu_m, A, \bar{n}) = 0, \quad m = 0 \tag{2.A.10f}$$

$$F(\mu_m, A, \bar{n}) = \frac{-J_0(\pi\mu_m) \prod_{n=1}^{\bar{n}-1} \left[1 - \frac{\mu_m^2}{\sigma^2 [A^2 + (n - \frac{1}{2})^2]} \right]}{\prod_{\substack{n=1 \\ n \neq m}}^{\bar{n}-1} \left[1 - \frac{\mu_m^2}{\mu_n^2} \right]} \tag{2.A.10g}$$

$$\sigma^2 = \frac{\mu_{\bar{n}}^2}{A^2 + (\bar{n} - \frac{1}{2})^2} \qquad SLL = -20 \log[\cosh(\pi A)] \tag{2.A.10h}$$

$\bar{n} \geq 2A^2 + \frac{1}{2}$

$$r' = \frac{r}{2r_0} \tag{2.A.10i}$$

Fig. 2.A.12 Tapered-Taylor aperture illumination functions. (After Barton and Ward, Prentice-Hall [20]; see also [13], [22], and [28].)

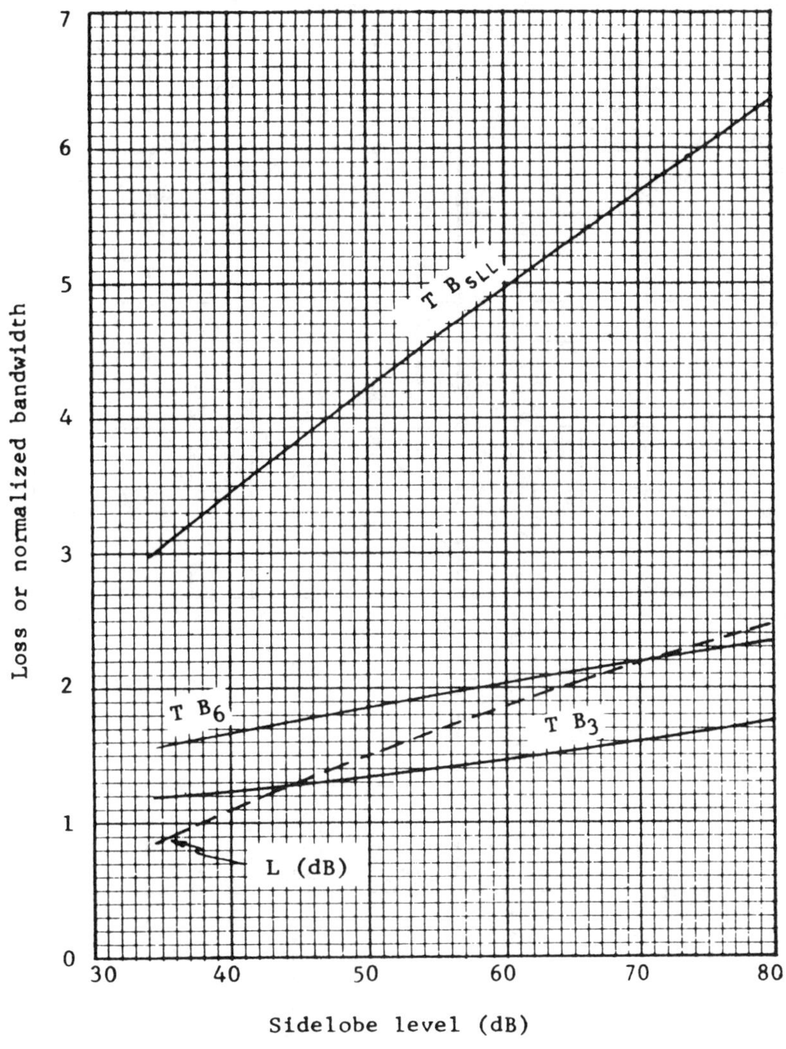

Loss or normalized bandwidth
7
6
5
4
3
2
1
0
30
40
50
60
70
80
Sidelobe level (dB)
T B_SLL
T B_6
T B_3
L (dB)

TABLE I
WINDOWS AND FIGURES OF MERIT

ANTENNA TAPER	HIGHEST SIDE-LOBE LEVEL (dB)	SIDE-LOBE FALL-OFF (dB/OCT)	COHERENT GAIN	$1/\eta_l$	3.0-dB BW (BINS)	SCALLOP LOSS (dB)	WORST CASE PROCESS LOSS (dB)	6.0-dB BW (BINS)	OVERLAP CORRELATION (PCNT)	
									75% OL	50% OL
RECTANGLE	−13	−6	1.00	1.00	0.89	3.92	3.92	1.21	75.0	50.0
TRIANGLE	−27	−12	0.50	1.33	1.28	1.82	3.07	1.78	71.9	25.0
$\cos^{\alpha}(x)$ α = 1.0	−23	−12	0.64	1.23	1.20	2.10	3.01	1.65	75.5	31.8
HANNING α = 2.0	−32	−18	0.50	1.50	1.44	1.42	3.18	2.00	65.9	16.7
α = 3.0	−39	−24	0.42	1.73	1.66	1.08	3.47	2.32	56.7	8.5
α = 4.0	−47	−30	0.38	1.94	1.86	0.86	3.75	2.59	48.6	4.3
HAMMING	−43	−6	0.54	1.36	1.30	1.78	3.10	1.81	70.7	23.5
RIESZ	−21	−12	0.67	1.20	1.16	2.22	3.01	1.59	76.5	34.4
RIEMANN	−26	−12	0.59	1.30	1.26	1.89	3.03	1.74	73.4	27.4
DE LA VALLE-POUSSIN	−53	−24	0.38	1.92	1.82	0.90	3.72	2.55	49.3	5.0
TUKEY α = 0.25	−14	−18	0.88	1.10	1.01	2.96	3.39	1.38	74.1	44.4
α = 0.50	−15	−18	0.75	1.22	1.15	2.24	3.11	1.57	72.7	36.4
α = 0.75	−19	−18	0.63	1.36	1.31	1.73	3.07	1.80	70.5	25.1
BOHMAN	−46	−24	0.41	1.79	1.71	1.02	3.54	2.38	54.5	7.4
POISSON α = 2.0	−19	−6	0.44	1.30	1.21	2.09	3.23	1.69	69.9	27.8
α = 3.0	−24	−6	0.32	1.65	1.45	1.46	3.64	2.08	54.8	15.1
α = 4.0	−31	−6	0.25	2.08	1.75	1.03	4.21	2.58	40.4	7.4
HANNING-POISSON α = 0.5	−35	−18	0.43	1.61	1.54	1.26	3.33	2.14	61.3	12.6
α = 1.0	−39	−18	0.38	1.73	1.64	1.11	3.50	2.30	56.0	9.2
α = 2.0	NONE	−18	0.29	2.02	1.87	0.87	3.94	2.65	44.6	4.7
CAUCHY α = 3.0	−31	−6	0.42	1.48	1.34	1.71	3.40	1.90	61.6	20.2
α = 4.0	−35	−6	0.33	1.76	1.50	1.36	3.83	2.20	48.8	13.2
α = 5.0	−30	−6	0.28	2.06	1.68	1.13	4.28	2.53	38.3	9.0
GAUSSIAN α = 2.5	−42	−6	0.51	1.39	1.33	1.69	3.14	1.86	67.7	20.0
α = 3.0	−55	−6	0.43	1.64	1.55	1.25	3.40	2.18	57.5	10.6
α = 3.5	−69	−6	0.37	1.90	1.79	0.94	3.73	2.52	47.2	4.9
DOLPH-CHEBYSHEV α = 2.5	−50	0	0.53	1.39	1.33	1.70	3.12	1.85	69.6	22.3
α = 3.0	−60	0	0.48	1.51	1.44	1.44	3.23	2.01	64.7	16.3
α = 3.5	−70	0	0.45	1.62	1.55	1.25	3.35	2.17	60.2	11.9
α = 4.0	−80	0	0.42	1.73	1.65	1.10	3.48	2.31	55.9	8.7
KAISER-BESSEL α = 2.0	−46	−6	0.49	1.50	1.43	1.46	3.20	1.99	65.7	16.9
α = 2.5	−57	−6	0.44	1.65	1.57	1.20	3.38	2.20	59.5	11.2
α = 3.0	−69	−6	0.40	1.80	1.71	1.02	3.56	2.39	53.9	7.4
α = 3.5	−82	−6	0.37	1.93	1.83	0.89	3.74	2.57	48.8	4.8
BARCILON-TEMES α = 3.0	−53	−6	0.47	1.56	1.49	1.34	3.27	2.07	63.0	14.2
α = 3.5	−58	−6	0.43	1.67	1.59	1.18	3.40	2.23	58.6	10.4
α = 4.0	−68	−6	0.41	1.77	1.69	1.05	3.52	2.36	54.4	7.6
EXACT BLACKMAN	−51	−6	0.46	1.57	1.52	1.33	3.29	2.13	62.7	14.0
BLACKMAN	−58	−18	0.42	1.73	1.68	1.10	3.47	2.35	56.7	9.0
MINIMUM 3-SAMPLE BLACKMAN-HARRIS	−67	−6	0.42	1.71	1.66	1.13	3.45	1.81	57.2	9.6
* MINIMUM 4-SAMPLE BLACKMAN-HARRIS	−92	−6	0.36	2.00	1.90	0.83	3.85	2.72	46.0	3.8
* 61 dB 3-SAMPLE BLACKMAN-HARRIS	−61	−6	0.45	1.61	1.56	1.27	3.34	2.19	61.0	12.6
74 dB 4-SAMPLE BLACKMAN-HARRIS	−74	−6	0.40	1.79	1.74	1.03	3.56	2.44	53.9	7.4
4-SAMPLE KAISER-BESSEL α = 3.0	−69	−6	0.40	1.80	1.74	1.02	3.56	2.44	53.9	7.4

NOTE: Filter mismatch loss $L_m = 10 \log_{10} \eta_l$

2.A Appendix: Additional Linear and Planar Aperture Data.

Useful addition design data for linear and planar apertures are given in Figs. 2.A.1 to 2.A.12.

2.6 References

[1] Allen, J. L., "The Theory of Array Antennas (with Emphasis on Radar Applications)", Massachusetts Institute of Technology, Lincoln Laboratory, Technical Report No. 323, 25 July 1963.

[2] Elliott, R. S., "Beamwidth and Directivity of Large Scanning Arrays", *Microwave Journal*, Part 1, Vol. 6, December 1963, pp. 53 - 60, and Part 2, Vol. 7, January 1964, pp. 74 - 82; see also *Significant Phased Array Papers*, R. C. Hansen (ed.) Artech House, Dedham, Mass., 1973.

[3] Blake, Lamont V., *Antennas*, John Wiley & Sons, New York, 1966.

[4] Skolnik, M., *Introduction to Radar Systems*, McGraw-Hill Book Comp., Inc., New York, 1962.

[5] Tai, Chen T., "The Nominal Directivity of Uniformly Spaced Arrays of Dipoles", *Microwave Journal*, Vol. 7, September 1964; pp. 51 - 55.

[6] Hansen, Robert C., *IEEE Trans. on Antennas and Propagation*, Vol. AP-20 No. 1, January 1972; pp. 100 - 102.

[7] Hansen, R. C. (ed.), "Microwave Scanning Antennas", Vol. II, *Antenna Theory and Practice*, Chapter 3, A. A. Oliner and R. G. Malech, *Mutual Coupling in Infinite Scanning Arrays*, Academic Press, New York, 1966; p. 303.

[8] Hannon, P. W., "The Element-Gain Paradox for a Phased-Array Antenna", *IEEE Trans.*, Vol. AP-12, July, 1964; pp. 423 - 433.

[9] Bach, Henning, "Directivity Diagrams for Uniform Linear Arrays", *Microwave Journal*, Vol. 15, December 1972; pp. 41 - 44.

[10] Hansen, *op. cit.*, Chapter 2, A. A. Oliner and R. G. Malech, "Radiating Elements and Mutual Coupling"; p. 77.

[11] Tai, C. T., "The Optimum Directivity of Uniformly Spaced Broadside Arrays of Dipoles", *IEEE Trans. on Antennas and Propagation*, Vol. AP-12, July 1964; pp. 447 - 454.

[12] Skolnik, M. I., *Radar Handbook*, McGraw-Hill Book Comp., New York, 1970.

[13] Ruze, J., "Antennas for Radar Astromony", in J. V. Evans and T. Hagfors (eds.) *Radar Astronomy*, McGraw-Hill Book Comp., New York, 1968.

[14] Dolph, C. L. "A Current Distribution for Broadside Arrays Which Optimizes the Relationship Between Beam Width and Side-Lobe Level", *Proc. of the I.R.E.*, Vol. 34, June 1946; pp. 335 - 348.

[15] Taylor, T. T., "Design of Line-Source Antennas for Narrow Beamwidth and Low Side Lobes", *I.R.E. Trans.*, Vol. AP-3, January 1955; pp. 16 - 28.

[16] Hansen, R. C., "Gain Limitations of Large Antennas", *I.R.E. Trans.*, Vol. AP-8, September 1960; pp. 490 - 495.

[17] Rudduck, R. C., and D. C. F. Wu, "Directive Gain of Circular Taylor Patterns", *Radio Science* Vol. 6, Number 12, December 1971; pp. 1117 - 1121.

[18] Mattingly, R. L., "Nonreciprocal Radar Antennas", *Proc. I.R.E.*, Vol. 48, April 1960; pp. 795 - 796.

[19] Kraus, J.D., *Antennas*, McGraw-Hill, New York, 1950.

[20] Von Aulock, W. H., "Properties of Phased Arrays", *Proc. I.R.E.*, Vol. 48, October 1960; pp. 1715 - 1727.

[21] Stegen, Robert J., "The Gain-Beamwidth Product of an Antenna", *IEEE Trans.*, Vol. AP-12, July 1964; pp. 505 - 506.

[22] Hannon, P. W., D. S. Lerner, and G. H. Knittel, "Impedance Matching a Phased-Array Antenna Over Wide Scan Angles by Connecting Circuits", *IEEE Trans.*, Vol. AP-13, January, 1965; pp. 28 - 34.

[23] Varon, D., and G. I. Zysman, "On the Mismatch of Electronically Steerable Phased-Array Antennas", *Radio Science*, Vol. 3 (New Series) No. 5, May, 1968; pp. 487 - 489.

[24] Knittel, G. H., "Wide-Angle Impedance Matching of Phased-Array Antennas — A Survey of Theory and Practice", in A. A. Oliner and G. H. Knittel (eds.) *Phased-Array Antennas*, Artech House, Inc., Dedham, Mass., 1972.

[25] Woodward, P. M., "A Method of Calculating the Field Over a Planar Aperture Required to Produce a Given Polar Diagram", *J. IEE* (London), Vol. 93, pt. 3A, pp. 1554 - 1558, 1946.

[26] Barton, D. K., and H. R. Ward, *Handbook of Radar Measurement*, Prentice-Hall, Inc., New Jersey, 1969.

[27] Tseng, Fung-I, and David K. Cheng, "Optimum Scannable Planar Arrays with an Invariant Sidelobe Level", *Proc. of the IEEE*, Vol. 56, No. 11, November 1968; pp. 1771 - 1778.

[28] Hansen, R. C., "Tables of Taylor Distributions for Circular Aperture Antennas", *IRE Trans.*, Vol. AP-8, January 1960; pp. 23 - 26.

[29] Sherman, J. W., "Aperture - Antenna Analysis", Chapter 9, *Radar Handbook*, M. I. Skolnik (ed.), McGraw-Hill Book Comp., New York, 1970.

[30] Jacquinot, P., and B. Roizen-Dossier "Apodisation," in E. Wolf (ed.), *Progress in Optics*, Vol. 3, Interscience Publishers, Inc., New York, 1963.

[31] Bayliss, E. T., "Design of Monopulse Antenna Difference Patterns with Low Sidelobes", *The Bell System Tech. J.*, Vol. 47, May-June 1968; pp. 623 - 650.

[32] Sanzgiri, S. M., J. K. Butler, and R. C. Voges, "Optimum Aperture Monopulse Excitation", *IEEE Trans.*, Vol. AP-20, May 1972; pp. 275 - 280.

[33] Skolnik, M. I., "Nonuniform Arrays", in R. E. Collin and F. J. Zucker (eds.) *Antenna Theory, Part I*, McGraw-Hill Book Comp., 1969.

[34] Kahrilas, P. J., "HAPDAR—An Operational Phased Array Radar", *Proc. of the IEEE*, Vol. 56, No. 11, November 1968; pp. 1967 - 1975.

[35] Reed, J. Emory, "The AN/FPS-85 Radar System", *Proc. of the IEEE*, Vol. 57, No. 3, March 1969; pp. 324 - 335.

[36] Daley, E., and F. Steudel "Modern Electronically-Scanned Array Antennas" *Electronic Progress*, Raytheon Comp., Vol. 16, No. 4, pp. 10 - 17, Winter 1974.

[37] Skolnik, M. I., J. W. Sherman III, and F. C. Ogg, Jr., "Statistically Designed Density-tapered Arrays, *IEEE Trans. Antennas Propagation*, Vol. AP-12, pp. 408 - 417, July 1964.

[38] Willey, R. E., "Space Tapering of Linear and Planar Arrays", *IRE Trans. Antennas Propagation*, Vol. AP-10, July 1962; pp. 369 - 377.

[39] Ruze, J., "Physical Limitations on Antennas", MIT Research Lab. Electronics Tech. Rept. 248, 30 October 1952.

[40] Steinberg, B. D., "The Peak Sidelobe of the Phased Array Having Randomly Located Elements", *IEEE Trans. on Antennas and Propagation*, Vol. AP-20, No. 2, March 1972; pp. 129 - 135.

[41] Lo, Y. T., "A Mathematical Theory of Antenna Arrays with Randomly Spaced Elements", Vol. AP-12, No. 3, May 1964; pp. 257 - 268.

[42] Lo, Y. T., "Random Periodic Arrays", *Radio Science*, Vol. 3 (New Series), No. 5, May 1968; pp. 425 - 436.

[43] Allen, J. L., "Beam Shaping by Omission of Elements" ("Space Tapering" or "Density Tapering") in J. L. Allen *et al.*, *Phased Array Radar Studies*, Tech Rept. No. 236, Lincoln Laboratory, M.I.T., 1 July 1960 to 1 July 1961.

[44] Cheston, T. C. and J. Frank, "Array Antennas", in M. I. Skolnik, *Radar Handbook*, McGraw-Hill Book Comp., 1970.

[45] White, W. D., "Pattern Limitations in Multiple-Beam Antennas", *IRE Trans.*, Vol. AP-10, July 1962; pp. 430 - 436.

[46] Van Der Maas, G. J., "A Simplified Calculation for Dolph-Tchebycheff Arrays", *Journal of Applied Physics*, Vol. 25, No. 1, January 1954; pp. 121 - 124.

[47] Salzer, H. E., "Calculating Fourier Coefficients for Chebyshev Patterns by Recurrence Formulas", *Proc. of IEEE*, Vol. 63, January 1975; pp. 195 - 196.

[48] Taylor, T. T., "Design of Circular Apertures for Narrow Beamwidth and Low Sidelobes", *IRE Trans.*, Vol. AP-8, January 1960; pp. 17 - 22.

APPENDIX B

$g(x')$	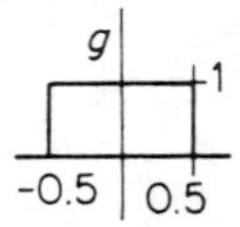	*Hamming* $0.54 + 0.46 \cos(2\pi x')$
Figure number	A.4	A.17
$A = \int_{-1/2}^{1/2} g(x')\,dx'$	1.000	0.536
$20 \log (A)$	0	-5.43
$C = \int_{-1/2}^{1/2} g^2(x')\,dx'$	1.000	0.394
$10 \log (C)$	0	-4.05
$D = \int_{-1/2}^{1/2} \left[\frac{dg(x')}{dx'}\right]^2 dx'$	–	–
$G = \int_{-1/2}^{1/2} [x'g(x')]^2\,dx'$	0.0850	0.0092
$H = \int_{-1/2}^{1/2} (x')^2 g(x')\,dx'$	0.0850	0.0216
$\eta_x = A^2/C$	1.000	0.729
$10 \log (\eta_x)$	0	-1.375
$\theta_n w/\lambda = C/A^2$	1.000	1.373
$\theta_3 w/\lambda$	0.886	1.296
$\Theta w/\lambda = \sqrt{D/C}$	–	–
$\mathscr{L}_s/w = 2\pi\sqrt{G/C}$	1.832	0.961
$10 \log [G_{sr}]$	+13.3	+42.8

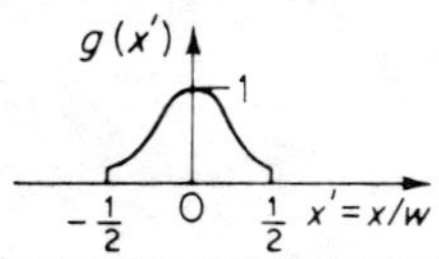

w = Aperture width
λ = Wavelength
η_x = Aperture efficiency
θ_n = Noise beamwidth (rad)
θ_3 = 3-db beamwidth (rad)
Θ = Rms beamwidth (rad)
$\mathscr{L}_s$ = Rms aperture width
$G_{sr} = \dfrac{\text{Gain of main beam}}{\text{Gain of highest sidelobe}}$

Figure 2.B.1

Uniform and Hamming Functions, g(x')*

*From Barton and Ward, Artech House [20]

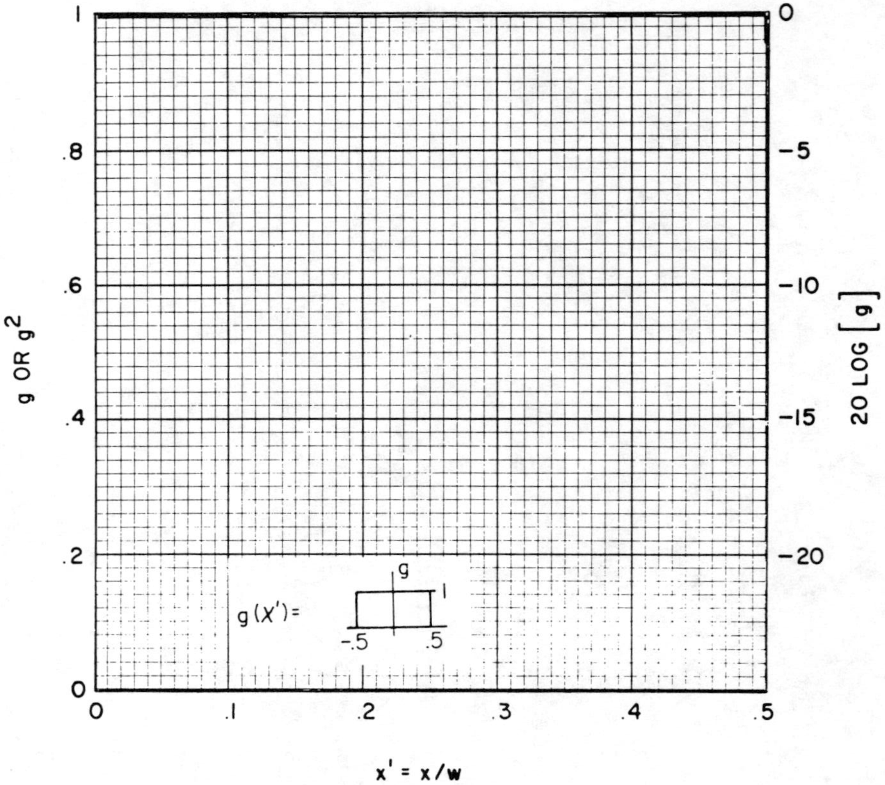

APERTURE ILLUMINATION FUNCTION

Figure 2.B.2

From Ward, H.R., "Antenna Pattern Handbook," Raytheon Tech. Rept. HRW-51, September 1968.

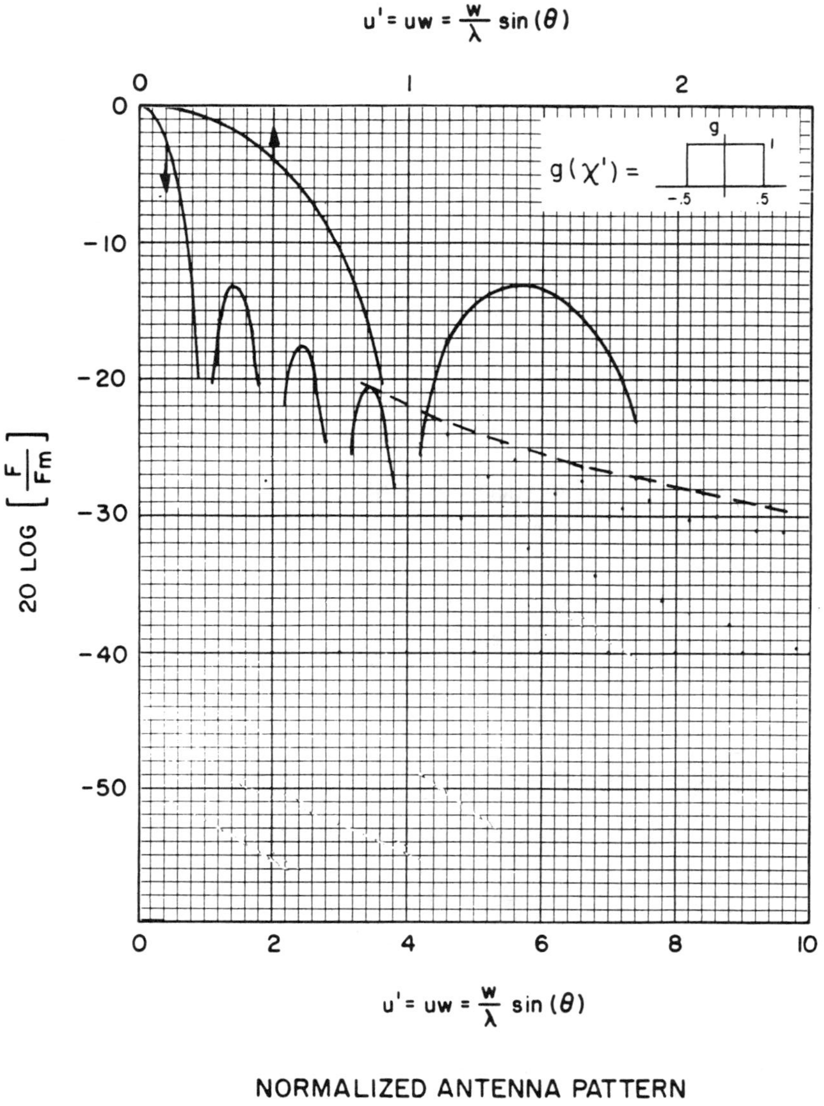

Figure 2.B.3

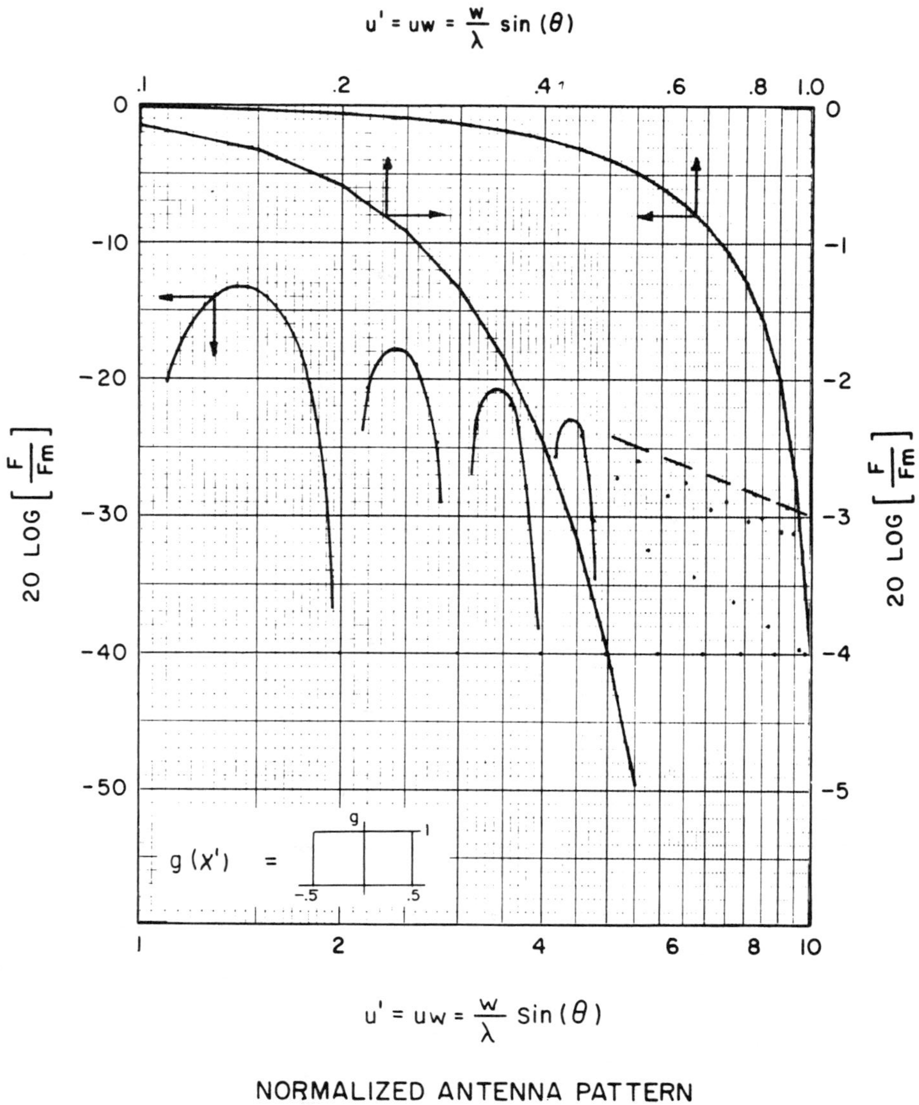

Figure 2.B.4

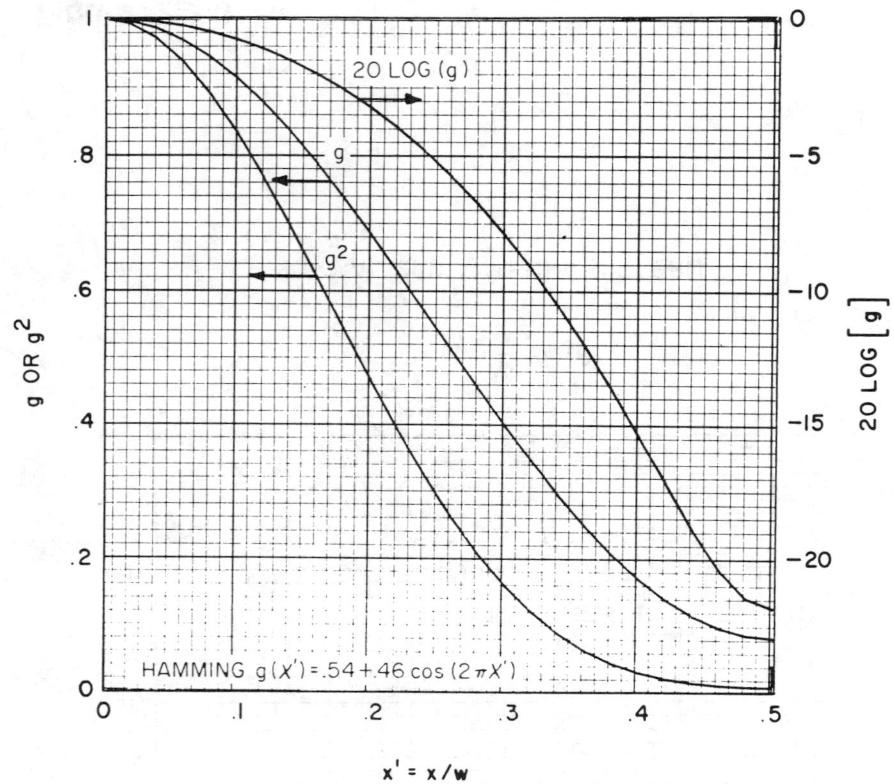

APERTURE ILLUMINATION FUNCTION

Figure 2.B.5

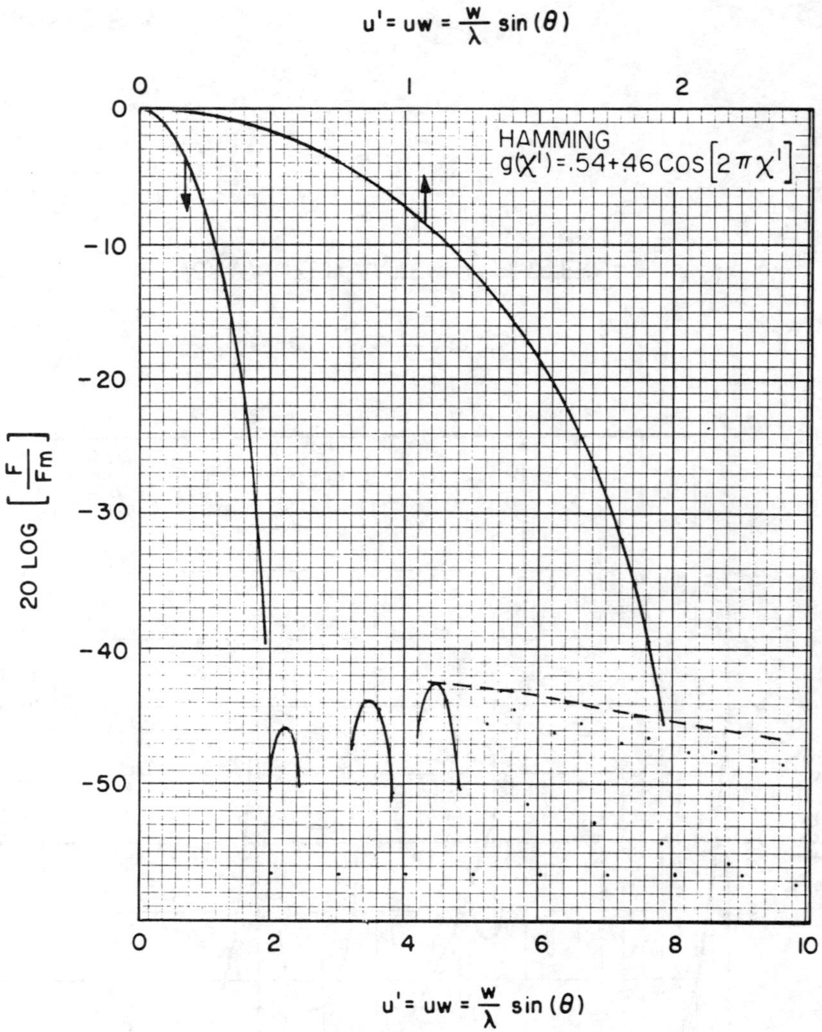

NORMALIZED ANTENNA PATTERN

Figure 2.B.6

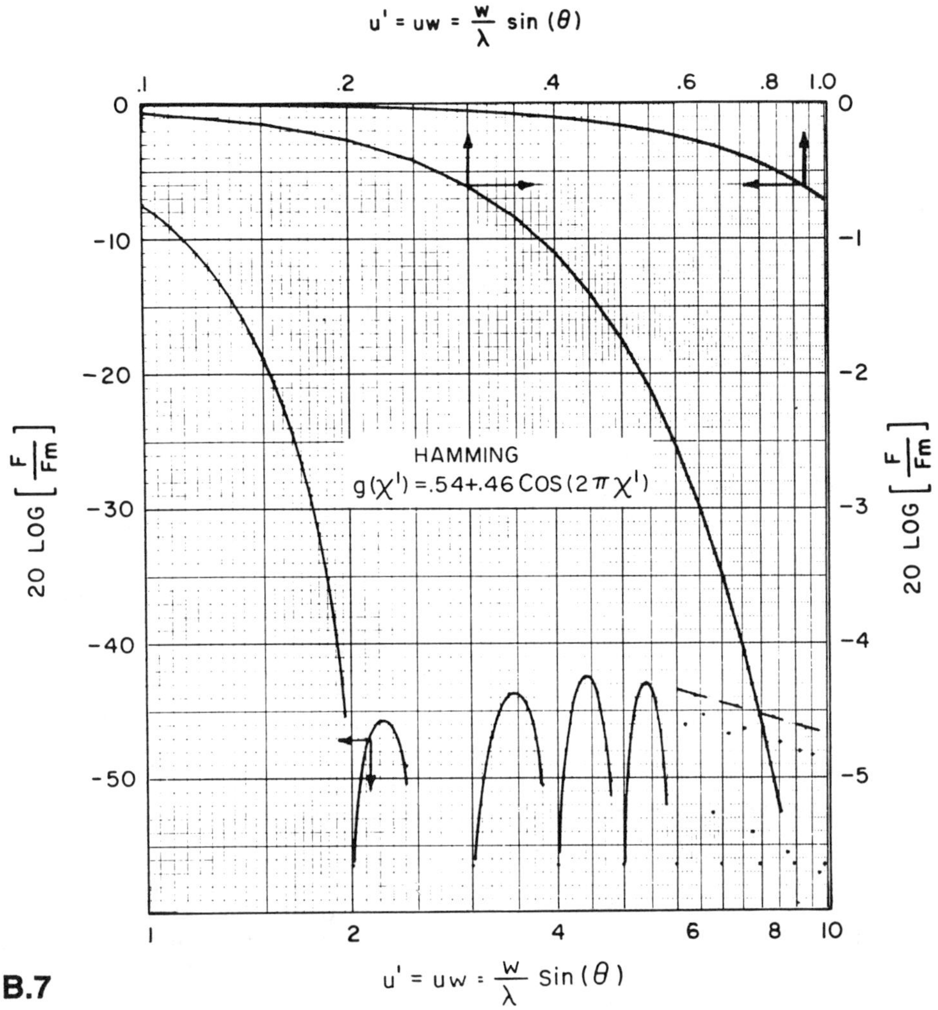

Figure 2.B.7

NORMALIZED ANTENNA PATTERN

Figure 2.B.8

CIRCULAR ILLUMINATION FUNCTIONS, $g(r') = [1 - (2r')^2]^n$

n	(*Uniform*) 0	1	2	3	4
Figure number	A.29	A.30			
$A = 8\int_0^{1/2} r'g(r')\,dr'$	1	0.490	0.327	0.245	0.1960
20 log (A)	0	−6.20	−9.72	−12.22	−14.15
$C = 8\int_0^{1/2} r'g^2(r')\,dr'$	1	0.327	0.1960	0.1399	0.1088
10 log (C)	0	−4.86	−7.08	−8.54	− 9.63
$\eta_a = A^2/C$	1	0.735	0.545	0.429	0.353
10 log (η_a)	0	−1.338	−2.64	−3.68	−4.52
$\theta_n D/\lambda = C/A^2$	1	1.361	1.836	2.33	2.83
$\theta_3 D/\lambda$	1.016	1.267	1.467	1.681	1.889
10 log [G_{sr}]	+17.6	+24.5	+30.5	+35.8	+40.6

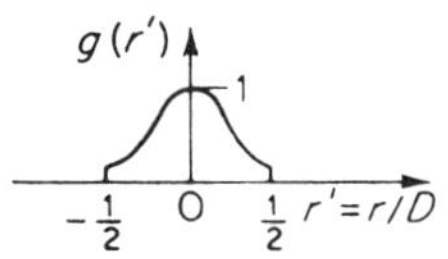

D = Aperture diameter
λ = Wavelength
η_a = Aperture efficiency

θ_n = Noise beamwidth (rad)
θ_3 = 3-db beamwidth (rad)
$G_{sr} = \dfrac{\text{Gain of main beam}}{\text{Gain of highest sidelobe}}$

Taylor* Linear Aperture Illumination Functions**

$\bar{n}$	2	3	4	5	6	8
SLL	20	25	30	35	40	45
Figure number	A.14		A.15		A.16	
$A = \int_{-1/2}^{1/2} g(x')\,dx'$	0.756	0.689	0.638	0.596	0.562	0.532
20 log (A)	−2.43	−3.23	−3.90	−4.49	−5.01	−5.48
$C = \int_{-1/2}^{1/2} g^2(x')\,dx'$	0.601	0.528	0.478	0.442	0.414	0.390
10 log (C)	−2.21	−2.78	−3.20	−3.55	−3.83	−4.09
$D = \int_{-1/2}^{1/2} \left[\frac{dg(x')}{dx'}\right]^2 dx'$	–	–	–	–	–	–
$G = \int_{-1/2}^{1/2} [x'g(x')]^2\,dx'$	0.0325	0.0224	0.0167	0.0131	0.0107	0.0090
$H = \int_{-1/2}^{1/2} (x')^2 g(x')\,dx'$	0.0517	0.0417	0.0344	0.0289	0.0247	0.0213
$\eta_x = A^2/C$	0.951	0.900	0.850	0.804	0.763	0.726
10 log (η_x)	−0.218	−0.455	−0.704	−0.948	−1.178	−1.390
$\theta_n w/\lambda = C/A^2$	1.051	1.111	1.176	1.244	1.311	1.377
$\theta_3 w/\lambda$	0.983	1.049	1.115	1.179	1.250	1.301
$\Theta w/\lambda = \sqrt{D/C}$	–	–	–	–	–	–
$\mathscr{L}_s/w = 2\pi\sqrt{G/C}$	1.462	1.297	1.176	1.085	1.015	0.959
10 log $[G_{sr}]$	+20.9	+25.9	+30.9	+35.9	+40.9	+45.9

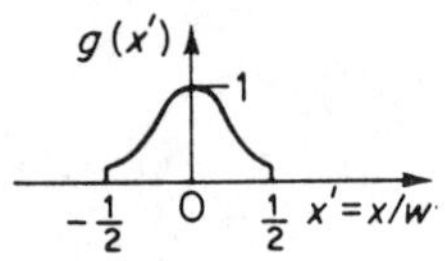

w = Aperture width
λ = Wavelength
η_x = Aperture efficiency
θ_n = Noise beamwidth (rad)
θ_3 = 3-db beamwidth (rad)
Θ = Rms beamwidth (rad)
$\mathscr{L}_s$ = Rms aperture width
$G_{sr} = \dfrac{\text{Gain of main beam}}{\text{Gain of highest sidelobe}}$

Figure 2.B.8 continued

*See Fig. 2.A.12 and Fig. 2.B.10
**From Barton and Ward, Artech House [20].

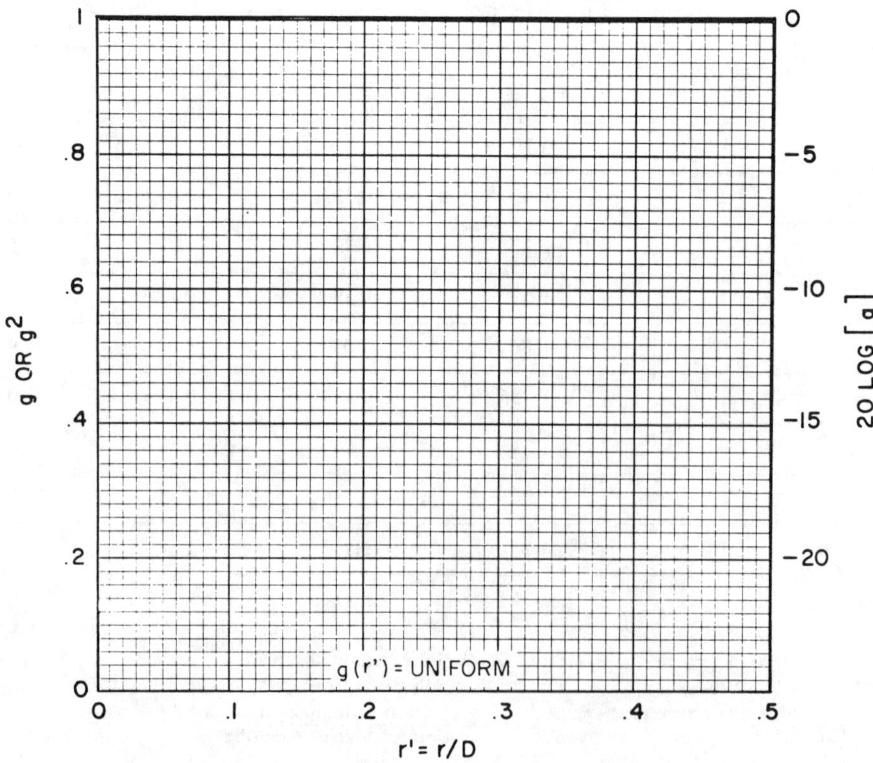

Figure 2.B.9 APERTURE ILLUMINATION FUNCTION

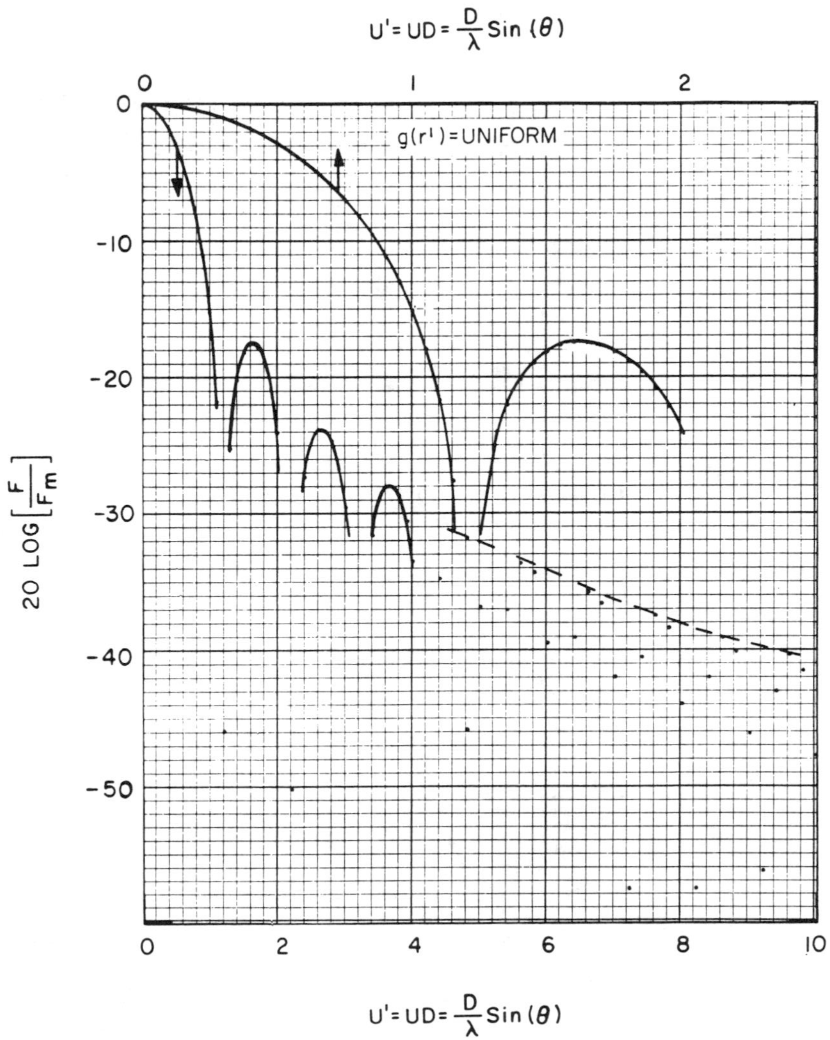

NORMALIZED ANTENNA PATTERN

Figure 2.B.10

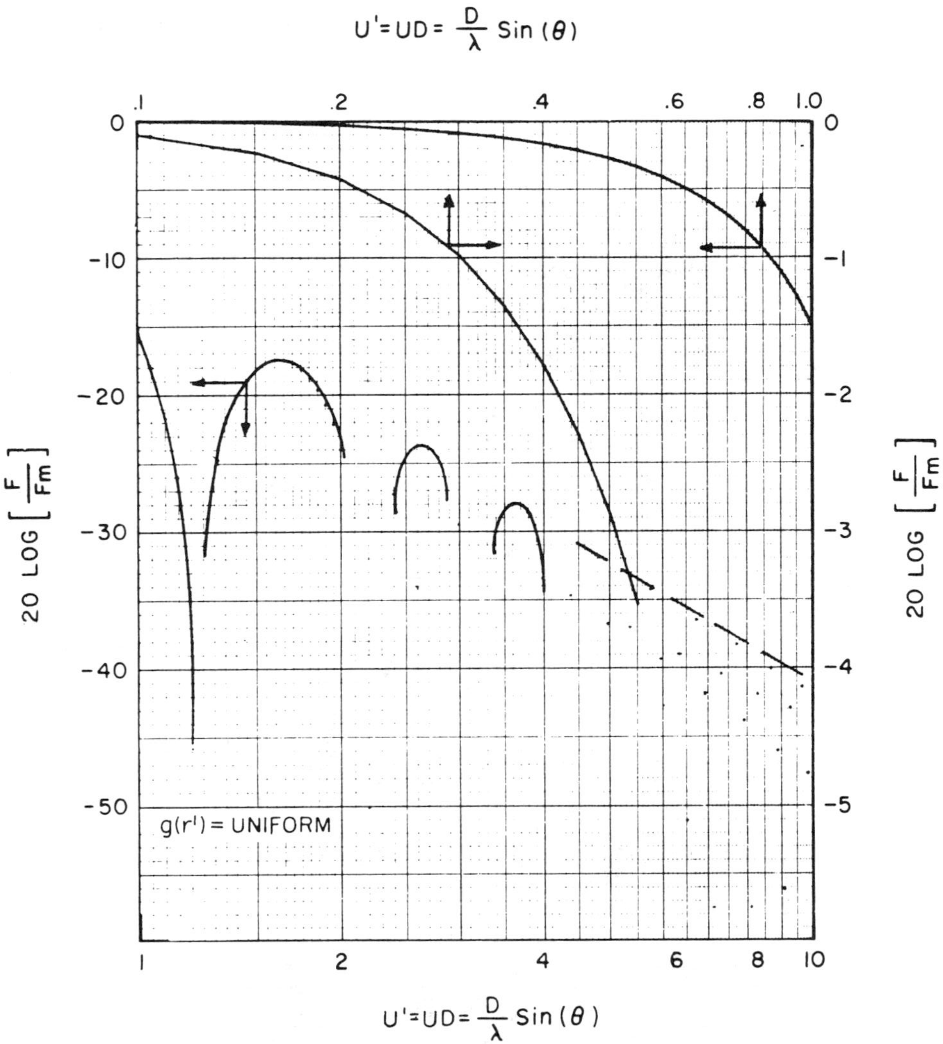

NORMALIZED ANTENNA PATTERN

Figure 2.B.11a

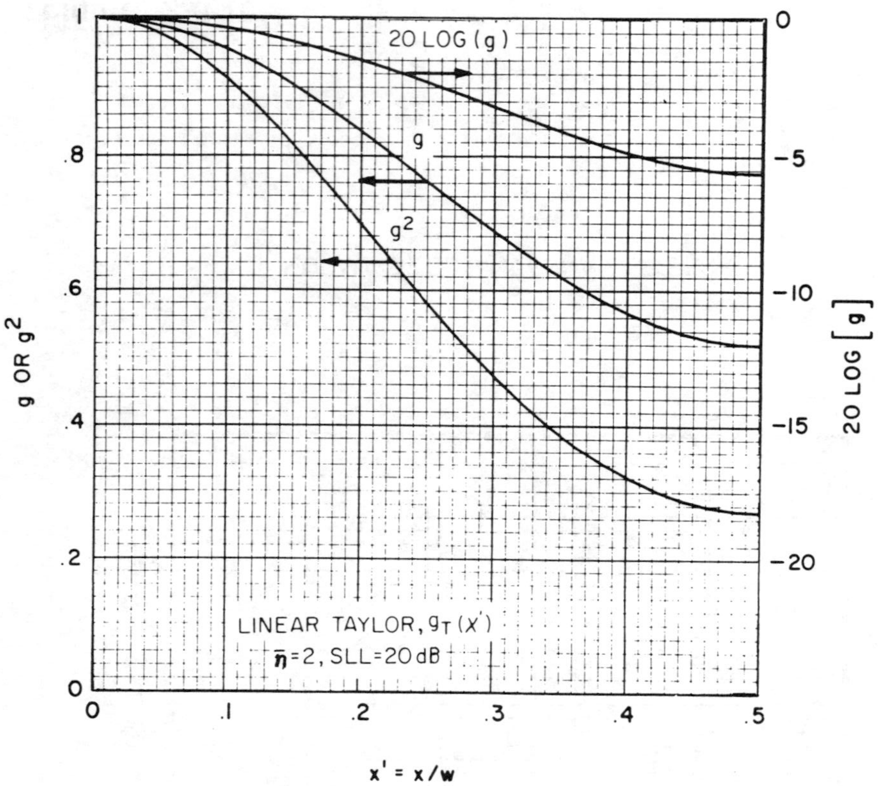

APERTURE ILLUMINATION FUNCTION

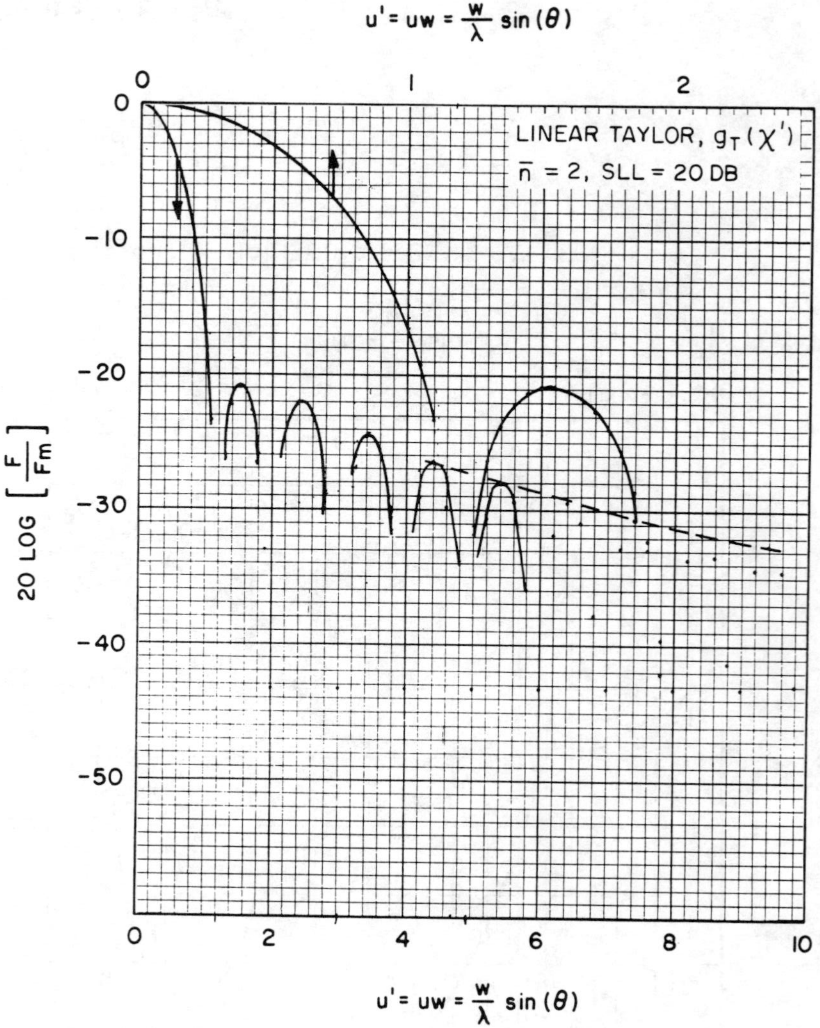

NORMALIZED ANTENNA PATTERN

From Ward, H.R., "Antenna Pattern Handbook," Raytheon Tech. Rept. HRW-11, September 1968.

Figure 2.B.11b

Figure 2.B.12

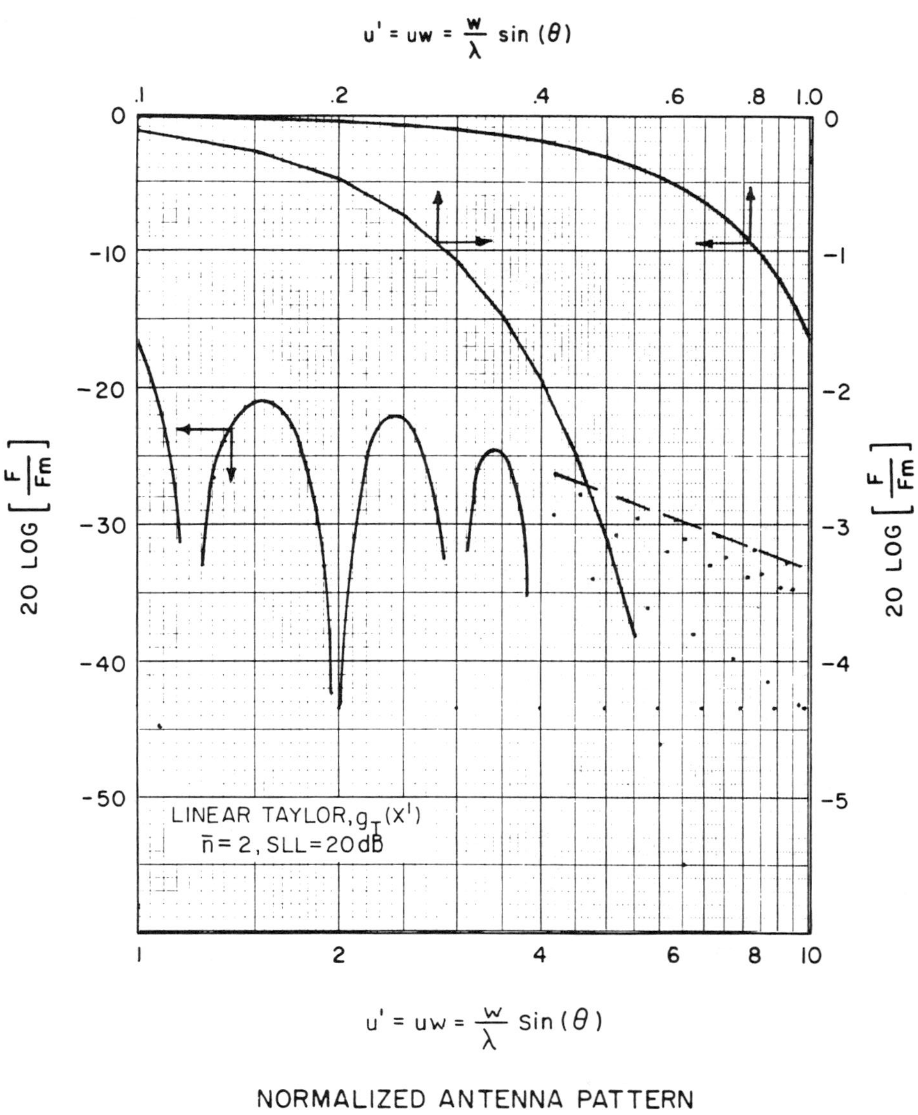

Figure 2.B.12 continued

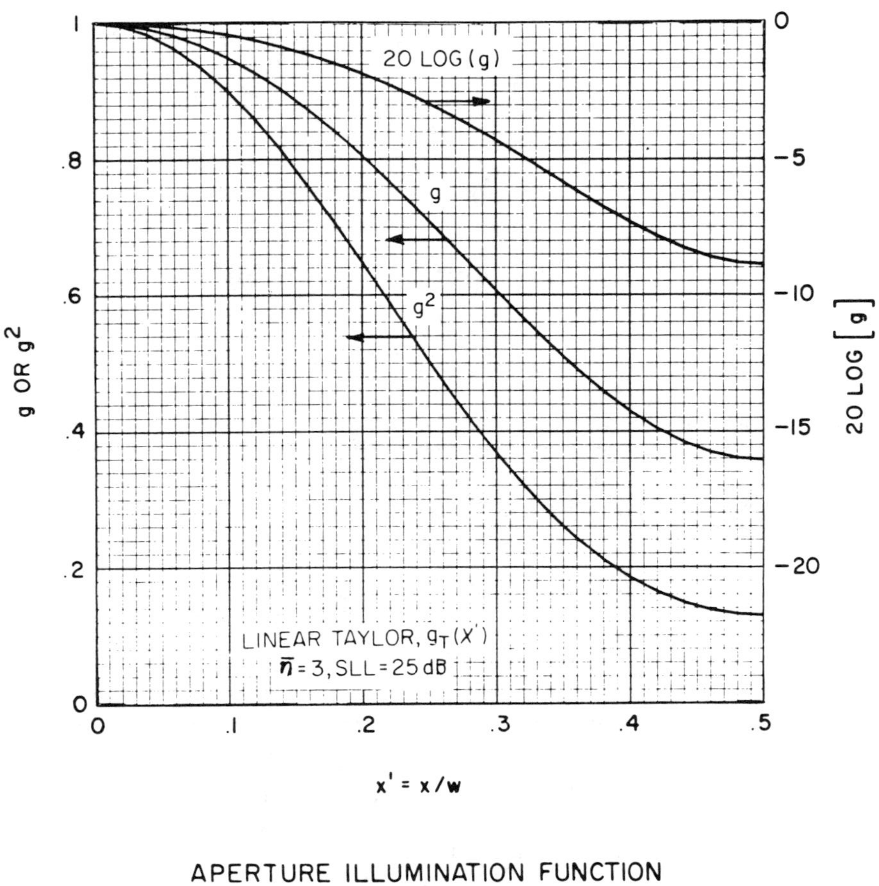

Figure 2.B.13

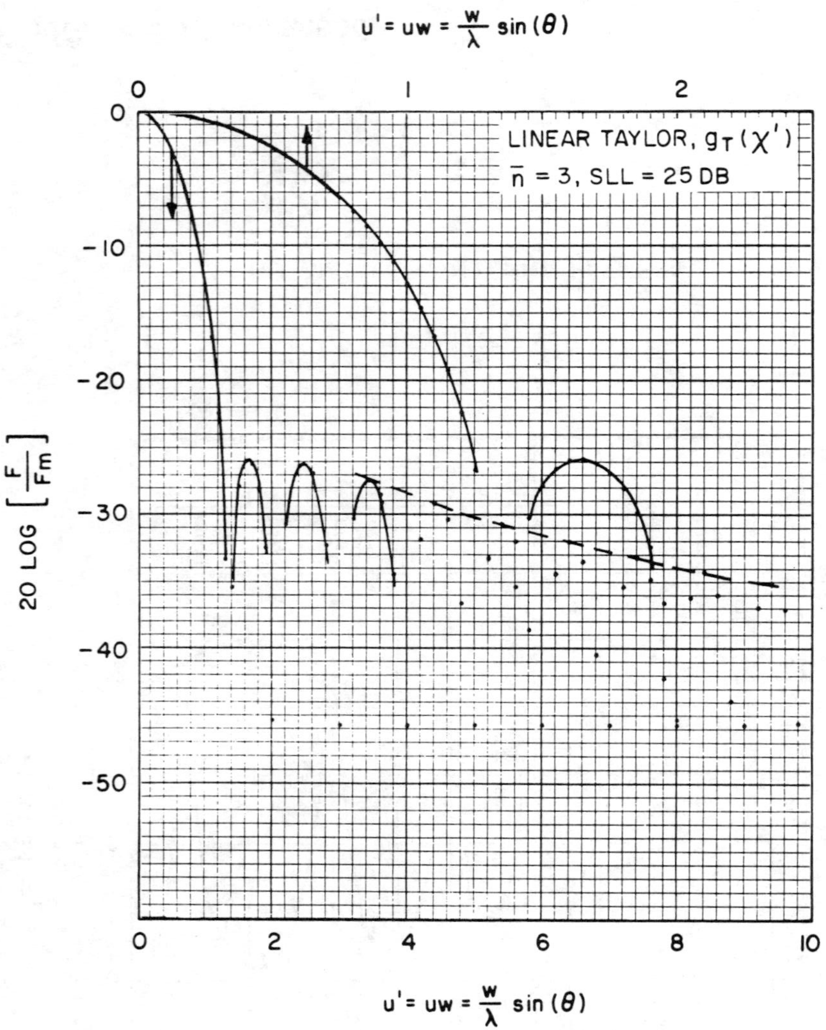

NORMALIZED ANTENNA PATTERN

Figure 2.B.14

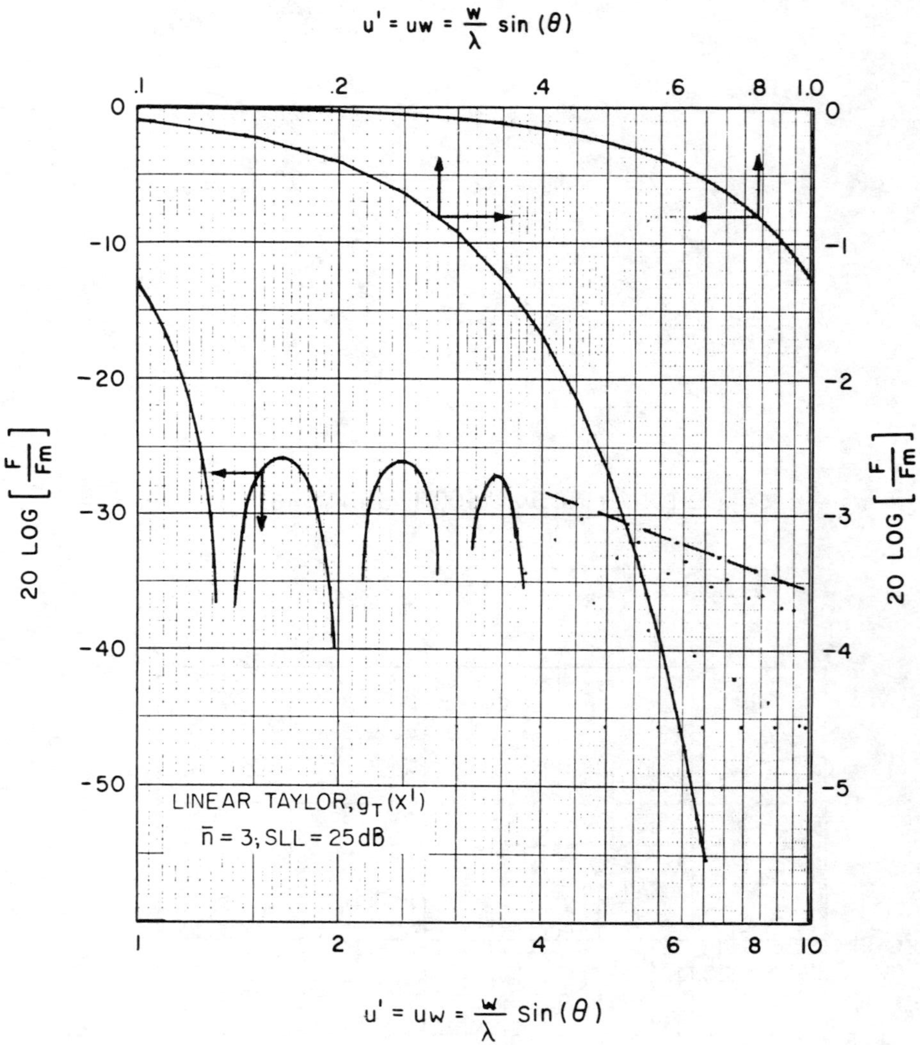

NORMALIZED ANTENNA PATTERN

Figure 2.B.15

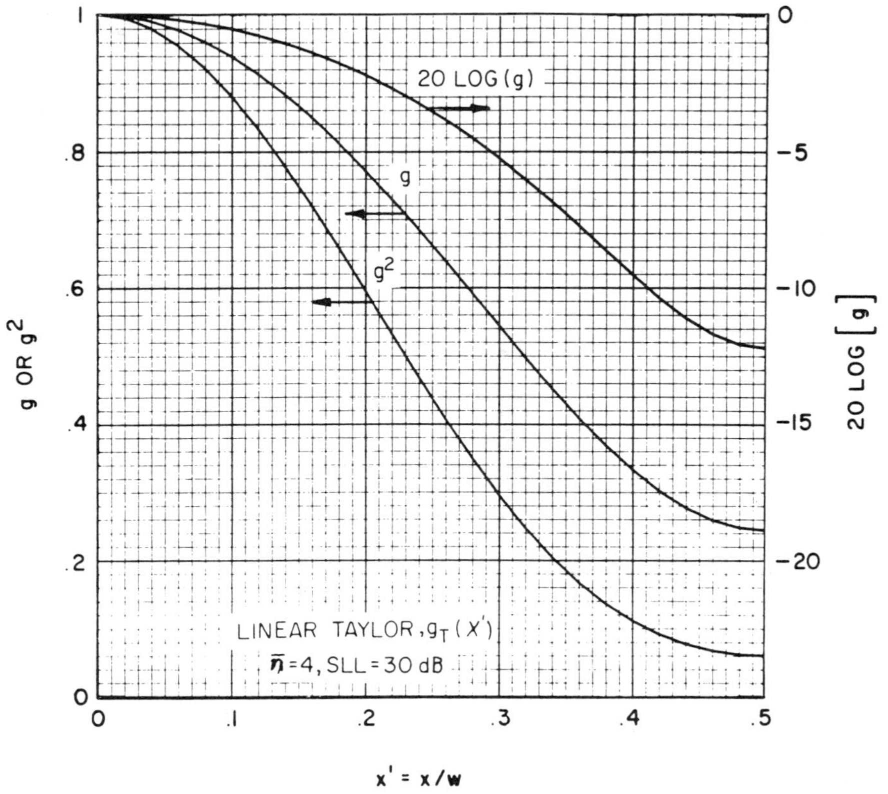

APERTURE ILLUMINATION FUNCTION

Figure 2.B.16

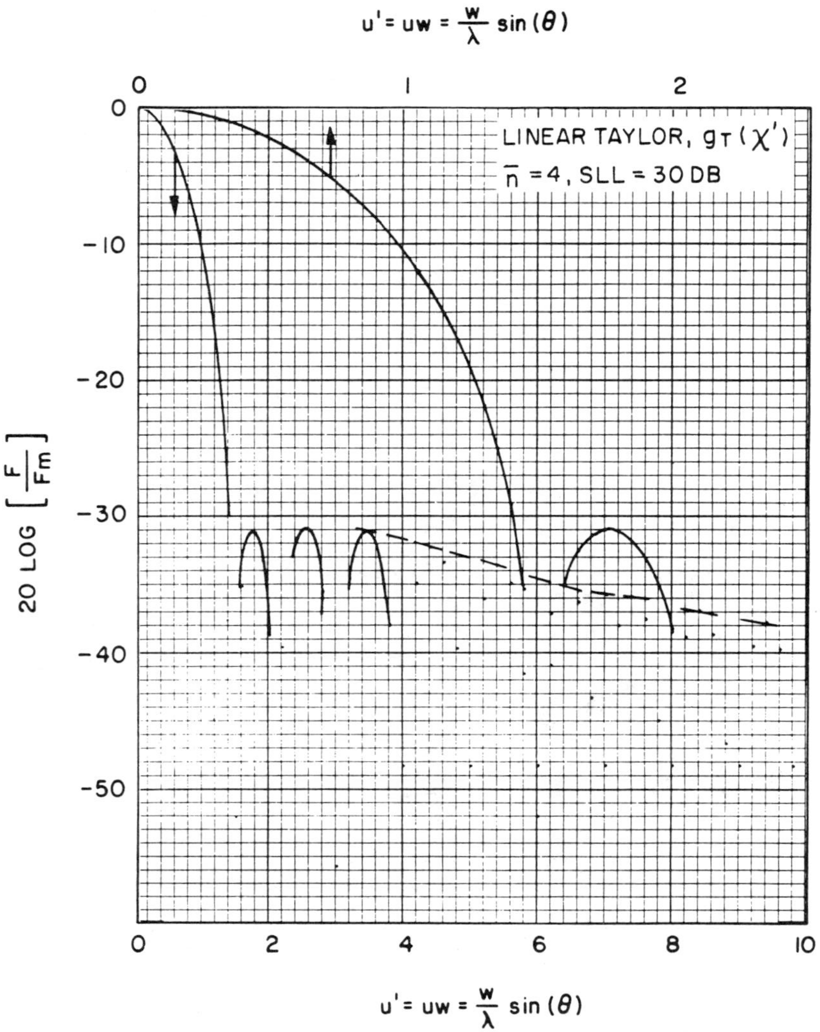

NORMALIZED ANTENNA PATTERN

Figure 2.B.17

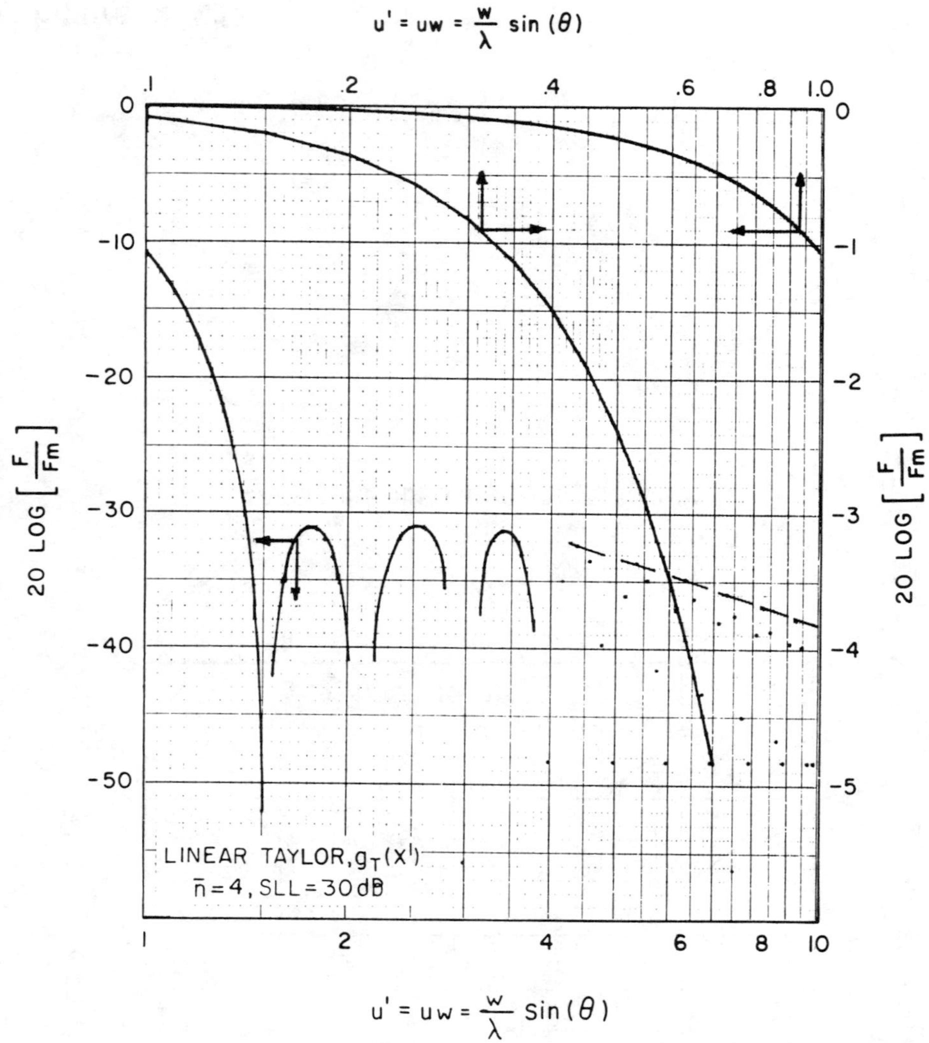

NORMALIZED ANTENNA PATTERN

Figure 2.B.18

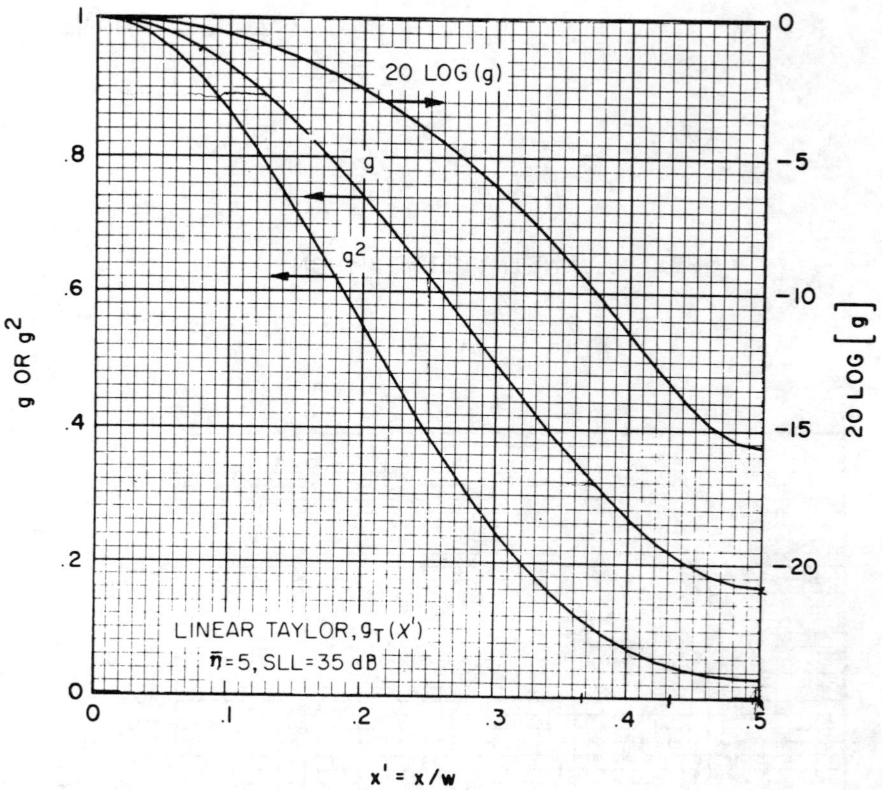

APERTURE ILLUMINATION FUNCTION

Figure 2.B.19

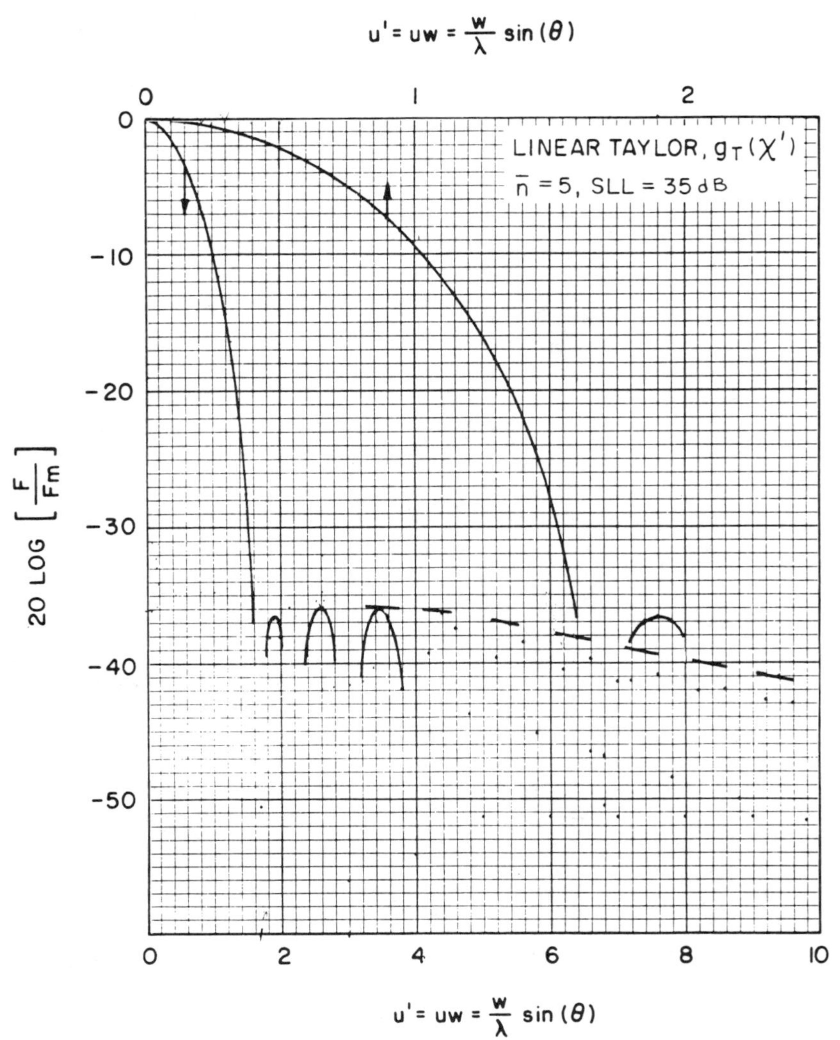

NORMALIZED ANTENNA PATTERN

Figure 2.B.20

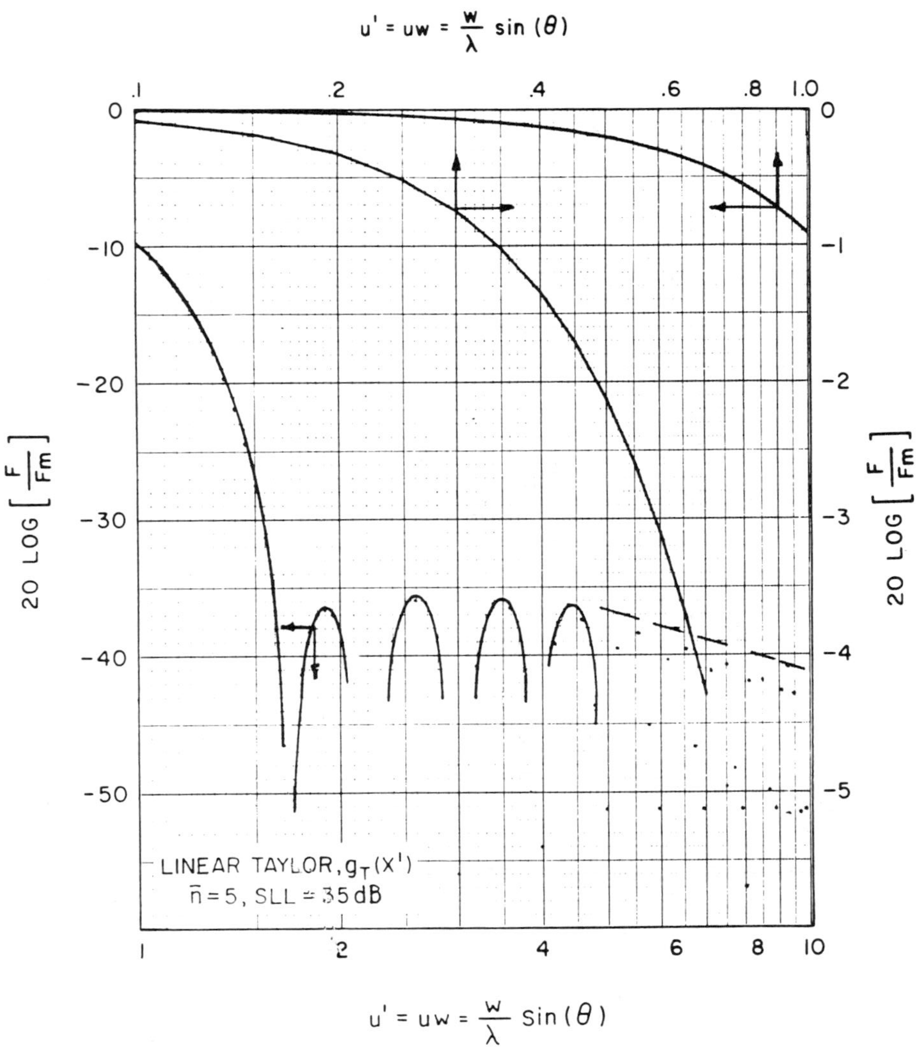

NORMALIZED ANTENNA PATTERN

Figure 2.B.21

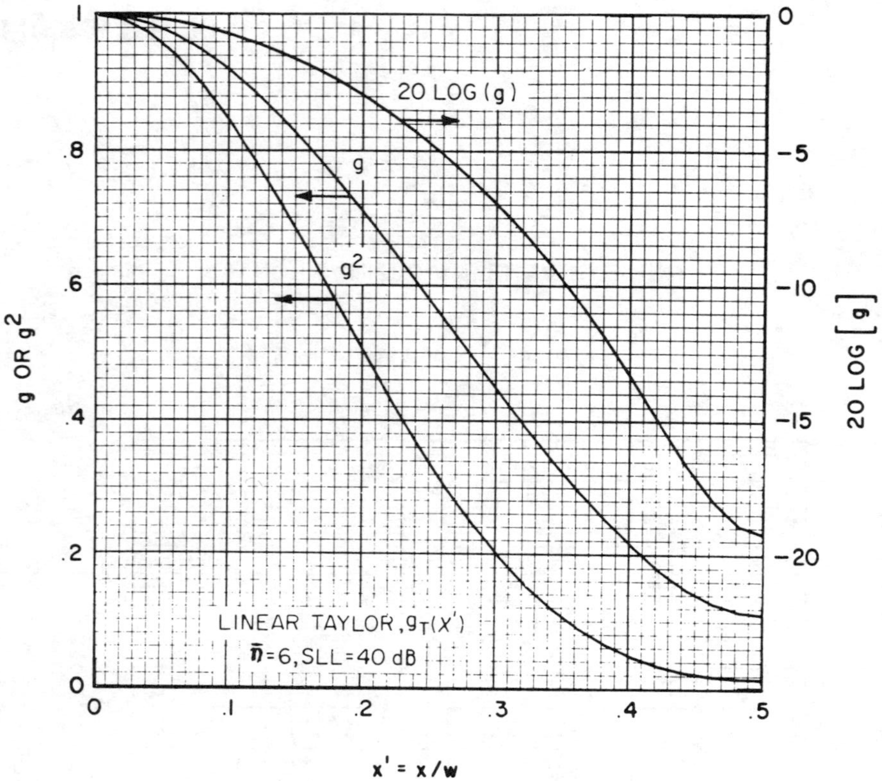

APERTURE ILLUMINATION FUNCTION

Figure 2.B.22

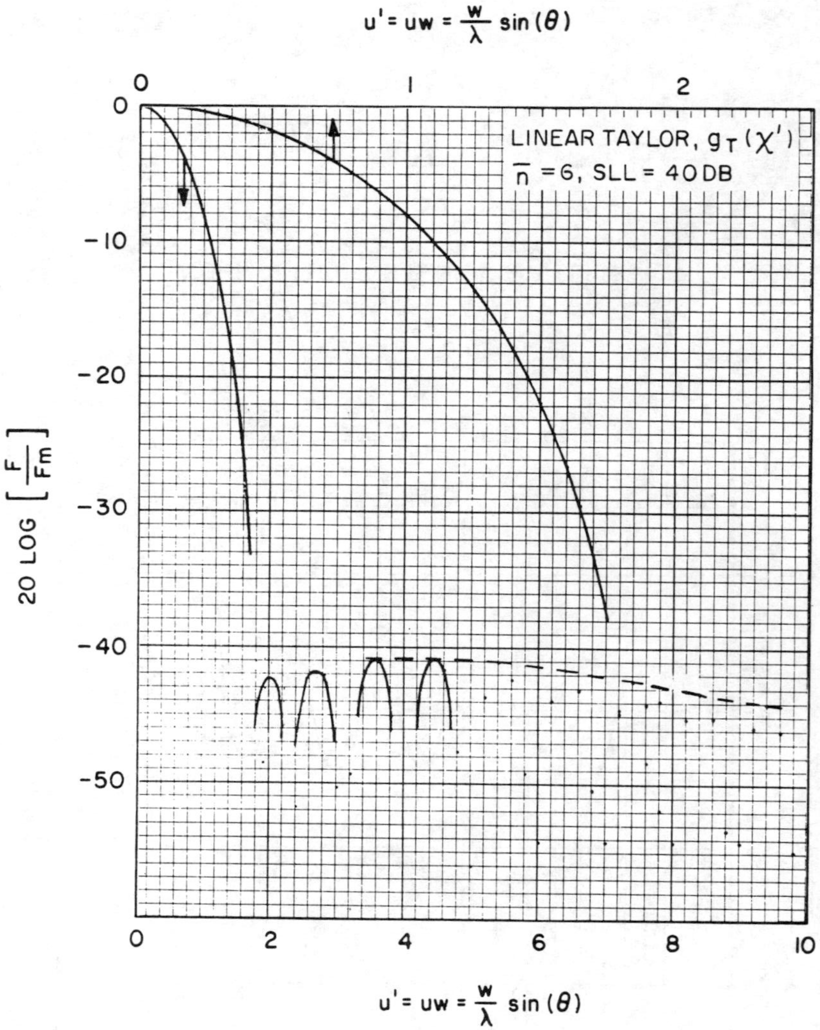

NORMALIZED ANTENNA PATTERN

Figure 2.B.23

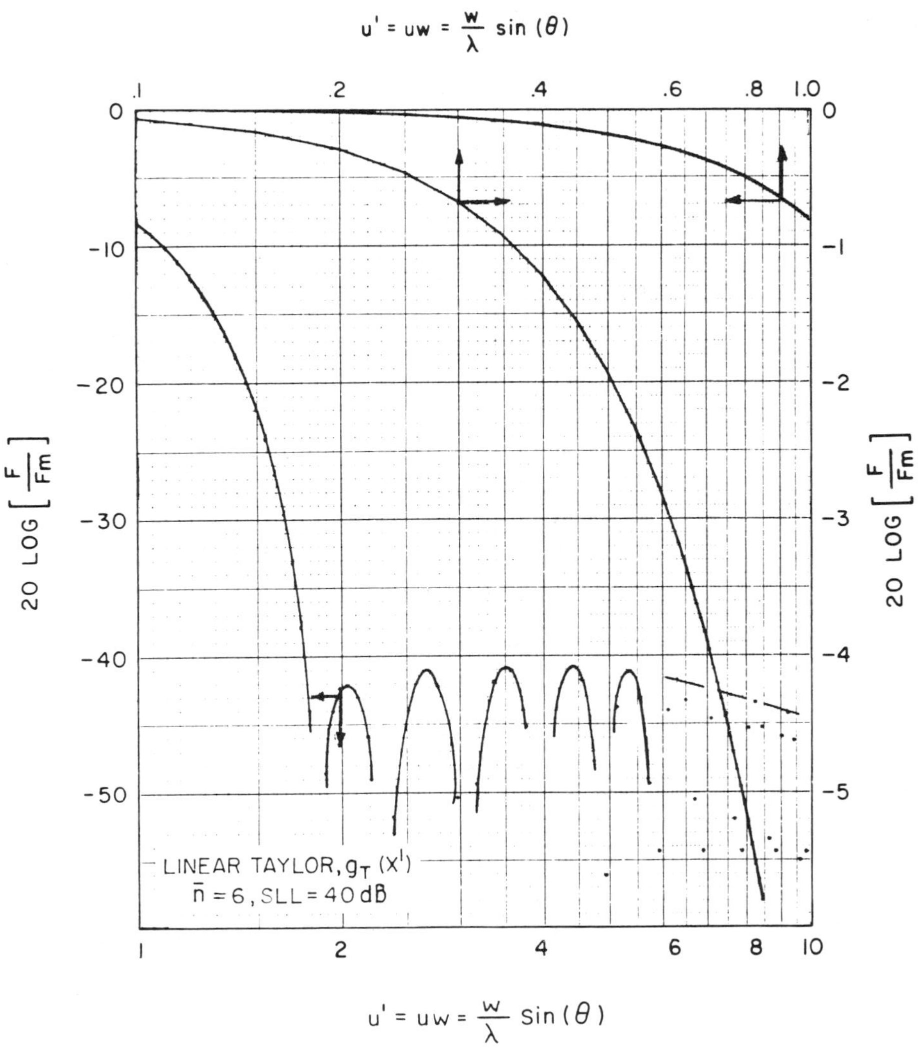

NORMALIZED ANTENNA PATTERN

Figure 2.B.24

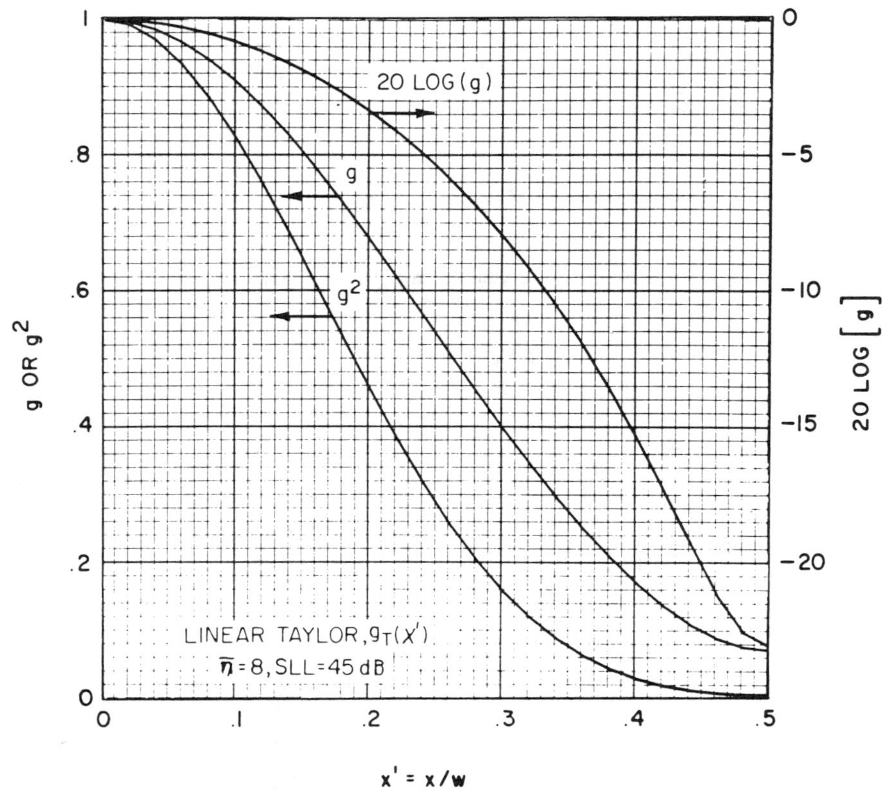

APERTURE ILLUMINATION FUNCTION

Figure 2.B.25

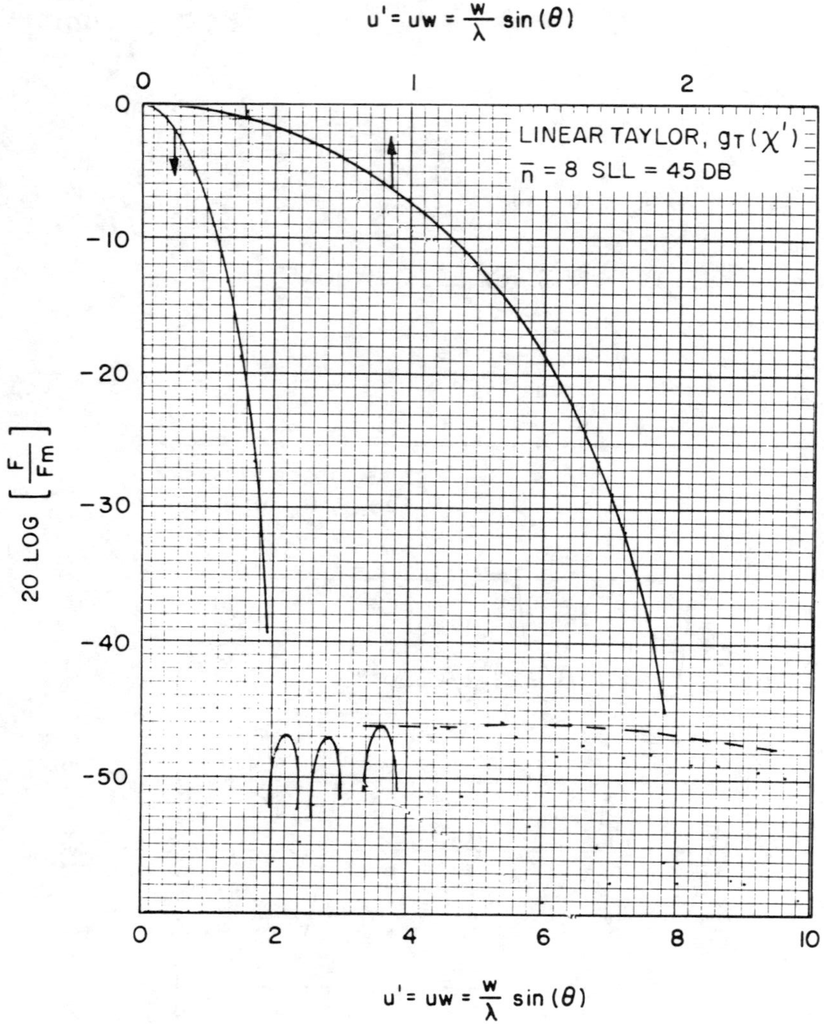

NORMALIZED ANTENNA PATTERN

Figure 2.B.26

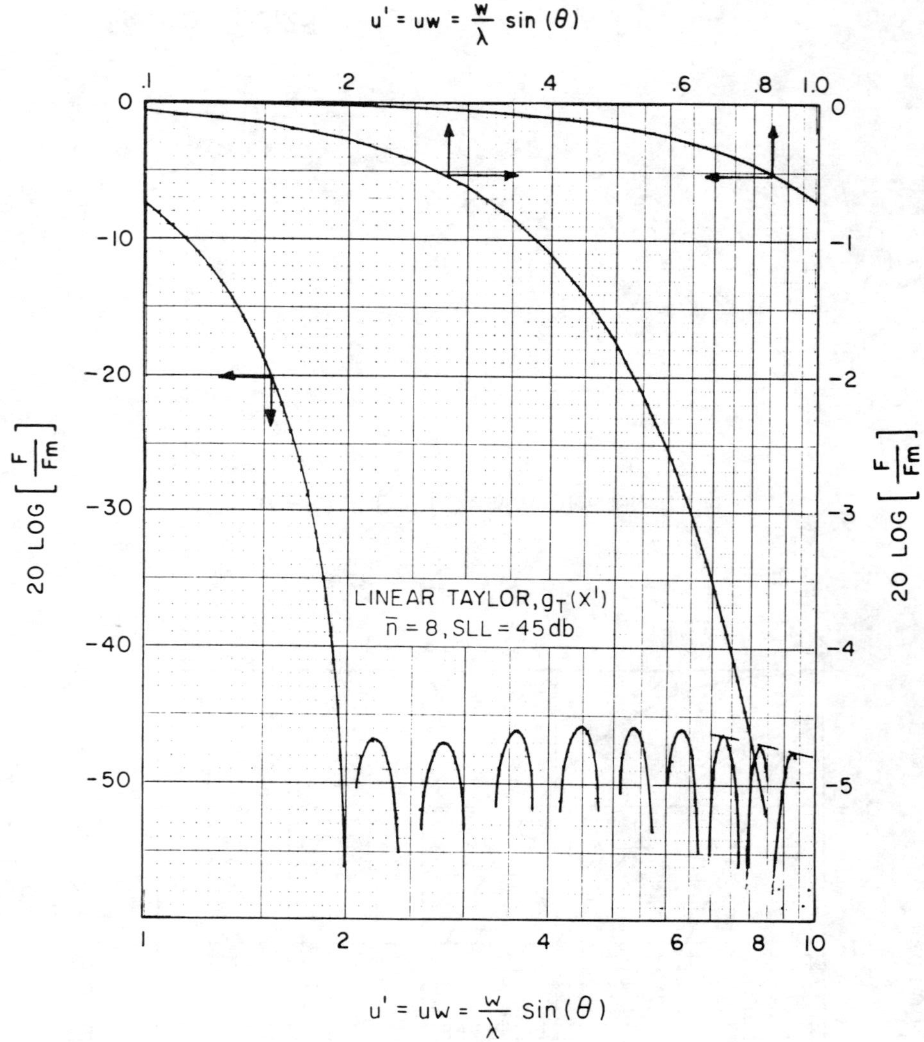

NORMALIZED ANTENNA PATTERN

Figure 2.B.27

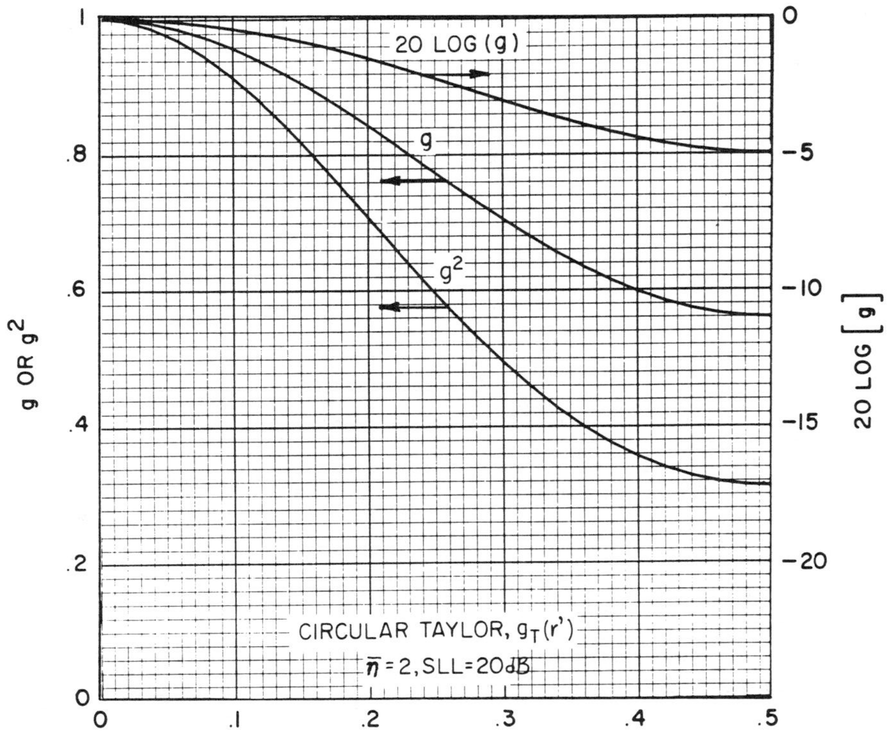

APERTURE ILLUMINATION FUNCTION

Figure 2.B.28

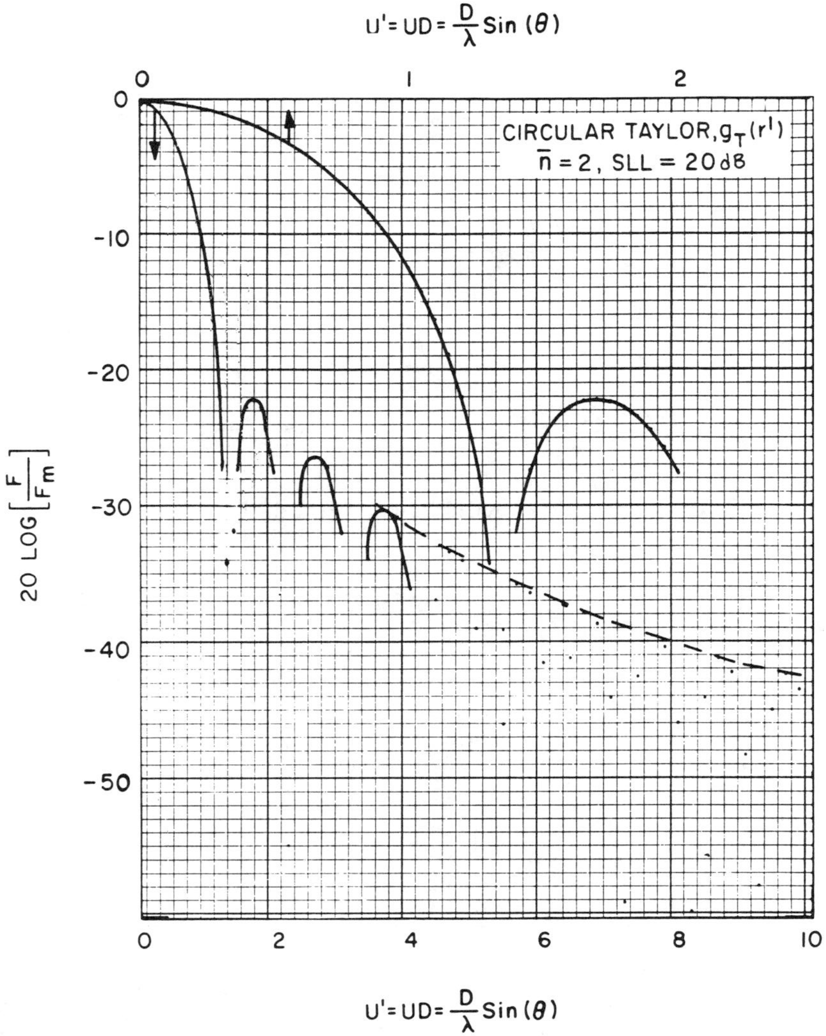

NORMALIZED ANTENNA PATTERN

Figure 2.B.29

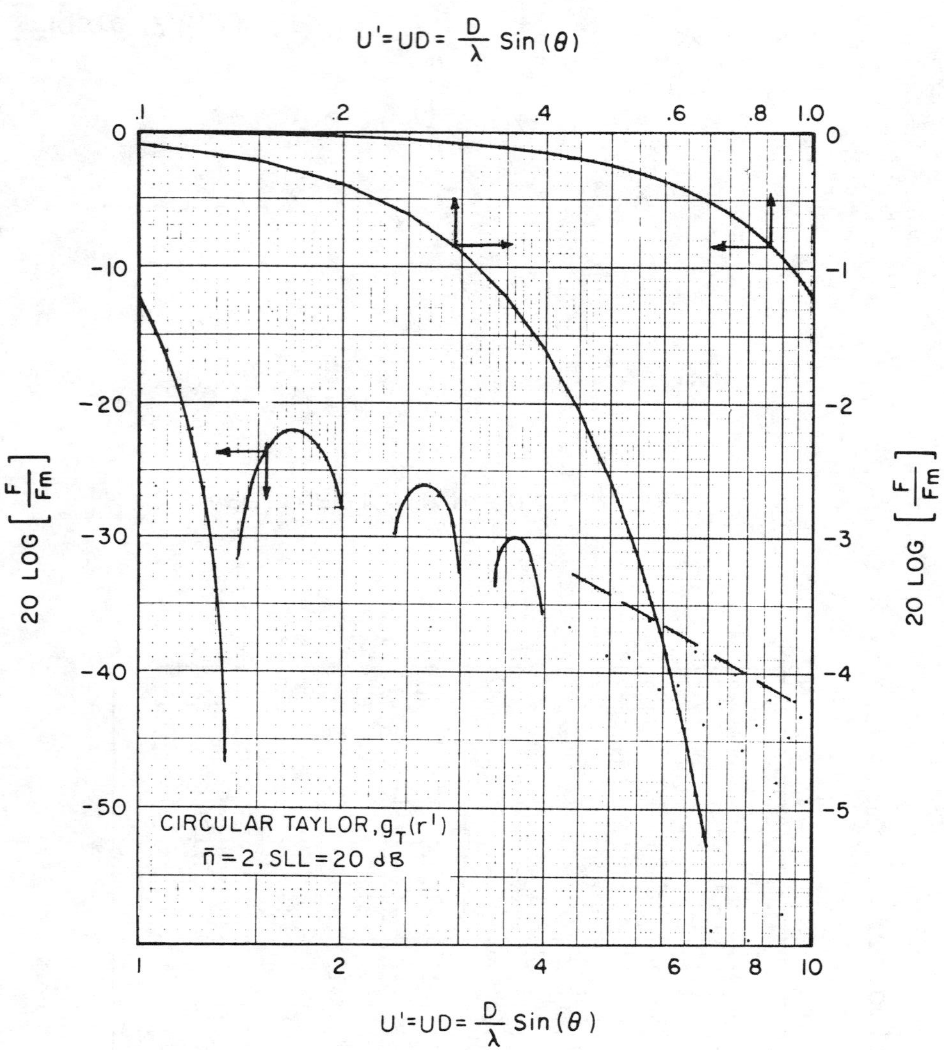

NORMALIZED ANTENNA PATTERN

Figure 2.B.30

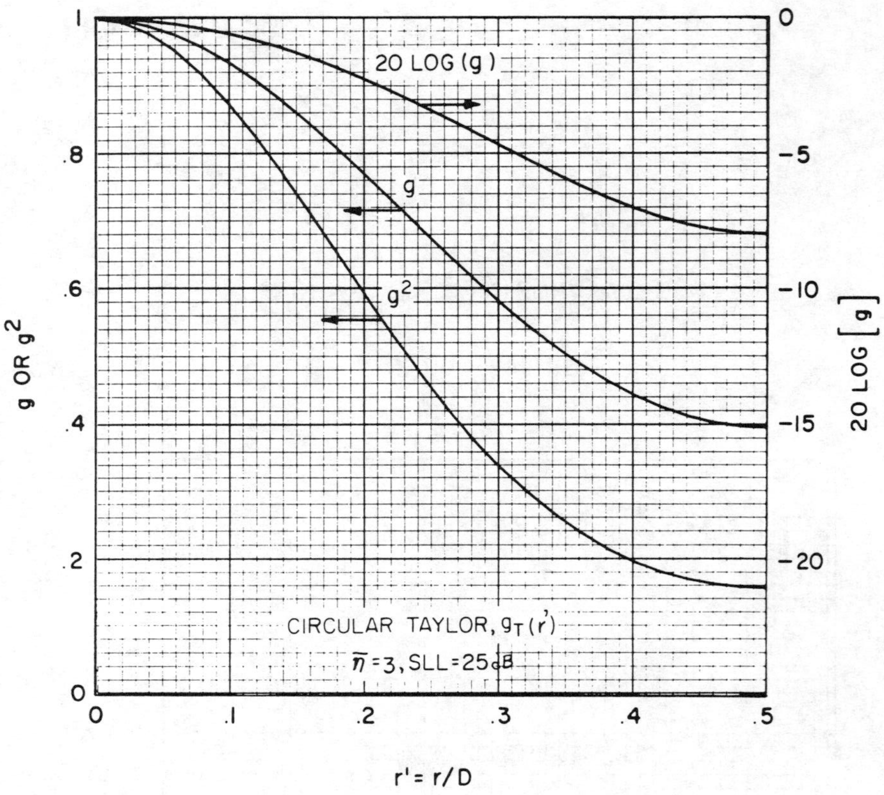

APERTURE ILLUMINATION FUNCTION

Figure 2.B.31

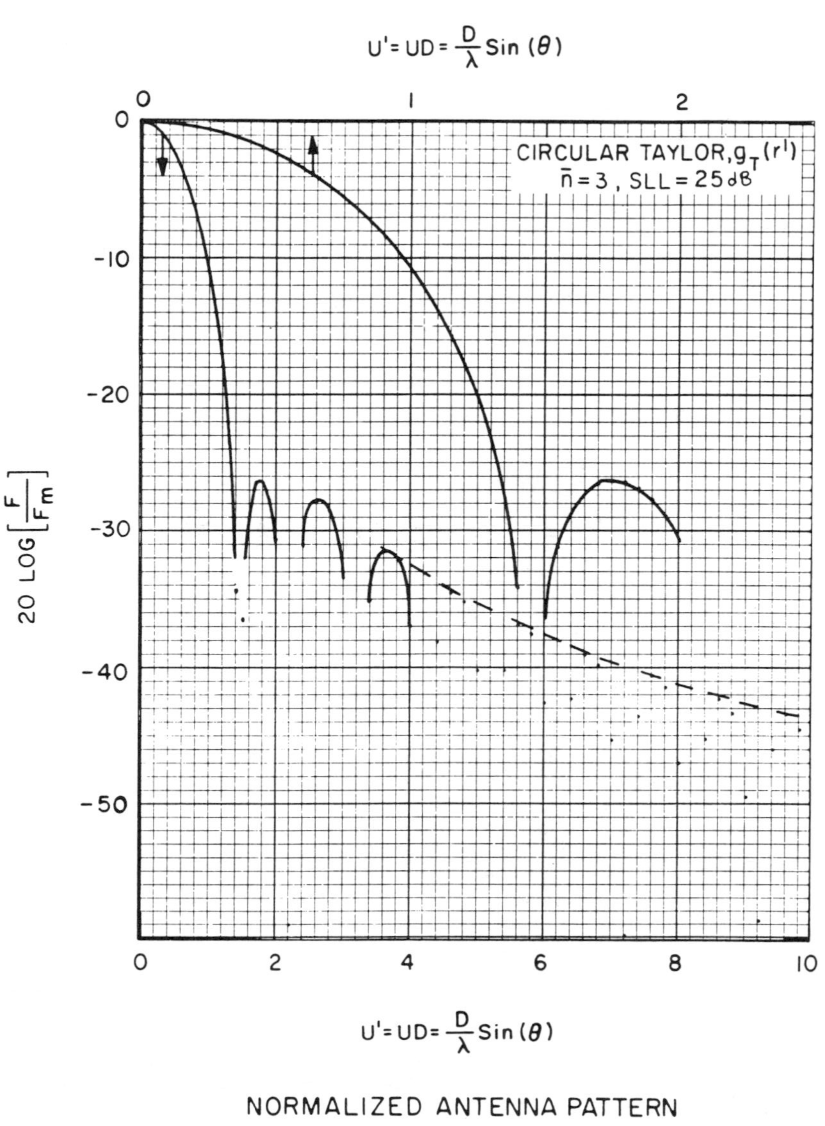

Figure 2.B.32

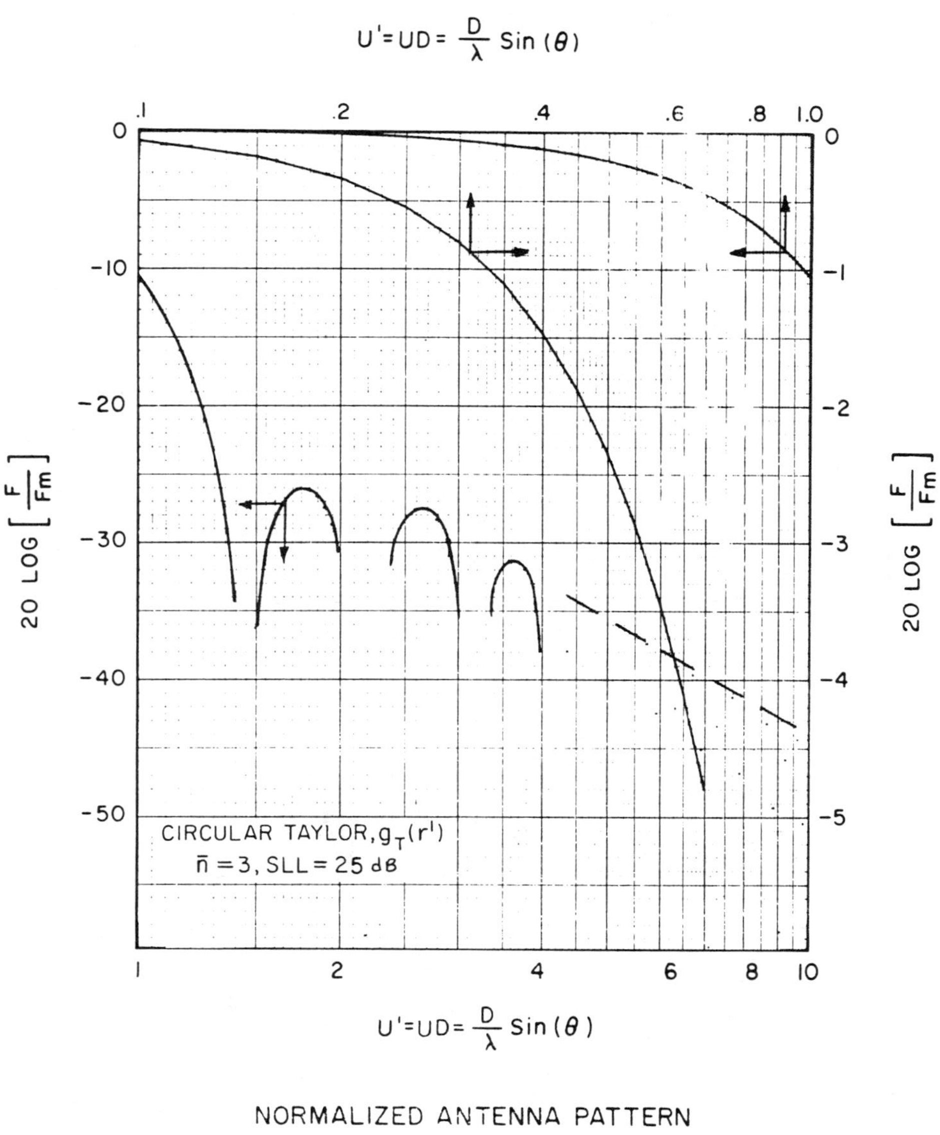

Figure 2.B.33

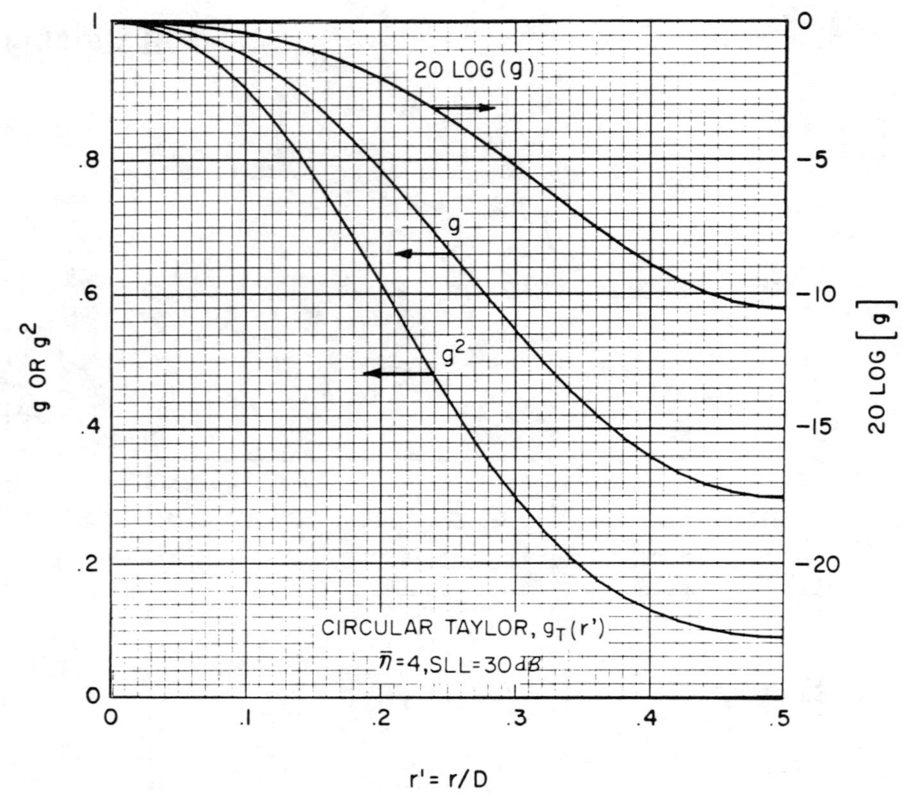

APERTURE ILLUMINATION FUNCTION

Figure 2.B.34

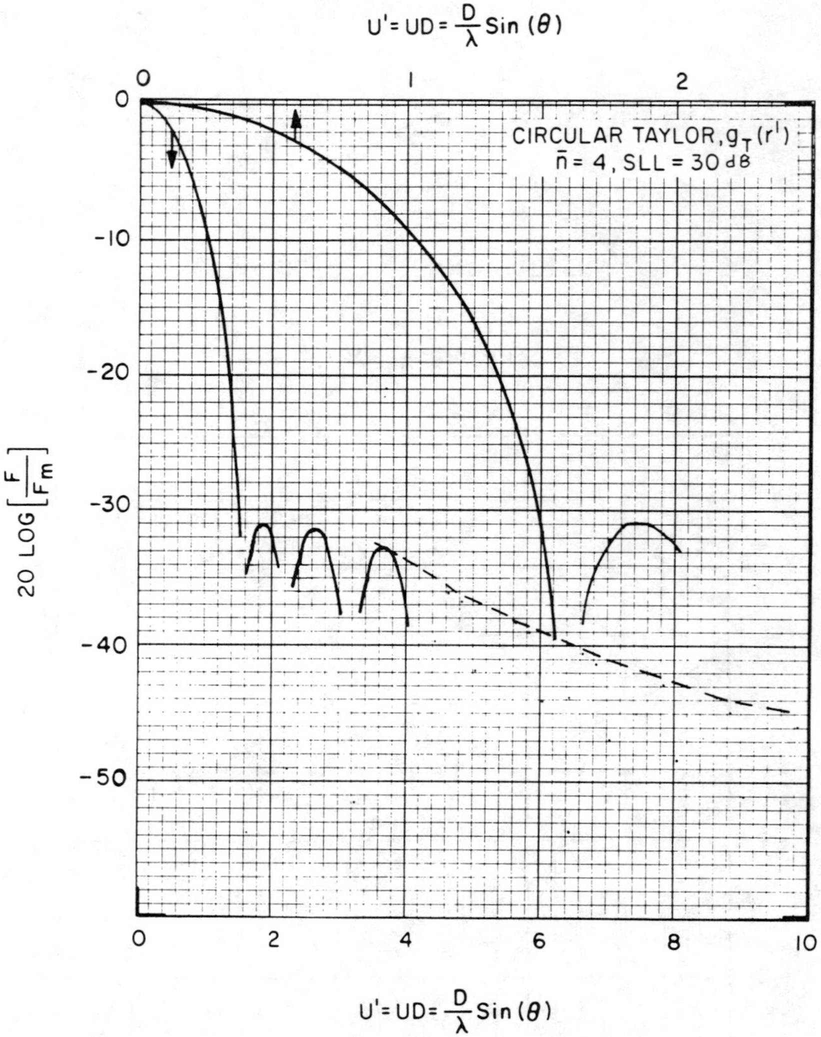

NORMALIZED ANTENNA PATTERN

Figure 2.B.35

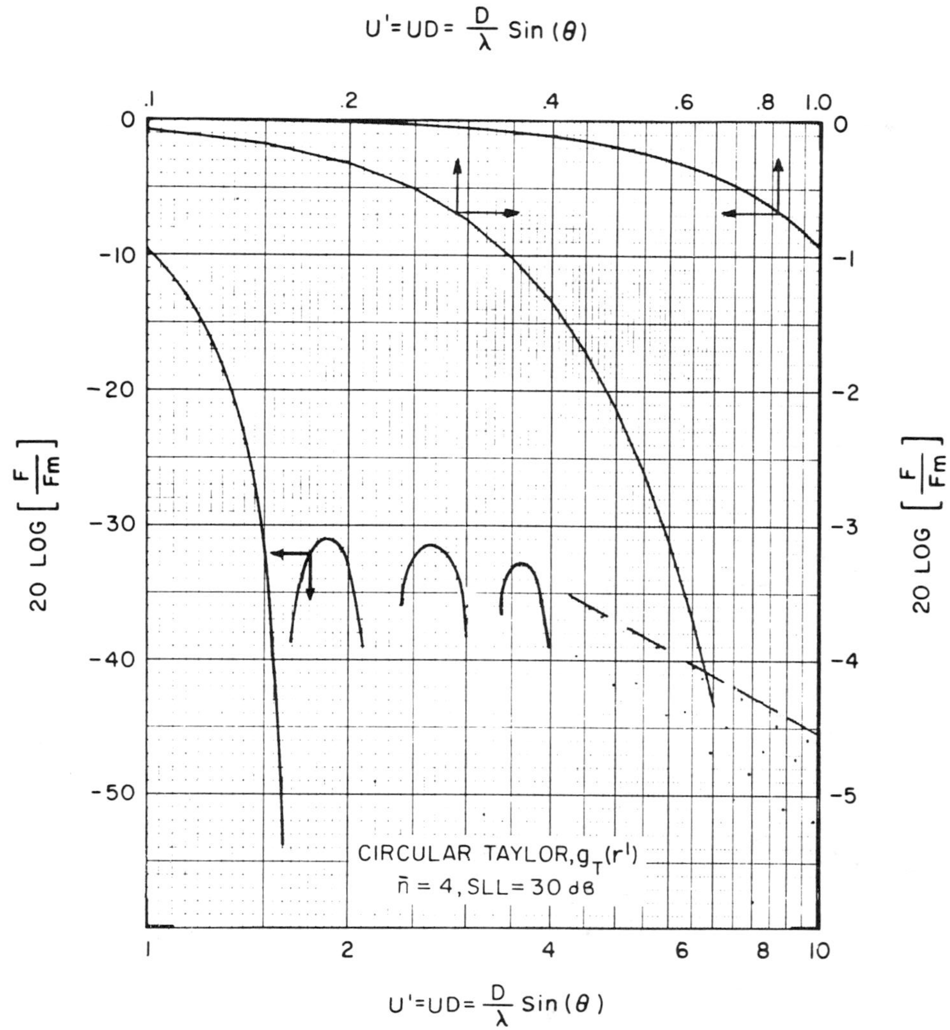

NORMALIZED ANTENNA PATTERN

Figure 2.B.36

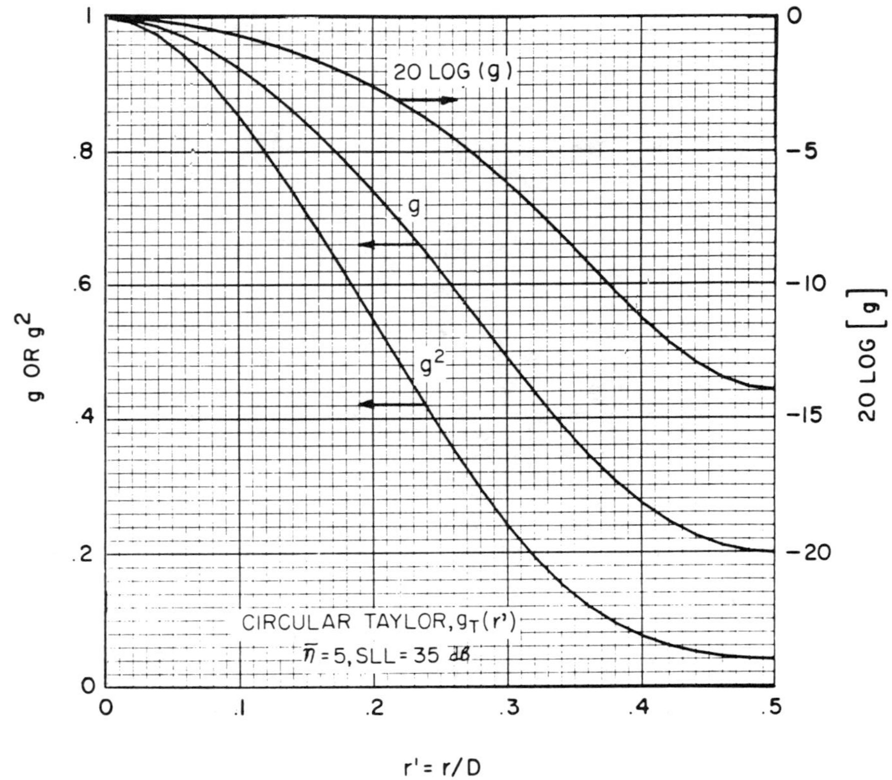

APERTURE ILLUMINATION FUNCTION

Figure 2.B.37

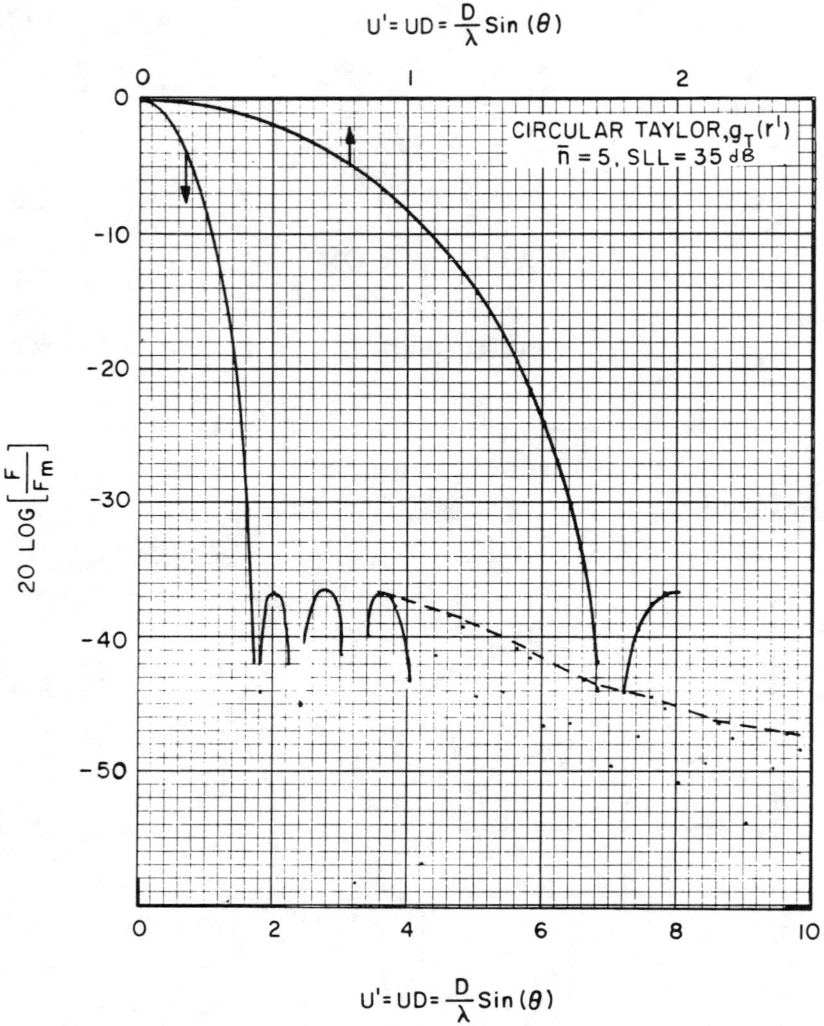

NORMALIZED ANTENNA PATTERN

Figure 2.B.38

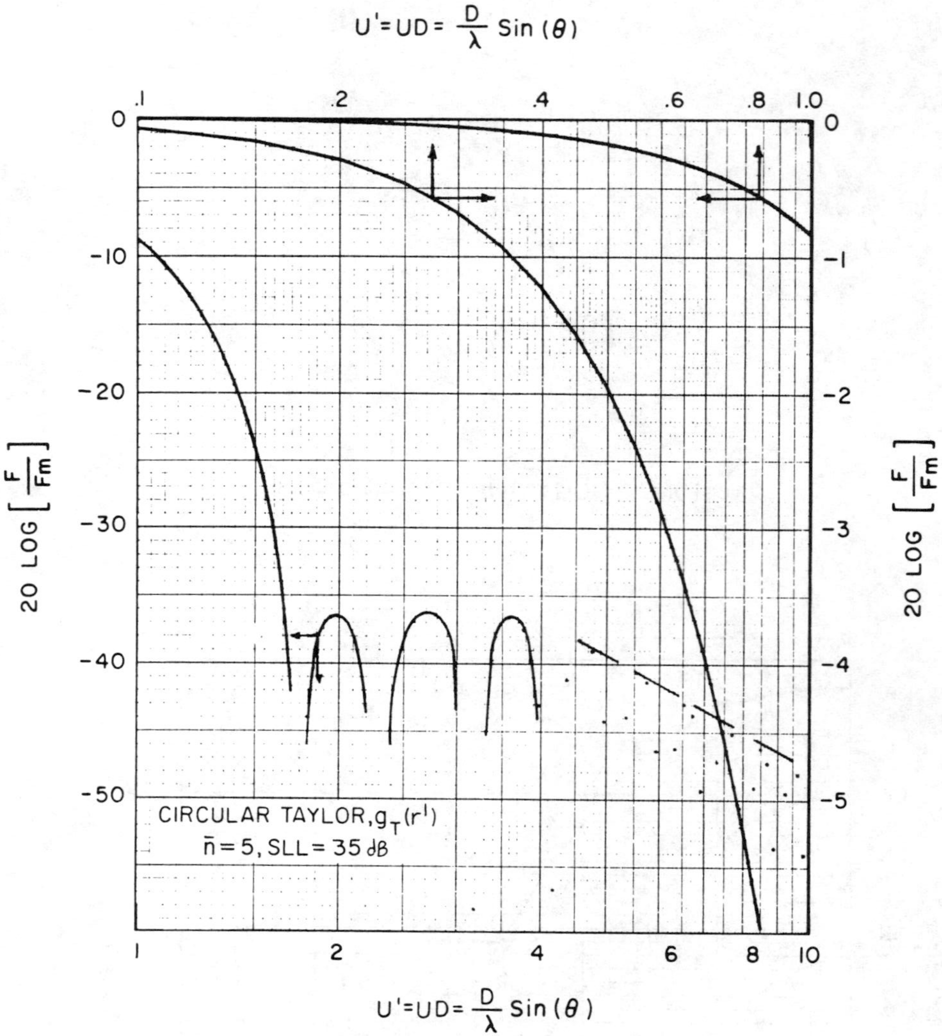

NORMALIZED ANTENNA PATTERN

Figure 2.B.39

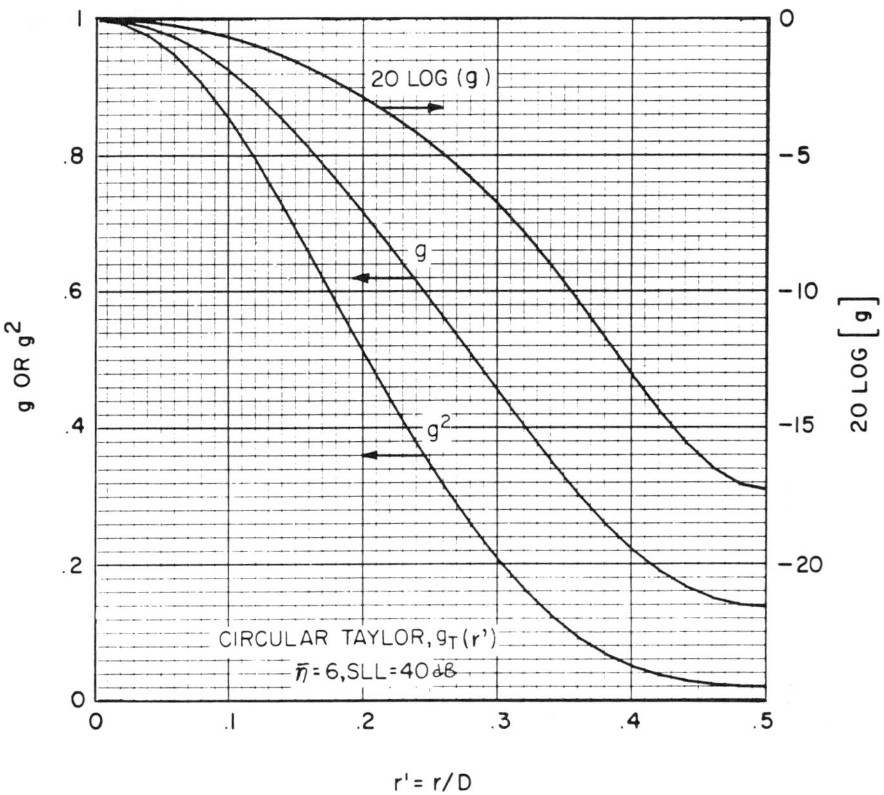

APERTURE ILLUMINATION FUNCTION

Figure 2.B.40

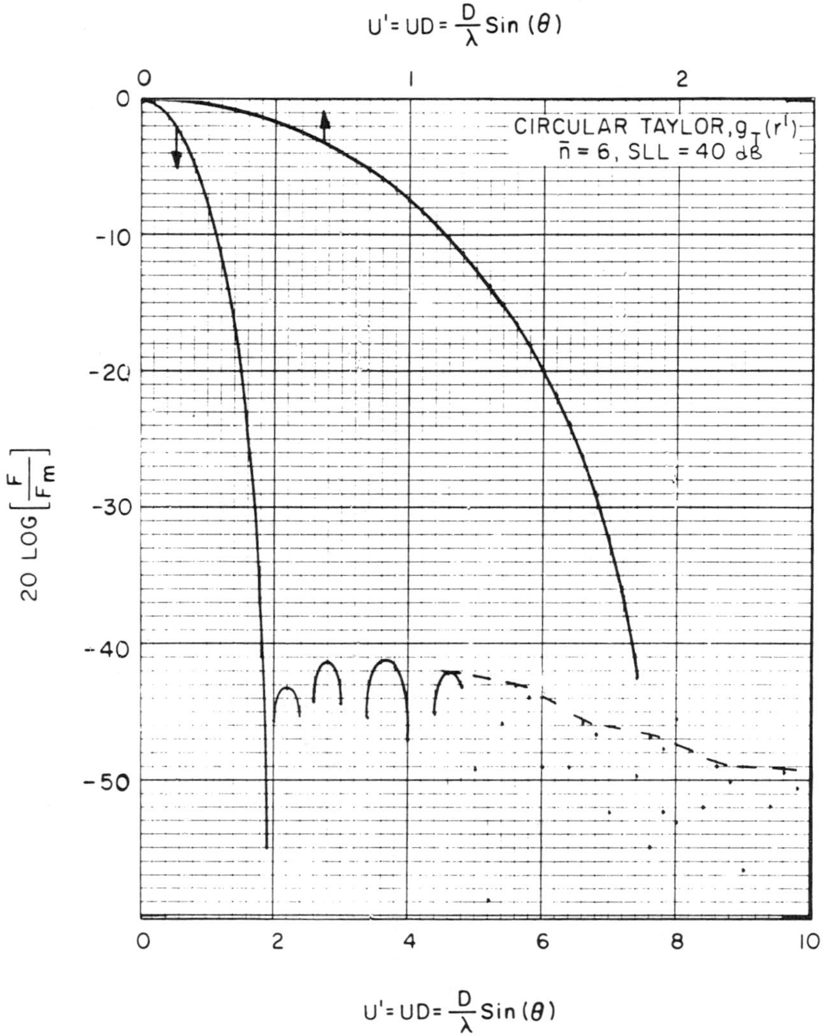

NORMALIZED ANTENNA PATTERN

Figure 2.B.41

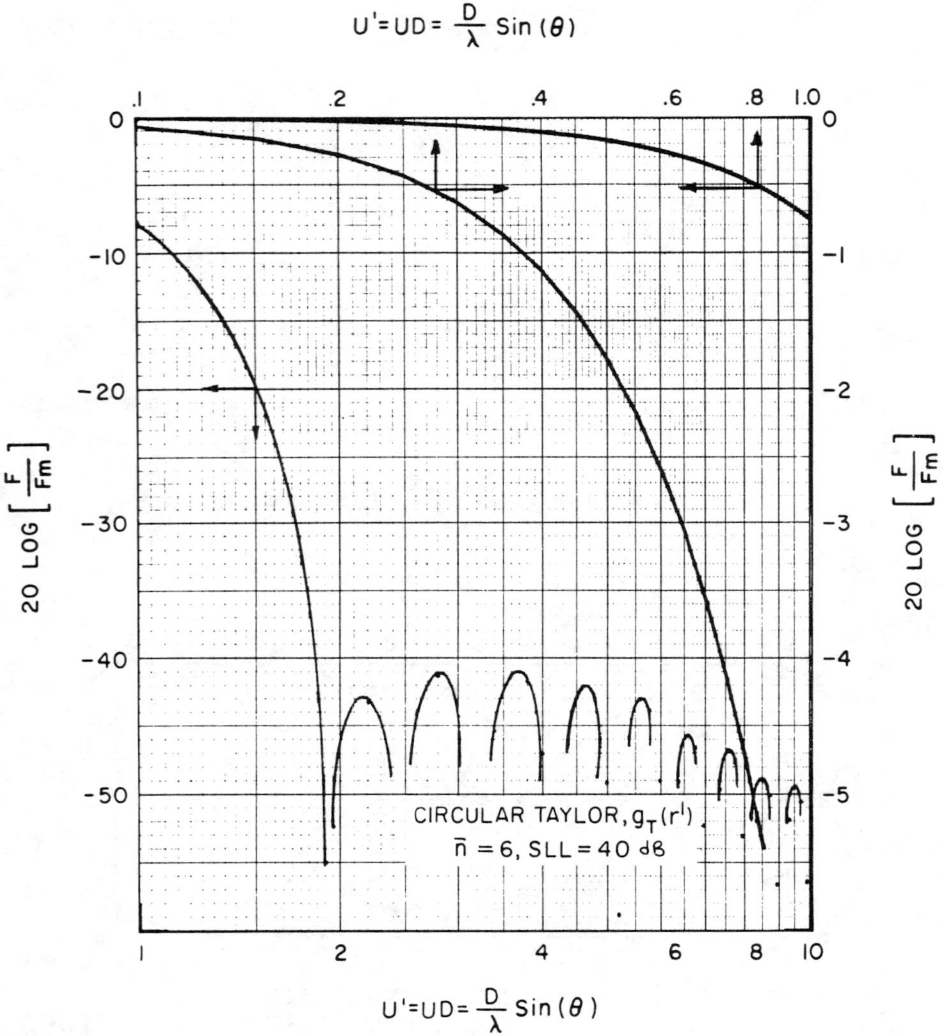

NORMALIZED ANTENNA PATTERN

Figure 2.B.42

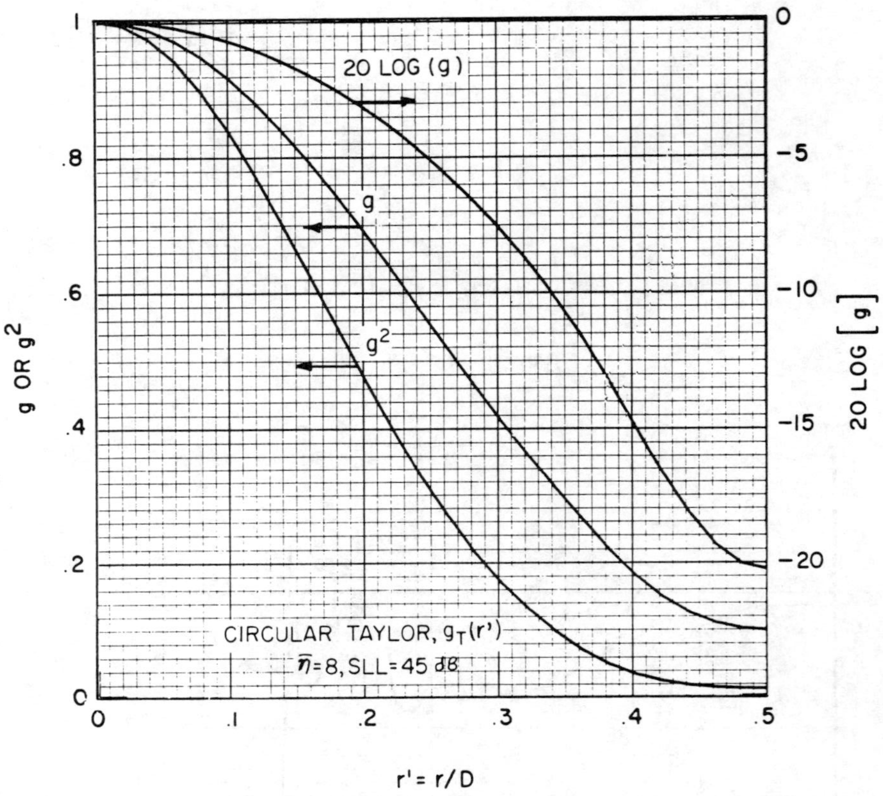

APERTURE ILLUMINATION FUNCTION

Figure 2.B.43

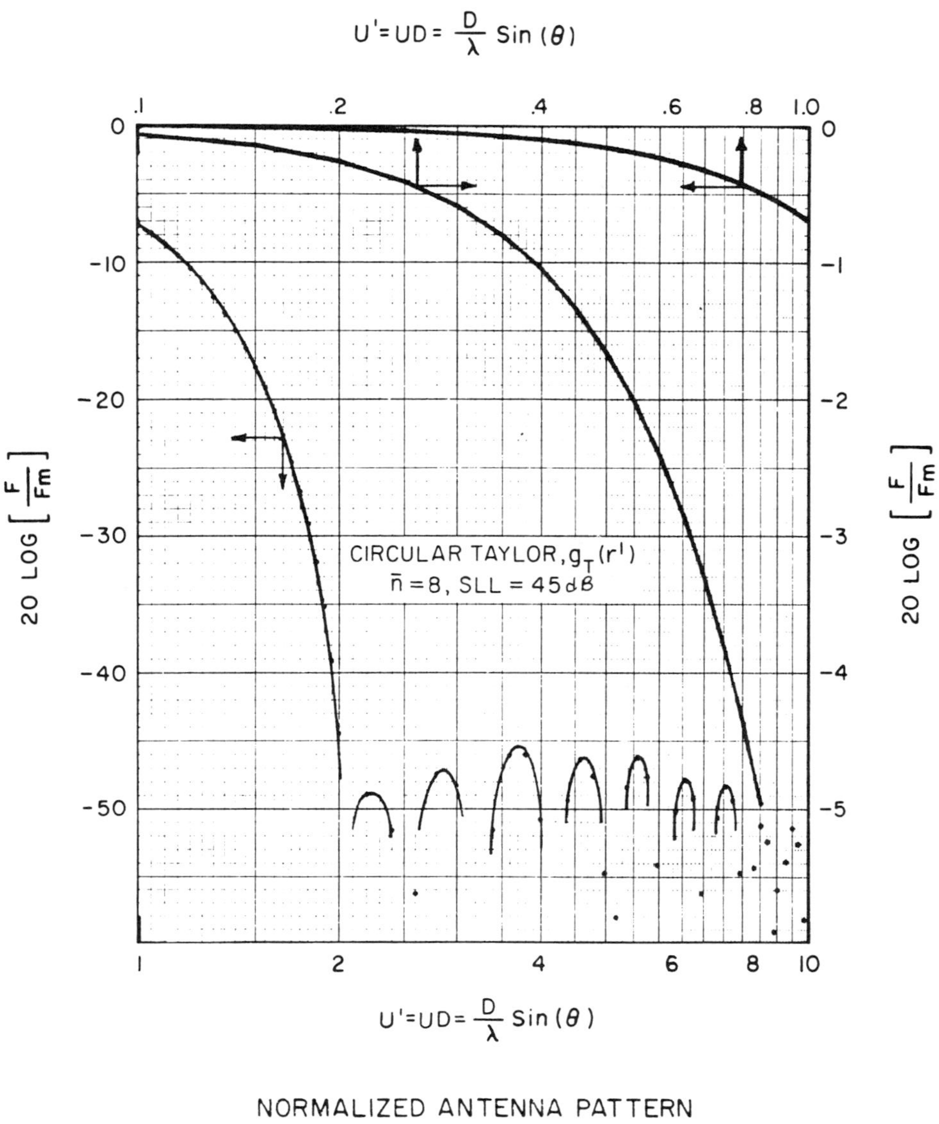

Figure 2.B.44

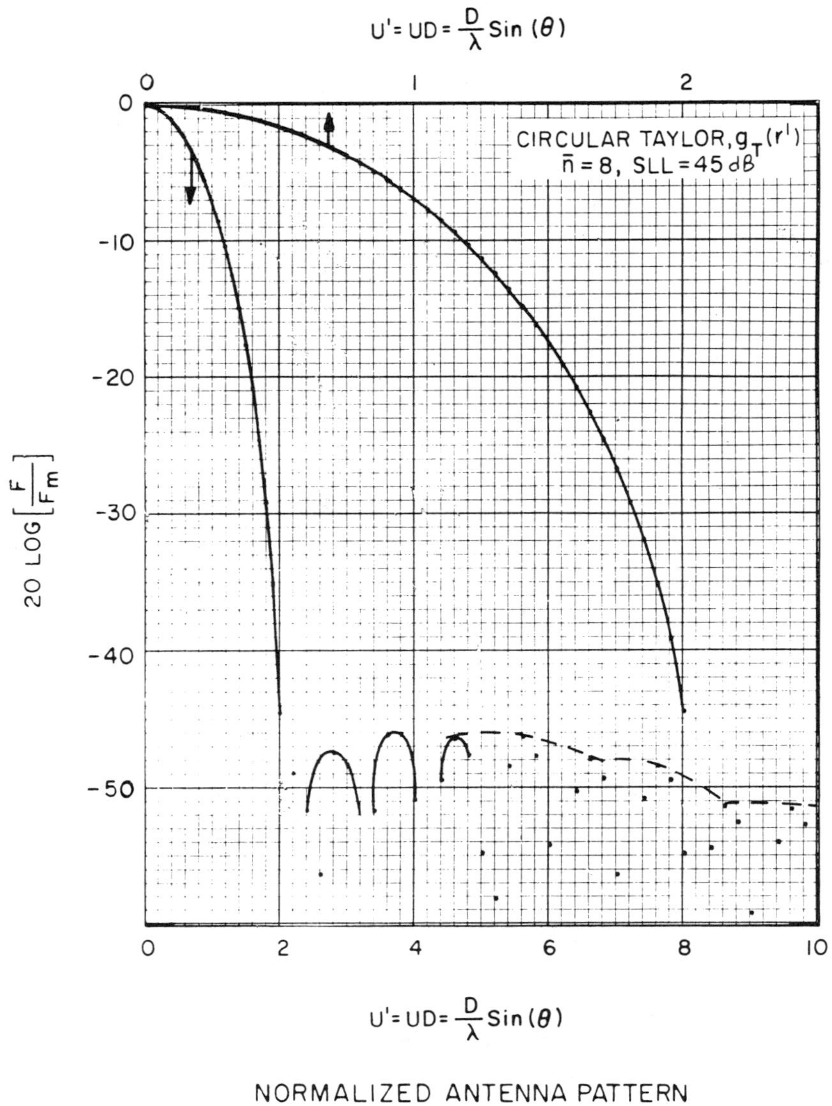

Figure 2.B.45

Part 2
Antenna Array Fundamentals

Dr. Eli Brookner

Raytheon Co., Wayland, MA

3.1 Effects of Errors on Antenna Array Characteristics.

Consideration is given to the effects of random errors on the sidelobe levels, directivity, and measurement accuracy capability of linear and planar arrays. The effects of periodic errors are also dealt with.

3.1.1 Effect of Errors on Antenna Sidelobe Levels and Directivity.

A. Random Errors

i. Errors Independent from Radiating Element to Radiating Element

Phase Errors

Consideration is given first to the effects of random phase errors. Assume a uniformly-weighted antenna linear array which is pointed broadside i.e., θ_0=o. If the currents of each radiator have no phase errors, then the current of each radiator can be represented by a unit vector of zero phase. When a phase error δ_i is present for the ith radiating element, then the current for the ith radiating element is a unit vector rotated by δ_i radians; (see Fig. 3-1a.) A determination is now made of the ratio of the average power at the peak of the main lobe to the average power in the sidelobes when phase errors are present.

The peak of the main-beam voltage pattern in the presence of phase errors is given by (3-1a) of Fig. 3-1. Squaring and averaging (3-1a) gives the average power at the peak of the main beam; see (3-1b).* The following three assumptions are made for the phase errors: (1) they have identical mean square values for all radiating elements, i.e., $\overline{\delta_i^2}=\overline{\delta^2}$ for all i, (2) they are independent of each other from radiating element to radiating element, and (3) they are small enough.#

For these assumptions, the mean square power at the peak of the main beam becomes (3-1d) of Fig. 3-1.

The contribution to the sidelobe voltage from the phase errors is determined by finding the average power of the sidelobe at the angle corresponding to where the beam pattern would have a null if no phase errors were present. The phase error shifted unit vector for the ith radiating element is given by

$$I_i = e^{j\delta_i}. \qquad (3\text{-}1k)$$

The above can be rewritten as the superposition of two vectors in quadrature;

$$I_i = \cos\delta_i + j\sin\delta_i \qquad (3\text{-}1l)$$

where for small enough phase errors (3-1l) becomes*

$$I_i = 1 + j\delta_i. \qquad (3\text{-}1m)$$

The above equation indicates that the currents can be considered as the superposition of the errorless unit current vector together with an error current vector $j\delta_i$. The contribution from the errorless part of the current vector is zero at the angles where the error-free sidelobes have a null. In the direction of a null, the far field vectors due to the radiating-element current errorless unit vectors have phases β_i, i = 1 to N, where the β_i are fairly uniformly distributed between 0 and 2π in order to add up to give a null at the location of an error-free pattern null. The error vector terms $j\delta_i$ will also have these same phases. Hence, $j\delta_i$ becomes $j\delta_i e^{j\beta_i}$ at the null of the error-free pattern. The actual values of the phase angles β_i are not important. They depend on the direction θ, ϕ of the null of the error-free pattern sidelobe. (β_i is given by exponent of (2-4) and (2-37) for linear and planar arrays, respectively.) Thus, the antenna voltage

* When determining the main-beam voltage in the presence of errors the contribution from the quadrature sin δ_i terms are negligible compared to that from the cos $\delta_i = 1-\frac{1}{2}\delta_i^2$ terms. The reverse is true for the contribution to the sidelobe power.

Quantitively the condition that the phase errors are small enough is that the following inequalities hold

$$\overline{\delta^2} << 4 \text{ and } 4N\overline{\delta^2} >> \overline{\delta^4} \qquad (3\text{-}1h)$$

when δ_i has a Gaussian probability density function the second of the above inequalities becomes

$$4N\overline{\delta^2} >> 3\overline{\delta^2}^2 \qquad (3\text{-}1i)$$

or

$$4N >> 3\overline{\delta^2}. \qquad (3\text{-}1j)$$

For the results obtained in this section, the assumption that the phase error has a Gaussian probability density function is not necessary. Moreover, no assumption is made relative to the probability density function of the phase errors. The only assumptions that need be made are that the phase errors are identically distributed for each radiating element and that the inequalities given by (3-1h) hold.

* Equations not given in text are given in Figures.

a) MAIN LOBE

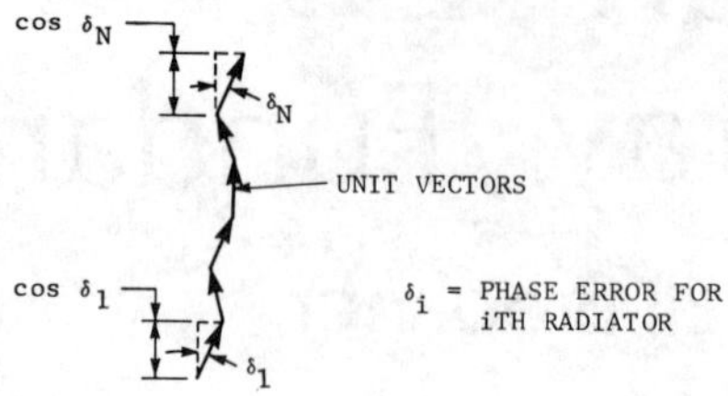

$$E_a(0,0) = \sum_{i=1}^{N} \cos\delta_i = \sum_{i=1}^{N}\left(1 - \frac{\delta_i^2}{2}\right) = N - \tfrac{1}{2}\sum_{i=1}^{N}\delta_i^2 \tag{3.1a}$$

$$\overline{|E_a(0,0)|^2} = \overline{\left(N - \tfrac{1}{2}\sum_{i=1}^{N}\delta_i^2\right)^2} = N^2 - N\sum_{i=1}^{N}\overline{\delta_i^2} + \tfrac{1}{4}\overline{\sum_{i,j}\delta_i^2\delta_j^2} \tag{3.1b}$$

ASSUME: $\overline{\delta_i^2} = \overline{\delta^2}$ (1c)

AND δ_i INDEPENDENT OF δ_j SO THAT

$$\overline{\delta_i^2\delta_j^2} = \overline{\delta_i^4} = \overline{\delta^4} \quad \text{FOR } i = j$$

$$= \overline{\delta^2}^2 \quad \text{OTHERWISE}$$

THEN FOR $4N\overline{\delta^2} \gg \overline{\delta^4}$ and $\overline{\delta^2} \ll 4$

$$|E_a(0,0)|^2 \doteq N^2 - N^2\overline{\delta^2} = N^2(1 - \overline{\delta^2}) \tag{3.1d}$$

Fig. 3.1 Simple derivation of expression for antenna sidelobe levels due to random antenna phase errors; uniformly-weighted linear array of isotropic elements.

b) AMPLITUDE OF PATTERN AT NULL OF ERROR-FREE SIDELOBE

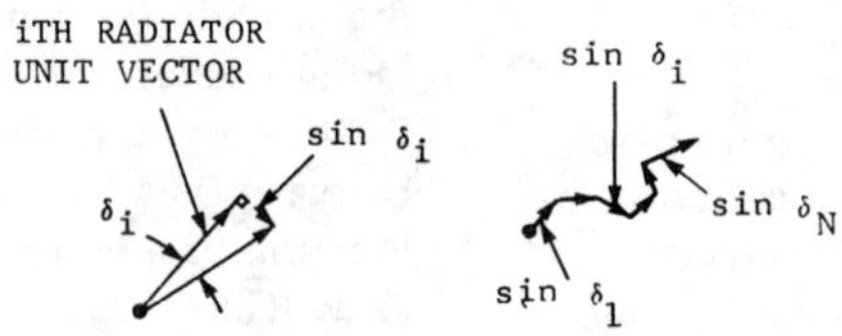

$$\Sigma_a(\theta,\phi) = \sum_{i=1}^{N} j(\sin\delta_i)e^{j\beta_i} = \sum_{i=1}^{N} j\delta_i e^{j\beta_i} \tag{3.1e}$$

$$\overline{|E_a(\theta,\phi)|^2} = \overline{\left(\sum_{i=1}^{N}\delta_i e^{j\beta_i}\right)\left(\sum_{i=1}^{N}\delta_i e^{-j\beta}\right)}$$

$$= \sum_{i=1}^{N}\overline{\delta_i^2} + \overline{\sum_{i\neq k}\delta_i\delta_k e^{j(\beta_i - \beta_k)}}$$

$$= N\overline{\delta^2} + 0 \tag{3.1f}$$

(c) SIDELOBE POWER RATIO

SIDELOBE POWER RATIO AT NULL OF ERROR-FREE PATTERN

$$\frac{N\overline{\delta^2}}{N^2(1-\overline{\delta^2})} = \frac{\overline{\delta^2}}{N(1-\overline{\delta^2})} \tag{3.1g}$$

Fig. 3.1 (con't) Simple derivation of expression for antenna sidelobe levels due to random antenna phase errors; uniformly-weighted linear array of isotropic elements. (Cont.)

pattern in the presence of phase errors in a direction at which the error-free pattern has a null is given by (3-1e). At other directions, this expression gives the contribution to the far field voltage pattern from the phase error components of the radiating currents. Squaring and averaging the expression (3-1e) gives the average power of the sidelobe; see (3-1f).

Taking the ratio of the average sidelobe power, given by (3-1f), to the average main lobe power, given by (3-1d), gives the average sidelobe power ratio at the null of the error-free pattern; see (3-1g). This ratio is referred to as $\overline{r^2_N}$. For the case of an aperture which has a tapered illumination pattern $\overline{r^2_N}$ becomes (3-2a) of Fig. 3-2.

The derivation given above applied for a linear array of isotropic elements for which no mutual coupling effects exist. For a linear array of non-isotropic elements the radiating element gain must be accounted for. Furthermore, to allow for mutual coupling effects the radiating element gain must be determined in the presence of the other elements, i.e., the gain must be determined for a centrally located element with all other elements terminated; see Sect. 2.1.1A and Fig. 2-11. Let $e(\theta,\phi)$ be the non-isolated radiating element gain. Then the power sidelobe ratio of (3-2a) should be multiplied by $[e(\theta_1,\phi)/e(\theta_0,\phi)]^2$ where θ_1 and θ_0 are the off-broadside angles for the sidelobe and main lobe, respectively.

The results apply as well for a planar array of non-isotropic elements. If the array is perfectly matched to free space for all scan angles then, as indicated in Sect. 2.2.1, the non-isolated element voltage gain pattern is given by

$$e(\theta,\phi) = \sqrt{\cos\theta} \tag{3-3}$$

Thus, for this case (3-2a) should be multiplied by the factor $\cos\theta_1/\cos\theta_0$. If the array is not ideally matched then the actual non-isolated element gain pattern should be used. Often it can be approximated by $\cos\theta$ to a fractional power greater than one.

The factor due to the antenna element pattern need not be included in (3-2a) except when the accuracy warrants it.

Phase and Amplitude Errors Together with Random Element Failures.

Following a similar simple analysis yields the result (3-2b) of Fig. 3-2 for the power sidelobe level ratio due to independent amplitude errors across the array. Similarly, one can derive the contribution to the average power sidelobe ratio due to random element failures as given by (3-2c), where P_e is the probability that the ith element does not fail. When random phase and amplitude errors and random element failures can exist simultaneously, the average power sidelobe ratio is given by (3-2d).[1] The loss of directivity due to the simultaneous existence of these errors is given by (3-4).[1]

No assumptions have been made relative to the probability density functions of the phase and amplitude errors δ_i and Δ_i. The only assumptions made was that they be identically distributed from element to element, independent from element to element, and small; see inequalities (3-1h). Similarly, the only requirement on the random failures are that they have identical probability of occurrence for all the radiating elements and that the probability of failure be independent from element to element. Nevertheless, the distribution of the sidelobe level voltage can be specified as long as the number of array elements N is large and the central limit theorem can be used. Under these conditions the voltage sidelobe level at the null of the error-free pattern is Rayleigh distributed (equivalently, the power of the sidelobe level at the null of the error free pattern has a chi-square distribution of 2 degrees of freedom).[2] As a result, r_N, the ratio of the sidelobe level voltage at the null of the error-free pattern to the main-lobe peak voltage, has approximately a Rayleigh distribution also. Eq. (3-5) gives the probability that the sidelobe level ratio is below some peak value r_p.

Consider now, the sidelobe level voltage for directions other than that at which the error-free pattern has a null. It consists of the sum of the error-free-pattern sidelobe level voltage plus the sidelobe level voltage due to the error currents. Thus, in the presence of phase errors, the sidelobe voltage is the sum of a Rayleigh variate* plus a constant term which is the amplitude of the error-free sidelobe voltage. This sum statistic has the well known Rice probability density function.[3] [4] The probability that r, the sidelobe level voltage ratio in the presence of phase errors, is less than some peak voltage r_p is given by (3-5a). Obviously, when $\overline{r^2_N} >> |E_o|$, where $|E_o|$ is the magnitude of the error-free voltage sidelobe ratio, the voltage sidelobe ratio is approximately Rayleigh distributed and (3-5a) becomes (3-5). When $\overline{r^2_N} << |E_o|^2$, the voltage sidelobe ratio is given by $|E_o|$.**

Figs. 3-5 and 3-6 give the increase in the peak sidelobe level due to array errors as a function of the ratio of the error-free sidelobe level power to the average power sidelobe level due to the errors alone, that is, the average power sidelobe level at the null of the error-free pattern. For example, when $\overline{r^2_N}$ is 10 dB below $|E_o^2|$ the peak sidelobe level is raised less than 2.3 dB 90% of the time, less than 3.8 dB 99% of the time, and less than 5.4 dB 99.99% of the time. Although the results of Figs. 3-5 and 3-6 give the peak sidelobe level degradation (or equivalently, peak sidelobe level) for a single sidelobe, the results can also be applied to determine the peak sidelobe level for m sidelobes. Use is made of the fact that if P is the probability that the peak of one sidelobe is less than some specified level, then the probability that m independent sidelobes having identical statistics are all less than that same specified level is given by $P^m = (1-Q)^m \doteq 1-mQ$ for $mQ<<1$ where $Q = 1-P$. For example, assume 1000 error-free sidelobes of equal peak amplitude. Let $\overline{r^2_N}$ be 10 dB below this peak error-free sidelobe level. Then from Figs. 3-5 and 3-6 the peak level of these 1,000 sidelobes will be degraded by ≤5.4 dB 90% of the time.

Fig. 3-7 works out in detail an example with respect to the effects of phase and amplitude errors and random element failures. In the example a determination is made of the allowable random phase and amplitude errors and random element failure which result in a degradation about 2 dB or less for each sidelobe level 90% of the time. The directiv-

* Actually, the sidelobe voltage due to the phase errors is a vector whose magnitude is Rayleigh distributed and whose phase is uniformly distributed.

** For $\overline{r^2_N} << |E_o|^2$, the voltage sidelobe ratio r is approximately Gaussianly distributed with mean $|E_o|$ and variance $\overline{r^2_N}/2$.[3]

1) PHASE ERRORS:

$$\overline{r_N^2} = \text{AVERAGE POWER SIDELOBE LEVEL AT NULL OF ERROR FREE PATTERN} = \frac{\overline{\delta^2}}{N\eta_t(1-\overline{\delta^2})} \quad (3.2a)$$

2) AMPLITUDE ERRORS:

$$\overline{r_N^2} = \frac{\overline{\Delta^2}}{N\eta_t} \quad (3.2b)$$

WHERE $\overline{\Delta^2}$ = MEAN SQUARE VALUE OF FRACTIONAL RADIATOR AMPLITUDE ERROR

3) ELEMENT FAILURE:

$$\overline{r_N^2} = \frac{1 - P_e}{N\eta_t P_e} \quad (3.2c)$$

WHERE P_e = PROBABILITY OF SURVIVAL OF iTH RADIATING ELEMENT

4) SIMULTANEOUS PHASE ERRORS, AMPLITUDE ERRORS, AND ELEMENT FAILURES:

$$\overline{r_N^2} = \frac{(1-P_e) + \overline{\Delta^2} + \overline{\delta^2}\, P_e}{P_e(1 - \overline{\delta^2})} \frac{1}{\eta_t N} \quad (3.2d)$$

Fig. 3.2 Antenna sidelobes due to random phase errors, amplitude errors, and element failures.

$$\frac{D_e}{D} \doteq \frac{P_e}{1 + \overline{\Delta^2} + \overline{\delta^2}} \quad (3.4)$$

WHERE D = DIRECTIVITY WITHOUT ERRORS

D_e = DIRECTIVITY WITH ERRORS

Fig. 3.3 Degradation in directivity due to phase errors, amplitude errors, and element failures.

1) AT NULL OF ERROR FREE PATTERN P(r) HAS RAYLEIGH DISTRIBUTION, THUS

$$P(r^2 \le r_p^2) = 1-\exp(- r_p^2/\overline{r_N^2}) \quad (3.5)$$

2) AWAY FROM NULL OF ERROR FREE PATTERN P(r) HAS A RICE DISTRIBUTION, THUS

$$P(r^2 \le r_p^2) = \int_0^{r_p} \frac{2r}{\overline{r_N^2}} \left[\exp - \frac{r^2 + |E_o|^2}{\overline{r_N^2}}\right] I_o\left(\frac{2r|E_o|}{\overline{r_N^2}}\right) dr \quad (3.5a)$$

WHERE $|E_o|$ = AMPLITUDE OF SIDELOBE IN DIRECTION θ, ϕ IN ABSENCE OF ERRORS

$= |E_a(\theta, \phi)_o|$

Fig. 3.4 Distribution of sidelobes in presence of errors.

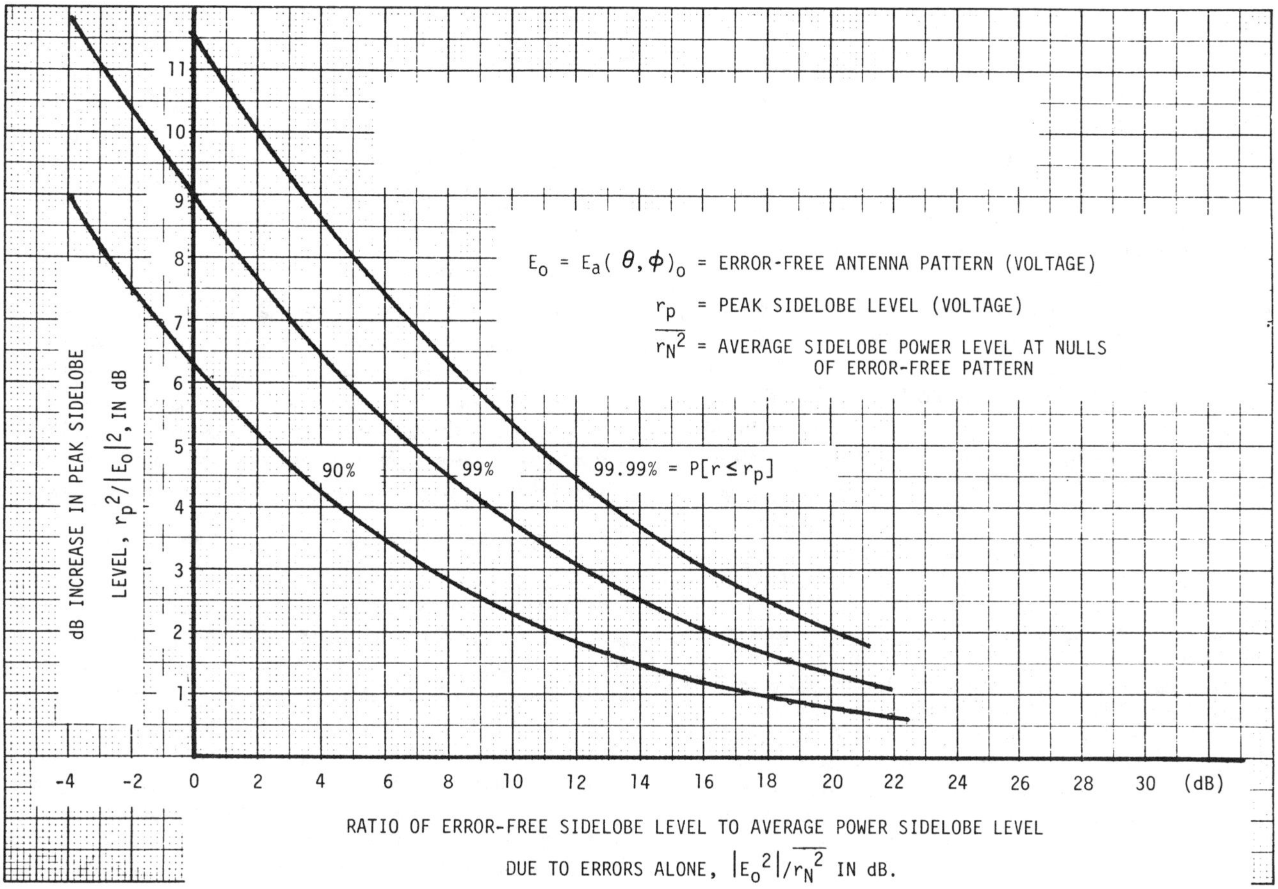

Fig. 3.5 Curves of increase in peak sidelobe level due to array errors.

ERROR-FREE SIDELOBE LEVEL TO AVERAGE SIDELOBE LEVEL DUE TO ERRORS ALONE dB	dB INCREASE IN PEAK SIDELOBE LEVEL		
	$P[r \le r_p]$ = 90%	99%	99.99%
0	6.3	9.0	11.5
5	3.8	5.9	8.0
10	2.3	3.8	5.4
20	0.8	1.3	2.0

Fig. 3.6 Tables of increase in peak sidelobe level due to array errors.

ty loss is found to be 2.4 dB for the values of $\overline{\delta^2}$, $\overline{\Delta^2}$, and P_e obtained; (see Fig. 3-8). Assuming that a 2.4 dB loss in directivity is too great for the system, Fig. 3-9 reworks the allowable amplitude and phase errors and allowable random element failure rates for the assumption that a maximum permissible directivity degradation is 1 dB. The values for δ^2, Δ^2, P_e obtained result in each 30 dB down sidelobe level being degraded by $\leqslant$ 1.5 dB 90% of the time.

The component contributors to the random phase errors and random amplitude errors are now elaborated on. Fig. 3-10 indicates the component contributors to the random phase errors. The magnitude of the mean square phase error $\overline{\delta^2}_Q$, due to the phase shifter random quantization errors for a P bit phase shifter, is given by (3-7).[5],[6] Substituting (3-7) into (3-2a) yields the average power sidelobe ratio due to the phase shifter random quantization errors $(\overline{r^2}_N)_Q$ given by (3-8). Fig. 3-12 gives a convenient plot of (3-8).

Fig. 3-13 gives an expression for the contribution to the phase error due to errors in the location of the radiating elements. [7] An expression for the contribution to the mean square amplitude error due to radiation element rotation errors is given in Fig. 3-14 [8]. Hence actually this term should have been discussed in connection with the amplitude errors of Fig. 3.16. The results of Fig. 3-14 are given for the assumption that the radiating elements consist of short dipoles aligned nominally along the x axis; see Fig. 2-31 for the coordinate system assumed. Usually the contribution to the mean square phase error from the rotation errors is negligible compared to the contribution from the other errors. The same applies to amplitude error Δ^2_{ROT} due to rotational misalignment.

In Fig. 3-9 a mean square phase error allotment of $\overline{\delta^2}$ = 0.0797 rad^2 was arrived at based on the constraint of a 1 dB directivity degradation due to the presence of amplitude and phase errors as well as random element failures. In the example of Fig. 3-15 this error is budgeted nearly equally among the contributions from the phase shifter random quantization error, the power amplifier phase errors, and the radiation element location errors. An attempt is made initially in Fig. 3-15 to divide the errors equally among these contributing factors; however, to make P and integer a slightly lower allotment is given to the contribution from the phase shifter random quantization error. It is found that a 4 bit phase shifter was required. After specifying a 4 bit phase shifter, the remaining part of the mean square phase error is divided equally between the power amplifier contribution and the radiation element location error contribution. The allotted errors found for the power amplifier is 10.5° rms and for the x, y, z position coordinates of each radiating element, 0.873 cm rms.

It is worth emphasizing again that although the 10.5° rms phase error is specified as arising from the power amplifier, as indicated in the definition on Fig. 3-10, it must also include any other sources of phase error arising in the generation of the ith radiation current.

The sources of the component amplitude errors are indicated in Fig. 3-16. Fig. 3-17 shows the typical type of amplitude-weighting errors that can arise for radiating elements.[7] Random amplitude weighting quantization errors will arise when the amplitude weighting is achieved using subarrays having pseudorandom shapes and locations. Each subarray is fed the same total power but has a different number of radiating elements in order to achieve amplitude tapering. Such an array would be used to remove the periodicity in the quantized approximation to the continuous amplitude

short dipoles aligned nominally along the x axis; see Fig. 2-31 for the coordinate system assumed. Usually the con-

weighting function desired.

A convenient plot giving the loss of directivity due to phase-shifter, random-quantization errors is given in Fig. 3-18.[5] This figure indicates that the 4 bit phase shifter of Fig. 3-15 contributes 0.055 dB to the total 1 dB loss in directivity for the example considered in Fig. 3-9.

ii. Correlated Random Errors

Consideration is now given to the case where the errors of the array are not independent from element-to-element. The errors are assumed to be identical over groups of radiating elements but independent from group-to-group. This is the case when the elements are grouped together with each group (of let us say, N_s elements) being fed by a single power amplifier. These small groups of elements are called subarrays.

Spatial Frequency and Paired Echo Approach.

To treat the case of a subarrayed antenna, spatial frequency and paired echo principles are used. Assume a uniformly-weighted linear array of isotropic elements having length L for which the main beam is pointed broadside. Then for the error-free case the radiator currents for all the radiating elements will be unit vectors of zero phase. Assume, however, that a phase error exists along the array that is described by a sinusoid having exactly n cycles as indicated in Fig. 3-19a. From simple discrete Fourier transform analysis [see (2-4], in the presence of this sinusoidal phase error the far field array pattern (which is the array factor) is found to consist of the error-free array pattern plus two additional components which are the error-free pattern shifted so as to be located at angles $\sin\theta = \pm n\,\lambda/L$ and weighted so as to have amplitudes $\pm a_n/2$; (see Fig. 3-19b). These shifted and weighted error-free array patterns yield a pair of sidelobes at the locations indicated. These two sidelobes are referred to as paired echos.[9],[10]

If the phase error across the aperture was cosinusoidal with exactly n periods across the aperture, (see Fig. 3-19c), then the array pattern would consist of the error-free array factor plus two weighted error-free array patterns having identical amplitudes $jb_n/2$ located at $\sin\theta = \pm n\,\lambda/L$, the same locations as for the case of the sinusoidal phase error; (see Fig. 3-19d). Having both these sinusoidal and cosinusoidal phase errors simultaneously results in the array pattern having sidelobes of complex amplitudes $-a_n/2 + jb_n/2$ and $a_n/2 + jb_n/2$ at the locations $\sin\theta = \pm n\,\lambda/L$; (see Fig. 3-19e). These sidelobes are of equal magnitude.

Now assume that the phase errors across the array are independent from element-to-element as shown in Fig. 3-20a. For simplicity of analysis the array radiating elements are assumed to have zero width. Then the resulting discrete sequence of phase errors along the x axis of the array can be expressed in terms of a discrete Fourier series with x as the variable; see (3-15a). The value of the phase error δ_i for the ith radiating element is given by the value of this series at the position x for the ith element. Because there are N elements, this discrete Fourier series will be comprised of the sum of N/2 spatial harmonics as indicated in (3-15a). The nth spatial harmonic is given by the superposition of the sinusoidal and cosinusoidal spatial phase error terms depicted by (3-14a) and (3-14b) of Fig. 3-19. As indicated there are N/2 of these harmonics with n going from 1 to N/2. These N/2 harmonics will yield N sidelobes located at the angles $\sin\theta = \pm n\,\lambda/L$, where n = 1, 2, . . . , N/2 as indicated

GIVEN:

CIRCULAR APERTURE WITH $2r_o = 50\lambda$

N = 7800, 30 dB TAPERED TAYLOR AMPLITUDE WEIGHTING (NO THINNING)

FIND:

$\overline{\delta^2}$, P_e, and $\overline{\Delta^2}$ WHICH RESULT IN ONLY ≤ 2 dB INCREASE IN EACH SIDELOBE 90% OF TIME.

FROM FIG. 2.41a OPTIMUM $\bar{n}=8$ AND $\eta_t = 0.8838$

FROM FIGS. 3.5 and 3.6 $\quad |E_o|^2/\overline{r_N^2} \doteq 10$ dB, HENCE $\overline{r_N^2} = -40$ dB $= 10^{-4}$.

BUDGET CONTRIBUTIONS TO SIDELOBES EQUALLY AMONG THE PHASE ERRORS, AMPLITUDE ERRORS, AND ELEMENT FAILURE CONTRIBUTIONS, i.e., EACH INDEPENDENTLY GIVES RISE TO $\overline{r_N^2} = \frac{1}{3}10^{-4}$.

FROM (3.2a) $\overline{r_N^2} = \dfrac{\overline{\delta^2}}{N\eta_t(1-\overline{\delta^2})} = \dfrac{1}{3}10^{-4} \Longrightarrow \overline{\delta^2} = 0.187 \text{ rad}^2$, $\boxed{\sqrt{\overline{\delta^2}} = 24.8^\circ}$

FROM (3.2c) $\overline{r_N^2} = \dfrac{1-P_e}{N\eta_t P_e} = \dfrac{1}{3}10^{-4} \Longrightarrow \boxed{P_e = 0.813}$

$\Longrightarrow$ $(1-P_e)N = 1459$ = NO. OF ELEMENTS THAT CAN FAIL

FROM (3.2b) $\overline{r_N^2} = \dfrac{\overline{\Delta^2}}{N\eta_t} = \dfrac{1}{3}10^{-4} \Longrightarrow \boxed{\overline{\Delta^2} = 0.230 = \sim 4.5\text{dB}}$ *

AMPLITUDE ERROR

CHECK: SUBSTITUTING VALUES OBTAINED ABOVE FOR $\overline{\delta^2}$, P_e, and $\overline{\Delta^2}$ IN (3.2d)

YIELDS: $\overline{r_N^2} = -39.0$ dB, SLIGHTLY TOO HIGH.

Fig. 3.7 Example of effects of errors.

EFFECT ON DIRECTIVITY:

FROM (3.4)

$$\frac{D_e}{D} = \frac{P_e}{1 + \overline{\Delta^2} + \overline{\delta^2}}$$

$$\boxed{\frac{D_e}{D} = -2.41 \text{ dB}}$$

Fig. 3.8 Example of effects of errors. (Cont.)

* $\sqrt{\overline{\Delta^2}} = 0.480$, $\quad 20\log_{10}(1+0.48) = 3.40$ dB, $\quad -20\log_{10}(1-0.480) = 5.67$ dB; $\quad \dfrac{3.40 + 5.67}{2} = 4.5$ dB

ASSUME -2.41 dB IS TO LARGE A LOSS.
ASSUME THAT MAXIMUM LOSS THAT CAN BE TOLERATED IN DIRECTIVITY IS 1dB.

PROBLEM: TO FIND NEW ALLOWABLE VALUES FOR $\overline{\delta^2}$, P_e, and $\overline{\Delta^2}$.
ASSUME CONTRIBUTION TO LOSS IN DIRECTIVITY BUDGETED EQUALLY AMONG $\overline{\delta^2}, P_e$, AND $\overline{\Delta^2}$.

1) $P_e = ?$ $\left(\frac{D_e}{D}\right)_{P_e} = P_e = -0.333 \text{ dB} = 0.926 \longrightarrow (1-P_e)N = \underline{576}$ ELEMENTS

2) $\overline{\Delta^2} = ?$ $\left(\frac{D_e}{D}\right)_{\overline{\Delta^2}} = \frac{1}{1+\overline{\Delta^2}} = -0.333 \text{ dB} = 0.926 \Rightarrow \overline{\Delta^2} = 0.0797 = \underline{\sim 2.5 \text{ dB}}$

3) $\overline{\delta^2} = ?$ $\left(\frac{D_e}{D}\right)_{\overline{\delta^2}} = \frac{1}{1+\overline{\delta^2}} = -0.333 \text{ dB} = 0.926 \Rightarrow \overline{\delta^2} = 0.0797, \sqrt{\overline{\delta^2}} = \underline{16.2^\circ}$

4) SIDELOBE LEVEL $\overline{r_N^2} = ?$

FROM (3.2d) $\boxed{\overline{r_N^2} = -44.1 \text{ dB}}$ NOW.

Fig. 3.5 THUS YIELDS THAT EACH 30 dB DOWN SIDELOBES IS DEGRADED BY $\leq$ 1.5 dB 90% OF THE TIME.

Fig. 3.9 Example of effects of errors. (Cont.).

$$\boxed{\overline{\delta^2} = \overline{\delta^2_{AMP}} + \overline{\delta^2_Q} + \overline{\delta^2_{MECH}} + \overline{\delta^2_{ROT}}} \qquad (3.6)$$

WHERE:

$\overline{\delta^2_{AMP}}$ = MEAN SQUARE PHASE ERROR DUE TO POWER AMPLIFIERS, MANUFACTURING ERRORS OF QUANTIZED PHASE SHIFTS (KNOWN AS DIFFERENTIAL PHASE SHIFT ERROR; SEE SECT. 3.1.2C, p 3-23).

$\overline{\delta^2_Q}$ = PHASE SHIFTER RANDOM QUANTIZATION ERROR

$\overline{\delta^2_{MECH}}$ = ERROR DUE TO INEXACT LOCATIONS OF RADIATING ELEMENTS

$\overline{\delta^2_{ROT}}$ = ERROR DUE TO ROTATIONAL MISSALIGNMENTS OF RADIATING ELEMENTS (GENERALLY SMALL RELATIVE TO OTHER ERRORS.)

Fig. 3.10 Sources of random phase errors.

1 BIT PHASE SHIFTER PHASES: 0°, 180°

2 BIT PHASE SHIFTER PHASES: 0°, 90°, 180°, 270°

P BIT PHASE SHIFTER PHASES: $s\left[360^\circ/2^P\right]$, $s = 0,1,..,2^P-1$

$$\boxed{\overline{\delta^2_Q} = \left(\frac{2\pi}{2^P}\right)^2 \frac{1}{12} = \frac{\pi^2}{3 \cdot 2^{2P}} \text{ rad.}^2} \qquad (3.7)$$

SUBSTITUTING (3.7) in (3.2a) YIELDS

$$\boxed{\left(\overline{r_N^2}\right)_Q = \frac{\overline{\delta^2_Q}}{N\eta_t\left(1-\overline{\delta^2_Q}\right)} \doteq \frac{\overline{\delta^2_Q}}{N\eta_t}} \quad \text{for } P \geq 3 \qquad (3.8)$$

WHERE $\left(\overline{r_N^2}\right)_Q$ = AVERAGE SIDELOBE LEVEL $\overline{r_N^2}$ DUE TO PHASE SHIFTER RANDOM QUANTIZATION ERROR.

Fig. 3.11 Phase shifter random quantization error.

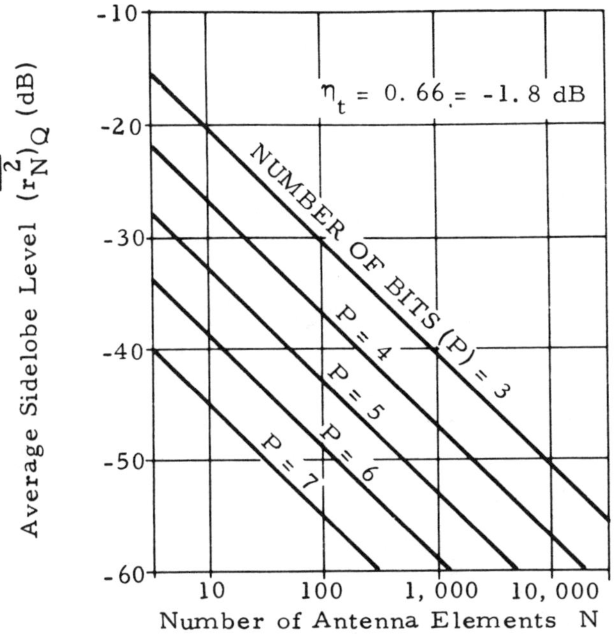

$(\overline{r_N^2})_Q$ = AVERAGE SIDELOBE LEVEL DUE TO PHASE SHIFTER RANDOM QUANTIZATION ERROR

$$\overline{(r_N^2)}_Q \doteq \frac{\overline{\delta_Q^2}}{N\eta_t} \quad \text{for } P \geq 3 \tag{3.8}$$

WHERE

$$\overline{\delta_Q^2} = \frac{\pi^2}{3 \cdot 2^{2P}} \tag{3.7}$$

Fig. 3.12 Average sidelobe level due to phase shifter random quantization error. (After Cheston and Frank in Skolnik, McGraw-Hill Book Comp. [6]; also Miller, RADC Symp. [5].)

$$\delta_{MECH} = \frac{2\pi}{\lambda}\left[\Delta x_n \sin\theta\cos\phi + \Delta y_n \sin\theta\sin\phi + \Delta z_n \cos\theta\right] \tag{3.9a}$$

WHERE $\Delta x_n, \Delta y_n, \Delta z_n$ = x, y, and z AXIS ERRORS FOR nth RADIATING ELEMENT; SEE Fig. 2.31

LET

$$\overline{(\Delta x_n)^2} = \overline{(\Delta y_n)^2} = \overline{(\Delta z_n)^2} = \sigma^2_{MECH} \tag{3.9b}$$

THEN

$$\overline{\delta^2_{MECH}} = \left(\frac{2\pi}{\lambda}\right)^2 \sigma^2_{MECH} \tag{3.9c}$$

Fig. 3.13 Phase error due to inexact location of radiating elements.

ASSUMPTION: RADIATING ELEMENTS ARE SHORT DIPOLES INTENDED TO BE PARALLEL TO X-AXIS; (SEE Fig. 2.31)

THEN

$$\overline{\Delta^2_{ROT}} \leq \tfrac{1}{2}\sigma_\alpha^2\left[\cos^2\phi + \cos^2\theta\sin^2\phi + \sin^2\theta\right] \tag{3.10a}$$

$$\overline{\Delta^2_{ROT}} \leq \sigma_\alpha^2 \tag{3.10b}$$

WHERE σ_α^2 = VARIANCE OF α, THE ROTATION ANGLE ERROR OF DIPOLE.

$= \sigma^2_{\alpha y} + \sigma^2_{\alpha z}$

$\sigma^2_{\alpha y}$ = VARIANCE OF α_y, THE ROTATION ANGLE ERROR ABOUT y AXIS

$\sigma^2_{\alpha z}$ = VARIANCE OF α_z, THE ROTATION ANGLE ERROR ABOUT z AXIS

Fig. 3.14 Amplitude error due to rotational missalignments of radiating elements.

ASSUME $\overline{\delta^2}$ BUDGETED EQUALLY BETWEEN $\overline{\delta^2_{AMP}}$, $\overline{\delta^2_Q}$, AND $\overline{\delta^2_{MECH}}$.

PHASE QUANTIZATION ERROR:

FROM Fig. 3.9 $\overline{\delta^2} = 0.0797$ SO THAT $\overline{\delta^2_Q} = 0.0797/3$. HENCE FROM (3.7) :

$$\overline{\delta^2_Q} = \frac{\pi^2}{3.2^{2P}} = \frac{0.0797}{3} \Longrightarrow P = 3.48 \longrightarrow \boxed{\text{4 BITS NEEDED.}}$$

FOR 4 BITS

$$\boxed{\overline{\delta^2_Q} = 0.01285}$$

$$\text{THUS} \quad \overline{\delta^2_{AMP}} + \overline{\delta^2_{MECH}} = \overline{\delta^2} - \overline{\delta^2_Q} = 0.0797 - 0.01285 = 0.0668 \qquad (3.11)$$

AMPLIFIER PHASE ERROR:

FROM ABOVE:

$$\overline{\delta^2_{AMP}} = \frac{0.0668}{2} = 0.0334 \longrightarrow \boxed{\sqrt{\overline{\delta^2_{AMP}}} = 10.5^{\circ} \text{ deg.}}$$

MECHANICAL ERROR: ASSUME $\lambda = 0.3$m (L-Band, $f_c = 1$ GHz) THEN

FROM (3.9c)

$$\overline{\delta^2_{MECH}} = \left(\frac{2\pi}{\lambda}\right)^2 \sigma^2_{MECH} = 0.0334 \qquad \boxed{\sigma_{MECH} = 0.873 \text{ cm}}$$

Fig. 3.15 Example of effects of errors (cont.)

$$\boxed{\overline{\Delta^2} = \overline{\Delta^2_{AMP}} + \overline{\Delta^2_{RAD}} + \overline{\Delta^2_Q}} \qquad (3.12)\ \overline{\Delta^2}_{ROT}$$

Where:

$\overline{\Delta^2_{AMP}}$ = MEAN SQUARE AMPLIFIER AND PHASE SHIFTER FRACTIONAL AMPLITUDE ERRORS

$\overline{\Delta^2_{RAD}}$ = MEAN SQUARE FRACTIONAL DEVIATION OF RADIATION ELEMENT VOLTAGE PATTERN FROM NOMINAL VALUE

$\overline{\Delta^2_Q}$ = MEAN SQUARE RANDOM FRACTIONAL DEVIATION OF ACTUAL ILLUMINATION FUNCTION FROM INTENDED ILLUMINATION FUNCTION DUE TO AMPLITUDE WEIGHTING QUANTIZATION

Fig. 3.16 Sources of random amplitude errors.

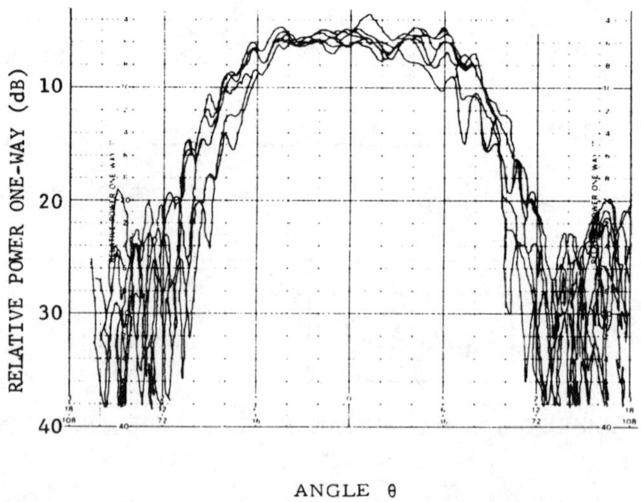

Fig. 3.17 Comparison of voltage patterns of log-periodic radiation elements of linear array. (After Allen, Lincoln Laboratory [7].)

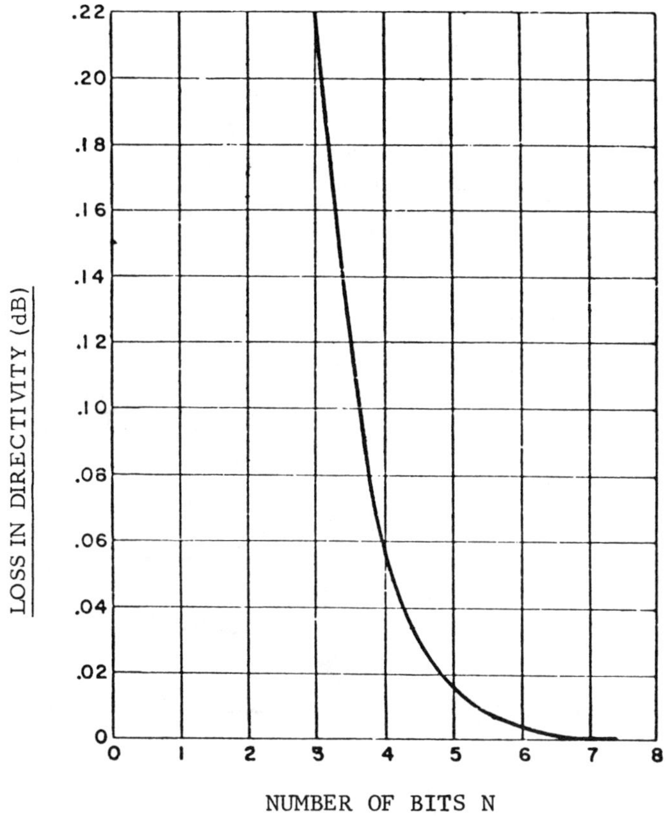

$$\frac{D_e}{D} = \frac{1}{1 + \overline{\delta_Q^2}} \qquad (3.13)$$

where

$$\overline{\delta_Q^2} = \frac{\pi^2}{3 . 2^{2P}} \text{ rad}^2 \qquad (3.7)$$

Fig. 3.18 Loss in directivity due to phase shifter quantization. (After Miller, RADC Symp. [5].)

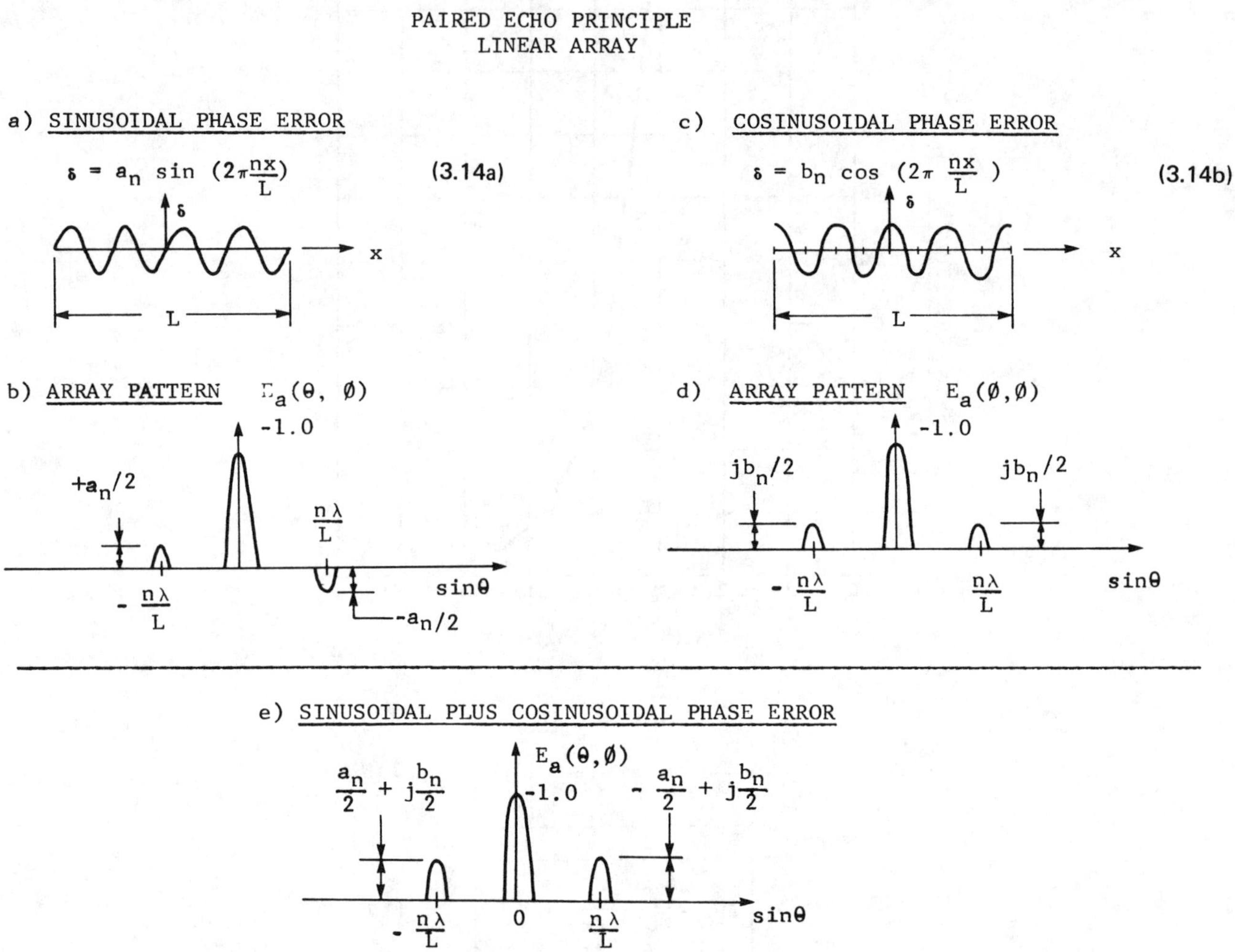

Fig. 3.19 Spatial frequency and paired echo approach to dealing with antenna array random phase errors; uniformly weighted linear array of isotropic elements.

a) PHASE ERROR FOR LINEAR ARRAY

$$\delta = \sum_{n=1}^{N/2} \left[a_n \sin 2\pi\frac{nx}{L} + b_n \cos 2\pi\frac{nx}{L} \right] \qquad (3.15a)$$

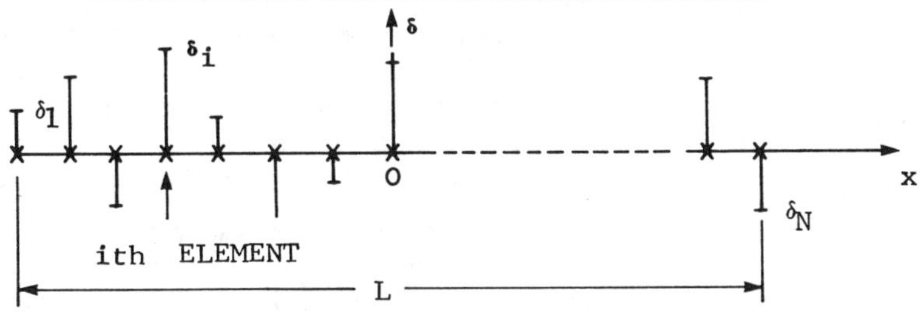

b) ARRAY ANTENNA SIDELOBES DUE TO PHASE ERRORS.

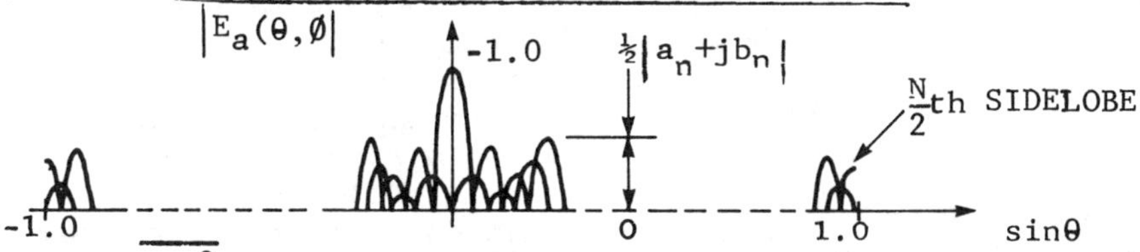

LET: $\overline{r_{No}^2}$ = AVERAGE POWER OF nth SIDELOBE.

FROM FIGURE THE AVERAGE POWER IN THE nth SIDELOBE IS GIVEN BY

$$\overline{r_{No}^2} = \left(\frac{\overline{a_n^2}}{4} + \frac{\overline{b_n^2}}{4} \right) \qquad (3.15b)$$

ASSUME $\overline{r_{No}^2}$ INDEPENDENT OF n.

$$\text{FROM (3.15a):}\ \overline{\delta^2} = \sum_{n=1}^{N/2} \left(\frac{\overline{a_n^2}}{2} + \frac{\overline{b_n^2}}{2} \right) = N\overline{r_{No}^2} \qquad (3.15c)$$

$$\boxed{\overline{r_{no}^2} = \frac{\overline{\delta^2}}{N}} \qquad (3.15d)$$

WHICH IS ESSENTIALLY THE SAME AS (3.2a) FOR $\eta_t=1$, THE ASSUMPTIONS MADE IN ABOVE DERIVATION. HENCEFORTH r_{No}^2 WILL THUS BE REPLACED BY SYMBOL $\overline{r_N^2}$ USED BEFORE.

Fig. 3.20 Spatial frequency and paired echo approach applied to case of independent phase errors from element-to-element; uniformly weighted linear array of isotropic elements.

in Fig. 3-20b. The amplitudes a_n and b_n of (3-15a), which determine the magnitudes of the sidelobes, are functions of the phase errors δ_i for the N radiating elements through standard discrete Fourier series relationships. Because the δ_i are random, it follows that a_n and b_n are random variables also with their values depending on the particular set of δ_i errors existing for a given situation.

From Fig. 3-20 it follows that the average power at the peak of the nth sidelobe is simply given by

$$\overline{r^2}_{No} = \frac{\overline{a^2}_n}{4} + \frac{\overline{b_n}^2}{4} \qquad (3\text{-}15b)$$

where the bar indicates an average of the possible r^2_{No}, a^2_n and b^2_n values over all possible sets of δ_i errors, that is, the bar indicates an ensemble average.

It is assumed that $\overline{r^2}_{No}$ is the same for all sidelobes. This assumption is equivalent to the assumption that the mean square phase error for all the radiating elements are the same, that is, $\overline{\delta^2}_i$ = constant = $\overline{\delta^2}$. From (3-15a) and (3-15b), (3-15c) follows. Solving (3-15c) for the average power sidelobe level yields (3-15d). Eq. (3-15d) is essentially the same as (3-2a) for a uniformly weighted aperture (that is, $\eta_t = 1$) when δ^2 is small. Thus (3-2a) has been derived using a different approach which gives added physical insight into the nature of the effects of phase errors on the antenna characteristics.

The derivation for (3-15d) and the results of Fig. 3-19 assumed a linear array of isotropic elements for which no mutual coupling effects exist. Accordingly as discussed previously relative to (3-2a) and (3-1g), a factor taking into account the non-isolated element gain pattern should be included for more exact results if the accuracy needed warrants it.

In Fig. 3-20 the N/2th sidelobe level is indicated as being located at $\sin\theta = 1$, that is, at $\theta = 90°$. This would be the case if the spacing between the radiating elements d was $\lambda/2$. If $d > \lambda/2$ then the N/2th sidelobe would occur to the left of $\sin\theta = 90°$. The magnitude of the antenna array factor is symmetric about the N/2th sidelobe as well about the mainlobe. Thus, for $d > \lambda/2$ there will be more than N/2 sidelobes on each side of the mainlobe even though there are only N/2 spatial harmonics. However, any sidelobe beyond the N/2th sidelobe will be identical to its symmetric sidelobe occurring before the N/2th sidelobe.

If $d < \lambda/2$, then the N/2th sidelobe occurs beyond $\sin\theta = 1$, that is, outside the visible space. In this case there are less than N/2 observable sidelobes. For the cases of linear and planar arrays, the actual number of sidelobes in visible space is given by m of Fig. 2-52, independent of d.

Subarrayed Antennas.

Again assume a uniformly-weighted linear array of isotropic elements having length L whose main beam is pointed broadside. Now, however, let the radiating elements be grouped together into groups of N_s elements with each such group being fed from one power amplifier as shown in Fig. 2-21. Each of these groups of N_s elements is referred to as a subarray. It is assumed that there are n_s subarrays with $n_s N_s = N$. Consideration is now given to the effects of phase errors introduced by the power amplifiers.*

A phase error in the ith power amplifier δ_i causes all the N_s radiating elements being fed from the ith power amplifier to have the same phase error δ_i; (see Fig. 2-21). The phase errors of each of the power amplifiers are assumed to be independent of each other. As before a spatial frequency representation of the phase error across the array can be derived. Because the array only has n_s independent phase errors spaced N_s array elements apart, there are now only $n_s/2$ spatial harmonics; see (3-16b) of Fig. 2-21. Each harmonic is composed of a sine and cosine term having exactly n complete periods across the length of the aperture L where n = 1 to = $n_s/2$. The sidelobes are located at $\sin\theta = \pm n\lambda/L$ where n = 1 to $n_s/2$ as depicted in Fig. 2-21. There are fewer sidelobes, n_s instead of N, $1/N_s$ th as many sidelobes. Also, the sidelobes do not generally cover the whole visible region from $\theta = -90°$ to 90°. Instead, they go from $\sin\theta = -n_s\lambda/2L$ to $n_s\lambda/2L$. Following arguments used to derive (3-15d) it is shown in Fig. 3-21 that the average sidelobe level is given by (3-16e). This expression is the same as for an array which does not use subarrays, that is, an array for which the phase errors are independent from element to element, see (3-2a), but with N replaced by n_s.

By similar arguments it is a simple matter to show that the expression derived for a linear array with n_s subarrays applies as well to a planar array with n_s subarrays; (see Fig. 3-22). For a square or circular planar array, however, the sidelobes due to the errors in the power amplifiers feeding the subarrays extend out only to $\sin\theta = \pm [(\sqrt{n_s}/2)(\lambda/2r_o)] = 1\sqrt{N_s}$#; (see Fig. 3.22). Equivalently, they exist only out to the $(\sqrt{n_2}/2$ th sidelobe. There are, however, about as many sidelobes (order-of-magnitude wise) produced by the power amplifier phase errors for the planar array as there are for the linear array when both consist of the same number of subarrays. For a planar square or circular array whose main beam is pointed broadside the sidelobes are distributed about the main beam in a cone angle having a half angle given by θ where $\sin\theta = 1/\sqrt{N_s}$ #; (see Fig. 3.22).

Similar results are obtained when the effects of power amplifier amplitude errors and random failures are considered. Specifically, the sidelobe ratio results for linear and planar arrays which do not use subarrays apply with N replaced n_s; (see Fig. 3-23), Eq. (3-17). The sidelobes produced by the power amplifier amplitude errors and random failures lie in the same region of space as do those produced by the power amplifier phase errors. The expression for the degradation in directivity when using subarrays is the same as when not using subarrays; see (3-18).

Fig. 3-24 gives an example for the case of a subarrayed antenna. The example is given for the antenna specified before in Fig. 3-7 except that now the 7800 elements of the array are divided into 40 subarrays of 195 elements each as indicated in Fig. 3-25. It is assumed that this subarraying is done on receive only for the antenna. The subarraying is achieved by the use of a corporate feed structure. The antenna is subarrayed in order to obtain the desired 30 dB Tapered-Taylor amplitude taper. (On transmit this antenna is assumed to have a uniform weighting.)

---

* In actuality not only the power amplifier contributes to the subarray error. There are in addition the subarray feed network phase errors and the subarray phase shifter quantization and differential phase shift errors.

Approximation best for $d = \lambda/2$.

LINEAR ARRAY OF SUBARRAYS

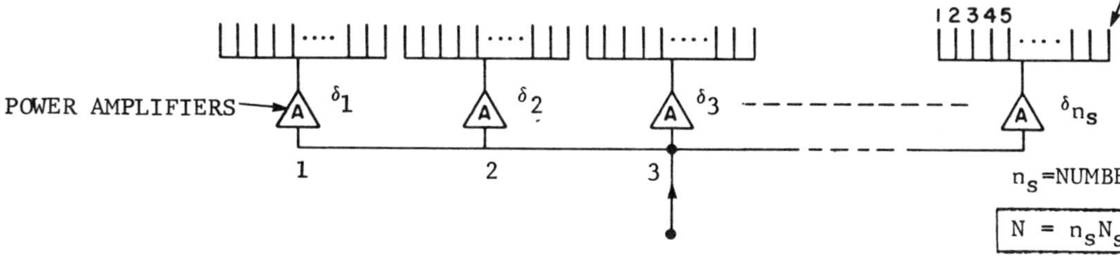

$$N = n_s N_s \quad (3.16a)$$

δ_i = PHASE ERROR FOR ith SUBARRAY DUE TO POWER AMPLIFIER DRIVING ith SUBARRAY.

NOW THE HIGHEST SPATIAL FREQUENCY PRESENT ACROSS THE ARRAY IS n_s/L, WHERE $n_s=N/N_s$= THE NUMBER OF INDEPENDENT PHASES ACROSS THE APERTURE.

THUS

$$\delta = \sum_{n=1}^{n_s/2} \left[a_n \sin 2\pi\frac{nx}{L} + b_n \cos 2\pi\frac{nx}{L} \right] \quad (3.16b)$$

ARRAY SIDELOBES DUE TO SUBARRAY PHASE ERRORS:

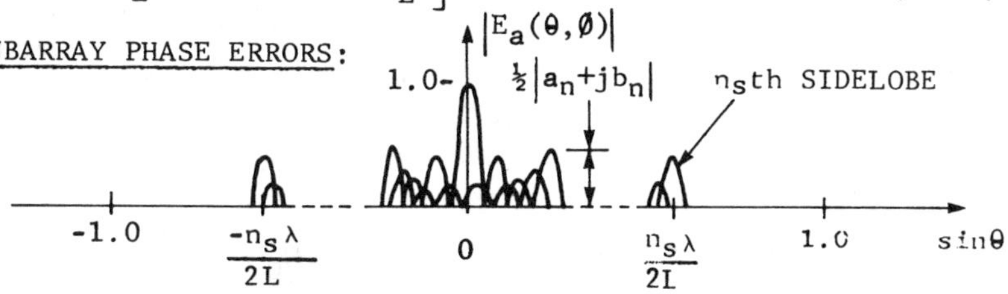

NOW: $\overline{\delta^2} = n_s\overline{r_N^2}$ (3.16c)

SO THAT FOR UNIFORMLY WEIGHTED APERTURE:

$$\overline{r_N^2} = \frac{\overline{\delta^2}}{n_s} \quad (3.16d)$$

AND FOR WEIGHTED APERTURE:

$$\overline{r_N^2} = \frac{\overline{\delta^2}}{n_s \eta_t (1-\overline{\delta^2})} \quad (3.16e)$$

Fig. 3.21 Average power sidelobe level due to random phase errors for linear array using subarrays; uniformly weighted linear array of isotropic elements.

EQUATION FOR LINEAR ARRAY APPLIES, THAT IS:

$$\overline{r_N^2} = \frac{\overline{\delta^2}}{n_s \eta_t (1-\overline{\delta^2})} \quad (3.16e)$$

HOWEVER NOW SIDELOBES EXTEND OUT TO $\frac{\sqrt{n_s}}{2}$ th SIDELOBE:

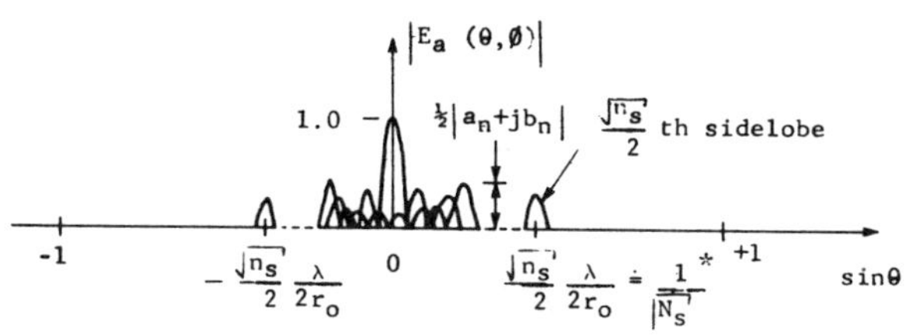

* APPROXIMATION BEST FOR $d = \lambda/2$.

Fig. 3.22 Average power sidelobe level due to random phase errors for planar square or circular arrays using subarrays.

SIDELOBE LEVEL:

AVERAGE POWER SIDELOBE LEVEL DUE TO SIMULTANEOUS RANDOM PHASE AND AMPLITUDE ERRORS AND RANDOM FAILURES; (2d) APPLIES WITH N REPLACED BY n_s, THE NUMBER OF SUBARRAYS; I.E.:

$$\overline{r_n^2} = \frac{(1 - P_e) + \Delta^2 + \delta^2 P_e}{P_e(1 - \delta^2)} \; \frac{1}{\eta_t n_s} \quad (3.17)$$

HERE δ AND Δ NOW ARE PHASE AND AMPLITUDE ERRORS WHICH ARE INDENTICAL FOR ALL RADIATING ELEMENTS OF ONE SUBARRAY BUT INDEPENDENT FROM SUBARRAY TO SUBARRAY, AND WHERE P_e = PROBABILITY OF SURVIVAL OF ith SUBARRAY

DIRECTIVITY:

$$\frac{D_e}{D} = \frac{P_e}{1 + \Delta^2 + \delta^2} \quad (3.18)$$

COMMENTS MADE FOR (3.17) ALSO APPLY FOR (3.18)

Fig. 3.23 Effects of errors on linear or planar arrays using subarrays: sidelobe level; directivity.

ASSUMPTIONS SAME AS MADE BEFORE (SEE Fig. 3.7) EXCEPT SUBARRAYS USED WITH:

$n_s = 40 \quad N_s = 195$

THIS SUBARRAYING ASSUMED TO BE ON RECEIVE ONLY WITH EACH SUBARRAY FEEDING A PARAMP. THE AMPLITUDE AND PHASE ERRORS INTRODUCED BY EACH PARAMP ARE SPECIFIED BY:

$$\overline{\Delta^2_{PAR}} = 1\text{dB}$$

$$\sqrt{\overline{\delta^2_{PAR}}} = 5^\circ$$

FROM (3.17)

$$\boxed{\overline{r^2_{N,PAR}} = -32.0 \text{ dB}} \qquad \text{FOR } \frac{-1}{\sqrt{N_s}} \le \theta \le \frac{1}{\sqrt{N_s}}$$

OR $-4.1^\circ \le \theta \le 4.1^\circ$ AT BROADSIDE.

FROM FIG. (2.40) FOR $\bar{n} = 8$:

$\sigma = 1.0827$, $K_{oo} = 60.55^\circ$ SO THAT AT BROADSIDE $\theta_{30} = 1.3^\circ$ FOR $2r_o = 50\lambda$. THUS $4.1^\circ/1.3^\circ = 3.15$ SO THAT THOSE SIDELOBE COMPONENTS WHICH ARE -32.0 dB DOWN EXTEND OUT TO $\sim 3\theta_{30}$ FROM PEAK OF MAIN LOBE. THERE ARE THREE CLOSE IN SIDELOBES DOWN 30 dB IN THE RANGE $\theta = 0^\circ$ TO 4.1° IN ANY ϕ = CONST. PLANE. ($\left[1/\sqrt{N_s}\right]/\left[\lambda/2r_o\right] \approx 3.6$)

FROM Fig. 3.5 EACH OF THESE CLOSE IN 30 dB DOWN SIDELOBES DEGRADED ≤ 5.2 dB 90% OF TIME, THAT IS, CLOSE IN SIDELOBES ≤ -24.8 dB 90% OF TIME.

Fig. 3.24 Example on effects of errors extended to antenna using subarrays.

$N = 7880$

$N_s = 195$

$n_s = 40$

$2r_o = 50\lambda$

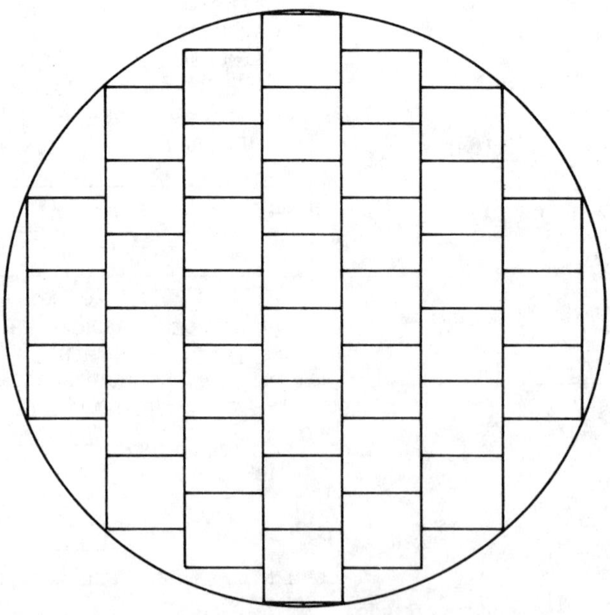

Fig. 3.25 Example antenna using subarrays.

Each subarray on receive is fed into a separate paramp. The independent amplitude and phase errors for these paramps are assumed to be 1 dB and 5°, respectively. Fig. 3-24 indicates that these amplitude and phase errors result in an error sidelobe at the null of the error free pattern which is 32 dB down from the peak of the main lobe. These sidelobes extend out to $\theta = 4.1^\circ$ from the main beam when it is pointed broadside. For the error-free pattern there are three sidelobes 30 dB down in the region from 0° to 4.1° in any constant ϕ plane. Fig. 3-5 indicates that these sidelobes would be degraded by 5.2 dB 90% of the time. Stated another way, the peak sidelobes degradation resulting from paramp phase and amplitude errors will be $\leqslant$-24.8 dB 90% of the time. Of course, to the paramp sidelobe degradation must be added the degradation from the other errors considered previously — the errors which are independent from element to element. The average power sidelobe level component from these other error factors must be added to the average power sidelobe produced by the power amplifiers to obtain the average power sidelobes from all errors.

B. Periodic Errors.

i. Phase Quantization Errors. [5], [6].

Fig. 3-26 gives the magnitude of the peak sidelobe level resulting from various types of periodic amplitude and phase errors across a planar array.[6] The factor $\sqrt{\cos\theta_1/\cos\theta_0}$ is included, where θ_0 and θ_1 are the directions of the main lobe and sidelobe (due to the periodic phase error), respectively. This factor arises from the assumption of the $\sqrt{\cos\theta}$ non-isolated voltage antenna-element gain pattern, see (3-3).

Two of the most common types of periodic phase errors that occur in phased arrays are the element-to-element phase ripple of Fig. 3-26f and the triangular phase modulation of Fig. 3-26g. Examples of how these occur are now given.

Recall that to point the beam of a linear aperture at some angle θ_o relative to broadside, successive elements across the array must have linearly increasing phase shifts; (see Fig. 2-1). Fig. 3-27 depicts this linearly increasing phase shift. When a P bit phase shifter is used to realize this linearly increasing phase shift across the aperture, a step approximation to the linearly increasing phase shift is obtained; (see Fig. 3-27). The difference between the desired linearly increasing phase shift and the step approximation to it represents the phase error resulting from the use of a P-bit phase shifter. This phase error is seen from Fig. 3-27 to be a periodic sawtooth wave. Fig. 3-27 also indicates the positions of the radiating elements. For the example illustrated in Fig. 3-27 the period of one of these sawtooths is equal to 8 element-to-element spacings, that is, the period s of Fig. 3-26g is equal to 8d. As a result, the phase shift between elements is equal to the $\pi/(4.2^P)$. For a 4-bit phase shifter, that is, P=4, and d=$\lambda/2$ it follows from (2-16) that this corresponds to a shifting of the main beam off broadside by 0.9°, that is, θ_o=0.9°. From (3-19) it follows that θ_1 = -13.6° and +15.4°. The peak phase error of Fig. 3-27 equals $\pi/2^P$. Hence, β of Fig. 3-26g equals $\pi/2^P$ and

$$\text{Peak quantization sidelobe level for triangular phase modulation} = \frac{\beta}{\pi}\sqrt{\frac{\cos\theta_1}{\cos\theta_o}} = \frac{1}{2^P}\sqrt{\frac{\cos\theta_1}{\cos\theta_o}}$$

$$= -24.2 \text{ dB} \tag{3-20}$$

for the 4-bit phase shifter with these peak lobes occurring both at θ_1 = +15.4° and -13.6°

It is possible also to specify a beam pointing angle for the antenna used in the example of Fig. 3-27 (d = $\lambda/2$, P = 4 bits) which results in the situation depicted in Fig. 3-28a.* The phase error is such as to alternate between $+\pi/2^{P+1}$, and $-\pi/2^{P+1}$, the situation of Fig. 3-26f. To achieve the situation depicted in Fig. 28a the differential phase shift between elements must be $\pi/2^P$. From (2-16) it follows that for a 4-bit phase shifter and $\lambda/2$ spacing between elements that this corresponds to a shifting of the main beam off broadside by θ_o = 3.58°, a beam pointing angle four times as large as for the example depicted in Fig. 3-27. From (3-19) it follows that θ_1 = -69.6° (the other lobe is not in visible space). For the case of element-to-element phase ripple illustrated in Fig. 3-28a the peak phase error β equals $\pi/2^{P+1}$ and

$$\text{Peak quantization sidelobe level for element-to-element phase ripple} = \beta\sqrt{\frac{\cos\theta_1}{\cos\theta_o}} = \frac{\pi}{2^{P+1}}\sqrt{\frac{\cos\theta_1}{\cos\theta_o}}$$

$$= -24.7 \text{ dB} \tag{3-21}$$

for a 4-bit phase shifter with the peak lobe occurring at θ_1= -69.6°.

For the two cases considered, involving θ_o = 0.9° and 3.58° in Figs. 3-27 and 3-28a, respectively, the element-to-element phase ripple case resulting when θ_o = 3.58° gives a slightly lower (0.5 dB) sidelobe level than the triangular phase modulation case which results when θ_o = 0.9°. However, in general, it is possible to find main beam scan angles which yield element-to-element phase ripple for which the peak sidelobes are worse than obtained for beam pointing angles that yield triangular phase modulation. This is because of the π factor present in (3-20) but not in (3-21). This π factor is partly offset by the fact that β for the element-to-element ripple cases is half that for the triangular modulation cases. As a result the worst peak sidelobe for the element-to-element ripple case will be $\pi/2$ = 3.9 dB worse than for the triangular modulation case.

A typically severe element-to-element ripple beam pointing case is illustrated for the same antenna (i.e., d= $\lambda/2$, P = 4 bits) in Fig. 3-28b. For this case the differential phase shift between elements is $13[2\pi/2^{P+1}]$. Hence, θ_o = 54.3° and θ_1 = -10.8° with the result that

$$\text{Peak quantization sidelobe level} = \beta\sqrt{\frac{\cos\theta_1}{\cos\theta_o}} = -17.9 \text{ dB}.$$

It appears that element-to-element phase ripple cases of the type obtained in Fig. 3-28b will in general give the highest peak sidelobe levels due to periodic phase errors arising from phase shifter quantization.

Fig. 3-29 gives a plot of (3-21) for the case of θ_o=30° and 60° with θ_1 = 0°, the worst possible value for θ_1. Also given is a plot of the peak sidelobe level obtained from (3-20) for the case of triangular phase modulation with

* The array in Fig. 3-27 actually differs from that of Fig. 3-28a in that the former has 2M+1 elements while the latter 2M; however, this difference is only one of convenience. The antennas can be thought of as being identical for the two figures for all practical purposes.

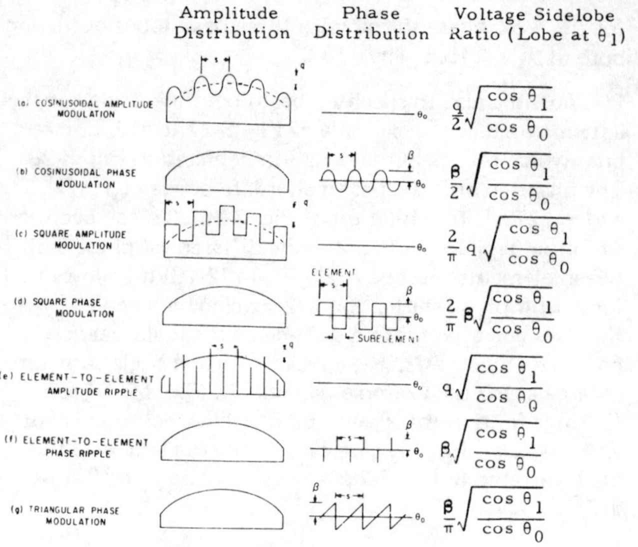

$$\sin\theta_1 = \sin\theta_0 \pm \frac{\lambda}{s} \quad (3.19)$$

Fig. 3.26 Effects of periodic amplitude and phase errors. (After Cheston and Frank in Skolnik, McGraw-Hill Book Comp. [6].)

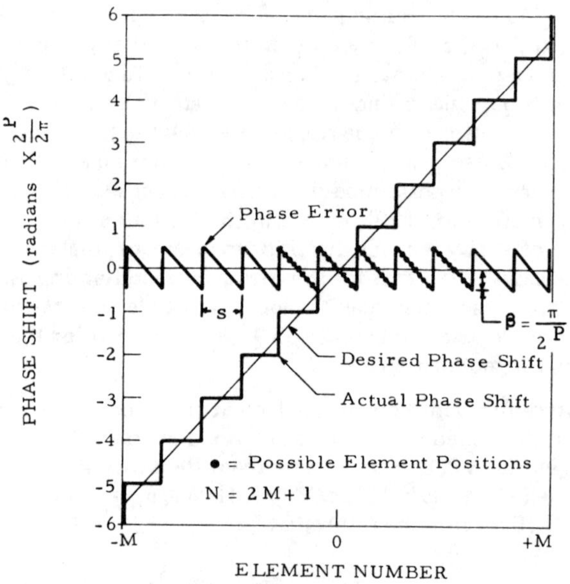

ASSUME: $d = \lambda/2$

THEN $s = 8d = 4\lambda$

$$\alpha = \left(\frac{2\pi}{2^P}\right) \div 8 = \frac{\pi}{4 \cdot 2^P}$$

FROM (2.16): $\theta_0 = 0.9°$ FOR P = 4 BITS

FROM (3.19): $\theta_1 = +15.4°$ and $-13.6°$

FROM (3.20): PEAK SIDELOBE LEVEL =

$$\frac{\beta}{\pi}\sqrt{\frac{\cos\theta_1}{\cos\theta_0}} = \underline{-24.2 \text{ dB}}$$

Fig. 3.27 Example case of triangular phase modulation due to phase quantization. (After Cheston and Frank in Skolnik, McGraw-Hill Book Comp. [6].)

EXAMPLE NO. 1

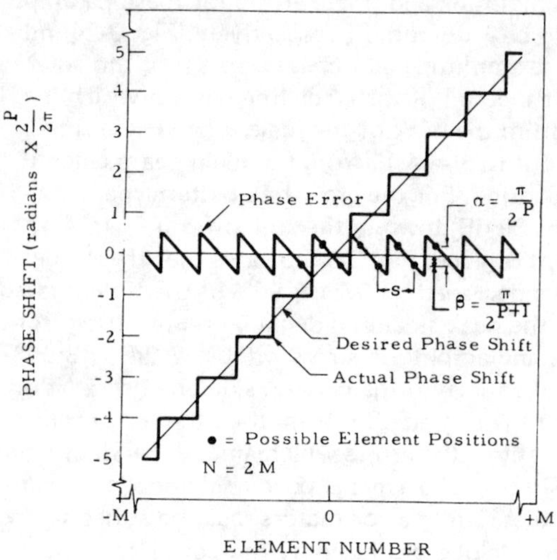

ASSUMPTIONS: SAME AS FOR Fig. 3.27 FOR ANTENNA SYSTEM; i.e., $d = \lambda/2$ AND P = 4 BITS. NOW THOUGH:

$s = 2d = \lambda$

$\alpha = \frac{\pi}{2^P}$

FROM (2.16): $\theta_0 = 3.58°$

FROM (3.19): $\theta_1 = -69.6°$

FROM (3.21): PEAK SIDELOBE LEVEL =

$$\beta\sqrt{\frac{\cos\theta_1}{\cos\theta_0}} = \underline{-24.7 \text{ dB}}$$

Fig. 3.28 Example cases of element-to-element phase ripple due to phase quantization; Example No. 1. (After Cheston and Frank in Skolnik, McGraw-Hill Book Comp. [6].)

EXAMPLE NO. 2

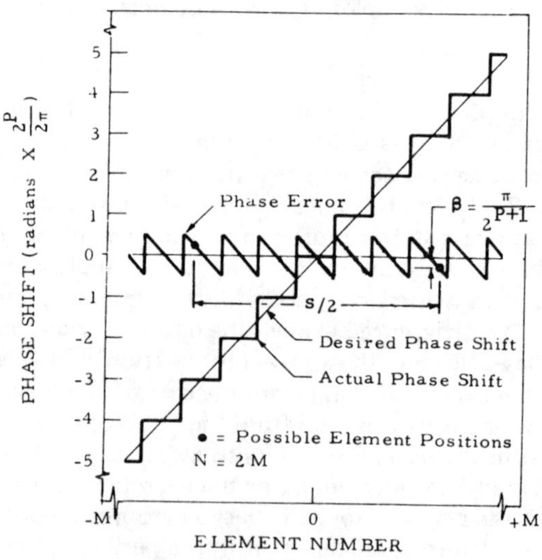

ASSUMPTIONS: SAME AS FOR FIG. 27 FOR ANTENNA SYSTEM; i.e., $d = \lambda/2$ AND P = 4 BITS. NOW THOUGH:

$s = \lambda$

$\alpha = \left(\frac{2\pi}{2 \cdot 2^P}\right)$ 13

FROM (2.16): $\theta_0 = 54.3°$

FROM (3.19): $\theta_1 = -10.8°$

FROM (3.21): PEAK SIDELOBE LEVEL =

$$\beta\sqrt{\frac{\cos\theta_1}{\cos\theta_0}} = \underline{-17.9 \text{ dB}}$$

Fig. 3.28 (con't.) Example cases of element-to-element phase ripple due to phase quantization; Example No. 2. (After Cheston and Frank in Skolnik, McGraw-Hill Book Comp. [6].)

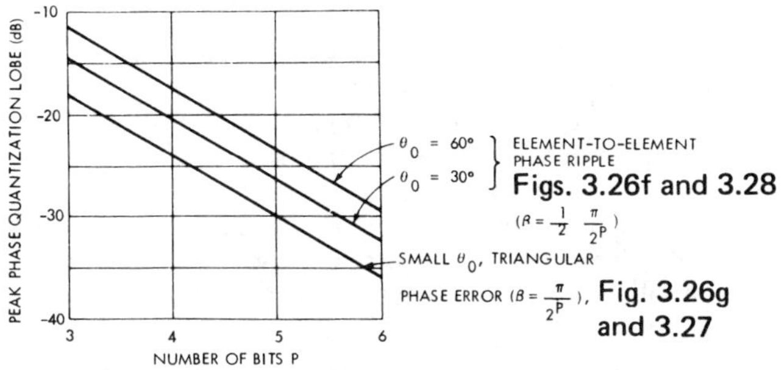

Fig. 3.29 Sidelobes due to periodic phase quantization errors. (After Cheston and Frank in Skolnik, McGraw-Hill Comp. [6].)

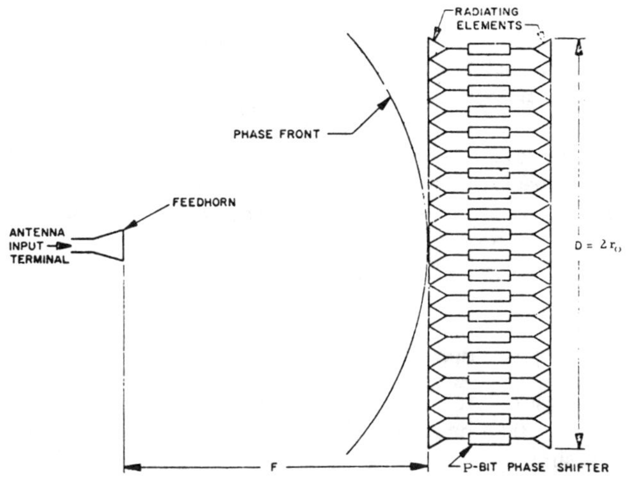

Fig. 3.31 Space fed array. (After Miller, RADC Symp. [5])

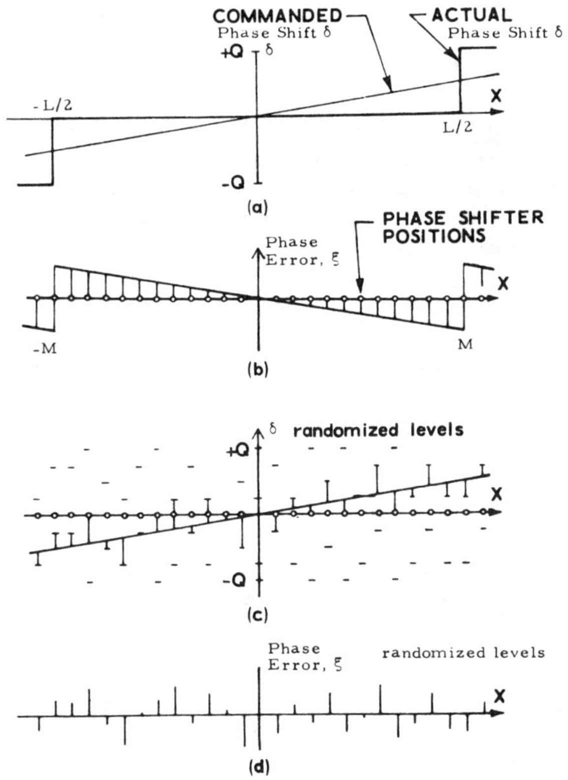

Q = PHASE OF LOWEST PHASE QUANTIZATION BIT

$= \frac{2\pi}{2^P}$

N = 2M + 1

Fig. 3.30 Randomization of phase quantization errors. (After Sahmel and Manasse, Aerospace Corp. [15]; also [14].)

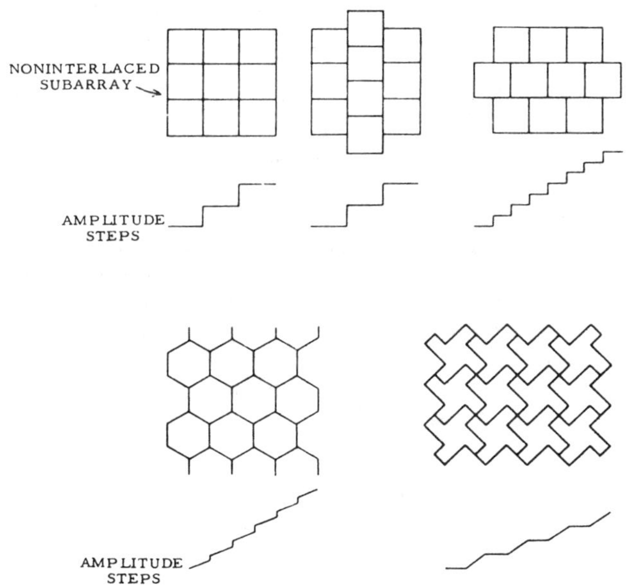

Fig. 3.32 Interlaced subarrays. (After Cheston and Frank in Skolnik, McGraw-Hill Book Comp. [6].)

θ_o and $\theta_1 << 1$ rad.

The peak sidelobe level obtained with a periodic element-to-element phase ripple error is to be compared with the peak sidelobe obtained when the phase shift quantization error is randomized as was assumed for Fig. 3-12. To make this comparison assume a linear array of isotropic elements as was done for Fig. 3-12. Then for the case of element-to-element phase ripple, a 4-bit phase shifter results in a peak sidelobe level of −20.2 dB for the situation depicted in Figs. 3-28a or 3-28b. For the case where the phase shifter quantization error is randomized, Fig. 3-12 and (3-8) indicate that for a 4-bit phase shifter the average sidelobe level would be −57.3 dB for a 30 dB sidelobe Tapered-Taylor 7800 element array (the example array used in Fig. 3-7). The peak sidelobe level due to a randomized phase quantization error should be about 10 dB larger (see Fig. 2-53) resulting in a value of −47.3 dB for the peak sidelobe level. Thus, if the phase quantization errors are randomized, the peak sidelobe level is about 27 dB lower than obtained when the phase quantization error is periodic.

The question that remains is: how does one randomize the phase quantization error? One way to achieve randomization is simply to add cables of known lengths in the path of each radiating element in series with its phase shifter.[11] These cables are to be of such lengths as to introduce values of phase shifts which are random between 0 and the phase shift of the lowest bit, which is $2\pi/2^P$ radians. In order to produce the linear phase gradiant across the aperture the phase shifters now have to subtract out the phase shifts introduced by the cables. This procedure results in random element-to-element phase errors as illustrated in Fig. 3-30.

Space feeding a phased array automatically introduces a randomization of the phase quantization error.[5] Fig. 3-31 depicts a space fed array. For such a system the signals arriving at the array from the feedhorn do not arrive in phase. This results from the fact that the distance from the feedhorn to different points on the array face differ. Specifically, the distance from the feedhorn to the center element of the array is much shorter than it is to the outer elements of the array. As an example assume an f number for the array of 1.6, that is, $F/2r_o$ of Fig. 3-31 equals 1.6, with $2r_o=50\lambda$. Then the distance to the outer elements of the array from the feedhorn is larger than it is to the center elements of the array by 3.82 wavelengths. The phase shifters of the array must correct for the fact that the signals arriving at different points on the array are not in phase. This phase correction is known as space feed collimation.[12] That space feed collimation phase correction results in a randomization of a phase quantization error has been demonstrated on the HAPDAR radar.[13]

ii. Amplitude Quantization Errors.

Amplitude quantization errors occur when the desired amplitude illumination taper over the aperture is approximated by a quantized version of the weighting through the use of subarrays. To achieve weighting on transmit each of these subarrays could have a different number of radiating elements with each subarray being fed by a single power amplifier. For a system which only uses amplitude tapering on receive, each subarray could have the same number of radiating elements with the output of each subarray being fed into a paramp followed by a suitable weighting network (as was the case of the example of Figs. 3-24 and 3-25). As mentioned previously the resulting amplitude quantization error can be randomized by the use of randomly shaped subarrays having centers which are randomly located. A technique for the reduction of the magnitude of the peak quantization error for the case where identical subarray shapes are used involves the use of subarray interlacing; (see Fig. 3-32).[6]

3.1.2 Angle Measurement Accuracy.[14],[15],[16]

A. Nonrandom Phase Errors. Fig. 3-30a shows the situation where one has the maximum angle measurement pointing error for the antenna system when no randomization of the quantization bits are used. It depicts the situation of maximum off broadside main beam pointing before the lowest significant phase shift bit is about to be shifted in for the antenna element at the edge of the antenna. It gives the maximum phase gradient one can have across the antenna aperture without any phase shifters being shifted in. For a 50 x 50 element rectangular antenna array using three bit phase shifters this maximum error at broadside is 5.1 ms; see Fig. 3-33. (The error is given here in terms of millisines, ms, which represents the error in $\sin\theta_o$ instead of θ_o. Thus, for this example the error in $\sin\theta_o$ equals 5.1×10^{-3}. For small errors, errors less than or equal to 6°, the error in ms approximately equals the error in milliradians, mr.) Although derived for broadside conditions, this maximum boresight angle error can occur at scans angles other than broadside.

B. Random Phase Errors. Fig. 3-30 also illustrates the situation where the phase error is randomized.[14],[15] This randomization could be achieved by the use of random cable lengths in series with each phase shifter.[11] For the case of randomized phase quantization errors the mean square boresight angle error is given by (3-25a) and (3-25b) of Fig. 3-34.[14],[15] The angles θ_{ox} and θ_{oy} correspond to the sine space angles of Fig. 2-32. As mentioned previously, these angles are referred to often* in the literature as α and β, respectively, and will be referred to as such in the remainder of this lesson. Note that the boresight angle measurement error is not dependent on the sum pattern weighting function or on any errors in the difference or sum pattern amplitude illumination functions, the boresight angle measurement error only depending on the errors in the phase shifts introduced for the monopulse excitation. For the special case of a rectangular antenna with a stepped-uniform monopulse weighting [see (3-26)] the boresight angle measurement error is given by (3-27).[16]

Using the example of Fig. 3-33 of a 50 x 50 rectangular array having a 3-bit phase shifter (3-27) yields an rms boresight angle measurement error of 0.116 ms at broadside. The peak error is about 10 times larger or ~1.2ms. This peak error is about 1/4th of the peak error obtained when no randomization of the phase shifter quantization error is used. [It is interesting to point out that the error for the case of no randomization is known as a function of beam pointing, (see Fig. 3-37). Consequently, although large, because it is known it can be subtracted out. Hence, the boresight angle measurement error even without randomization can be made comparable to that with randomization in principle. Nevertheless, randomization would be used generally to avoid high sidelobes].

---

*Although unfortunately not universally as is apparent from the other usages adopted for α and β by the other lecturers; see Figs. 7-6 and 8-45 and definitions at beginning of lecture notes.

FIG. 3.30a DEPICTS THE SITUATION WHICH GIVES THE WORST POINTING ERROR. THE MAXIMUM POINTING ERROR SHALL BE EXPRESSED IN TERMS OF THE MAXIMUM ERROR IN $\sin\theta_o$.

LET $\Delta(\sin\theta_o)_{max.}$ = MAXIMUM ERROR IN $\sin\theta_o$.

THE VALUE OF $\sin\theta_o$ FOR THE SITUATION ILLUSTRATED IN FIG. 3.30a GIVES $\Delta(\sin\theta_o)_{max.}$ RECALL FROM (2.16) THAT:

$$\sin\theta_o = \frac{\lambda\alpha}{2\pi d} \tag{3.22}$$

FROM FIG. 3.30a

$$\alpha = \alpha_{max.} = \frac{Q}{N-1} \doteq \frac{Q}{N} = \frac{2\pi}{N2^P} \tag{3.23}$$

SUBSTITUTING (3.23) IN (3.22) YIELDS

$$\Delta(\sin\theta_o) = \frac{\lambda\alpha_{max.}}{2\pi d} \doteq \frac{\lambda}{L2^P} \tag{3.24}$$

EXAMPLE GIVEN: 50 BY 50 ELEMENT SQUARE ARRAY

$d = \lambda/2$

P = 3 BITS

FROM: (3.24) $\Delta(\sin\theta_o) = 5.1$ ms

Fig. 3.33 Equations of this figure apply for linear array. For rectangular array N should be replaced by N_x or N_y depending on whether error in $\sin\theta_{ox}$ or $\sin\theta_{oy}$ of Fig. 2.32 is being calculated. Similarly, replace L by L_x or L_y, θ_o by θ_{ox} or θ_{oy}, $\Delta(\sin\theta_o)$ by $\Delta(\sin\theta_{ox})$ or $\Delta(\sin\theta_{oy})$.

VARIANCE OF ERROR IN $u = \sin\theta_{ox} = \sin\alpha$:

$$\sigma_u^2 = \frac{\sigma_\xi^2 \sum_{m,n} \Delta_{xmn}^2}{k^2\left(\sum_{m,n} \Delta_{xmn} x_{mn}\right)^2} \tag{3.25a}$$

VARIANCE OF ERROR IN $v = \sin\theta_{oy} = \sin\beta$:

$$\sigma_v^2 = \frac{\sigma_\xi^2 \sum_{m,n} \Delta_{ymn}^2}{k^2\left(\sum_{m,n} \Delta_{ymn} y_{mn}\right)^2} \tag{3.25b}$$

WHERE:

σ_ξ^2 = variance of difference pattern phase error of a radiating element (assumed the same for all radiating elements)

$\Delta_{xmn} = \Delta_x(x_{mn}, y_{mn})$

$\Delta_{ymn} = \Delta_y(x_{mn}, y_{mn})$

$\Delta_x(x,y)$ = x monopulse difference pattern amplitude weighting (illumination) function at coordinates x, y

$\Delta_y(x,y)$ = y monopulse difference pattern amplitude weighting (illumination) function at coordinates x, y

k = propagation constant $= \frac{2\pi}{\lambda}$.

Fig. 3.34 Boresight errors for randomized phase quantization errors. (After Sahmel and Manasse, Aerospace Corp. [15]; also [14].)

RECTANGULAR ANTENNA WITH STEPPED-UNIFORM MONOPULSE WEIGHTING THAT IS:

$$\Delta_x(x,y) = -1 \text{ for } x \le 0, \quad = +1 \text{ for } x > 0 \tag{3.26}$$

STANDARD DEVIATION OF ERROR IN $\theta_{ox} = \alpha$ GIVEN BY:

$$\sigma_{\theta_{ox}} = \sigma_\alpha = \frac{0.64\,\sigma_\xi}{\sqrt{N}} \; \frac{\lambda}{L_x \cos\alpha} \tag{3.27}$$

WHERE L_x = ARRAY DIMENSION ALONG X AXIS.

N = NUMBER OF ELEMENTS IN ARRAY (WHEN PHASE ERRORS ARE INDEPENDENT FROM ELEMENT-TO-ELEMENT; OTHERWISE EQUALS NUMBER OF INDEPENDENT PHASE ERRORS ACROSS THE ARRAY).

EXAMPLE: SAME AS FOR Fig. 3.33, I.E.,

50 BY 50 ELEMENT SQUARE ARRAY

$d = \lambda/2$

P = 3 BITS

$\theta_{ox} = \alpha = 0$ (BROADSIDE MAIN BEAM)

FOR RANDOMIZED PHASE QUANTIZATION ERRORS, FROM (3.7)

$$\sigma_\xi = \sqrt{\overline{\delta_Q^2}} = \sqrt{\frac{\pi^2}{3 \cdot 2^{2P}}} = \frac{\pi}{\sqrt{3}\;2^P}$$

THUS FROM (3.27)

$$\sigma_{\theta_{ox}} = \sigma_\alpha = 0.116\text{ms}$$

Fig. 3.35 Boresight error for randomized phase quantization errors: special case.

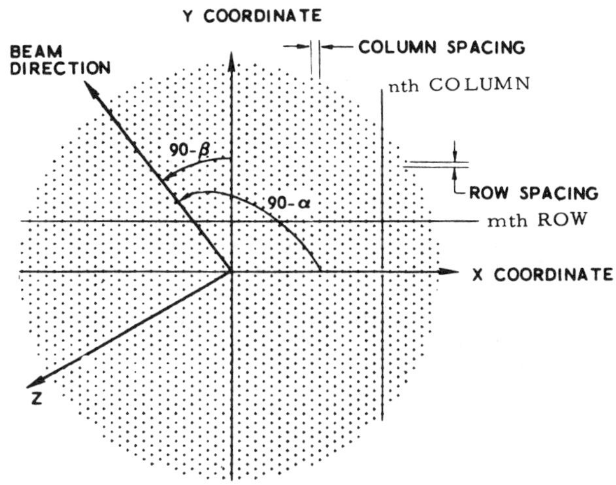

(a) CIRCULAR ANTENNA ARRAY AND ITS COORDINATE SYSTEM

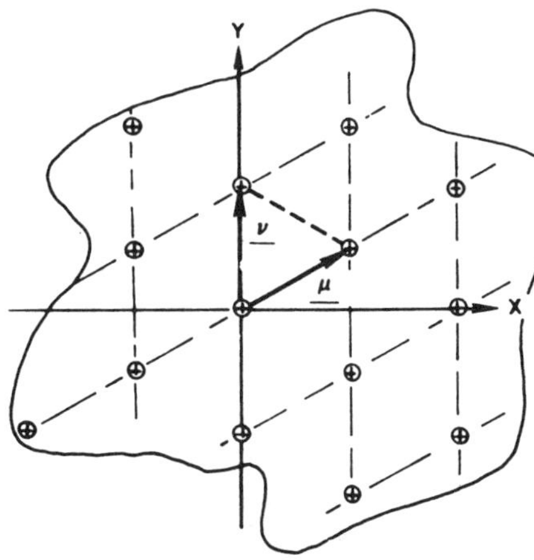

(b) EXPANDED VIEW NEAR ORIGIN SHOWING EQUILATERAL TRIANGLE LATTICE

Fig. 3.36 Simulation example antenna array parameters. (After Sahmel and Manasse, Aerospace Corp. [15]; also [14].)

(c) PARAMETERS:

d_x	(column spacing)	=	0.500 λ
d_y	(row spacing)	=	0.289 λ
μ_x	(lattice parameter)*	=	0.500 λ
μ_y	(lattice parameter)*	=	0.289 λ
ν_x	(lattice parameter)#	=	0.000λ
ν_y	(lattice parameter)#	=	0.578 λ
$2r_o = D$	(aperture diameter)	=	26.4 λ
r_o	(aperture radius)	=	D/2 = 13.2 λ
N	(number of elements)	=	1891
Δ_x	(difference pattern amplitude weighting function)	=	sin πx/R
Δ_y	(difference pattern amplitude weighting function)	=	sin πy/R
F	(focal length in the space feed mode)	=	D
P	(no. of phase shifter bits)	=	3
θ_{30}	boresight beamwidth	=	2.6°

*μ_x and μ_y are projections of $\underline{\mu}$ of Fig. 3.36b on x and y coordinates, respectively.

#ν_x and ν_y are projections of $\underline{\nu}$ of Fig. 3.36b on x and y coordinates, respectively.

Fig. 3.36 (con't.) Simulation example antenna array parameters (Cont.). (After Sahmel and Manasse, Aerospace [15]; also [14].)

PRINCIPAL PLANE CUT ($\beta = 0°$; SEE Fig. 3.36a

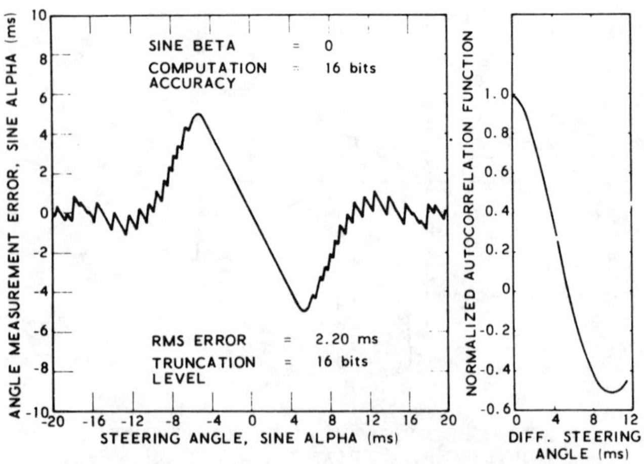

Fig. 3.37 Boresight error for non-randomized phase quantization errors. (After Sahmel and Manasse, Aerospace Corp. [15]; also [14].)

PRINCIPAL PLANE CUT ($\beta = 0°$; SEE Fig. 3.36a

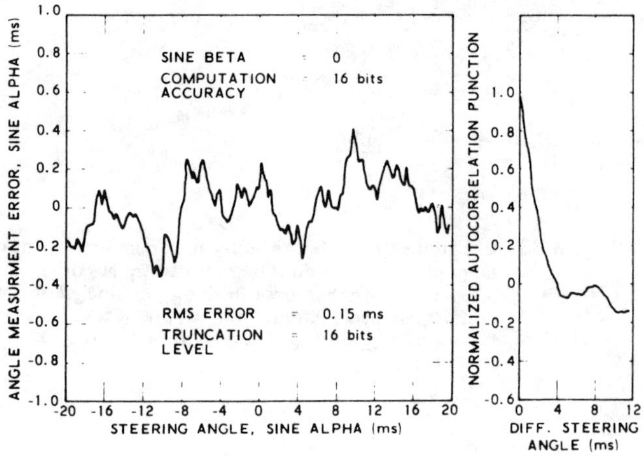

Fig. 3.38 Boresight error for randomized phase quantization errors. (After Sahmel and Manasse, Aerospace Corp. [15]; also [14].)

SPACE FEED COLLIMATION

NUMBER OF BITS r TO WHICH ROW AND COLUMN COMPUTED PHASE SHIFTS ARE TRUNCATED TO*	BORESIGHT ACCURACY AT BROADSIDE IN PRINCIPAL PLANE, $\beta = 0$ (MS)
16	0.149
6	0.23
4	0.45

*COMPUTATION DONE TO 16 BITS BEFORE TRUNCATION; I.E., REQUIRED ROW STEERING AND COLLIMATION COMPUTED TO 16-BIT ACCURACY, THEN ADDED WITH 16-BIT ACCURACY, AND FINALLY TRUNCATED TO r BITS. SAME DONE FOR COLUMN PHASE SHIFT.

Fig. 3.39 Effects of computer truncation errors on boresight error.

C. Simulation Results. [14], [15] A simulation had been carried out to determine the effect on the boresight angle measurement errors of 1) randomization of the phase shift quantization errors, 2) space feed collimation, 3) computer truncation errors, and 4) differential phase shift errors. [14], [15] This simulation was carried out on the antenna whose parameters are given in Fig. 3-36. [Note the whereas before a rectangular lattice arrangement was used for the radiating elements (see Figs. 2-31 and 2-47) for this example a triangular lattice arrangement is assumed]. The parameters for the circular array antenna used in the simulation are similar to those given in Fig. 3-33 for the rectangular aperture, hence, a comparison of the results can be made.

Fig. 3-37 shows the results of the computer simulation obtained for the case where no randomization of the phase shifter quantization error is used. The figure indicates that large angle measurement errors are obtained. These peak angle measurement errors are called quantization lobes because they are due to the system quantization phase error. As expected, the peak of these quantization lobes is about ±5 ms which compares well with the value of 5.1 ms computed in the example of Fig. 3-33. Furthermore, these peak angle measurement errors occur when the main beam is pointed off broadside by about ±5 ms, just at the point where the phase shifter lowest bit is about to be shifted in at the edge of the antenna as discussed for the example of Fig. 3-33. The rms angle measurement error is 2.2 ms. The angle measurement error quantization lobes given in Fig. 3-37 repeat themselves several times in the visible space.

Also shown in Fig. 3-37 is the autocorrelation function of the angle measurement error. The 3 dB width of this autocorrelation function is about 4.6 ms.

In obtaining the results of Fig. 3-37, the row and column computations performed in the computer were done to 16 bit accuracy, then added with a 16 bit accuracy to provide the required phase shift for the mn th radiating element before finally being truncated to 3 bit accuracy, the number of bits used for the phase shifters.

When the phase quantization errors were randomized the results given in Fig. 3-38 were obtained. In this case, the rms boresite angle measurement error was found to be 0.15 ms. This value is close to the 0.116 ms value obtained for the rectangular aperture in Fig. 3-35. A calculation using the simulation antenna parameters of Fig. 3-36 yields an rms boresite angle error of 0.1486 ms. The 3 dB width of the autocorrelation function of the measurement angle error is found to be about 2 ms.

The simulation was also carried out for the case where no randomization of the phase shift quantization error was used, but instead space feed collimation was employed. For this case, the rms error was found to be 0.149 ms for the α coordinate and 0.203 ms for the β coordinate. The poorer angle measurement accuracy for the β coordinate. may be due to the fact that either (or combination of both) the simulation is not averaged over a large enough sample or space feed collimation does not completely randomize the phase shift quantization errors.

For the case where space feed collimation was used, consideration was given to the effect of using a smaller number of bits in the computation of a phase shift required for each radiating element. Specifically, consideration was given to calculating the row and column phase shifts required to 16 bit accuracy, then truncating the results obtained to r bits of accuracy, adding the results to obtain the phase shifts for an individual radiating element to r bit accuracy, and finally truncating this number to 3 bit, the accuracy needed by the 3 bit phase shifter. Results were obtained for r = 16, 6 and 4 bits and are summarized in Fig. 3-39. [15] The boresight angle accuracies obtained show the pronounced increase in the error when truncating to r = 4 bits. This large error results from the truncation to a number of bits nearly equal to the number of phase shifter levels. When this is done the phase shift errors for the various radiating elements are no longer independent of each other. A strong row and column correlation exists due to the truncation process. As a result, the number of independent phase errors reduces from N, the total number of radiating elements, to roughly $2\sqrt{N}$. Such a reduction in N for (3-27) results in an increase in the boresight angle measurement error by the factor $\sqrt[4]{N}/\sqrt{2} = \sqrt[4]{1891}/\sqrt{2} = 4.7$ resulting in the original 0.149 ms error increasing to about 0.7 ms. Actually, however, the errors are not completely correlated along the rows and columns because the final 3 bits of truncation introduces some randomization. Hence, the boresight angle measurement angle error is somewhat smaller than 0.7 ms, the value 0.45 ms being obtained in the simulation.

Up to now, the phase shifts introduced by each bit of the phase shifter were assumed to be exact, that is, the manufacturing inaccuracies of each bit phase shifter were not accounted for. This manufacturing error is referred to as the differential phase shift error. (In Fig. 3-10 this error was grouped with the amplifier error.) Generally, the effect of such an error is not severe. Assume a 10° rms differential phase shift error, that is, each bit of the 3 bit phase shifter has a manufacturing error of 10° rms with this error being different and independent from bit to bit. Also assume that randomized phase shift quantization is used. Then the simulation indicates that the rms boresight angle error increases from 0.15 ms to about 0.195 ms due to the presence of a 10° rms differential phase shift error. This is about as expected. A 10° rms error per bit results in the rms phase error per bit increasing by $\sqrt{3/2}\,(10°) = 12.25°$. [Each bit is present half the time on the average, thus the mean square differential phase error contribution, $\sigma_{\xi d}^2$, to σ_ξ^2 of (3-25a) and (3-25b) from the three bit phase shifter is $\sigma_{\xi d}^2 = \frac{1}{2} \cdot 3(10°)^2 = (12.25°)^2$]. The contribution to σ_ξ from the randomized bit error is from (3-7) 13° rms for a 3 bit phase shifter. Thus σ_ξ increases from 13° when no differential phase shift error is present to 17.85° when it is present, an increase of a factor of 1.37. The boresight error should increase by the same amount, from 0.15 ms to 0.206 ms which is in fair agreement with the value of 0.195 ms obtained in the simulation.

Fig. 3-40 shows the effect on the angle boresight error of the use of different monopulse difference amplitude weighting patterns for the antenna array system of Fig. 3-36. The results are given for the assumption of randomized phase quantization errors. Eqs. (3-25a) and (3-25b) were used to obtain the results given.

D. Decorrelation of Boresight Errors Due to Phase Quantization Errors. [14], [15] For the computer simulation example, as mentioned before, the autocorrelation function of the boresight error is about 2ms for the case of randomized phase shift quantization errors and about the same for the case of space feed collimation. Thus the target angle must change by a few millisines before an independent phase error measurement due to phase quantization errors

ASSUMPTIONS:

RANDOMIZED PHASE SHIFT QUANTIZATION ERRORS.

PRINCIPAL PLANE CUT: $\beta = 0^\circ$ FOR σ_u

$\alpha = 0^\circ$ FOR σ_v

ALL OTHER ASSUMPTIONS SAME AS IN Fig. 3.36

WEIGHTING FUNCTION		ONE-SIGMA ERROR
SINE	$\sin \frac{\pi x}{r_o}$	$\sigma_u = 0.1486$ ms $\sigma_v = 0.1485$ ms
LINEAR	$\left(\frac{x}{r_o}\right)$	$\sigma_u = 0.1258$ ms $\sigma_v = 0.1258$ ms
CUBIC	$\frac{x}{r_o}\left[1 - \left(\frac{x}{r_o}\right)^2\right]$	$\sigma_u = 0.1406$ ms $\sigma_v = 0.1404$ ms
TRUNCATED RAYLEIGH	$\frac{x}{r_o} \exp - \left(\frac{2x}{r_o}\right)^2$	$\sigma_u = 0.1775$ ms $\sigma_v = 0.1775$ ms
UNIFORM	$\begin{cases} 1, & x > 0 \\ -1, & x < 0 \end{cases}$	$\sigma_u = 0.1464$ ms $\sigma_v = 0.1482$ ms

Fig. 3.40 Effect of monopulse difference amplitude pattern on boresight error. (After Sahmel and Manasse, Aerospace Corp. [15].)

is obtained. For many situations of interest the target angle will not change by more than 2 or more ms between looks.

Yet it is desirable to have this error independent from look-to-look in order to further reduce (by look-to-look smoothing) the boresight angle error due to phase quantization errors. An application where such a reduction is desirable is that of command guidance intercept. A number of ways are available for achieving this independence from look-to-look. They are now briefly described.

Phase Dithering

By simply adding the same arbitrary phase (0 to 2π) to the phase commands for all the array elements, the beam steering command angle is not changed; however, the phase quantization errors of Fig. 3-30 are changed because the individual phase shifts specified for each array element have changed. Consequently, the boresight error will be different every time a new arbitrary phase shift is used for any observation of the same target at the same location. That is, the boresight error will be independent from look-to-look as a result of adding a new arbitrary phase to the phase shifters between looks.

The addition of different random phases between looks is not always desirable. For example when phase coherent clutter rejection techniques are used. In this case second time around (or higher) returns would not be rejected.

Frequency Dithering

Frequency dithering involves changing the radiated frequency from look-to-look by some specified amount, such as, 5-10%. For frequency dithering to work it is necessary that there be a frequency dependent phase shift for each radiating element, such as, due to the cable length from the transmitter to the radiating element in a corporate feed network or due to the different space paths for a space fed array. In calculating the phase commands for each array element, the computer must supply a phase correction which compensates for the frequency dependent phase path delays seen by the different array elements. When a change in frequency is made, the correction required changes. Hence, a randomization is in effect achieved equivalent to that of adding a random phase shift as was done with phase dithering.

For the space fed array example of Fig. 3-36, an 8% change in the carrier frequency results in most of the phase shifters in the array changing by Q/2, that is, by an amount equal to one-half the phase shift of the least significant bit's phase shift.

When using frequency dithering, pulse-to-pulse coherence techniques such as MTI and pulse doppler processing for clutter rejection may not be useable at all.

Beam Dithering.

The autocorrelation function of the angle boresight error was about 2 ms for the case where randomized quantization phase shift errors was used and the case where space feed collimation was used. Consequently, a natural technique for decorrelation is to steer the beam slightly off boresight a different amount between looks. For example, dithering the beam uniformly between ±5 ms (about ±10% of the beam width) produces effective decorrelation of the boresight angle error due to phase quantization errors for the antenna system of Fig. 3-36.

E. Thermal Noise Errors. The standard equation for determining the target angle measurement error due to thermal noise in a monopulse boresight tracking system is [17]

$$\frac{\sigma_\theta}{\theta_3} = \frac{1}{k_m \sqrt{2n(S/N)_1}} \tag{3-28}$$

where

σ_θ = rms error in measurement of θ

$(S/N)_1$ = power signal-to-noise ratio per pulse at output of IF sum channel

n = number of pulses integrated.

θ_3 = 3 dB beamwidth

k_m = normalized monopulse slope

$$= \left[\left. \frac{\partial(\Delta/\Sigma)}{\partial(\theta/\theta_3)} \right|_{\theta=0} \right] (\sigma_\Sigma/\sigma_\Delta) \tag{3-28a}$$

Δ = magnitude of signal voltage out of IF difference channel

Σ = magnitude of signal voltage out of IF sum channel

θ = target off boresight angle error

σ_Δ = rms noise out of IF difference channel.

σ_Σ = rms noise out of IF sum channel.

The first factor in (3-28a) for k_m gives the system angular sensitivity, i.e., the sensitivity of the signal error voltage Δ/Σ to the target off boresight angle error. The factor $\sigma_\Sigma/\sigma_\Delta$ accounts for the use of the sum instead of the difference channel power signal-to-noise ration in (3-28).

For an antenna using a Tapered-Taylor sum pattern and Tapered-Bayliss difference pattern, typically

$$\frac{\partial(\Delta/\Sigma)}{\partial(\theta/\theta_3)} = 1.1$$

$$\sigma_\Sigma/\sigma_\Delta = 1.4$$

so that

$$k_m = 1.54 \quad .$$

Assume the antenna of Fig. 3-36. Then $\theta_3 = 2.6° = 45.4$ mr at boresight and for $(S/N)_1$ = 16dB and n = 10, (3-28) gives

$$\sigma_\theta = 1.04 \text{ mr.} = 1.04 \text{ ms.}$$

The above angle error due to thermal noise is considerably bigger than the randomized quantization error of 0.15 ms of Fig. 3-38.

3.2 Detection Considerations for Phased Arrays.

It is the purpose of this lesson to just give some exposure to the subjects of beam shape loss, beam packing loss, optimum beam spacing, and sequential detection.

3.2.1 Beam Shape Loss, Beam Packing Loss, and Optimum Beam Spacing.[18]

Fig. 3-41 shows two possible packings (rectangular and triangular lattice packings) of the beam for searching over the volume of space to be covered by the phased array. In Fig. 3-41, the beams

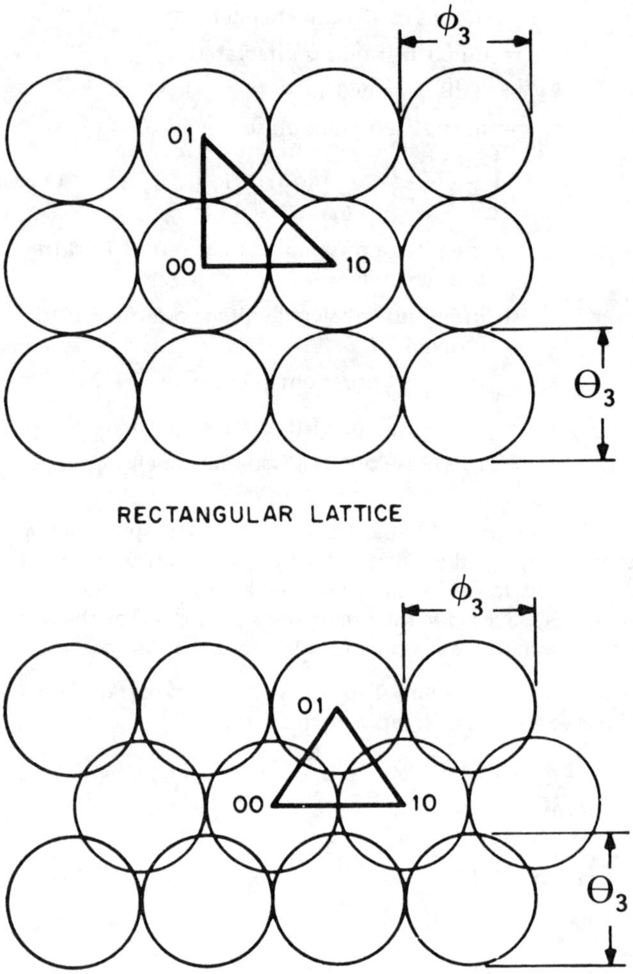

Fig. 3.41 A geometrical representation of beam positions. Circles represent the 6 dB two-way beam contours. (After Hahn and Gross, IEEE AES [1].)

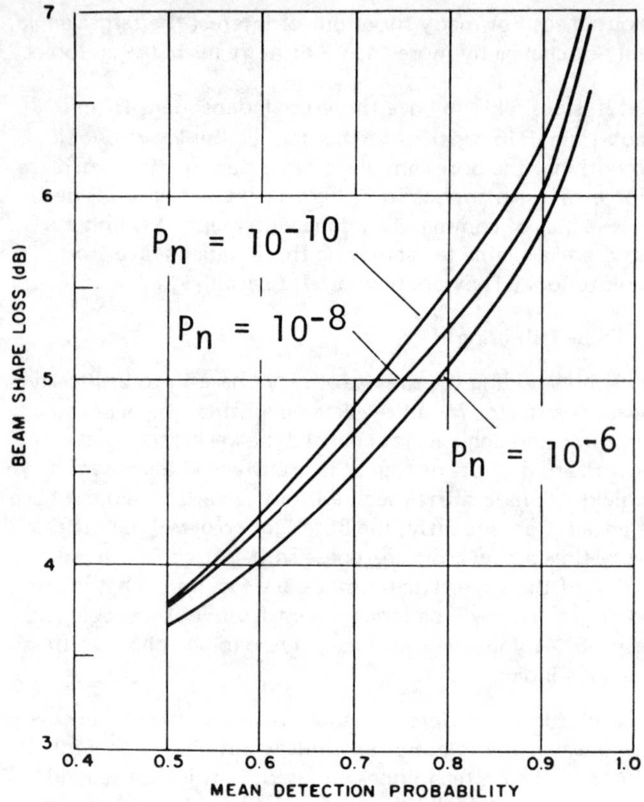

Fig. 3.42 Beam shape loss versus mean detection probability for Gaussian beam shape on a rectangular lattice with 6 dB contours touching, one pulse per beam position and steady target type. P_n= false alarm probability. (After Hahn and Gross, IEEE AES [1].)

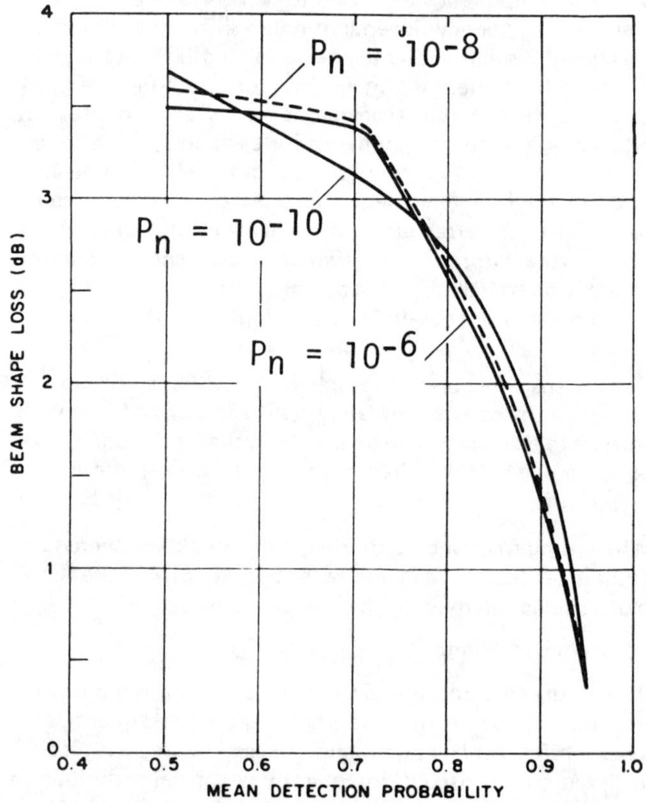

Fig. 3.43 Beam shape loss versus mean detection probability for Gaussian beam shape on a rectangular lattice with 6 dB contours touching, one pulse per beam position and Swerling case I/II target. P_n= false alarm probability. (After Hahn and Gross, IEEE AES [1].)

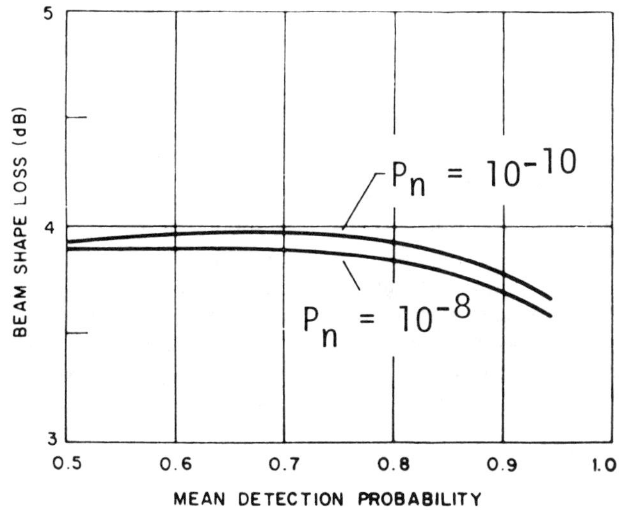

Fig. 3.44 Beam shape loss versus mean detection probability for Gaussian beam shape on a rectangular lattice with 6 dB contours touching, one pulse per beam position, and Swerling case III/IV target. P_n = false alarm probability. (After Hahn and Gross, IEEE AES [1].)

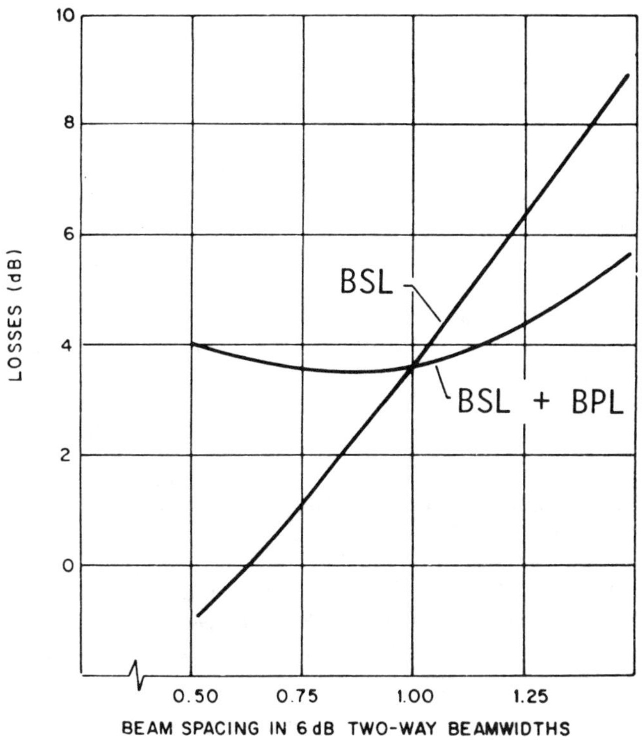

Fig. 3.46 Beam losses versus beam spacing for Gaussian beam shape on a rectangular lattice with one pulse per beam position, Swerling Case I/II target, mean detection probability = 0.5, and false alarm probability = 10^{-8}. (After Hahn and Gross, IEEE AES [1].)

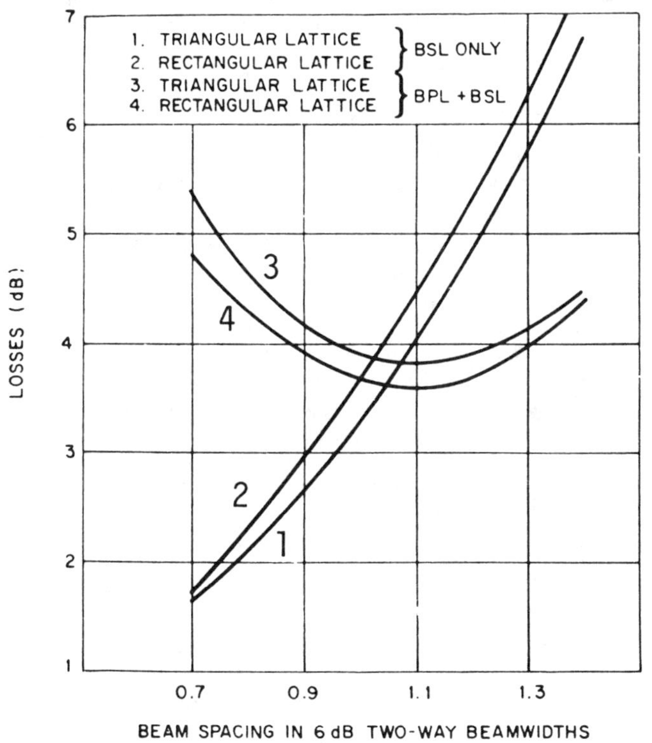

Fig. 3.45 Beam losses versus beam spacing for Gaussian beam shape with one pulse per beam position, steady target type, mean detection probability = 0.5, and false alarm probability = 10^{-8}. (After Hahn and Gross, IEEE AES [1].)

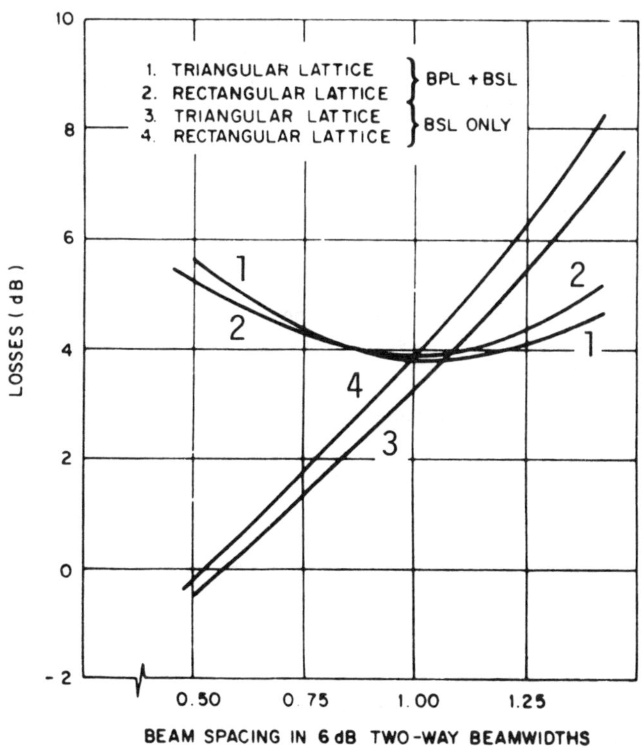

Fig. 3.47 Beam losses versus beam spacing for Gaussian beam shape with one pulse per beam position, Swerling case III/IV target, mean detection probability = 0.5, and false alarm probability = 10^{-8}. (After Hahn and Gross, IEEE AES [1].)

are arranged so that their horizontal center-to-center spacing is equal to the 6 dB 2-way beamwidths. Note that the beams are tangent at the 6 dB 2-way contours. Assume a Gaussian shaped beam pattern. The probability of detecting a target of a specified cross section will depend on where the target is relative to the peak of the beam, or more precisely, relative to where it lies in one of the triangles shown. One need only consider the possible locations in one of the triangles because the beam patterns are repetitive. The average probability of detection over all possible target locations in one of these triangular regions is denoted as the mean detection probability (MDP). The beam shape loss (BSL) is defined as the ratio of the power required to achieve a specified MDP when beam shape and beam overlap are taken into account to the beam power required when ideal non-overlapping rectangular beams are used.

Figs. 3-42 to 3-44 show the BSL for the rectangular lattice case of Fig. 3-41 for various target models of interest as a function of the MDP and the false alarm probability. For MDP = 0.5, the BSL's are within 0.5 dB of each other for all the target models considered. However, as MDP is increased from 0.5 to 0.95, the BSL's vary quite differently for the various target models considered.

As the beams are put closer together, the BSL decreases. However, more beam positions are needed to scan out the search volume and hence more total energy is needed for search. This increase in the energy required for search is a loss factor designated as the beam packing loss (BPL). It is assumed that the BPL is 0 for the situations depicted in Fig. 3-41. Any decrease or increase in the total power required to search out the volume due to the use of a different packing arrangement results in a minus or positive BPL, respectively. Fig. 3-45, 3-46, and 3-47 show the BSL and the sum of BPL and BSL versus the spacing between the beams for both rectangular and triangular lattices. The sum curve shows a broad minimum near the situations depicted in Fig. 3-41, that is, for the cases where the beams are spaced apart by the 6 dB 2-way beamwidth. Note that neither the rectangular or triangular lattice have a great advantage over one another.

Although scan loss was not considered in the above analysis, the results and conclusions drawn should be applicable in general.

Simple procedures for calculating the mean detection probabilities and required signal-to-noise ratios in the presence of BSL and BPL are given in Ref. [4].

3.2.2 Sequential Detection. Mechanically scanning antennas are generally committed in advance to a specific fixed amount of energy per beam position for determining the presence or absence of a target at any and all regions of space. A more efficient detection procedure would involve the radiation initially of a small amount of energy at a beam location. This energy is used to determine if it is likely that a target is present at that particular beam location. Only if the low energy probing signal indicates that a target is likely to be present is more energy radiated at a given beam position. Such sequential detection schemes are feasible with electronically scanned phase arrays.

The initial low energy probing could indicate one of three things: (1) a target is present (this would be the case if the target is very large and the received signal exceeds a very high threshold level set a priori); (2) a target is not present (the case where the received signal is very weak or else noise alone is present, so that a very low threshold level is not exceded); and (3) it is not certain whether the target is present (the received signal voltage is for this case between the low and high level settings) and more energy is needed to decide the presence or absence of the target. For cases (1) and (2) no more energy than the initial small probing energy is needed to decide the presence or absence of the target. Hence a savings in search power required is realized. Only for case (3) is more energy used in order to decide if the signal is present.

There are many forms of sequential detection schemes.[19], [20] Usually they are designed to come to a decision after the transmissions of the second pulse for case (3). This simple two-stage sequential detector can provide savings of from 2 to 4 dB in search energy typically. The savings depends on the number of range bins being searched in one beam position. The larger the number of bins the smaller the savings — about 2 dB being realized typically when there are 1000 range cells and about 4 dB when there are about 10 cells.[19]

3.2.3 Acknowledgement. The author is indebted to Fritz Steudel of the Raytheon Company for educating him on the simple physical-spatial-frequency and sum-of-random-vectors approaches for dealing with array errors.

The author has benefited from discussions with a number of other people at the Raytheon Company. These are: Fernando Beltran, James Howell, and Charles Winter.

Finally, the author is grateful to the Raytheon Company management for supporting participation in this seminar.

3.3 References.

[1] Skolnik, M. I., "Nonuniform Arrays", in R. E. Collin and F. J. Zucker (eds.) *Antenna Theory, Part 1*, McGraw-Hill Book Comp., 1969.

[2] Ruze, J., "Physical Limitations on Antennas", MIT Research Lab. Electronics Tech. Rept. 248, October 30, 1952.

[3] Rice, S. O., "Mathematical Analysis of Random Noise", *Bell System Tech. J.*, Vol. 23, pp. 282 - 332, 1944 and Vol. 24, pp. 46 - 156, 1945; also N. Wax (ed.), *Selected Papers on Noise and Stochastic Processes*, Dover Pub., 1954.

[4] Brookner, Eli, "How to Look Like a Genius in Detection Without Really Trying", *IEEE Northeast Electronics Research and Engineering Meeting (NEREM)*, Part 4, "Radar Systems and Components", pp. 37 - 64, October 28 - 31, 1974.

[5] Miller, C. J., "Minimizing the Effects of Phase Quantization Errors in an Electronically Scanned Array", *Proc. 1964 Symp. Electronically Scanned Array Techniques and Applications*, RADC-TDR-64-225, Vol. 1, pp. 17 - 38 (AD448481).

[6] Cheeston, T. C. and J. Frank, "Array Antennas", in M. I. Skolnik, *Radar Handbook*, McGraw-Hill Book Comp., 1970.

[7] Allen, J. L., "Some Extensions of the Theory of Random Error Effects on Array Patterns", in J. L. Allen *et al.*, *Phased Array Radar Studies*, Tech. Rept. No. 236, Lincoln Laboratory, M.I.T., 1 July, 1960 to 1 July, 1961.

[8] Elliott, R. S., "Mechanical and Electrical Tolerances for Two-dimensional Scanning Antenna Arrays", *IRE Trans.*, Vol. AP-6, pp. 114 - 120, January 1958.

[9] Cook, C. E. and M. Bernfeld, *Radar Signals*, Academic Press, New York 1967.

[10] Farnett, E.C., T.B. Howard and G.H. Stevens, "Pulse-Compression Radar", in M. I. Skolnik (ed.), *Radar Handbook*, McGraw-Hill Book Comp., 1970.

[11] Steudel, F., "Phased Array", U.S. Patent 3,680,109, July 25, 1972.

[12] Hatcher, B. R., "Collimation of Row-and-Column Steered Phased Arrays", *Proc. of the IEEE*, Vol. 56, pp. 1787 - 1968, November 1968.

[13] Kahrilas, P. J., "HAPDAR—An Operational Phased Array Radar", *Proc. of the IEEE*, Vol. 56, No. 11, pp. 1967 - 1975, November 1968.

[14] Sahmel, R. H. and R. Manasse, "Spatial Statistics of Instrument-Limited Angular Measurement Errors in Phased Array Radars", *IEEE Trans.*, Vol. AP-21, No. 4, pp. 524 - 532, July 1973.

[15] Sahmel, R. H., and R. Manasse, "Spatial Statistics of Instrument - Limited Angular Measurement Error in Phased Array Radars", Aerospace Rept. No. TR-0073 (3901-02)-1, AF Rept. No. SAMSO-TR-72-309, The Aerospace Comp., El Segundo, Ca., October 2, 1972.

[16] Nester, W. H., "A Study of Tracking Accuracy in Monopulse Phased Arrays", Vol.

[16] Nester, W. H., "A Study of Tracking Accuracy in Monopulse Phased Arrays", *IRE Trans. on Antennas and Propagation*, Vol. AP-10, pp. 237 - 246, May 1962.

[17] Barton, D. K. and H. R. Ward, *Handbook of Radar Measurements*, Prentice-Hall, Inc., New Jersey, 1969.

[18] Hahn, P. M. and S. D. Gross, "Beam Shape Loss and Surveillance Optimization for Pencil Beam Arrays", *IEEE Trans. on Aerospace and Electronic Systems,* Vol. AES-5, pp. 674-675, July, 1969.

[19] Capers, J. W., "Automatic-Detection Theory", in M. I. Skolnik, "Radar Handbook", McGraw-Hill Book Comp., 1970.

[20] Nathanson, F. E., Radar Design Principles, New York: McGraw-Hill, 1969.

Phased Array Elements - Part 1

Dr. L.R. Lewis
Raytheon Company, Bedford, MA

4.1 General Considerations of Array Elements.

The phased array element is usually defined as consisting of a phase shifter and radiator. The following discussion is primarily concerned with the radiator; the transition from feed and phase shifters to free space. Here the term element is also considered to be the radiator. Extensive discussions of array elements, from which much of these lectures are derived, appear in references [1], [2], and [3].

The important considerations of radiator design may be classified as electrical and non-electrical. Both classifications are ultimately related to performance and involve sets of constraints which will dictate design choices. Those considered as electrical are: impedance matching, grating lobe suppression, polarization control and power capacity. Impedance matching corresponds to maximizing the power radiated, usually considered for some frequency range and scanning sector. Grating lobe suppression means having only one beam in space. This avoids a transmission and receive ambiguity with corresponding gain loss; as power is shared among these radiated beams. Multiple beams arise from exceeding an electrical distance between elements due to lattice spacing and/or phasing by the phase shifters. As a further discriminant a unique specification of polarization is imposed. A polarization selection is used to enhance radar returns or communication link efficiency. Power capacity is usually more closely associated with the phase shifters rather than radiator. However, excessive power levels 1) may cause arcing in very thin element aperture arrays; and 2) may limit choices of suitable element material.

Of equal importance are the so-called non-electrical considerations of element design. Both environment (weather, location, and mission) and cost are major influences in choice of array type and both dictate and limit performance. An array designed to operate in a rainy environment requires special features on the aperture plane to prevent a rain sheet build up; one that can be utilized in a high performance aircraft will require lightweight and low volume, etc.

These days perhaps the most significant design consideration for both array system concept and component choice is cost. The cost of the large planar phased array is nearly directly related to, and consists mainly of, the number of elements. This is primarily because of phase shifter costs. A primary design objective is to utilize the minimum number of phase shifters, either by reduced number of elements, or by "thinning" (load termination of an element rather than active excitation through a phase shifter). As will be proven later, the minimum number of elements consistant with grating lobe suppression for a conical scan sector is achieved using an equilateral triangular grid.

Shown in Fig. 4.2 are two arrays of rectangular waveguides in a triangular grid. The orthogonal orientations of the waveguides will produce spacially orthogonal polarizations. Identical scanning for either array will produce the same beam locations; however, beam polarizations, matching conditions, beam shapes and sidelobe levels will be different. The similarities in the resulting patterns will arise from properties of the lattice, which is called array factor. The differences are associated with the particular radiator and its orientation, and are related to an element factor. For example, in the case of these rectangular waveguide arrays the patterns in the x - z plane for the "isolated" single waveguides for each configuration is considerably different. The patterns for elements without an array environment are called the "isolated" element patterns, and with all neighbors passively terminated are called array element patterns. It is also instructive to note that orthogonal waveguide orientations (i.e. orthogonal polarizations) used in the same lattice requires the use of different waveguide sizes for each choise. Furthermore, the different waveguide size possesses a different waveguide characteristic impedance and aperture impedance, resulting in a different element match. In addition, the dis-similarity of configurations produces different mutual coupling between elements, also effecting the impedance properties.

For a polarization diverse array, square or circular waveguide apertures are commonly used. All combinations of polarizations can be transmitted or received efficiently for these geometries depending on amplitude and relative phase controls of a pair of spatially orthogonal excitations. As an example one might choose these arrays with a single linear and one sense circular polarization, with discrimination against the orthogonal circular and linear.

The choice of square or circular waveguides, or more generally for the case of linear polarization rectangular or circular would probably be governed by the assembly configuration and machining methods. For example, elements mounted into a ground plane would suggest the use of circular holes for machining of the ground plane.

IMPORTANT CONSIDERATIONS IN ELEMENT DESIGN

I. ELECTRICAL

a. IMPEDANCE MATCHING

b. GRATING-LOBE SUPPRESSION

c. POLARIZATION CONTROL

d. POWER CAPACITY

II. NON-ELECTRICAL

a. ENVIRONMENT

b. COST

Figure 4.1

DUAL-POLARIZATION WAVEGUIDE ARRAYS

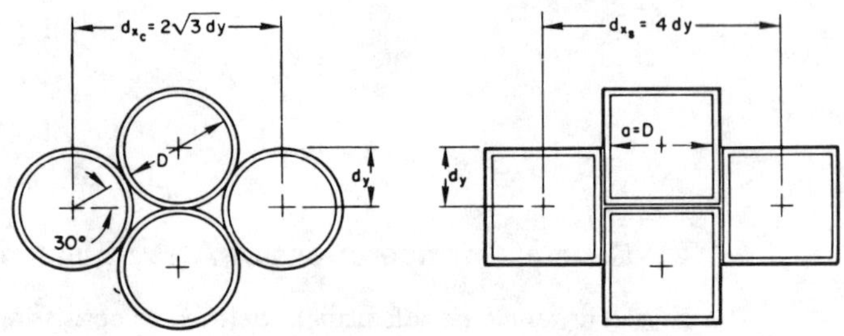

Figure 4.3

RECTANGULAR WAVEGUIDES IN TRIANGULAR LATTICE SINGLE POLARIZATION

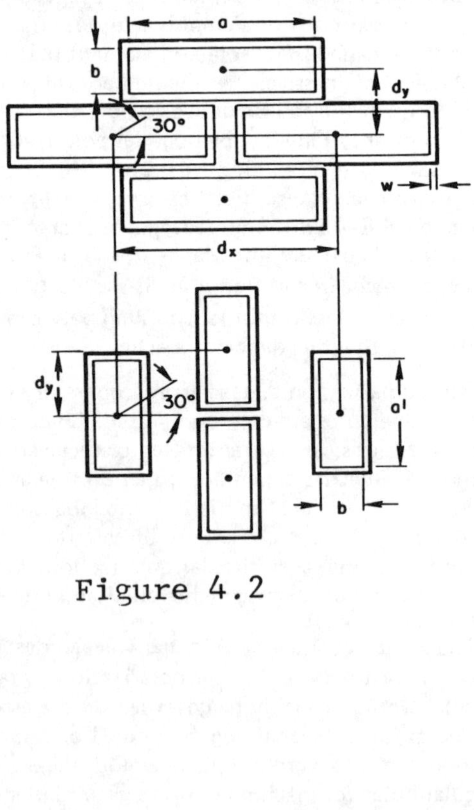

Figure 4.2

DIPOLES WITH BALUNS

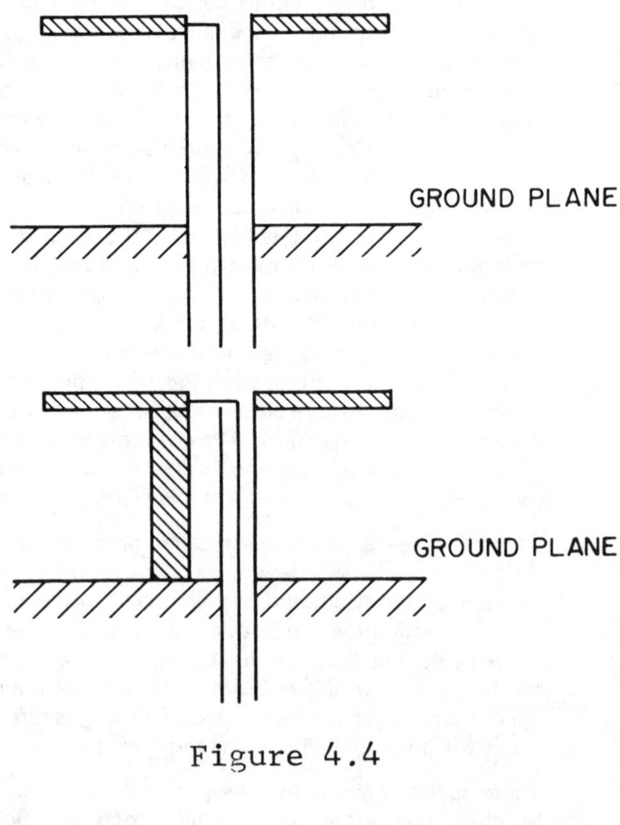

Figure 4.4

In addition to waveguide aperture arrays, dipole elements are also used. Examples of dipole elements are shown in Fig. 4.4. A summary comparison table is given below for the dipole and waveguide arrays:

Dipoles	– low frequencies $<$1 GHz (not strongly dependent on tolerances) – power handling limitation – difficult analysis to include balun effects – good transition from stripline and coaxial feeds
waveguide apertures	– generally used in $>$1 GHz arrays – waveguide radiators natural transition from waveguide cross section phase shifters – analysis very refined and accurate – experimental verification for large array sample – good track record

In addition to dipoles and aperture type elements, traveling wave radiators are also used in phased arrays. Polyrods are used as dielectric rod elements extending from waveguide apertures ($> 1\ \lambda$) with the purpose of increasing element gain or bandwidth. For large arrays no gain benefits would be forthcoming, however, some inexpensive matching can be accomplished with the polyrod antenna. The matching benefits may be offset by resulting "blind" regions of the scan sector, (total reflection regions). This phenomena of blindness is discussed in detail in section B of 4.2.1.

Examples of structures that have demonstrated broadband (octave or better) behavior are arrays of spiral and notch elements. A detailed discussion of a version of the notch element will be presented in Lecture 5.

4.2 The Large Phased Array.

We will separate arrays into the category of large and small. Large arrays are approximated in analysis by an infinite array where impedance properties are readily found. Small array analysis is considerably more difficult, in particular the extent of computations. Moreover, clear-cut design methods for the small array are not as yet available.

For large arrays, linear or planar, the *far* electric field vector $E(\theta,\phi)$ may be written as a product of two terms: an element factor $e(\theta,\phi)$ and scalar array factor $E_a(\theta,\phi)$. The element factor includes those effects associated with the specific choice of element, while the array factor is determined by lattice only. The design of element factor and array factor for the large array may be done independently. This will be a good approximation if there are relatively few edge elements (good demonstrated agreement in element patterns of a planar array in one plane was for a 3 x 39 element array), the grating lobes are not near "radiating space" and no "blindness" (nulls in the element patterns or total reflection condition) occurs within the scan sector.

The design is then carried out by assuming a continuous or discrete aperture distribution for the array factor to predict array patterns and assumes an infinite array environment to calculate or measure element patterns. Specifically, the infinite array possesses a special symmetry for linear phase taper excitations where the field distribution in any cell, (which can reproduce the entire array) is the same as that in any other within a multiplicative phase factor which depends only on the relative cell locations. As a result of this property, the consideration of the infinite array reduces to that of a single cell waveguide with phase-shifting type walls. The field distributions (or modes) for waveguides of this type, resulting from infinite periodic boundries are well known and facilitate the field matching problem required to solve for the element pattern. One example of these properties of periodic structures as applied to phased arrays is given in [4].

As an example of comparing the continuous and discrete distributions for predicting array factor the far field of a uniformly excited 10 element array is compared to that resulting from a corresponding size continuous uniform illumination. The agreement as shown in Fig. 4.7 remains under 1dB out to and beyond the second sidelobe, which in this case is out to nearly 45°. The principal pattern features; namely beamwidth and first sidelobe are nearly identical. However, grating lobes, or spatial harmonics, that result from the periodicity of the array are not predicted. The patterns of the continuous distribution should not be considered for the array case beyond the second sidelobe. The array lattice produces these secondary maximums which must be determined separately.

4.2.1 Array Factor.

There are certain properties of the array factor which can be determined directly from a knowledge of the array lattice and frequency. Perhaps most important is the locations of the grating lobes. Depending on the scan conditions, grating lobes can become radiating beams. There are simple relations for determining main beam and grating lobe locations as a function of lattice and inter-element phasing.

A. Linear Array – Array Factor. Initially an N element linear array is considered. The elements are spaced d apart and operate at a frequency f with wavelength λ. The factor k is the free space wavenumber given as $2\pi/\lambda$. For no scanning (all elements are excited in phase) the array factor is

$$E_a(u) = \sum_{n=0}^{n-1} a_n e^{jnkdu} \tag{4.1}$$

a_n is the amplitude of the nth element, and the parameter u is a generalization of a $\sin\theta$ term that appears from a geometric interpretation of the radiated (far) field, where θ is measured from the array normal.

For a given spacing and frequency d/λ is specified and u is then the only parameter. From the array factor expression we see that beam maximums occur for $u = \dfrac{2p\pi}{kd}$; for p all positive and negative integers including zero. We define "real space" or radiating space as the regime of real angles θ which correspond to the range of u between +1 and −1. Each value of p for which u falls in this range defines a separate radiating beam. The p=0 beam is usually called the main lobe, all other p values are the grating lobes.

For a prescribed linear phase taper array excitation we can write the array factor as

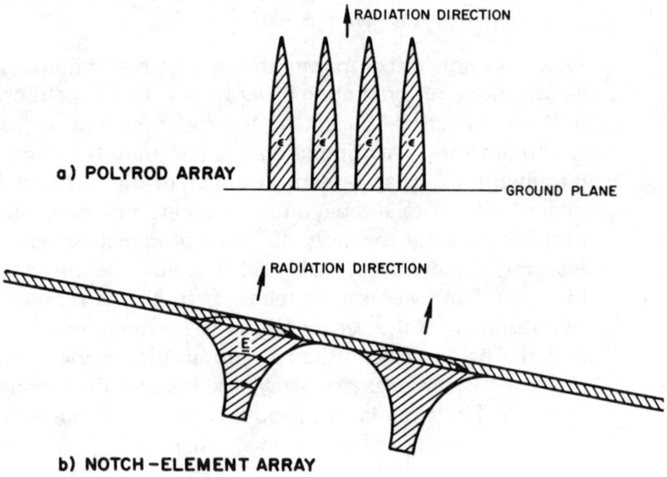

Figure 4.5

DESIGN OF LARGE ARRAY ANTENNAS

FOR LARGE LINEAR OR PLANAR ARRAYS

$$\underline{E}(\Theta,\varphi) = \underline{e}(\Theta,\varphi)\, E_a(\Theta,\varphi)$$

DESIGN $\underline{e}$ AND E_a INDEPENDENTLY

GOOD APPROXIMATION IF:

1. RELATIVELY - FEW EDGE ELEMENTS
2. GRATING LOBES NOT NEAR "RADIATING SPACE"
3. NO BLINDNESS CONDITION WITHIN SCAN SECTOR

$\underline{e}$	E_a
ASSUME INFINITE ARRAY	ASSUME CONTINUOUS APERTURE

Figure 4.6

COMPARISON OF RADIATION PATTERNS

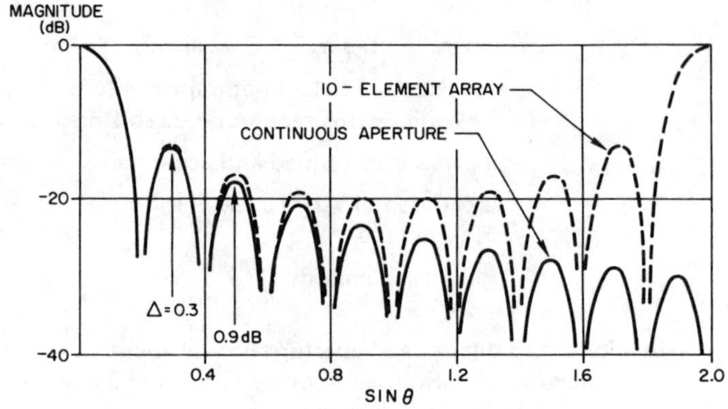

Figure 4.7

LINEAR ARRAY-REAL SPACE

Θ

Y ←— d —→ Y Y . . . . . . Y_{N-1}

n = 0

NON-SCAN ARRAY FACTOR: $E_a(u) = \sum_{n=0}^{N-1} a_n e^{Jnkdu}$

$u = \sin\theta$

BEAM MAXIMUMS: $u = \frac{2p\pi}{kd} = \frac{p\lambda}{d}$; $p = 0, \pm 1, \ldots$

REAL SPACE: $-1 \le u \le 1$

SCANNING ARRAY FACTOR: $E_a(\bar{u}) = \sum_{n=0}^{N-1} \left| a_n \right| e^{Jnkd\bar{u}}$

$a_n = \left| a_n \right| e^{-jnkdu_o}$

$\bar{u} = u - u_o = \sin\Theta - \sin\Theta_o$

BEAM MAXIMUMS: $\bar{u} = \frac{p\lambda}{d}$; $p = 0, \pm 1, \ldots$

REAL SPACE $-1 - u_o \le \bar{u} \le 1 - u_o$

Figure 4.8

$$E_a(\bar{u}) = \sum_{n=0}^{N-1} |a_n| e^{jnkd\bar{u}} \quad (4.2)$$

where the array illumination is $a_n = |a_n| e^{-jnkdu_o}$ and the parameter $\bar{u}$ is $u - u_o$. The inter-element phasing is kdu_o with u_o a dimensionless factor that is a way of specifying the the phasing. Extending the geometric interpretation of $u = \sin\theta$ to $\bar{u} = \sin\theta - u_o$, we can identify an array factor maximum (e.g. $\bar{u} = 0$) at $\theta = \theta_o$ and write,

$$\bar{u} = \sin\theta - \sin\theta_o. \quad (4.3)$$

The beam maximums are found from the array factor as before, with $\bar{u} = \frac{2p\pi}{kd}$ for all p integers. The real angle extent is still $-1 \leqslant u \leqslant 1$, thus real space for the parameter $\bar{u}$ lies between $-1 - u_o$ and $1 - u_o$.

We can further examine resulting beam locations of this linear array for particular examples of spacing and phasing. Beam maximums lie within and beyond real space depending on spacing and scan parameter.

For a broadside beam (non-scan) the scan parameter u_o equals zero. The main lobe (ML) is at $\theta = 0°$ and corresponds to the $p = 0$ beam. The grating lobes (GL) are identified as those beam maximums where $p \neq 0$. For the case of $p = \pm 1$, these are the grating lobes with the lowest modulus value of u. For element spacing $d/\lambda < 1$ the magnitude of u for the GL are greater than 1 and thus these beams are outside real space with all the other higher order grating lobes. The grating lobes outside real space are non-radiating and are the higher order modes of the unit cell waveguide (associated with stored energy at the interface). When the spacing becomes one wavelength both the $p = \pm 1$ grating lobes are at the edge of real space, or $\theta = \pm 90°$. There are then three beams that will share the radiated power, the amount of which depends on the element factor.

When the array is scanned the ML is at $\theta = \theta_o = \sin^{-1} u_o$. For a positive scan angle θ the $p = -1$ GL moves toward real space for increasing θ. The (−1)GL is at endfire when

$$\bar{u}(p = -1) = \bar{u}_{-1} = \sin(\theta = 90°) - \sin\theta_o = -\frac{\lambda}{d} \quad (4.4)$$

Thus, $\sin\theta_{oGL} = \frac{\lambda}{d} - 1$; and the scan angle at which the grating lobe is incipient to real space is specified. This equation will be generalized for the planar array.

As the elements are closer spaced the grating lobes are moved further out from radiating space. The spacing $d/\lambda = 0.5$ is the widest spacing (fewest elements for a given array size and frequency) to maintain a grating lobe free scanned hemisphere. Typical array spacings lie between $0.5 \leqslant d/\lambda \leqslant 1$, and are determined by a consideration of fewest number of elements for a grating lobe free scan volume.

B. Planar Array — Array Factor. In a manner similar to that describing the linear array, the array factor for the planar array case can be described. The array factor for a rectangular grid, spaced d_x and d_y in the x and y directions respectively with elements located at each of the grid crossings, is

$$E_a(u, v) = \sum_{n,m}^{N,M} a_{nm} e^{j[mkd_x(u - \frac{\delta x}{kdx}) + nkdy(v - \frac{\delta y}{kdy})]} \quad (4.5)$$

where u, v are related to direction angles from the x and y axis, or the usual spherical coordinates as,

$$u = \sin\theta \cos\phi$$

and

$$v = \sin\theta \sin\phi$$

There are two scanning parameters here, $\frac{x}{kdx} = u_o$ and $\frac{y}{kdy} = v_o$ associated with the inter-element phasing in x and y. A non-scan beam is where u_o and v_o are zero.

We first consider the broadside beam (ML at $\theta = 0°$). The array factor maximum that corresponds to the main lobe (ML) for the planar array case is $u = v = 0$. The locations of the pq grating lobes are:

$$u = p\frac{\lambda}{d_x} \quad \text{and} \quad v = q\frac{\lambda}{d_y} \quad (4.6)$$

for p and q all integers. Equations (4.6) satisfy all the multiplicities of $2l\pi$ in the exponential of array factor that yield unique beam maximums, (l an integer).

Real space is again the regime of the real angle θ or $\sin\theta \leqslant 1$. In terms of the variables u and v we find that $u^2 + v^2 = \sin^2\theta$, hence real space is defined as,

$$u^2 + v^2 \leqslant 1. \quad (4.7)$$

A graphical representation of all beam locations, ML and GL, is conveniently shown in a coordinate system of u - v. This plot is called the 'grating lobe grid', 'reciprocal lattice', 'T-space', and the 'u - v plane'.

A plot of beam locations in the u - v plane is shown in Fig. 4.11 for the rectangular lattice of d_x by d_y. The main lobe is located by an x at $u = v = 0$, with grating lobes at all the $p\frac{\lambda}{d_x}$ and $q\frac{\lambda}{d_y}$ (p, q integer) points shown as circles. The labeling of the (p, q) GL here is the integer values of λ/d_x and λ/d_y respectively. The rectangular lattice of d_x and d_y has a rectangular grid in the u - v plane, with dimensions proportional to the reciprocals of the physical dimensions.

Real space on this plot is the area within a circle of radius 1. The grating lobe locations in this figure indicate that $\frac{d_x}{\lambda}$ and $\frac{d_x}{\lambda}$ are less than one since the GL are out of real space for this non-scan condition.

LINEAR ARRAY-BEAM LOCATIONS

NO SCANNING: $(u_o = 0)$ MAINLOBE (ML) AT $\theta = 0°$

SPACING: $\frac{d}{\lambda}$	GRATING LOBES: $u_{\pm 1} = \frac{\lambda}{d}$	
$\frac{1}{4}$	±4	ML
$\frac{1}{2}$	±2	ML
1	±1	GL ← → GL

SCANNING: $(u_o = \sin\theta_o \neq 0)$

ML AT $\theta = \theta_o = \sin^{-1} u_o$

END FIRE GRATING LOBE:

$$\underbrace{-1 - \sin\Theta_o}_{\text{END FIRE } \Theta = -90°} = \bar{u} = \underbrace{\frac{-2\pi}{kd}}_{-1 \text{ BEAM}}$$

$$\boxed{\sin\Theta_{oGL} = \frac{\lambda}{d} - 1}$$

AT $\frac{d}{\lambda} = \frac{1}{2}$ GL ← → ML

TYPICAL $\frac{1}{2} < \frac{d}{\lambda} < 1$

Figure 4.9

PLANAR ARRAY-DIRECT LATTICE

RECTANGULAR GRID

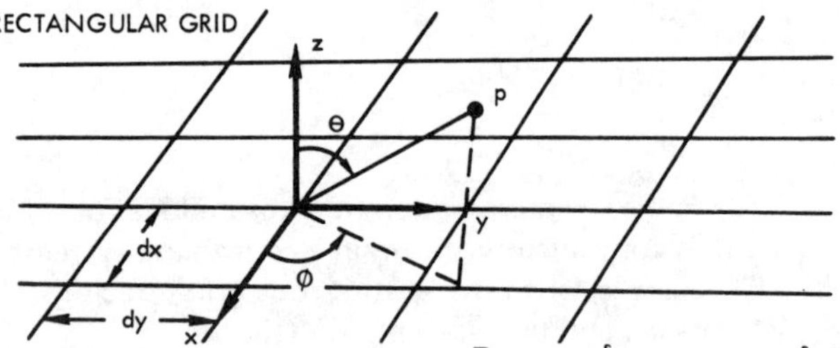

ARRAY FACTOR: $E_a(u, v) = \sum_{n,m} a_{nm} e^{j\left[mkd_x\left(u - \frac{\delta_x}{kdx}\right) + nkdy\left(v - \frac{\delta_y}{kdy}\right)\right]}$

$u = \sin\Theta\cos\phi$

$v = \sin\Theta\sin\phi$

NON SCAN: $\frac{\delta_x}{kdx} = u_o = 0; \frac{\delta_y}{kdy} = v_o = 0$

BEAM MAXIMUMS:

$u = v = 0$ ML

$$\left.\begin{array}{l} u_{pq} = p\frac{\lambda}{dx} \\ v_{pq} = q\frac{\lambda}{dy} \end{array}\right\} \text{GL}$$

ALL p, q INTEGERS

Figure 4.10

PLANAR ARRAY-GRATING LOBE GRID

REAL SPACE: $u^2 + v^2 \quad \sin^2\theta \leq 1$

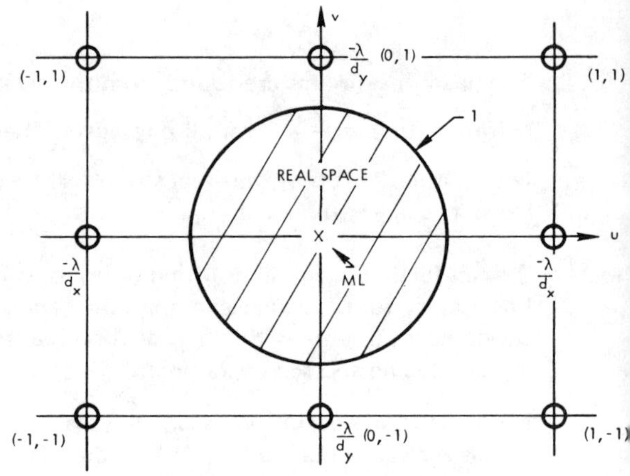

RECTANGULAR LATTICE - RECTANGULAR GL GRID

Figure 4.11

PLANAR ARRAY-SCANNED

EXPONENT: $mkd_x\left(u - \frac{\delta_x}{kd_x}\right) + nkdy\left(v - \frac{\delta_y}{kd_y}\right)$

ML: $u_o = \sin\theta_o\cos\phi_o = \frac{\delta_x}{kd_x}$

$v_o = \sin\theta_o\sin\phi_o = \frac{\delta_y}{kd_y}$

LOCATION OF ML

$\frac{v_o}{u_o} = \tan\phi_o$

$u_o^2 + v_o^2 = \sin^2\theta_o = \left(\frac{\delta_x}{kd_x}\right)^2 + \left(\frac{\delta_y}{kd_y}\right)^2$

GRATING LOBE LOCATIONS

$u_{pq} = u_o - \frac{p\lambda}{d_x}$

$v_{pq} = v_o - \frac{q\lambda}{d_y}$ p,q = 0, ±1, ...

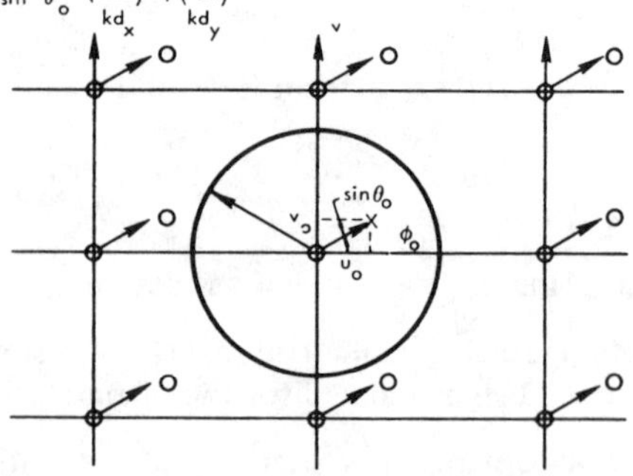

Figure 4.12

Scanning of the rectangular grid planar array can be examined in terms of the exponent of the array factor. From the definition of planar array factor, this term is,

$$mkd_x(u - \frac{\delta_x}{kd_x}) + nkd_y\,(v - \frac{\delta_y}{kd_y}) \quad . \qquad (4.8)$$

The main lobe is at,

$$u_o = \sin\theta_o \cos\phi_o = \frac{\delta_x}{kd_x} \qquad (4.9)$$

and

$$v_o = \sin\theta_o \sin\phi_o = \frac{\delta_y}{kd_y} \qquad (4.10)$$

Dividing the two equations we get,

$$\frac{v_o}{u_o} = \tan\phi \quad . \qquad (4.11)$$

From ϕ_o the angle θ_o is obtained from either of the u_o or v_o equations; or we have,

$$u_o^2 + v_o^2 = \sin^2\theta_o = \left(\frac{\delta_x}{kd_x}\right)^2 + \left(\frac{\delta_y}{kd_y}\right)^2 \quad . \qquad (4.12)$$

These equations locate the main beam in both θ, ϕ coordinates and the u - v plane.

On the grating lobe plot in the u - v plane we see the ML has moved in modulus equal to $\sin\theta_o$ and rotated an angle ϕ_o off the u axis.

The conditions for grating lobes for this rectangular grid array are;

$$kd_x\,(u_{pq} - \frac{\delta_x}{kd_x}) = kd_x(u_{pq} - u_o) = 2p\pi$$

or

$$u_{pq} - u_o = p\frac{\lambda}{d_x} \;;\; p,q = 0, \pm1, \pm2, \ldots$$

and similarily

$$v_{pq} - v_o = q\frac{\lambda}{d_y} \;;\; p,q = 0, \pm1, \pm2, \ldots \qquad (4.13)$$

We observe that the grating lobe locations are shifted in u, v from their non scan locations by an amount identical to the main beam scan. Thus, scanned GL positions are specified directly as shown in Fig. 4.12. Let us look at a simple example. Our array is a rectangular grid and planar with $\frac{d_x}{\lambda} = 2/3$ and $\frac{d_y}{\lambda} = 1$. We desire to scan to $\theta_o = 45°$ in the plane $\phi_o = 0°$ and would like to specify the inter-element phasing and determine the resulting grating lobe structure.

$$v_o = \sin\theta_o \sin\phi_o = 0 = \frac{\delta_y\lambda}{2\pi d_y} \Rightarrow \delta_y = 0$$

$$u_o = \sin\theta_o \cos\phi_o = \frac{1}{\sqrt{2}} = \frac{\delta_x\lambda}{2\pi d_x} \Rightarrow \delta_x = \frac{2\sqrt{2}}{3}\pi \quad .$$

The inter-element phasing is $\frac{2\sqrt{2}\pi}{3}$ radians or $\sim 167°$. Then $|u_{-1}| = 0.8 < 1$ and this grating lobe has entered real space!

This would not be a good design. For this scan sector the x spacing should be changed. Using the grating lobe relations $u_{-1} - u_o = -\frac{\lambda}{d_x}$ we specify the GL at $u_{-1} = 1$ at the scan angle u_o and determine d_x/λ.

$-1 - 1/\sqrt{2} = -\lambda/d_x$ then $d_x/\lambda = \frac{\sqrt{2}}{\sqrt{2}+1}$ for GL free scanning. Another example in Fig. 4.13 is a grating lobe grid for a square array of spacings equal to one wavelength. Here there are 4 GL symmetrically located at the edge of real space. Any scanning will bring a grating lobe in.

In a practical design we take into account the finite width of the beam. For this array although all four of the beam peaks may be just outside real space essentially half the beams are inside. Thus if we assumed no reflection in the elements with isotropic element patterns, then for a finite width beam only one third of the power will be going into the main beam. A good design would always exclude at least the full width of the GL beams from real space at full scan extent.

The more general case of a planar array lattice is called skew. This lattice may be described by a pair of vectors $\underline{d}_1$ and $\underline{d}_2$ which upon translation defines all lattice sites and will fill the space and form the array. The vectors are base vectors from which we construct the direct lattice. The angle between $\underline{d}_1$ and $\underline{d}_2$ we have shown as α. The phasing terms δ_1 and δ_2 are inter-row and column phasings.

There is a corresponding pair of vectors $\underline{b}_1$ and $\underline{b}_2$ that will describe the lattice sites in the u - v plane (or reciprocal lattice.). These relations are;

$$\underline{b}_1 = \frac{\underline{d}_2 \times \underline{z}_o}{A_d} \;;\; \underline{b}_2 = \frac{\underline{z}_o \times \underline{d}_1}{A_d} \;, \text{ where} \qquad (4.14)$$

A_d is the area of the direct lattice. The base vectors $\underline{b}_1$ and $\underline{b}_2$ are used to form the reciprocal lattice. In the u - v plane these vectors are $\lambda\underline{b}_1$ and $\lambda\underline{b}_2$.

The main beam location in terms of the phasings δ_1 and δ_2 are given by

$$u_o = \frac{\delta_1}{k} b_1 \sin\alpha$$

$$v_o = -\frac{\delta_1}{k} b_1 \cos\alpha + \frac{\delta_2}{k} b_2 \quad . \qquad (4.15)$$

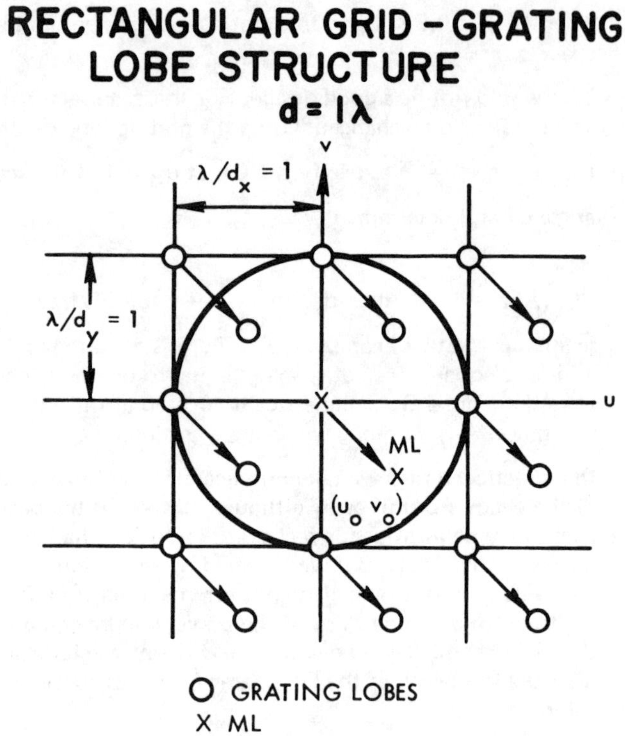

Figure 4.13

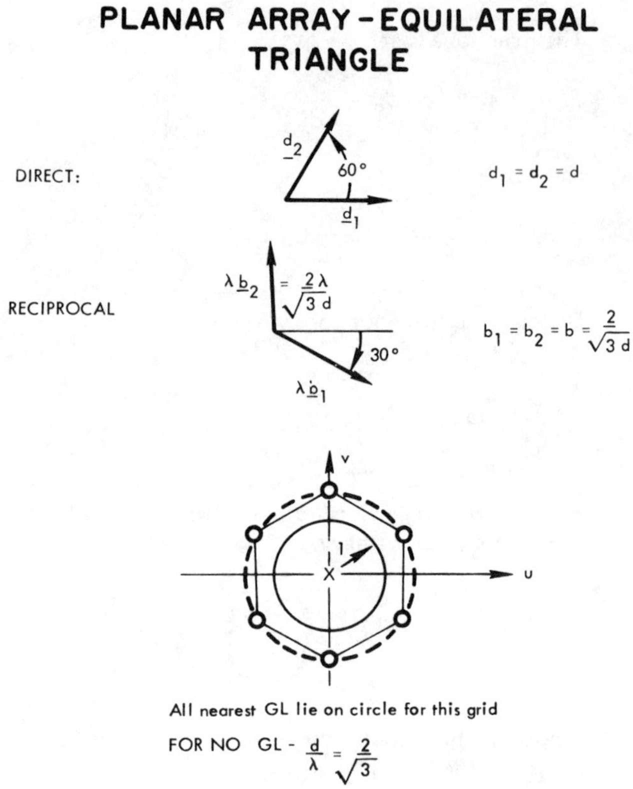

Figure 4.15

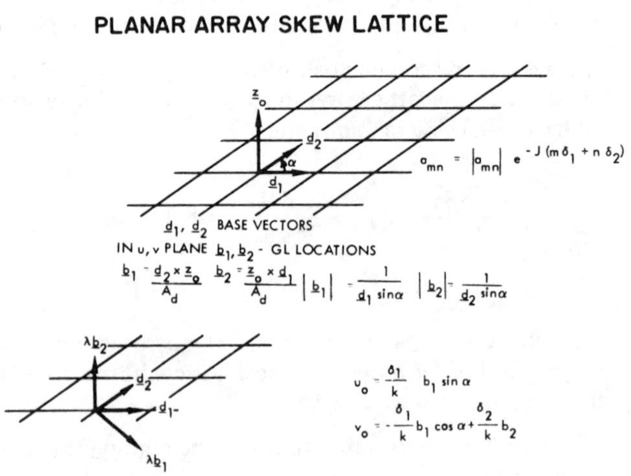

Figure 4.14

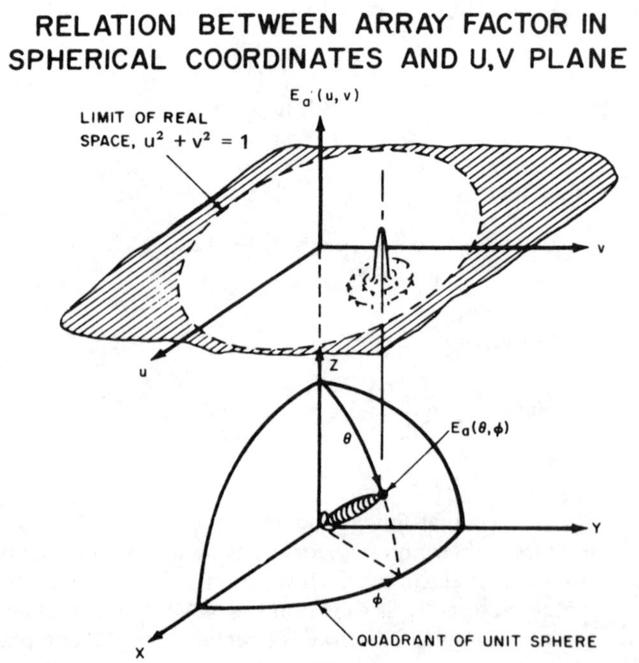

Figure 4.16

As an example we consider a planar array with an equalateral triangular grid (a case of significant interest). The base vectors are equal in length and separated by 60° in the direct lattice; in the reciprocal lattice the base vectors are also equal and seperated by 120°.

For this grid a regular hexagon is the geometry formed by the nearest grating lobe sites. These GL points are thus on a circle for this grid. GL points of a square lattice will form a square where also a circle passes through all points, however, the circle for the triangular grid is $\frac{\sqrt{3}}{2}$ wider for the same length of base vectors in the direct lattice.

Thus 12% fewer element are needed for this triangular grid than a square grid using the criteria of GL incipience to establish grid size. This is proven in the following manner. We select a spacing d for a square lattice array which sets a grating lobe circle at $\frac{\lambda}{d}$ in the reciprocal lattice. The direct lattice area for this square array is d^2. A triangular grid array with a grating lobe circle at $\frac{\lambda}{d}$ is associated with an array whose base vectors are $\frac{2}{\sqrt{3}}$ d long. The direct lattice area for this triangular grid array is $4/3d^2 \sin 60° = d^2\, 2/\sqrt{3}$. The cell area ratio of triangular to square grid is $2/\sqrt{3}$ or 1.12. Associating the same number of elements per cell in each case results in 12% fewer elements using the triangular grid.

Fig. 4.16 is a pictorialization of the array factor in the spherical coordinate system and in the u - v plane.

The beam or beams in real space can be projected onto a unit sphere to present a visualization of the beam scanned in the spherical coordinate system.

One important characteristic of the beam representation in the u, v plane is the invariance of beam shape and size with beam angle (u_o, v_o). For example, the array factor exponential for the rectangular lattice is of the form;

$$mkd_x\,(u - u_o) + nkd_y\,(v - v_o)$$

where the behavior of the function, in u and v, about u_o and v_o is independent of the values of u_o and v_o (the exponential depends only on the difference). This result is also true for the arbitrary skew lattice.

An example of a circular beam scanned in v (or $\phi = 90°$) is shown in Fig. 4.17. Here the u - v projection and the spherical coordinate representation are combined. As the beam is scanned its shape in the u - v plane is invarient while the projections of the beam on the unit sphere change.

We can find an approximate expression for the beamwidth as the array scans [5]. For no scanning we have at the half power beamwidth $E_a(\theta = \frac{\Theta_{30}}{2})$ in the plane v = 0, u = sin $\frac{\Theta_{30}}{2}$ (Θ_{30} is full beamwidth). From the beam maximum to the 3dB beamwidth we call $\Delta u \equiv \sin\frac{\Theta_{30}}{2}$ which remains constant with scan. The beamwidth is thus Θ_{30} (rad) = $2\mathrm{SIN}^{-1}(\Delta u)$. The beam is scanned to θ_s, where $u - u_s = \mathrm{SIN}\theta - \mathrm{SIN}\theta_s$. At the angle closest to broadside $u_1 - u_s = \mathrm{SIN}\theta_1 - \mathrm{SIN}\theta_s$ and at the side furthest $u_2 - u_s = \mathrm{SIN}\theta_2 - \mathrm{SIN}\theta_s$. We want to find $\theta_2 - \theta_1$ as the beamwidth that corresponds to θ_s where Δu has remained constant. In this fashion,

$$\Delta u = u_2 - u_1 = (u_2 - u_s) - (u_1 - u_s) = \sin\theta_2 - \sin\theta_1 \quad (4.16)$$

Solving for $\Theta_3 = \theta_2 - \theta_1 = 2\sin^{-1}\dfrac{\Delta u}{\cos\frac{1}{2}(\theta_s + \theta_1)}$ (4.17)

For narrow beam antennas,

$$\Theta_{30} \approx 2\Delta u \quad (4.18)$$

then

$$\Theta_3 = \frac{2\Delta u}{\cos\frac{1}{2}(\theta_2 + \theta_1)} \approx \frac{\Theta_{30}}{\cos\theta_s} \quad (4.19)$$

A simple graphic representation can be used to obtain the beamwidth relations for different scan angles. Using the invarient parameter Δu the beam angles θ_1 and θ_2 are readily obtained by projections of the beamwidth on the u axis into the unit circle. (See Fig. 4.18).

4.2.2 Element Factor. The "large" array approximation permits the removal of the element factor from the field pattern expressions. Each element of the large array is then assumed to have an identical pattern; this of course is not the case because the different environments for edge and central array elements give rise to different element patterns. However, we recall again the assumption that there are relatively few edge elements compared to the total number.

The element pattern is defined as the pattern of one element in its array environment with all other elements terminated in their generator impedances. The element gain pattern puts an absolute gain level on this element pattern and includes the array reflection properties in addition to the element directivity. For a small array the patterns and gains are different for all elements in the array. The total field and gain of the array can be determined by superposition, imposing all the different excitations and computating the far fields and reflections one at a time. The large array approximation simplifies the mathematics as well as the number of computations to be made.

When we use the infinite array approximation and factor the element pattern out of the field expression the realized gain of the array can be written as

$$G_r(\theta, \phi) = Ng_r(\theta, \phi) \quad (4.20)$$

where $G_r(\theta, \phi)$ is the realized array gain for a linear phase taper excitation to scan to θ, ϕ. N is the number of array elements and $g_r(\theta, \phi)$ is the realized element gain pattern. The array gain for a particular scan is thus proportional to the element pattern. The coupling between array elements and subsequent effect on the gain and reflection properties are both imbedded in the element gain pattern $g_r(\theta, \phi)$. To give some idea of how strong the effects of mutual coupling are on these element patterns, Fig. 4.19 shown is the H-plane element gain patterns (reflections are excluded) for the cen-

BEAMWIDTH AND ECCENTRICITY OF THE SCANNED BEAM

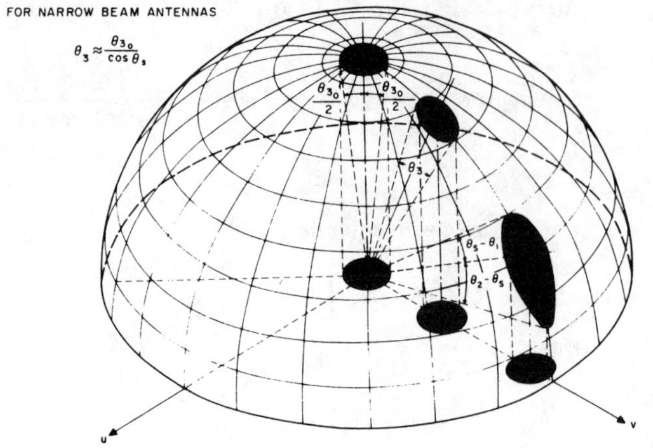

Figure 4.17

BEAMWIDTH RELATIONS FOR SCANNING

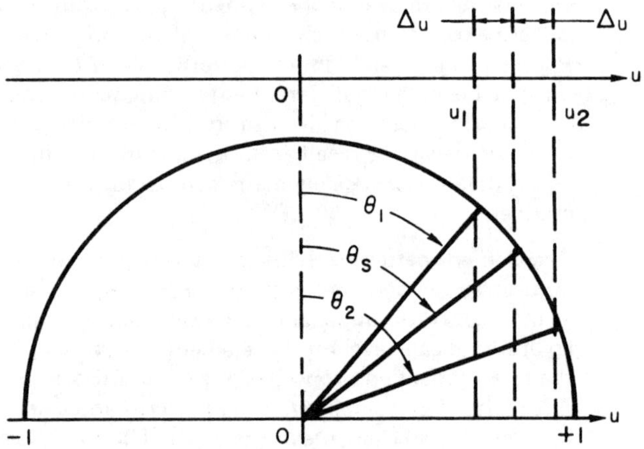

Figure 4.18

FREE SPACE DIPOLE ELEMENT PATTERNS

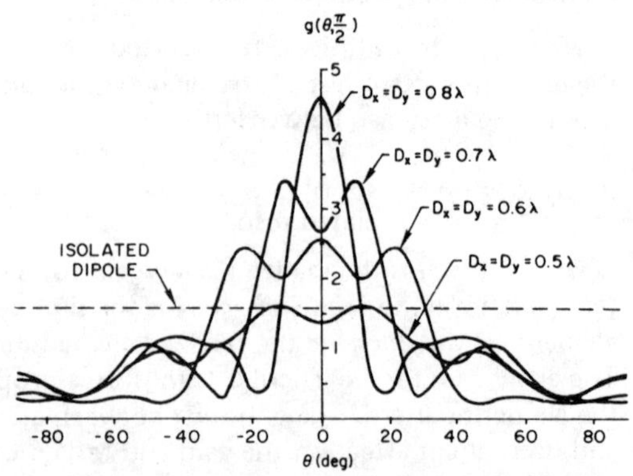

H-PLANE ELEMENT-GAIN FUNCTIONS FOR THE CENTER ELEMENT OF A 7-BY-9 ARRAY OF λ/2 DIPOLES IN FREE SPACE, FOR VARIOUS ARRAY SPACINGS. THE CORRESPONDING CURVE FOR THE ISOLATED DIPOLE IS INCLUDED FOR COMPARISON (FROM [6])

Figure 4.19

TWO WAYS TO VIEW PHASED ARRAYS

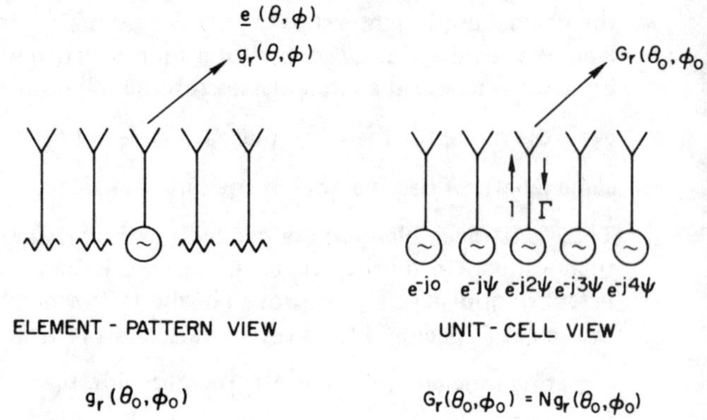

$$G_r(\theta_o,\phi_o) = \frac{4\pi N A_d}{\lambda^2} \cos\theta_o \left[1 - |\Gamma(\theta_o,\phi_o)|^2\right] \text{ (no GL)}$$

Figure 4.20

EFFECT OF MUTUAL COUPLING

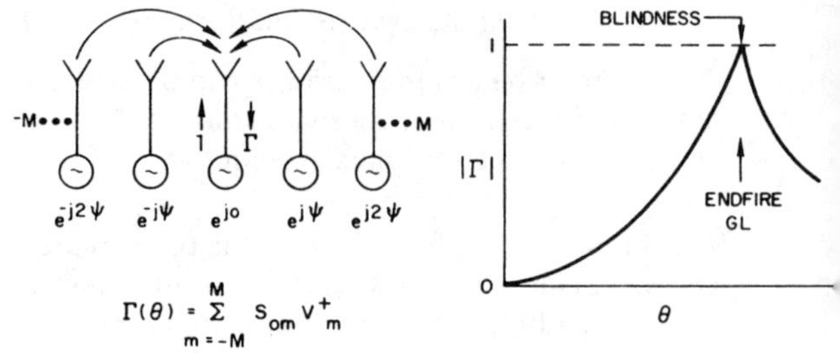

Figure 4.21

EXAMPLES OF RADIATING ELEMENTS FOR WHICH EXTERNAL RESONANCES MAY OCCUR

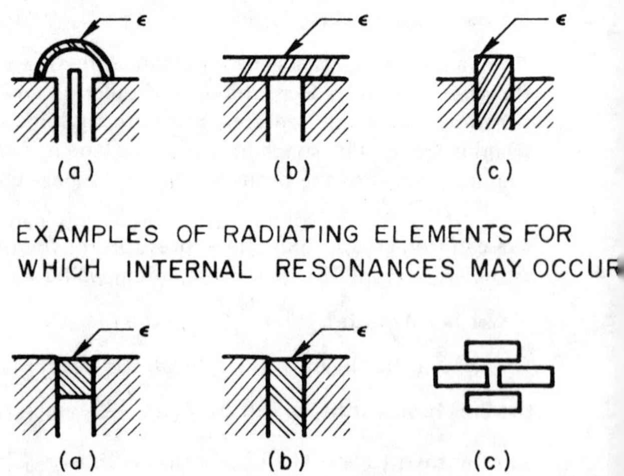

Figure 4.22

ter element of a 7 x 9 array of $\lambda/2$ dipiles in free space, as a function of element spacing. A curve of the "isolated" dipole pattern is also shown for reference, (the isolated element removes all neighbors to infinity). The different array spacings provide a different environment for the driven element, and as seen significantly change the pattern shape, and gain. The equation relating the element gain pattern to the gain of the fully excited array is significant because it ties together two different operational forms of the phased array: the singly excited element, and the fully excited uniform illuminated linear phase taper array. Measurements or calculations of element patterns can thus be used as a design basis for the fully excited array.

There are certain relations for a large array which will help further to tie the element pattern and array gain factors together. The array is assumed lossless; then we may write the ratio of a realized gain to directive gain as *

$$\frac{G_r(\theta,\phi)}{D(\theta,\phi)} = 1 - |\Gamma(\theta,\phi)|^2 \qquad (4.21)$$

where $|\Gamma(\theta,\phi)|^2$ is the modulus squared of the active reflection coefficient in any array element for an array scanned to θ, ϕ. This reflection coefficient is termed "active" because it corresponds to the fully excited array. The directive gain of a large uniformly illuminated planar array in the direction of the peak when no grating lobes are present is related to the physical array area by,

$$D(\theta,\phi) = \frac{4\pi N A_d}{\lambda^2}\cos\theta \qquad (4.22)$$

A_d is the cell area or area of direct lattice; in the case of a rectangular grid A_d is $d_x\, d_y$.

Using the relations between directive and realized gain (4.21), (4.22) with the equation for element and array gain (4.20) we have the realized element gain factor as,

$$g_r(\theta,\phi) = \frac{4\pi A_d}{\lambda^2}\cos\theta\,\left[1 - |\Gamma(\theta,\phi)|^2\right] \qquad (4.23)$$

To maximize the realized gain of the array, the designers task is in matching the active element over both a scan volume and frequency bandwidth. The relations presented not only point out the element design problem but relate two different operational conditions from which design approaches have evolved.

A. Element Pattern Nulls or Blindness. We see that the effect of mutual coupling for the large array can be expressed in a form where it is manifest in the reflection coefficient. The reflection coefficient is both frequency and scan dependent; and for the large array is approximately the same for all elements.

Shown in Fig. 4.21 is a segment of a large array uniformly excited with a linear phase taper. The reflection coefficient of, lets say, the 0th element consists of the self-reflection and the coupled terms from all other elements. We can write the scan dependent reflection coefficient of this element in terms of the self and mutual coupling for arbitrary array excitation as,

$$\Gamma_o(\theta) = \sum_{m=-M}^{M} S_{om}\, V_m^+ \qquad (4.24)$$

where θ is the scan variable and S_{om} is a scan independent scattering coefficient which is the coupling of the mth element into the 0th with all neighbors passively terminated. The incident voltage excitation in the mth element is V_m^+ (complex; includes phase excitation and implicitly specifies θ).

One of the phenomena of mutual coupling that we have briefly mentioned is array "blindness" or element pattern null, or resonance. The blindness corresponds to a total reflection condition for some scan in the fully excited array. An array null ($|\Gamma| = 1$) shown in Fig. 4.21 is associated with scanning to where the endfire grating lobe emerges into real space. This may be more generally thought to be the point where a radiating beam (or space mode) changes from non-propagating to propagating (cut-off condition) and hence passes through a zero or infinity of admittance for H or E modes. At the grating lobe condition $\sin\theta_{o\,GL} = \frac{\lambda}{d} - 1$; thus $d/\lambda = 1/(\sin\theta_o + 1)$, and

$$\frac{2\pi d}{\lambda}\sin\theta_o + \frac{2\pi d}{\lambda} = 2\pi \qquad (4.25)$$

We identify $\frac{2\pi d}{\lambda}\sin\theta_o$ as the inter-element phasing to scan to θ_o and $\frac{2\pi d}{\lambda}$ as the spacial phase path difference between elements along the array. Thus at the GL condition the inter-element coupling (combining the fixed path difference and the excitation phasing) is in-phase; which in part substantiates the large mismatch.

Shown in Fig. 4.22 are examples of radiating elements for which resonances can occur, [7]. The resonances are distinguished by two types, *external* and *internal,* according to where in the structure a higher mode cut-off condition exists. This mode cut-off condition we will show is related to the resonance phenemona.

The elements of Fig. 4.22A can give rise to external resonances. These configurations can support the propagation of a higher external mode in a region just beyond the ground plane, but will be cut-off in free space just beyond.

Internal resonances can occur for elements of Fig. 4.22B. In (a) and (b) the cut-off condition of a *higher mode* in the dielectrics has associated resonances [8], [9]. In (c) a brick array is shown where a resonance has been observed when no dielectrics were loading the elements [9], [10]. However, a higher order waveguide mode was at the cut-off condition in the air loaded guide.

An element pattern from a "brick" array which displayed the blindness phenomena is shown in Fig. 4.23. The resonance was clearly distinct from the GL condition and it has been verified theoretically that this blindness is associated with the cut-off condition of the waveguide TE_{20} mode.

* No grating lobes.

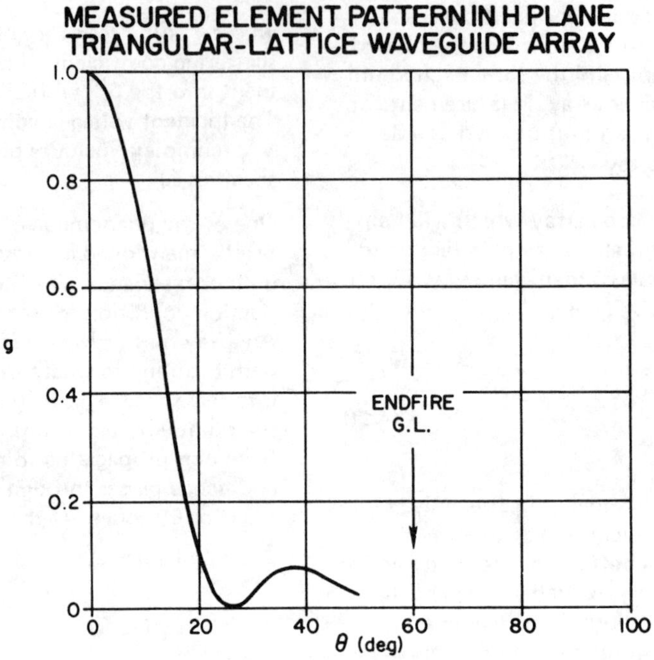

Figure 4.23

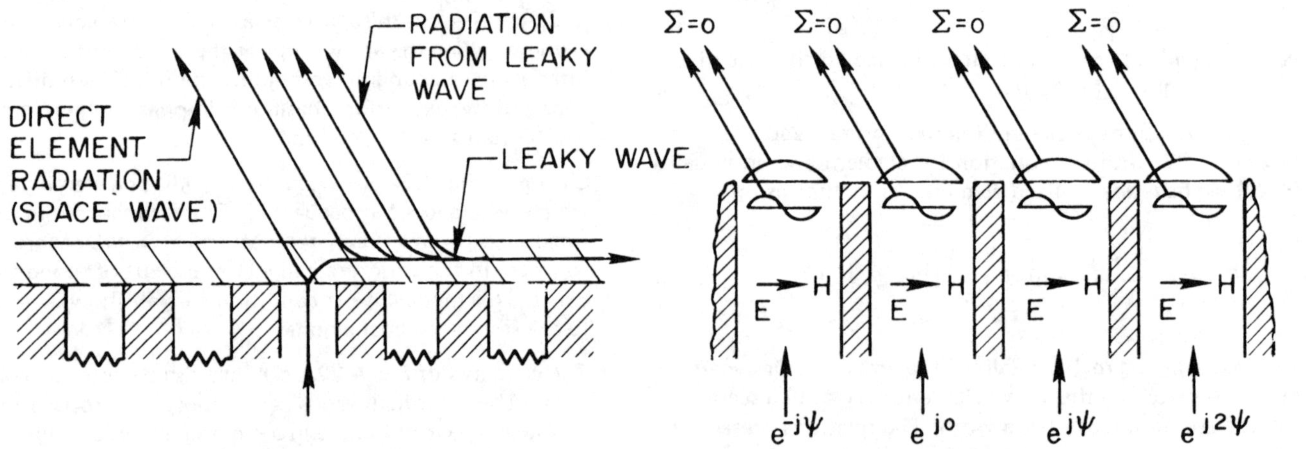

(A)

SKETCH SHOWING INTERFERENCE BETWEEN DIRECT ELEMENT RADIATION AND THAT FROM LEAKY WAVE TO PRODUCE ELEMENT PATTERN NULL ON LEFT SIDE OF PATTERN. (FROM KNITTEL, HESSEL AND OLINER)

(B)

SKETCH SHOWING THE CANCELLATIONS OF RADIATION FROM THE TWO LOWEST-ORDER WAVEGUIDE MODES TO PRODUCE THE COMPLETE-REFLECTION CONDITION. (COURTESY OF G. H. KNITTEL)

Figure 4.24

There are two points of view which are used to physically interpret the blindness phenomena, [7]. These are the same points of view we used to describe element patterns: one element excited in the passively terminated array environment, and each element of the fully excited array. Both blindness explanations may be related to radiation properties of the two array structures.

In Fig. 4.24A the element null is shown to be caused by a cancellation of direct element radiation (space wave) by radiation from a leaky wave excited along the array. The single element radiates a space wave and in addition couples some of its energy into a leaky wave which is a guided wave mode of the structure along the array surface. The leaky wave wave will radiate its energy in some direction in space and it turns out to be of a particular amplitude and phase to cancel the space wave; hence there is a null in the element patterns.

The other point of view uses the fully excited array to facilitate explanation of the anomalies. Fig. 4.24B shows a cutaway of a rectangular waveguide array, sectioned along the H-plane. For a particular scan condition a field asymmetry in the apertures is needed to match the field asymmetry for a scanned beam. This asymmetry in field forces an aperture excitation of the TE_{20} waveguide mode (excitation not by a generator but by requirements of the boundary conditions). A condition can exist where the radiation from the sum of these two modes in the scan direction is zero.

The relation of blindness in phased arrays to higher mode cut-off conditions have been well correlated by Knittel [12]. A dispersion diagram relating frequency and inter-element phasing (a k-β diagram) is used to graphically relate the resonances to the mode cut-off conditions.

A k-β diagram is shown in Fig. 4.25 where k is the free space wave number (proportional to frequency) and k_x is the element phasing per period to provide scanning at angle $\theta = \sin^{-1} k_x/k$. A configuration of a triangular grid waveguide array appropriately dimensioned is also shown on the figure. The curve axes have been normalized with A/π for convenience to this array geometry (where TE_{20} waveguide mode cut-off is at $k\,A/\pi = 1$).

Horizontal lines on this curve are constant frequency (scan angle varies) while straight lines from the origin are of constant scan angle; $\sin\theta = k_x/k$. The scan angles $\theta = 0^\circ$ and 90° correspond to straight lines along the $k\,A/\pi$ axis and orientated 45° in the plane respectively. The 45° sector from the $k\,A/\pi$ axis is real space for positive scan angles.

Fig. 4.26 shows a locus of $|\Gamma| = 1$ (blindness) for a triangular grid array as chosen by Farrell and Kuhn [10] who observed a resonance for this geometry with no dielectric plug, (t = 0); and the frequency used by these authors is labelled FK. There are three cut-off condition loci shown on this curve: The TE_{20} mode cut-off in the unloaded waveguide, the TE_{20} mode cut-off in this same waveguide filled with dielectric of $\epsilon = 2.0$, and the cut-off condition of one of the higher order space modes (GL).

The following observations are made: with no dielectric plug in the waveguide (t = 0) the blindness curve begins and ends at one of the higher order mode conditions and never deviates substantially from these lines; as the thickness of a dielectric plug in the waveguides is increased ($t > 0$ and $\epsilon = 2.0$) the blindness curve moves down in frequency toward the cut-off condition of the TE_{20} mode in the dielectric loaded waveguide. The blindness curve for the thickened dielectric plug is asymptotic to the TE_{20} mode cut-off line and a second blindness condition is formed which also moves downward toward the first, and so on (multiple resonances).

Fig. 4.27 shows the same kind of behavior for resonances in a dielectric slab covered parallel plate guide array scanned in the E-plane. The curves here are asymptotic to the cut-off conditions of the next higher order GL (unit cell modes) in the dielectric slab.

For increasing slab thickness the resonance curve as before moves downward. Thus for a constant frequency the blindness occurs closer to broadside. This curve migrates toward the unit cell mode cut-off condition in the dielectric slab, and a second curve is formed. This second resonance condition, whose slope is opposite the -1 grating lobe mode cut-off is associated with the +1 unit cell mode cut-off.

To give some insight into both the methods of solution for active reflection coefficient in the infinite phased array and a good example for comparison of theory and measurements for an array which exchibits a blindness the array configuration shown in Fig. 4.28 is examined [13]. The array is made up of protruding-dielectric elements fed by parallel plate waveguides in the TEM mode for E-plane scan. The progressive inter-element phasing is $\mathcal{H}_o\,d$, and there is no variation in the y dimension.

The active array reflection coefficient (i.e. element patterns) was determined in the following manner:

- The transverse fields in each region were expressed in a modal representation.
- An approximate field matching technique determines the mode amplitude level in each of the regions; namely the scattering matrix for each interface is solved, the feed guide-dielectric slab interface and the free space-dielectric slab interface.

An equivalent network of the unit cell is constructed based on the scattering matrices of the two interfaces. This equivalent network is shown in Fig. 4.29. The travelling wave modal voltages are indicated for each region. The voltage V_f^+ is prescribed and the active array reflection coefficient is V_f^- / V_f^+. The network N connecting the two junctions is composed of equal lengths of uniform modal transmission lines uncoupled except at the junctions. The solution for network voltages or currents is rather straightforward once the modes and scattering matrices have been determined. The evaluation of a phased array of feed waveguides with a single discontinuity at the free space-feedguide interface is determined from a simplified form of this same circuit. Only one scattering matrix needs to be determined; actually only one scattering parameter since the array is viewed on transmit. The dielectric sheath covered array (of x = h thickness) also has the equivalent circuit shown. In the sheathed array case the interface at x = h keeps the modes decoupled and the circuit can be simplified.

One can see that this circuit shown in Fig. 4.29 is rather general for a large class of phased array problems and suggests a useful representation for the array interface (or radiator) as a scan dependent circuit. The difficulties in these problems, as with most waveguide discontinuities, is in determining the circuit element values; here the scattering matrix.

k-β DIAGRAM SHOWING PARAMETERS AND DIMENSIONS

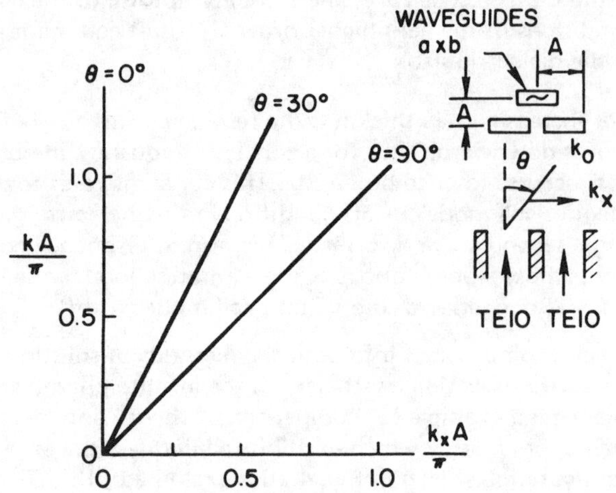

Figure 4.25

|Γ| = 1 CURVES (SOLID LINES) FOR FARRELL-KUHN GEOMETRY SHOWING RELATION TO HIGHER-MODE CUTOFF CONDITIONS. FK IS THE FREQUENCY USED BY FARREL-KUHN.

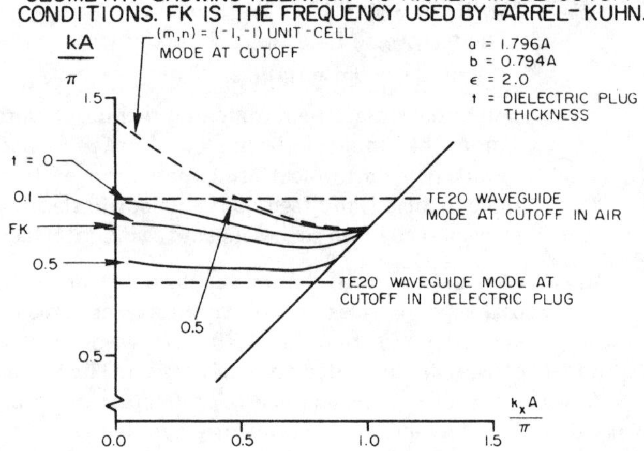

Figure 4.26

|Γ| = 1 CURVES FOR A TEM-FED PARALLEL-PLATE ARRAY COVERED BY A DIELECTRIC SLAB, E SCAN. SLAB PARAMETERS t, ε = 2.0; UNIT CELL = A; WAVEGUIDE HEIGHT = 0.5A.

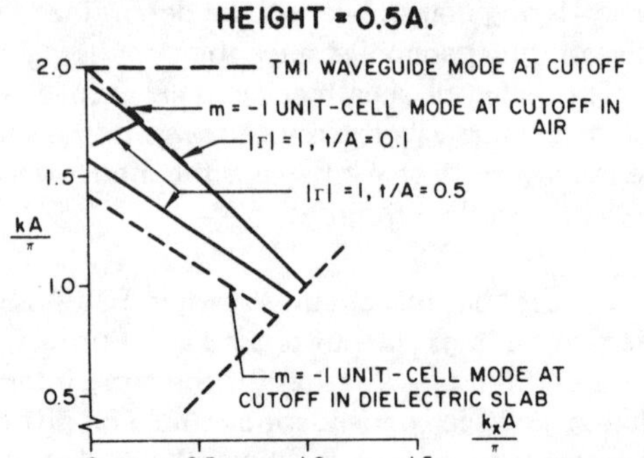

Figure 4.27

PHASED ARRAY OF PROTRUDING-DIELECTRIC ELEMENTS FED BY PARALLEL-PLATE-WAVEGUIDES IN TEM MODE FOR E-PLANE SCAN

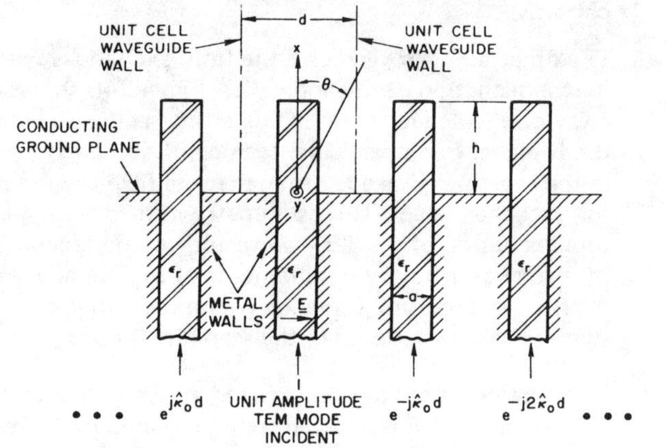

Figure 4.28

EQUIVALENT NETWORK OF UNIT CELL

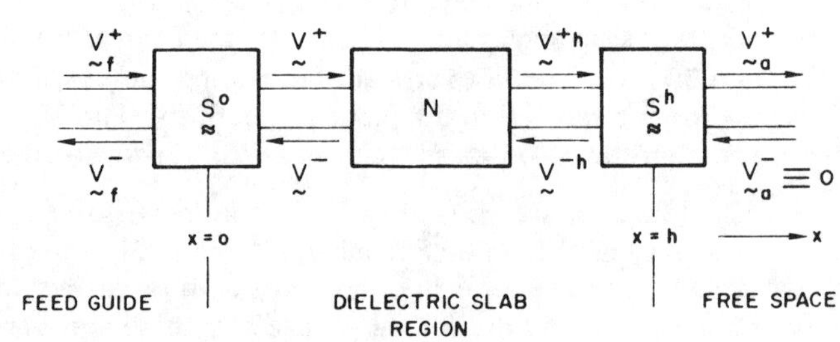

Figure 4.29

The next two curves are computed E-plane element patterns for this geometry for dielectric slab protrusions of $\lambda/2$ and λ, where the element spacing $d/\lambda = 0.535$ (grating lobe emerges into real space at 60° scan). In Fig. 4.30 resonance (blindness) occurs at 46.5° which is an angle between where the second mode propagates in the slab region and the GL is endfire. Similarly, in Fig. 4.31 two resonances appear for this λ protrusion slab array case; again occuring between the angle where the second mode propagates in the dielectric slab and the GL enters in real space.

General Trends of this Blindness.

- For small protrusions a resonance is formed at an angle closer to broadside than the GL angle. As the protrusion is increased the resonance moves toward broadside.
- A second resonance is formed for protrusion $0.5\lambda \leqslant h \leqslant \lambda$.
- A third resonance for $\lambda \leqslant h \leqslant \frac{3}{2}\lambda$, and so on.
- All resonances lie between the GL angle in free space and the angle where the second mode propagates in the slab region.
- This same behavior is observed for dielectric sheathed arrays.

Two test arrays were built to verify these resonances. An 11 x 11 element array was not large enough to clearly show this blindness. Subsequently, a 3 row x 39 column array was designed to approximate the two dimensional structure. A front view of this array is shown in Fig. 4.32.

Measured patterns of Figs. 4.33 and 4.34 are for the 3 x 39 test array and compare very well to the theoretical curves of Figs. 4.30 and 4.31. The resonance locations are very nearly the same. This array has very sharp fall off at $\theta = 90^\circ$ which one expects for the large array as indicated by the $\cos\theta$ term of the element pattern.

ACKNOWLEDGEMENT

A special thanks to Professor Alexander Hessel of the Polytechnic Institute of New York for many insights from his lectures, private discussions and joint projects included herein.

My appreciation to Raytheon Company for permitting me to participate in this lecture series.

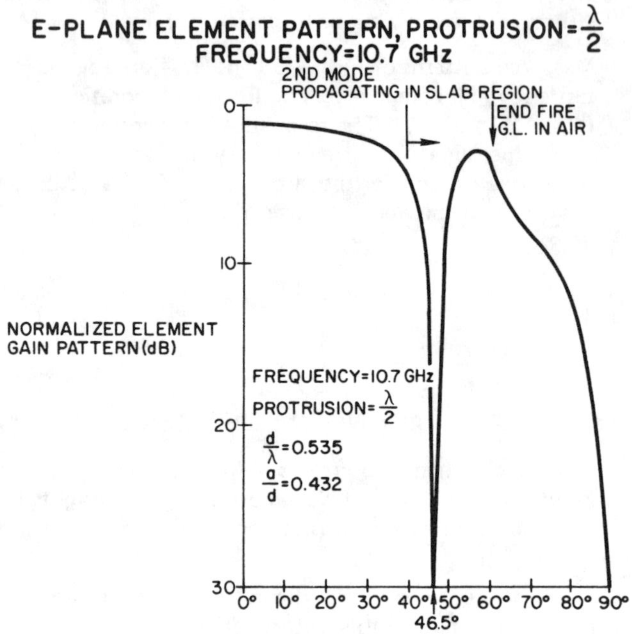

Figure 4.30

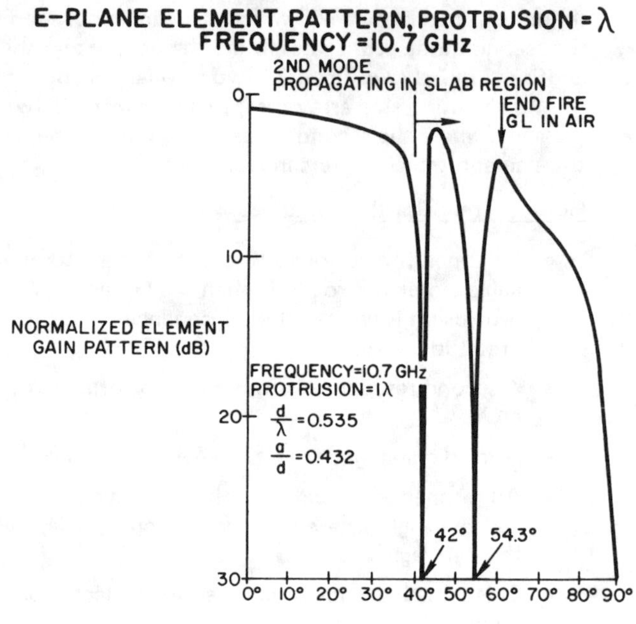

Figure 4.31

FRONT VIEW OF THE 3x39 TEST ARRAY OF PROTRUDING-DIELECTRIC RECTANGULAR WAVEGUIDE ELEMENTS

Figure 4.32

Phased Array Elements - Part 2

Dr. L.R. Lewis

Raytheon Company, Bedford, MA

5.1 Element Matching. We have seen that effects of array environment (or mutual coupling) can significantly change the reflection properties of an element as a function of both scan and spacing. The variation of the magnitude square of reflection coefficient with scan angle for short dipoles above a ground plane is shown in Fig. 5.1. With increased element density the element impedance variation tends toward the continuous distribution of a resistance sheet of 120π ohms (perfect match for normal incidence). In addition to showing the strong effects of mutual coupling, an interesting characteristic of these curves is the decrease in variation of impedance with scan for closer element spacing. This tends to dispell the notion that closer spacing, and more coupling, will increase element impedance variation.

5.1.1 Matching with Frequency. Having selected an element type and lattice we have some perspective into the remaining problems of the array element designer. The impedance as a function of scan and frequency is to be determined and subsequently matched.

Methods of element design for the large array are based on three approaches;

1. infinite array calculations for the aperture impedance,
2. small array measurements for element patterns and mutual coupling, and
3. waveguide simulator measurements (a simulation of an infinite array by a small number of array elements in a waveguide by imaging in the waveguide walls).

An analysis program for use in design of large rectangular waveguide arrays in rectangular or triangular lattices and has been developed at Lincoln Labs. The designs use multiple disc loading of rectangular waveguides with arbitrary step cross-sections. There are some limitations but generally most network configurations of this form can be evaluated. This program does not provide a step by step design procedure, but relies heavily on circuit designers ingenuity and experience to use the program to evolve a design.

Recently a systhesis of a matching network for a waveguide array element [14] in the infinite array has been carried out by a systematic design procedure at Raytheon Co. A double tuned type of response is achieved and physically realized by a combination of dielectric and below cut-off waveguide sections. This is an easily designed element with low complexity and high performance. The method is discussed here is for linear polarization but also been extended for use in circularly polarized arrays.

Basic to the analytic approach for the design of an element in an infinite array is the network representation of the discontinuity between the element aperture and free space. As indicated earlier, this interface can be represented by a scattering matrix relating the coupling between feedguide modes and those of the unit cell (space harmonics). The designer seeks a matching network that can be inserted into each of the feed waveguides, leaving the aperture interface unchanged, such that the dominant mode transmission line is matched, as viewed at the generator (or in practice in the phase shifter line). The addition of the matching circuit in the feedguides results in the equivalent circuit of Fig. 5.4. Here $\underline{\underline{S}}$ is the scattering matrix of the unmatched aperture interface and $\underline{\underline{S}}_N$ is that of the matching circuit.

To design, or synthesize $\underline{\underline{S}}_N$ the design matching network, considering many modes (or truncated to a few) appears unwieldy and devoid of physical insight. The network considered initially is thus simplified to a dominant mode approximation. The synthesis is based on a two port model, where $\underline{\underline{S}}$ can be transformed to a scalar admittance Y_T, the active aperture admittance of the dominant feedguide mode with all other modes terminated in their characteristic admittances

The design based on the two port is extended to the multimode case, but not considered here.

Consider a linearly polarized infinite array of circular waveguide elements in a triangular grid. The grating lobe free, active element admittance variation for TE_{11} mode excitation, over a 12 percent frequency band, has been computed for an array of this geometry. Fig. 5.5 shows this admittance plot for $0.37\lambda_M$ diameter elements spaced at $0.614\lambda_M$ with dielectrix loading $\epsilon = 6.6$ (λ_M is the wavelength corresponding to mid-frequency). The angle ϕ is measured from the H-plane of the array in the array plane, and θ is the scan direction from the array normal. The labeling ω_L, ω_M, and ω_H correspond to the low-, mid-, and high-frequency spiders that characterize the array admitance at a single frequency with scan.

One first selects a scan condition where (over the frequency band) matching is expected to result in the best overall match throughout the scan volume. At this angle θ_o, ϕ_o one can approximately achieve with lumped circuits, for the dominant feedguide mode, the matching of Fig. 5.6 (the double-tuned response) assuming 1) the unmatched aperture conductance remains essentially constant over the frequency band, and 2) the susceptance values vary as ω or $1/\omega$. In Fig. 5.6, $Y_{in}(\omega, \theta_o, \phi_o)$ is the resulting input admittance, and S is the maximum VSWR, over frequency at θ_o, ϕ_o.

The response of Fig. 5.6 is realized from the double-tuned lumped circuit of Fig. 5.7(a). A susceptance $B(\omega)$ is added to the unmatched aperture admittance Y_T to provide mid-frequency resonance. The tank circuit shown in Fig. 5.7(b) is the secondary resonator with admittance Y_s, whose variation with frequency is shown in Fig. 5.8(a). This admittance passed through the ideal J-inverter [15] (with an appropriate normalization transformer) is shown in Fig. 5.8(b). The J-inverter forms the reciprocal of the admittance Y_s, thus making it appear as if a series tuned circuit is present. The transformer normalized the high and low frequency responses to the equator of the admittance chart, where $|Z| = |Y| = 1$. The primary resonator L_p, C_p is shown in Fig. 5.7(a), serves to bring together on the conductance axis the high and low frequency admittance values, and produces the desired response of Fig. 5.6.

For this ideal circuit, there is an independent specification of lumped circuit parameter values which are easily found by imposing the resonance conditions at ω_L, ω_M, and ω_H, and by requiring the conductance values at ω_L and ω_H to be reciprocal to that at ω_M. The distributed circuit realization that will be discussed departs from this idealization. Circuit values are not independently specified and a modified synthesis is used.

The resultant admittance locus is the element performance over frequency and scan for a design initiated at θ_o, ϕ_o. An iterative procedure is used to select the design scan angle that results in a "best match" condition. If the "best match" condition is the minimum reflection coefficient circle containing all transformed admittance values then the trends observed using different values of θ_o, ϕ_o suggest that a match angle near the edge of the unmatched spiders be used.

A simple structure that exhibits similar behavior to the ideal double-tuned lumped circuit of Fig. 5.7(a) is shown for circular feedguides in Fig. 5.9. A pair of short circular dielectric discs are coupled via a section of below cut-off waveguide.

The short length of propagating guide adjoining the aperture provides a shunt capacitance. The below cut-off waveguide can be represented as a J-inverter with symmetric shunt inductances. The second dielectric disc again approximates a shunt capacitor. The circuit is clearly in the desired form for double-tuned matching. The principal departure from the ideal lumped circuit arises from the dispersive nature of the circuit values and the interdependence of the primary and secondary resonators via the shunt inductors L. Although these limitations exist, resonance for both the primary and secondary circuits can be achieved at ω_M, the needed inversion can be carried out, and some wrapping with the primary is possible while keeping the secondary resonator Q low. The resulting response will be considerably better than single tuning though not as good as ideal double tuned.

A matching network similar to that shown in Fig. 5.9 was designed for the element example whose admittance was plotted in Fig. 5.5. The design scan range was limited to 60° in the E-plane and 45° in the H-plane. The design scan angle here was $\theta_o = 30°$ and $\phi_o = 0°$.

The resulting admittance plot is shown in Fig. 5.10 for the array over the 12 percent frequency band. This element results in a maximum VSWR of 4:1, and typifying the performance is an average VSWR of 2:1, corresponding to 10 percent reflected power (averaging over frequency and scan angle).

5.1.2 Wide Angle Impedance Matching (WAIM).

Alternative, or additional, approaches to matching stress the scan dependent parameter rather than frequency dependence. These schemes are commonly called Wide Angle Impedance Matching (WAIM) for phased arrays.

Figure 5.11 defines what is meant by WAIM. Fig. 5.11(a) shows a typical immittance locus for a planar array scanned in the E- and H-planes at a single frequency. The unmatched curve is symmetrized in some fashion (like we have just described) about the center of the chart by use of a scan-independent matching structure. By adding a scan-dependent matching structure one can change the size and/or shape of the locus as shown in Fig. 5.11(b) and 5.11(c); this is WAIM.

There are many WAIM schemes that have been used with relative degrees of success. Two distinct type of schemes can be used as a classification; those modifying the transmission line region (feedguides) and those altering the free space region beyond the aperture discontinuity. An extensive summary is presented in reference. [16]

The transmission line schemes rely on inter-connecting curcuits or multi-mode apertures. The free space WAIM schemes use baffles for dipole arrays and for aperture elements use dielectric sheets adjacent to the array plane and denser element packing.

One very useful approach was proposed by Magill and Wheeler in 1965 [17] and uses a thin high dielectric constant sheet in front of the array aperture plane. The scan dependent equivalent susceptance introduced by the sheet is used to partially cancel the admittance variation of the array aperture.

The dielectric sheet is spaced a short distance in front of the array aperture plane, as shown in the Fig. 5.12. The reflection coefficient with and without the sheet was measured for several discrete scan points. If the impedance of the three scan conditions are centered on the reflection chart using single tuned matching, one gets $|\Gamma| \leq 0.38$ (VSWR $< 2.2:1$) without the sheet and $|\Gamma| \leq 0.2$ (VSWR $< 1.5:1$) with the sheet.

WAIM by use of denser packing is shown by comparing the next figures, 5.13(a) and 5.13(b), [18]. An equilateral triangular grid array of circular apertures with element diameters $.356\lambda$ loaded with $\epsilon = 4.1$ is investigated. The top figure shows the admittance locus for a 10 percent frequency bandwidth where the inter-element midband spac-

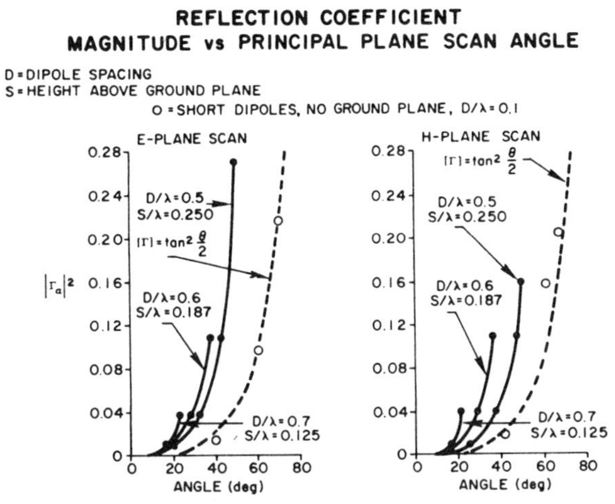

Figure 5.1

APPROACHES TO ELEMENT DESIGN

- INFINITE ARRAY CALCULATIONS — VERIFY BY SIMULATOR MEASUREMENTS
- SMALL ARRAY MEASUREMENTS
- WAVEGUIDE SIMULATOR MEASUREMENTS

(SMALL ARRAY MEASUREMENTS and WAVEGUIDE SIMULATOR MEASUREMENTS: USE IN CONJUNCTION WITH ANALYSIS)

Figure 5.2

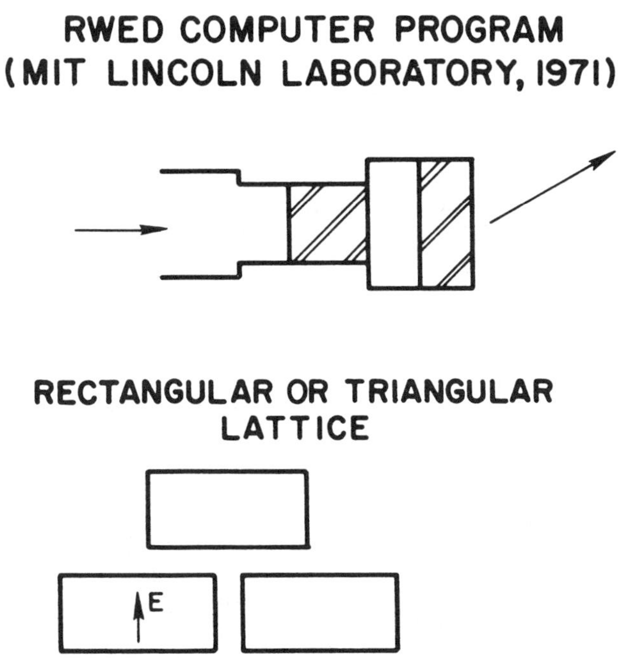

Figure 5.3

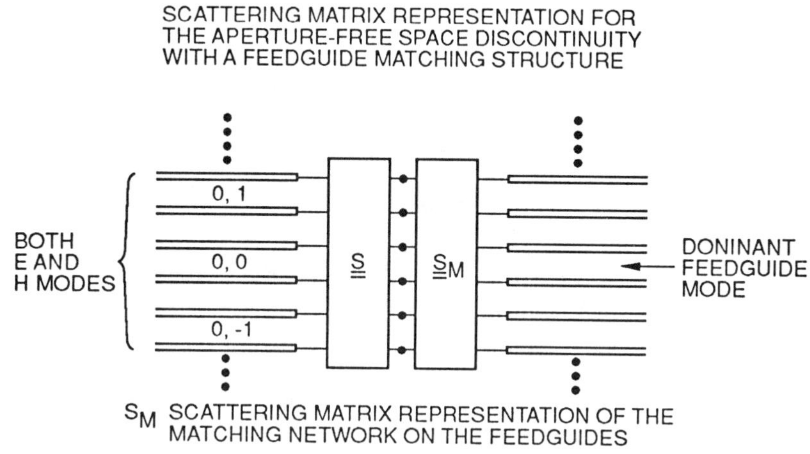

Figure 5.4

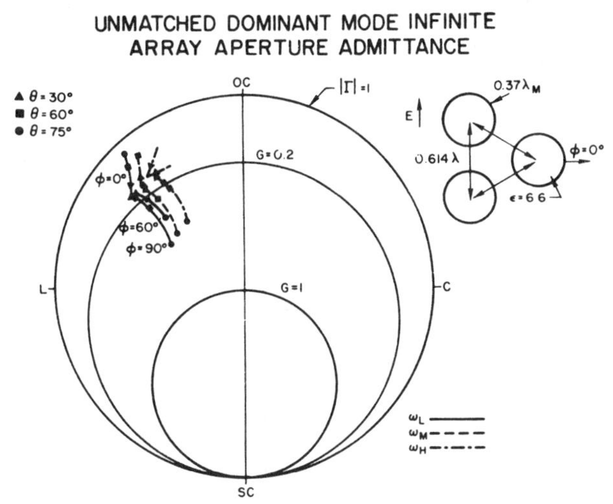

Figure 5.5

DOUBLE - TUNED ADMITTANCE RESPONSE AT SELECTED SCAN CONDITION (Θ_o, ϕ_o)

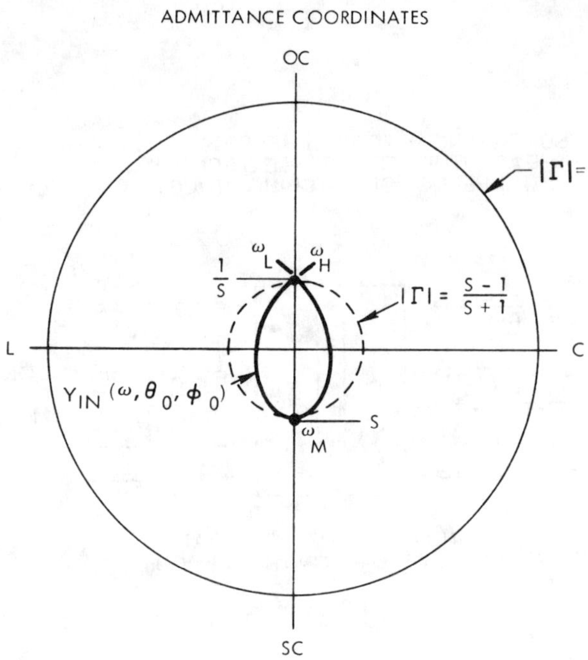

Figure 5.6

DOUBLE TUNED NETWORK REPRESENTATION

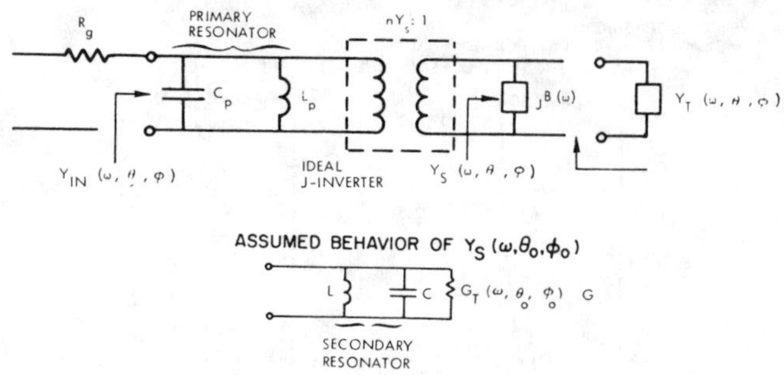

Figure 5.7

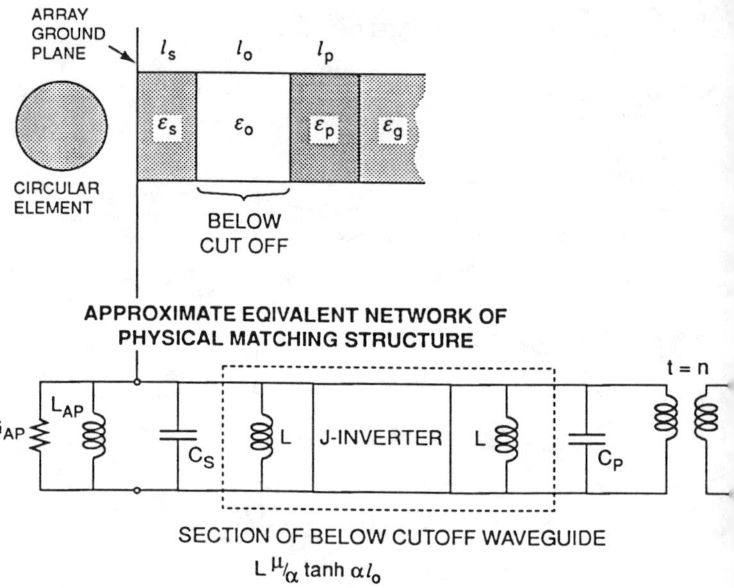

Figure 5.9

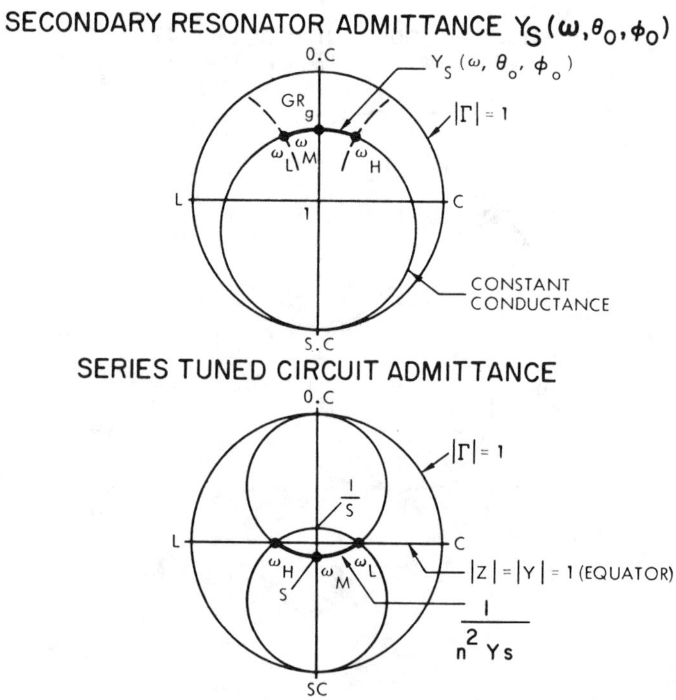

Figure 5.8

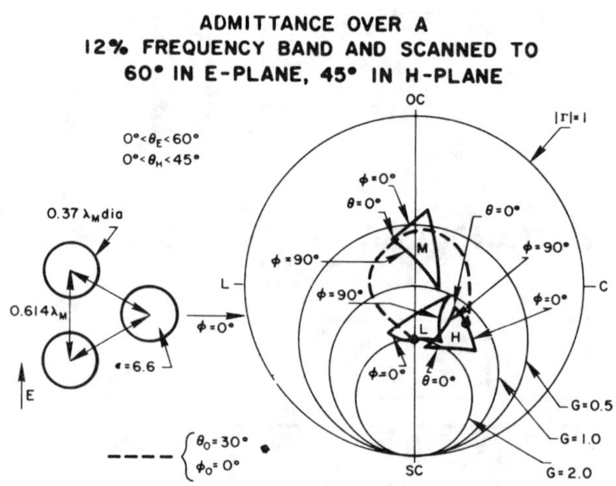

Figure 5.10

DEFINITION OF WIDE-ANGLE IMPEDANCE MATCHING (WAIM); E=E-PLANE SCAN, H=H-PLANE SCAN (REF. [16])

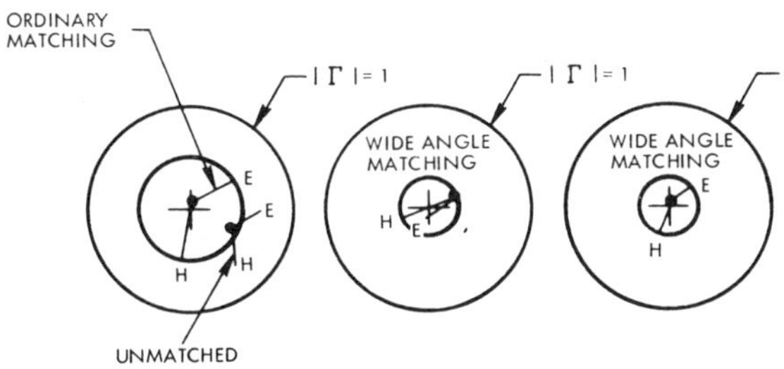

(a) ORDINARY MATCHING WITH SCAN-INDEPENDENT STRUCTURES IN ELEMENT LINES
(b) WAIM BY MOVING THE SCAN LOCI CLOSER WITH A SCAN-DEPENDENT MATCHING STRUCTURE
(c) WAIM BY REDUCING THE SIZE OF THE SCAN LOCI WITH A SCAN-DEPENDENT MATCHING STRUCTURE

Figure 5.11

WAIM BENEFIT WITH A THIN, HIGH-ε DIELECTRIC SHEET, FROM SIMULATOR MEASUREMENTS (FROM MAGILL AND WHEELER [17])

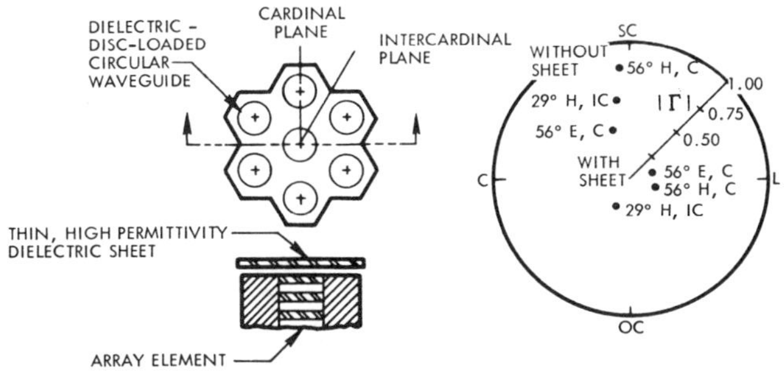

Figure 5.12

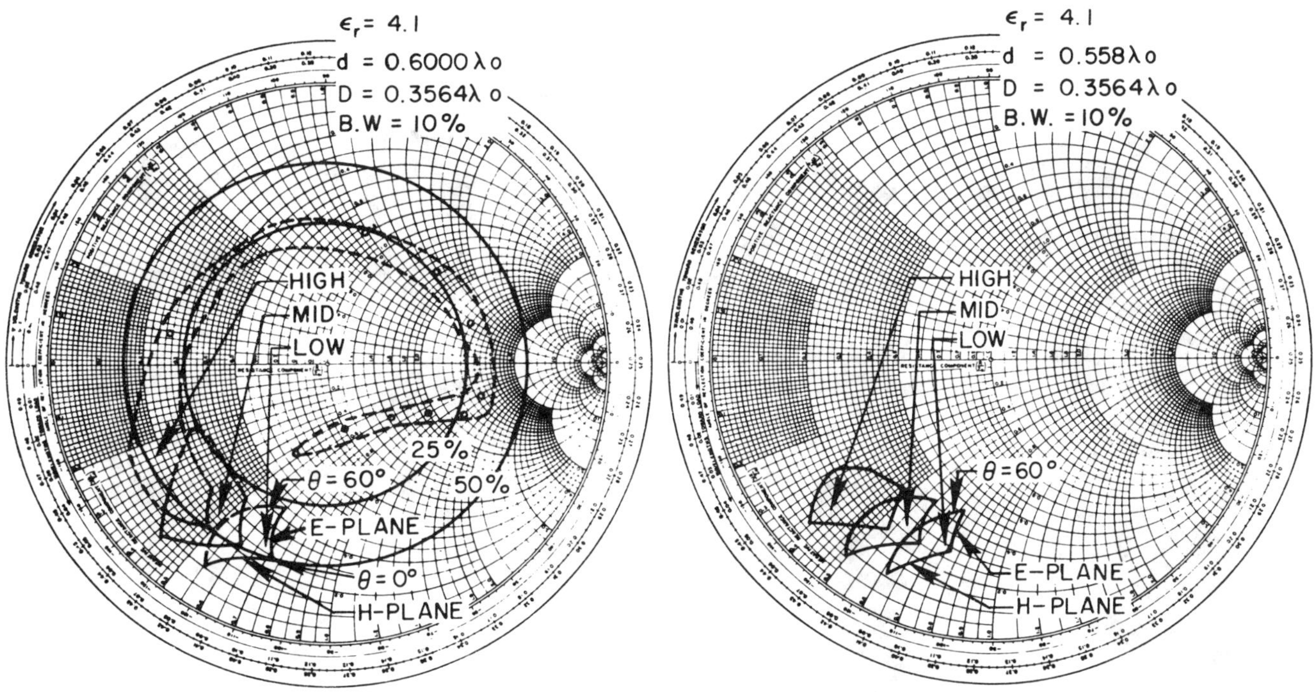

ACTIVE RADIATION ADMITTANCE REFERRED TO APERTURE PLANE

ACTIVE RADIATION ADMITTANCE REFERRED TO APERTURE PLANE WITH REDUCED INTERELEMENT SPACING

Figure 5.13

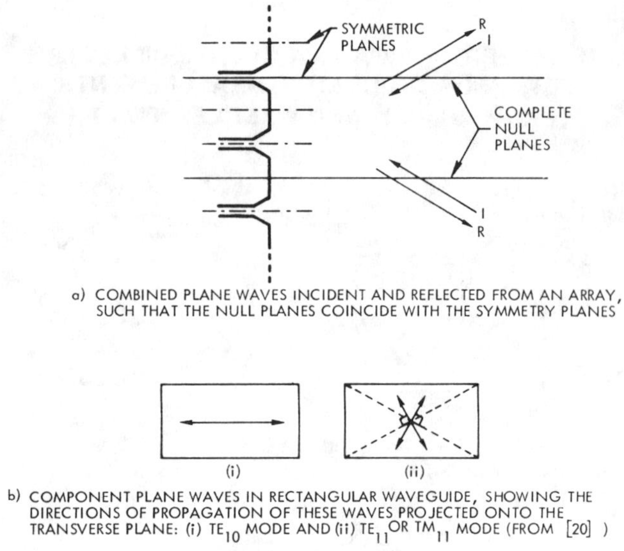

Figure 5.14

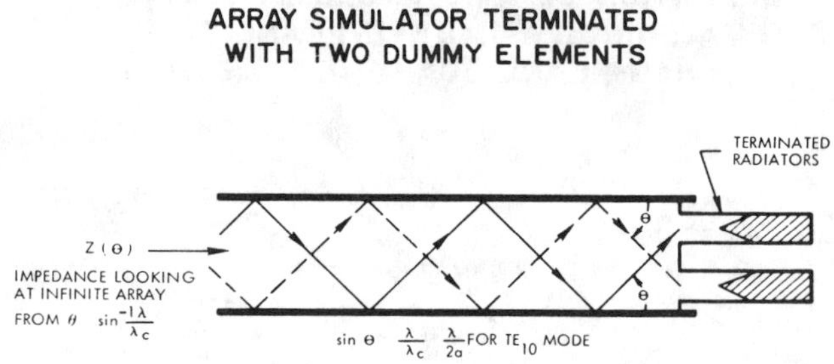

Figure 5.15

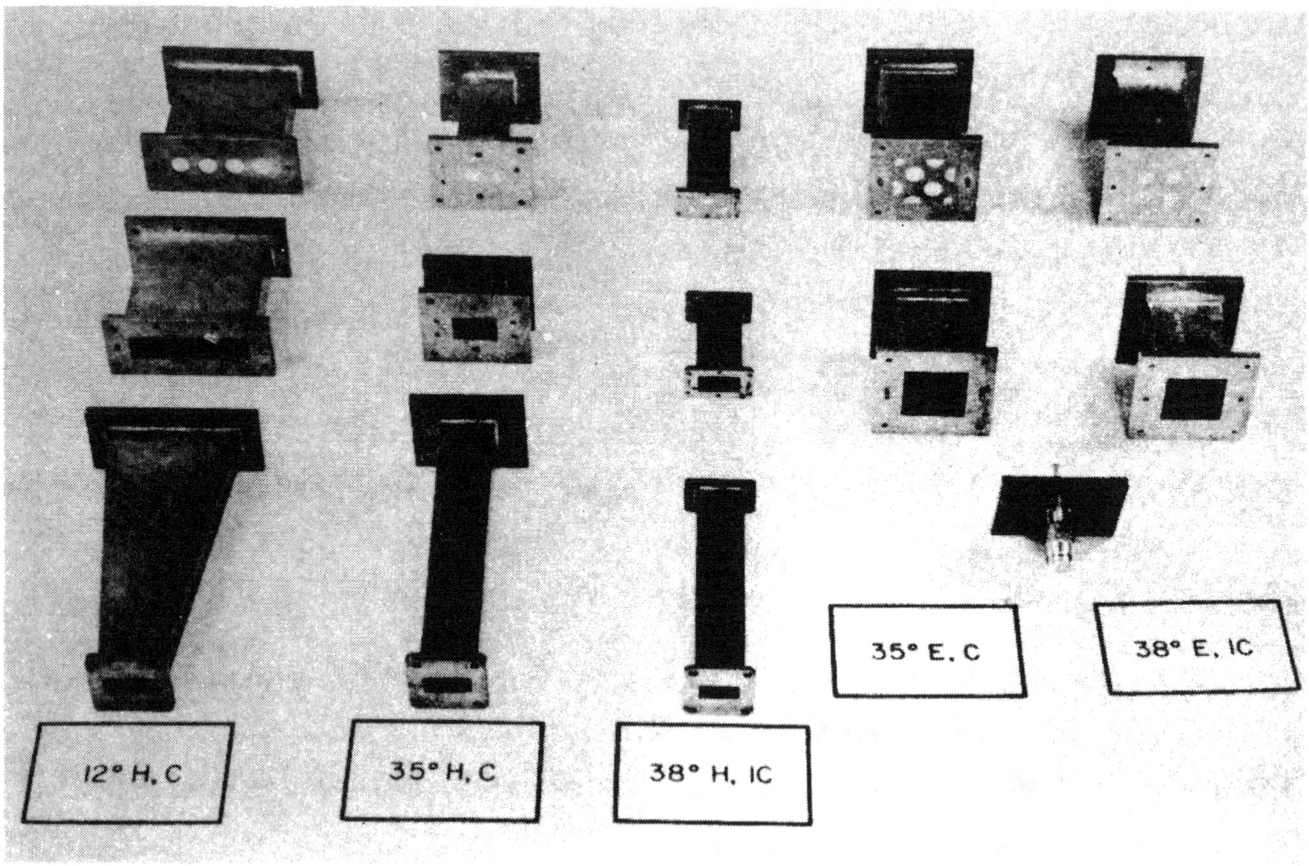

Figure 5.16

Simulators for five conditions of beam radiation.

ing is 0.6λ. Very large admittance variations are observed, and thus the array would be difficult to match. The admittance variation where the array spacing was reduced to 0.558λ is shown in the admittance diagram of Fig. 5.13(b). These tightly packed spiders can be matched considerably better than the former case.

5.2 Waveguide Simulators [19,20]. We have mentioned waveguide simulation of the infinite array in connection with element design. The procedure utilizes the unit cell concept, or infinite array property that the electromagnetic fields in any cell are related to any other only by a phase shift.

The basis of the simulator waveguide is that a set of conducting planes can be placed along symmetry planes of the infinite array in such a way that a few elements enclosed in this (waveguide simulator) boundary will correctly image all other array elements for a particular range of scan angles. The simulator waveguide modes have a plane wave decomposition that corresponds to the radiation properties of the infinite array. In this fashion the reflection and transmission properties of this waveguide terminated in a few array elements "simulates" the plane wave reflection and transmission of the infinite phased array.

To graphically indicate the concept we show in Fig. 5.14(a) a pair of equal amplitude plane waves incident at equal and opposite angles on a phased array. The reflected plane waves have identical reflection coefficients Γ, which is the same (via superposition) as the reflection for each plane wave alone. The pair of plane waves produce an interference pattern that results in sets of null planes. We choose to associate these null planes with the electric field, such that conducting walls can be placed along the null planes without disturbing the fields. Locating the null planes coincident with the symmetry planes correctly images the array.

Consider the rectangular waveguide for use as a simulator waveguide. Fig. 5.14(b) shows the projections of plane wave propagation onto the transverse waveguide plane for the (1) TE_{10} and the (ii) TE_{11} and TM_{11} modes. The field orientations, scan plane and incident direction are a unique operating array condition for a specified frequency.

Figure 5.15 shows a rectangular waveguide simulator excited in the TE_{10} mode and terminated with two dummy elements. The plane waves propagate from side to side in the waveguide (H-plane) and make an angle θ with respect to array face normal. This angle θ is well known to be;

$$\theta = \mathrm{SIN}^{-1}\ (\lambda/\lambda c) \tag{5.1}$$

where λc is the cut-off wavelength.

As an example, we assume the element spacing in the figure is 0.6λ; then this two element simulator is 1.2λ wide. The cut-off wavelength for this TE_{10} mode in rectangular waveguide is 2(1.2λ) = 2.4λ. Therefore this simulator scans this array $\theta = \mathrm{SIN}^{-1}$ (1/2.4) = 24.6 degrees in the H-plane; and the reflection properties are equal to that of the active infinite array scanned to this angle at frequency f = c/λ.

There are numerous simulators corresponding to many of the array symmetries, however only discrete scan conditions can be simulated using different size and shape waveguides operating in different modes.

Examples of several simulators that have been built are shown in Fig. 5.16 for a rectangular grid array of circular elements. [20]

Extensive discussions and element design approaches using simulators are presented by H.A. Wheeler in references [2] and [3]. The methods proposed are generally combinations of computations of reflection coefficient in conjunction with measurements on simple simulators.

A simple simulator that has been used to verify computed resonances is shown in the next figure. [22] The array elements are dielectric loaded rectangular waveguides feeding narrow slot apertures covered by a dielectric slab. The computed and measured admittance is shown in Fig. 5.18.

5.3 The Small Array. Methods of element design using the small array are extensive [23], [24] but are generally more costly and approximate than the analysis and simulator approaches, and would probably only be used for difficult geometries and in very special cases. It is perhaps more informative to examine and contrast the performance of the small (finite) array as compared to the large (infinite) array.

Fig. 5.19 shows calculated reflection coefficient modulus for a dielectric slab covered parallel plate guide array as a function of number of elements. For the infinite array two pronounced resonances appear. The one closest to broadside scan ($\sin\theta \approx .12$) is higher Q than the other at $\sin\theta = 0.59$, similar to the resonance width trends observed for the protruding eielectric elements. A reduction to forty nine elements dramatically alters the near broadside resonance, but the one further out still possesses a high reflection; similarly for 15 elements. However, for 5 elements all correlation to the infinite array pattern is lost. These curves show that depending on the resonance width and location the "large enough" approximation for the infinite array is uncertain. For the protruding element cases good correlation was obtained for measurements on 39 columns; here for the slab covered array more than 49 are needed.

We will now look at some interesting comparisons of element gains and patterns in a small array that does not have resonances. [25]

A triangular grid array of 59 rectangular aperture elements is shown in Fig. 5.20. The apertures are cavity backed rectangular slots fed by strip transmission line, where the isolated element has a VSWR of 2.4. A comparison between computed and measured reflection coefficients at the generators are shown in Fig. 5.21. Typical agreement is within 15 percent, and a departure from the isolated case is noted.

The table of Fig. 5.22 summarizes broadside gain performance. The isolated element gain is 5.3dB. The array cell area gain for this particular choice of lattice configuration is 7.3dB. Thus, in order to achieve full array aperture gain the element gains need to increase approximately 2dB, on the average.

The array area gain defined as the cell area gain times the number of elements is 25.0dB. The predicted gain is 23.7dB. If we take into account 0.4dB due to taper loss for the assumed illumination, and 0.5dB reflection loss, 0.4 differential is still present. This difference is attributed to direc-

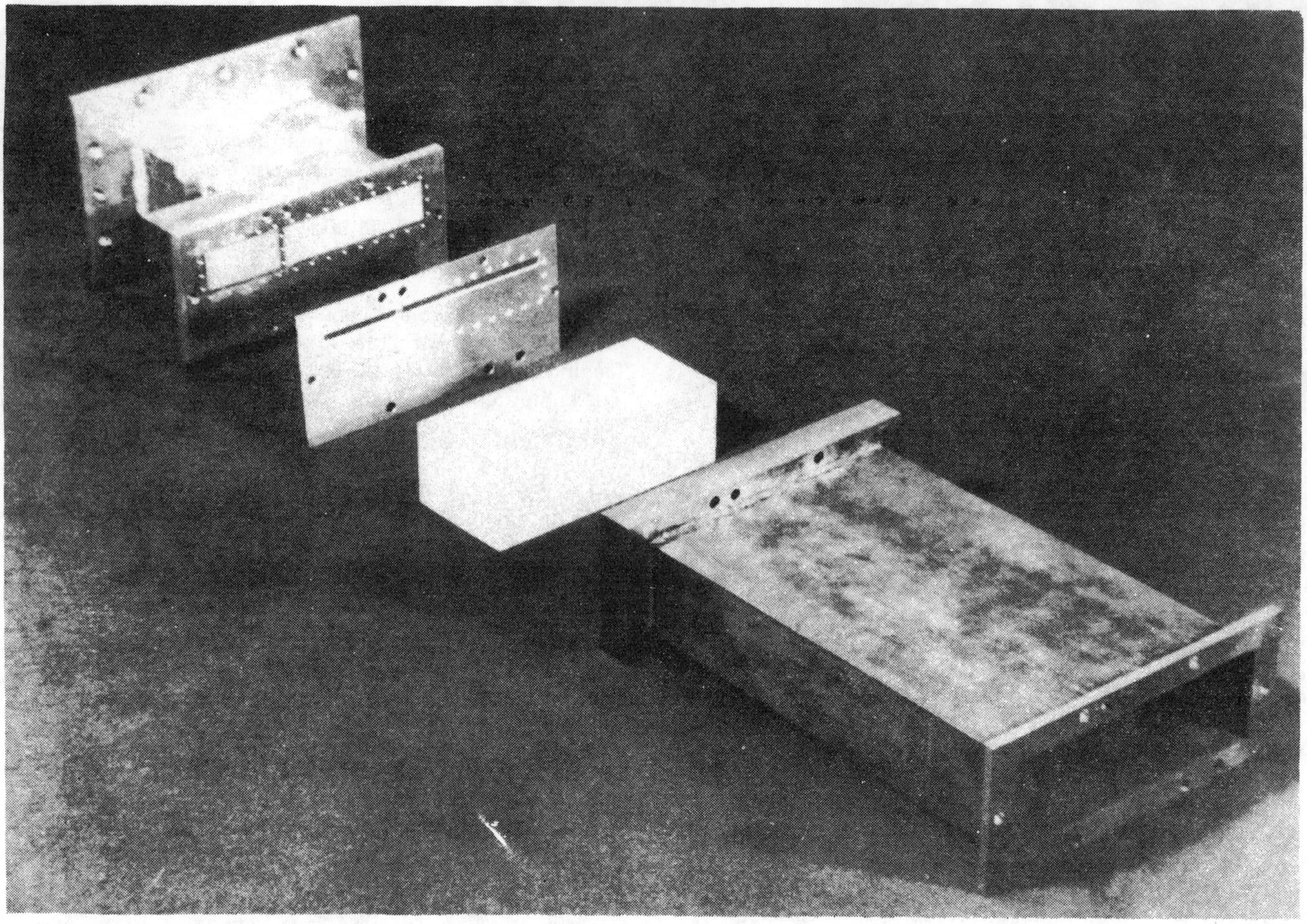

Figure 5.17

Exploded view of waveguide simulator showing rexolite-loaded feed waveguides, slot plate, rexolite slab, and simulator waveguide

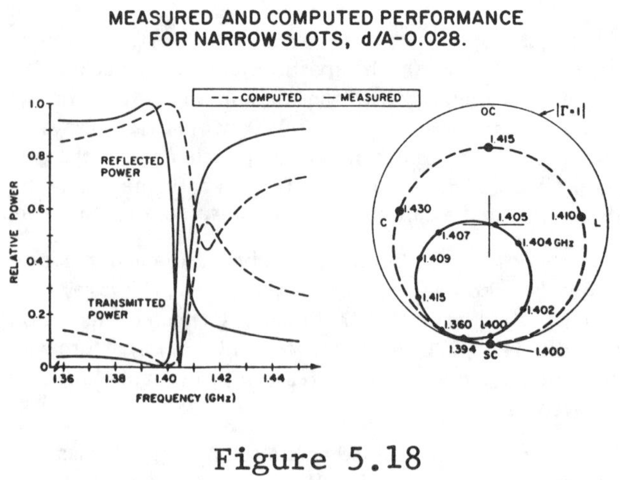

Figure 5.18

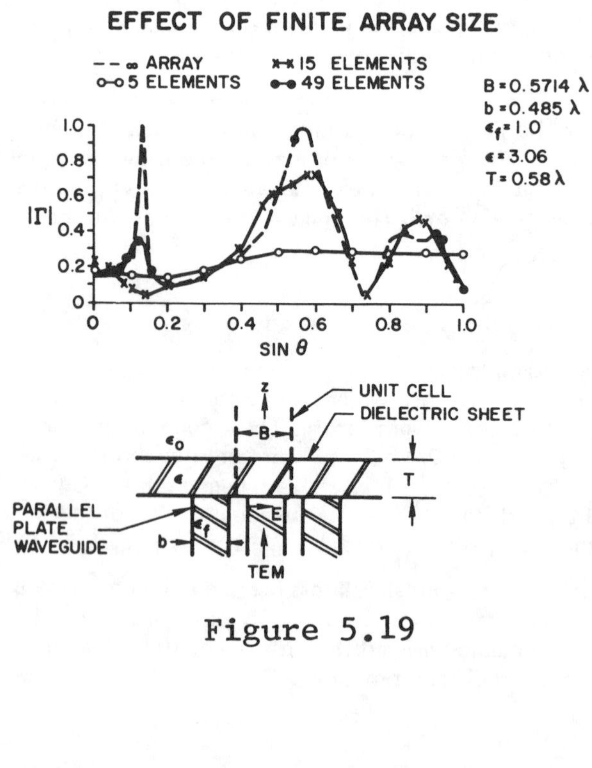

Figure 5.19

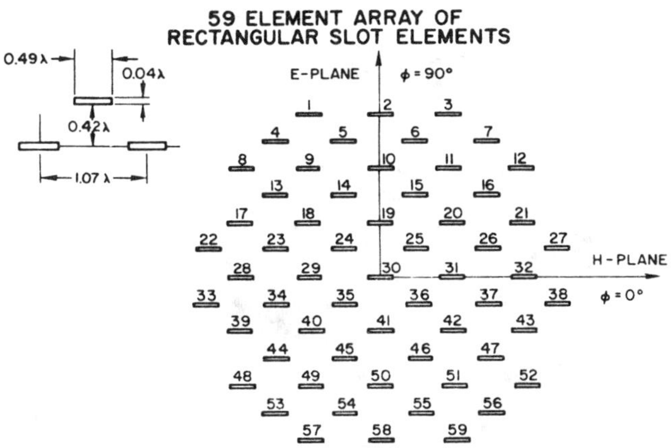

Figure 5.20

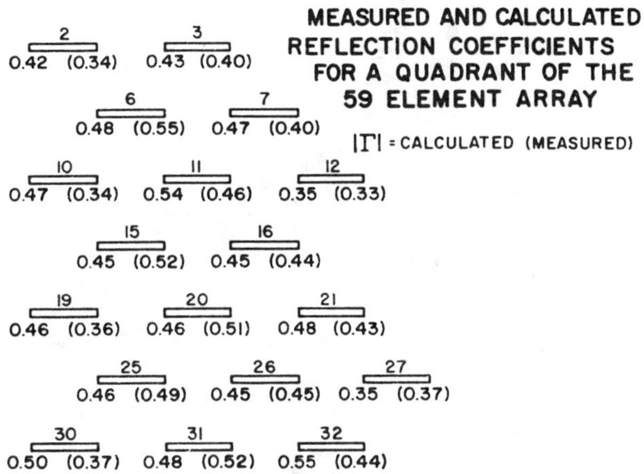

Figure 5.21

BROADSIDE GAIN

ISOLATED ELEMENT GAIN (MEASURED MATCH)	= 5.3 dB
CELL AREA GAIN	= 7.3 dB
ARRAY AREA GAIN	= 25.0 dB
PREDICTED ARRAY GAIN	= 23.7 dB
TAPER LOSS = 0.4 dB	
REFLECTION LOSS = 0.5 dB	
DIRECTIVITY LOSS = 0.4 dB	
ELEMENT GAIN IN ARRAY ENVIRONMENT:	
ELEMENT 30, ARRAY CENTER	=7.5 dB
ELEMENT 2, E-PLANE EDGE	= 5.7 dB
ELEMENT 25, CLOSEST CENTER NEIGHBOR	= 6.2 dB
ELEMENT 27, EDGE ELEMENT	= 5.6 dB
ELEMENT 32, H-PLANE	= 7.9 dB

Figure 5.22

DUAL-POLARIZED DIPOLE ARRAY

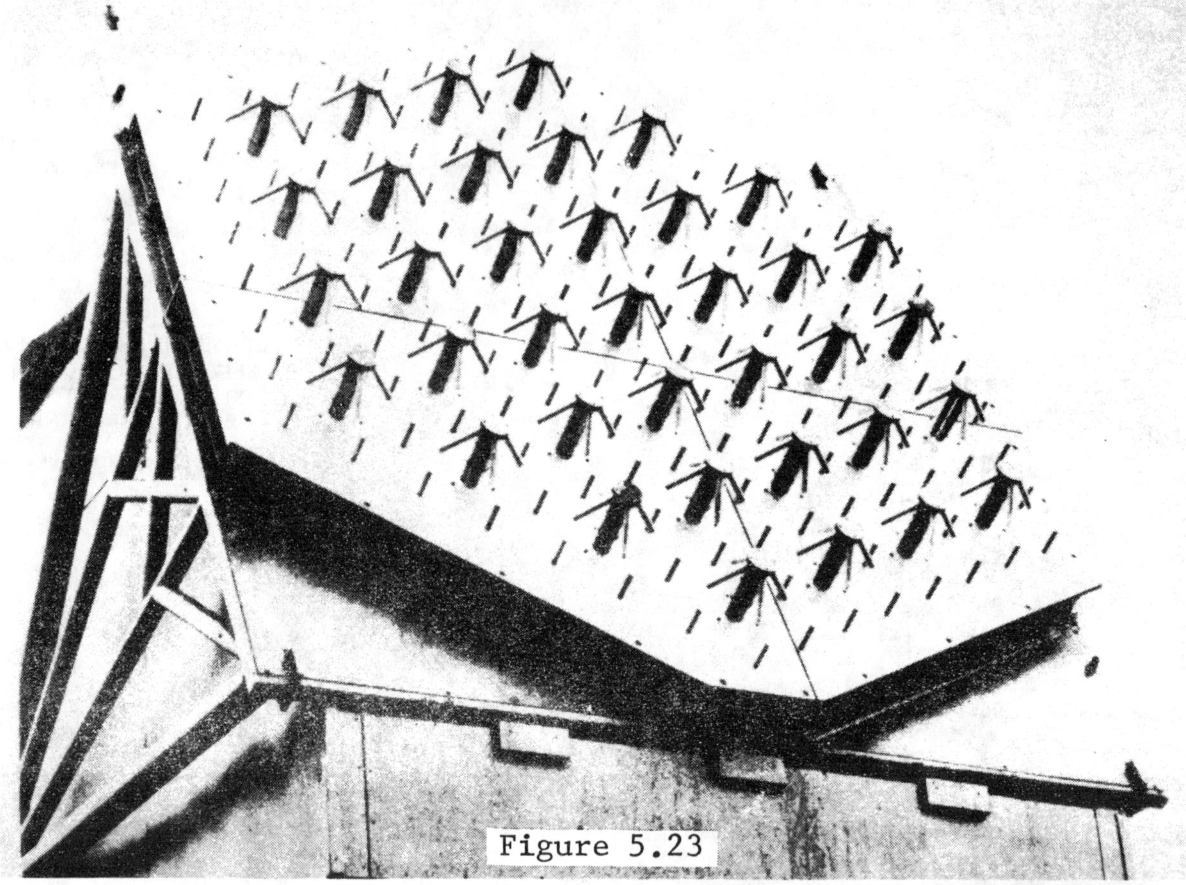

Figure 5.23

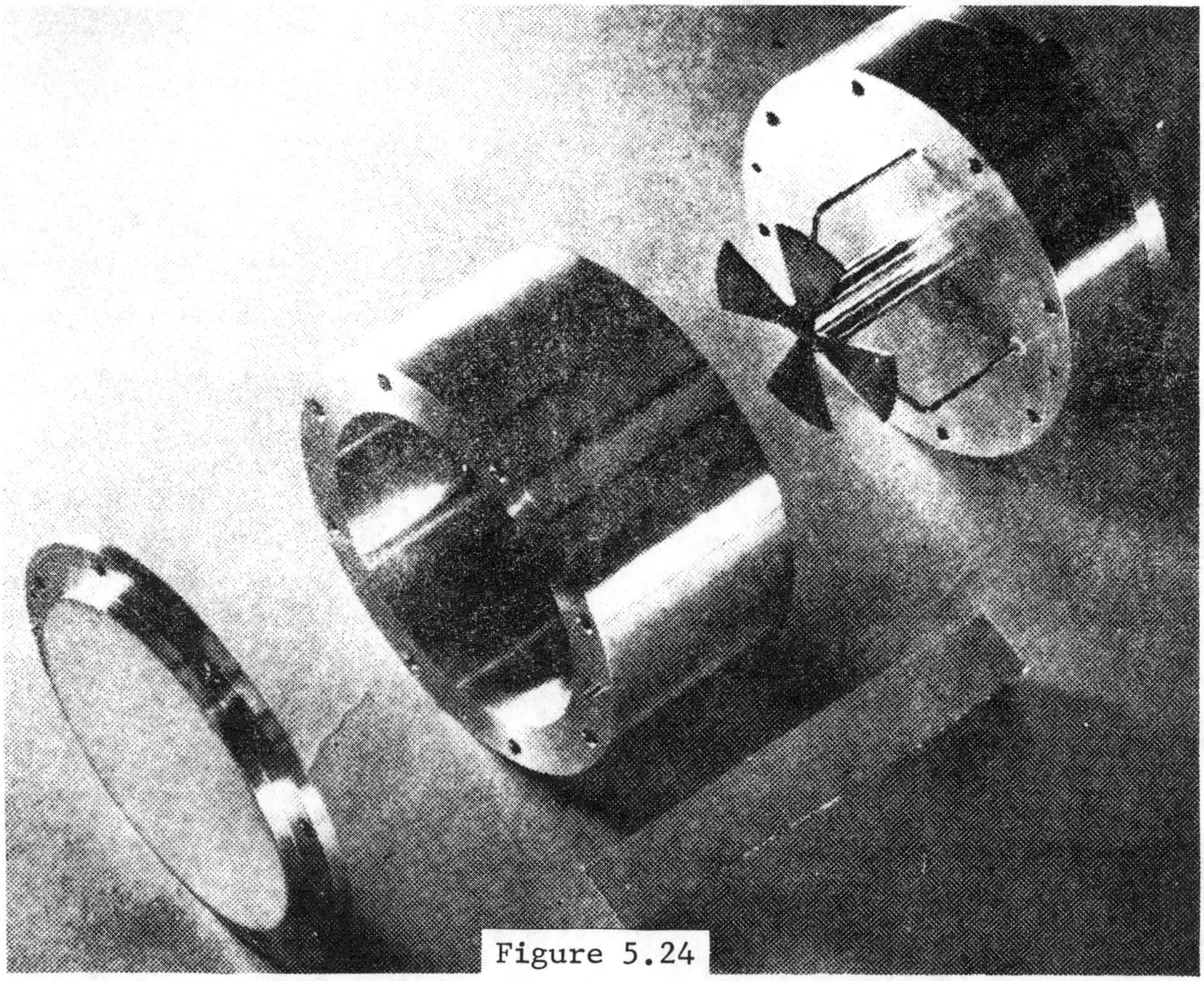

Figure 5.24

tivity loss in the element patterns (i.e. they do not all achieve their peaks at broadside) and to a change in the illumination function. However, we do see that on the average the elements do achieve a gain of 6.9dB in the array environment.

Typical element gains in the array environment are tabulated. The center element achieves a gain of 7.5dB, higher than the cell area gain. The E- and H-plane edge elements achieve gains of 5.7 and 5.6dB respectively, close to that of the isolated element. The highest value of 7.9dB is that of element 32, located on the H-plane axis.

5.4 Some Designed Elements.

5.4.1 Narrow Band Designs. Fig. 5.23 is an array of dual linearly polarized dipoles above a conducting ground plane plane. This array is a test section of an array containing many more elements, developed at Bell Telephone Laboratories. A waveguide example of a dual polarized element is shown in Fig. 5.24; a probe pair feeds a section of below cut-off waveguide coupled to a front (aperture plane) disc. This element was an early version (Wheeler laboratories) of a dual polarized, double tuned type element.

Fig. 5.25 shows an array radiator with an integrated non-reciprol toroidal phase shifter and flux driver (Raytheon Co.) The element is used in a wide angle scanned space fed phased array (C-band). Linearly polarized energy from the space feed enters the element at one end via a rear radiator, passes through the phase shifter receiving proper excitation phase command and then is radiated again linearly polarized. For space fed transmit type arrays two free space-element interfaces are present. Each has a distinct requirement: the "rear" radiator subtends a small angular region of incident radiation (plane wave excitation near broadside from feed horns), while the "front" radiator requires a wide scan element design. For this element a rear design is facilitated by a slightly protruding dielectric loaded waveguide element (no resonance for this small protrusion with limited incidence angles) and the front radiator is a double tuned disc type previously discussed. This design has an average dissipation loss of 1.0dB and can handle 3.5Kw of peak power.

5.4.2 Broadband Designs. Phased arrays are capable of broadband behavior. Limitations are imposed by the elements rather than mutual coupling. However, for grating lobe free operation most broadband elements have distinct low frequency performance degradation. Spirals and helices have diameters approximately equal to 2/k (i.e. λ/π) at the lowest frequency. Thus about 1.5:1 (wide scan) to 3:1 (no scan) bandwidth arrays are possible, with these type of elements.

An example of a phased array of spiral elements is shown in Fig. 5.26. A corporate feed serves as the power distirbution network. The spiral elements are circularly polarized and will operate over a broad bandwidth, limited here by the feed for a high performance application (low sidelobes and good circularity). The spiral elements are etched from copper clad foam, forming a light weight easily fabricated design. (Etched foam elements are not without problems: high temperature failure, tolerances and machining; etc.)

A low cost, small lightweight broadband design has recently been presented in the literature. [26] A stripline fed tapered notch is a broadband radiator capable of multi-octave bandwidths and multiple polarizations. Shown in Fig. 5.27 is the linearly polarized flared notch that will also be shown in a polarization diverse configuration. This antenna is capable of multi-octave bandwidth.

The notch region is formed by etching away the stripline outer conductors and is excited by the stripline TEM mode. The line remains balanced since the notch is etched symmetrically from both outer conductors.

The excited voltage in the gap couples to a leaky wave of the notch and travels toward the stripline board edge, with the polarization lying in the plane of the board.

The transition to free space at the board edge remains well matched because the impedance discontinuity is small. The wavenumber of the leaky mode lies between that of a parallel plate TEM mode in the dielectric constant and that in air. The dielectric constant is 2.2.

The boadband character of the stripline-notch transition arises because of the low Q of the transition.

(i) It has been found empirically that the length of extension of the center conductor beyond the notch should be about a quarter wavelength. Alternative designs using a short circuit pin beyond the gap provided similar results. In both cases a low frequency limitation arises from a large increase of gap susceptance.

(ii) The general curvature of the notch was found to not be that important, although the width at the board edge appeared to be best at greater than $\lambda/2$, (recently narrower aperture width designs have been carried out making possible an increased array bandwidth) similarly the gap width at the transition region was found to be small for best matching. (Typically one-two linewidths).

Fig. 5.28 is a curve of VSWR versus frequency for a single linearly polarized notch antenna. The frequency spread is for an octave. This data corresponds to an early design where the high frequency dimensions were:

a) 0.6λ *gap width* across the board edge.
b) 0.37λ notch height
c) 0.05λ *gap width*
d) 0.12λ center conductor height above notch bottom
e) 0.2λ extension of center conductor
f) 70 Ω line

The VSWR typically remains under 2:1 except at the lower frequency where 3:1 VSWRs occur. In these designs fine adjustments of line width and notch width results in further matching benefits. The element gain varied between 1-3dB over this octave. A non-scan linearly polarized 64 element square array of these notch elements was built. The primary objective was to determine how close to full aperture gain could be achieved over a 5:1 frequency band.

The array as shown in Fig. 5.29 is formed by 8 columns of 8 elements each, with metal plates forming the rows in a square grid. The spacing was selected at .953λ at the highest frequency. Broadband reactive power dividers are used for the corporate feed.

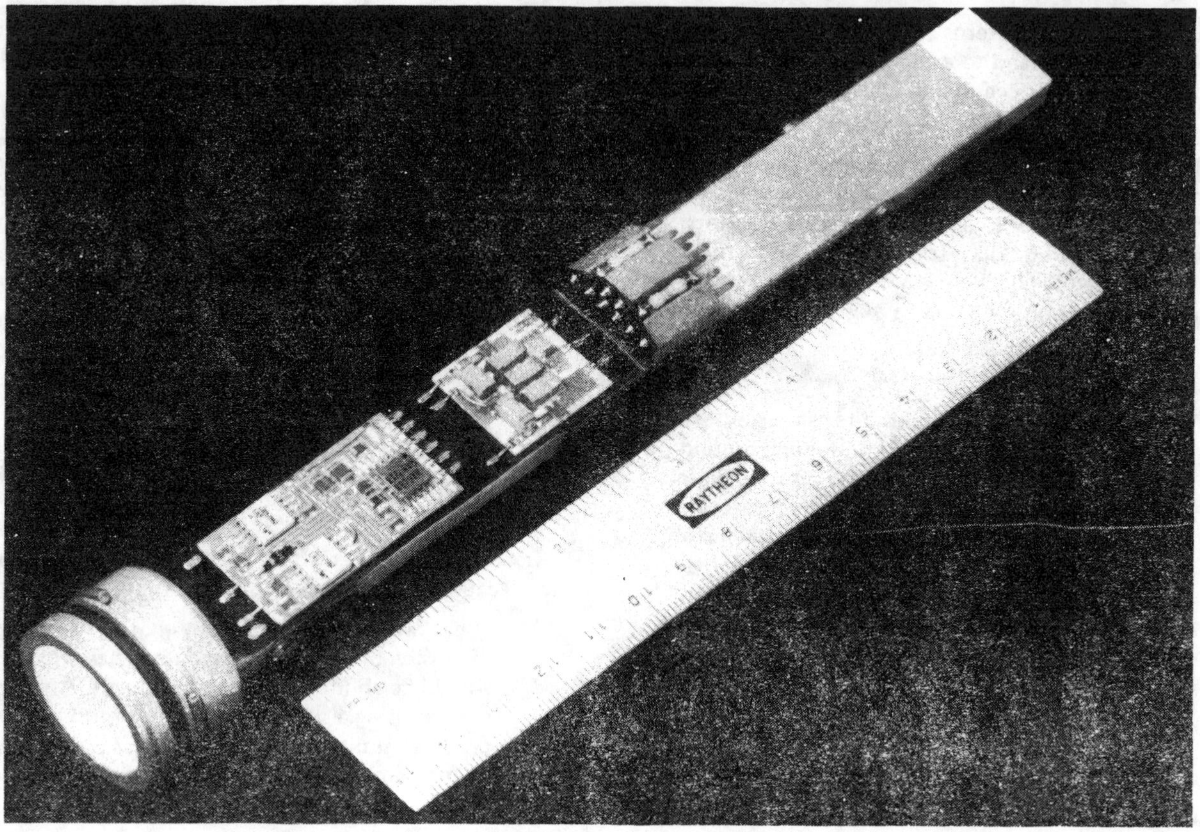

Figure 5.25

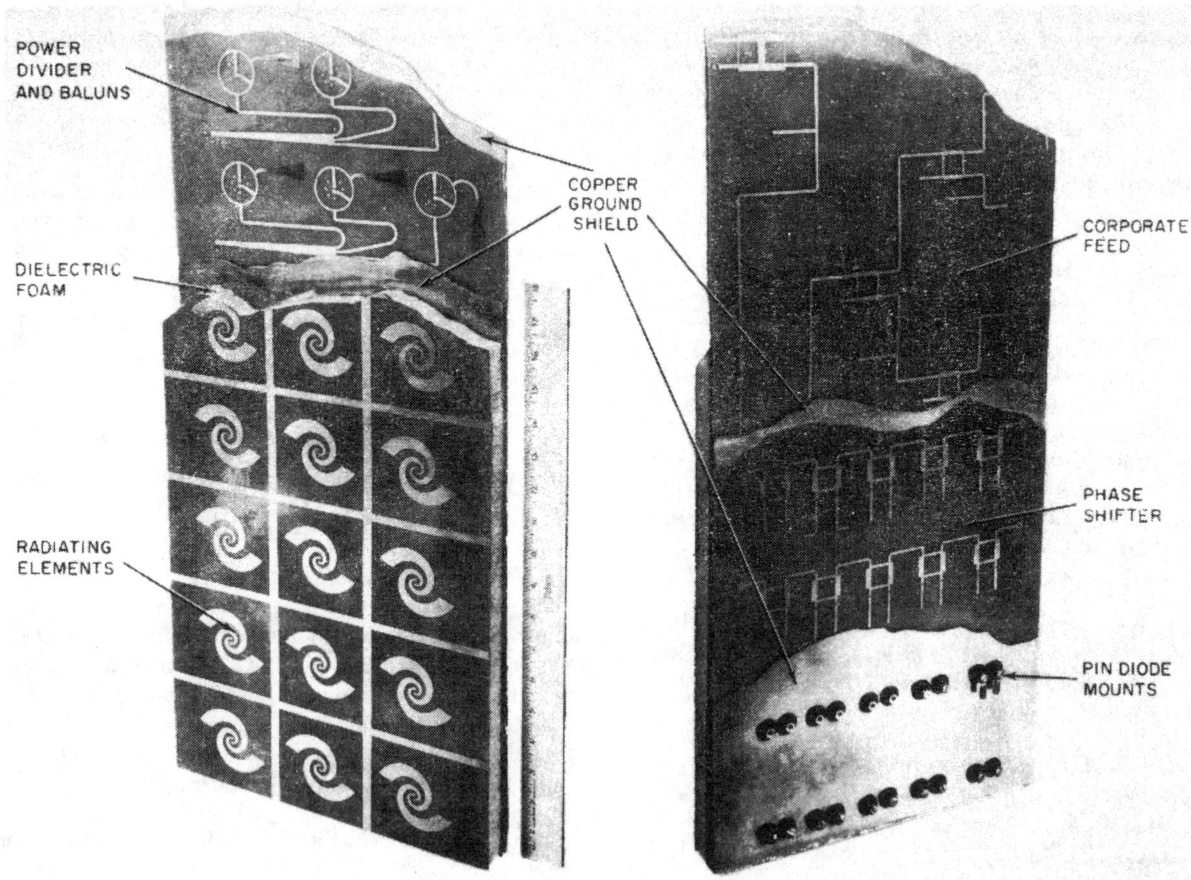

Figure 5.26

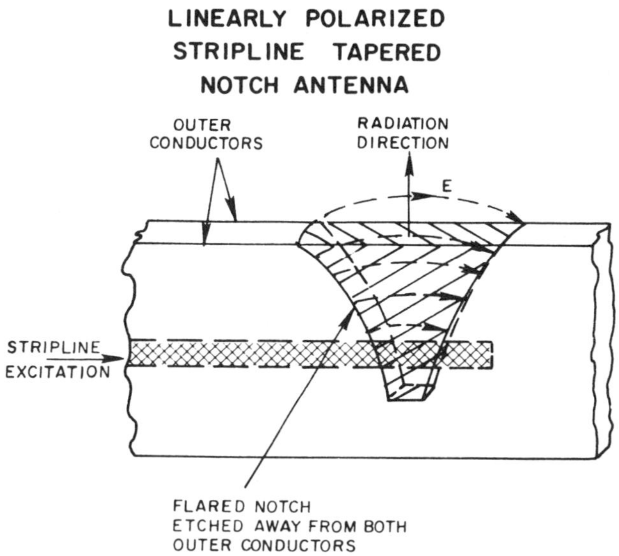

Figure 5.27

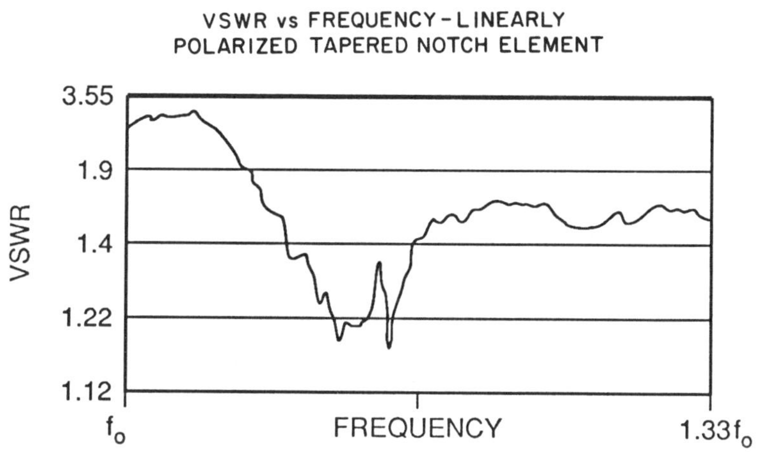

Figure 5.28a

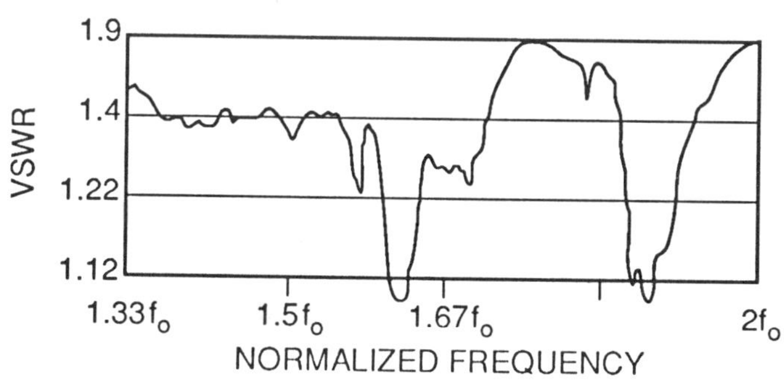

Figure 5.28b

Figure 5.29

Nonscan Tapered Notch - 64 Element Array

The columns of elements were formed by etching a series of 8 equi-spaced notches on a stripline board. These were then formed into an "egg crate" type construction whose cross rows were aluminum strips.

These elements were designed as isolated elements in a free space environment and in waveguide simulators. Fig 5.31 shows the measured gain and theoretical directive gain. These gains are within 2-3dB over most of the 5:1 band. All feed system and aperture mismatch is included in the measured gain. Losses primarily arise in the feed network. Typically 1-2dB, with an increase at the high frequency. The dips that appear are attributed to several sources.

- feed mismatch
- aperture edge effects (finiteness)
- back radiation from array ground plane

Fig. 5.32 is the antenna VSWR versus frequency. Typically the VSWR remains under 2:1 except at the low end of the band. This is primarily due to the increase in gap susceptance at the low frequencies. The several other points reaching 3:1 VSWR's again arise from the same effects as the dips that occured in the gain.

Figs. 5.33(a), (b) and (c) are array patterns for uniform amplitude and phase excitation. All patterns are generally well focused and have correct sidelobe and beamwidth values. At the lower frequency there is some defocusing here arising from the larger mismatches and subsequent influence on aperture excitation.

The next investigations were to incorporate these antennas, as elements of a scanning array. Fig. 5.34 shows the directive gain (losses and mismatch removed) versus frequency for a 64 element array, square spaced at .657λ at 2.25 f_0, with E-plane scan angle as a parameter. The frequency range here is $f_0 \rightarrow 2.25\ f_0$. Static scanning was accomplished by extending the appropriate line lengths in the feed. The scan loss is generally consistant with the aperture foreshortening (cosθ). No peculair scan losses appear although finer scanning is needed for adequate verification. The dips at 1.5 f_0 were a connector problem, not an effect associated with the elements or aperture. H-plane scan results are about the same.

Shown in Fig. 5.35 is a sketch of a broadband polarization diverse radiator formed by two linearly polarized notches.

a) One element is groved from the board edge to just above the stripline center conductor feedline.

b) The other element is grooved just below the center conductor feed such that the joined elements essentially retain all the characteristics of the individual linearly polarized elements.

The top grooved element has shorting pins at the notch bottom to form the short circuit plane for the other element.

its obvious that in order to retain the external notches as identical, the stripline center conductor feeds have to be slightly offset. This offset is typically one line width.

This element demonstrated the broadband characteristics of the linearly polarized version.

5.5 Instantaneous Array Bandwidth [27],[28]

5.5.1 Aperture Effects. Although phased array elements can be matched over very broad bandwidths, there are instantaneous bandwidth limitations imposed by phase shifters and feeds. A scanned beam will change in direction as frequency is changed for arrays steered by phase rather than time delay. For series feeds, bandlimiting arises from a resulting change in inter-element electrical spacing.

To separate the aperture from feed effects a parallel feed is considered. In this fashion the equal path lengths to each of the elements eliminate the scanning effects of the feed. Fig. 5.36 shows a equal line length parallel feed with element spacing d scanned to θ_o with respect to the array normal.

The phase of the ith element for scanning to θ_o at λ is,

$$\psi_i = \frac{2\pi\ i\ d}{\lambda} \sin\theta_o \qquad (5.2)$$

The frequency derivative of this equation with θ_o is,

$$-\frac{c}{f^2}\ df = \frac{2\pi\ i\ d}{\psi_i} \cos\theta_o d\theta_o \qquad (5.3)$$

where c = fλ = speed of light.

Using $\lambda = \frac{c}{f} = \frac{2\pi\ i\ d}{\psi_i} \sin\theta_o$ one gets,

$$-\frac{df}{f}\frac{2\pi\ i\ d}{\psi_i} \sin\theta_o = \frac{2\pi\ i\ d}{\psi_i} \cos\theta_o d\theta_o$$

Whence,

$$\Delta\theta_o = -\Delta f/f \tan\theta_o \qquad (5.4)$$

is the change in scan angle with a change in frequency. The quantity Δf/f is termed fractional bandwidth.

The equation for θ_o indicates no scanning for a broadside array with changes in frequency. Although antenna size and beamwidth are not explicitly stated in the equation it is evident from the next figure that a bandwidth definition needs to incorporate a beamwidth parameter in light of the frequency scanning expression.

As the array is scanned the power in the direction θ_o is decreased. For example (see Fig. 5.37(a)) if the array is frequency scanned a half beamwidth then the power in the θ_o direction is down 3dB. For a ¼ beamwidth scan there is approximately 0.7dB gain loss in the desired direction θ_o. An approximate curve of loss of gain in the θ_o direction is shown in Fig. 5.37(b), where fractional beamwidth scanned at the band edge is varied. If at the lowest frequency in the band a half beamwidth shift is noted then at *that* frequency there is a 3dB loss. For the CW case this is a 3dB loss; however for a pulse with this same low frequency scan (symmetric about θ_o at center frequency f_0) and corresponding high frequency shift, the gain loss, as interpreted as total energy on target, is about 1dB.

A reasonable criteria for bandwidth has been suggested by Frank. [27] One can permit, or design to, a quarter beamwidth scan from frequency at a scan angle of 60°. The corresponding gain loss will be about 0.7dB. This criteria as applied in the previous beam shift equations leads to a convenient rule of thumb,

$$\text{Bandwidth (\%)} \approx \text{Beamwidth (deg)} \qquad (5.5)$$

NOTCH ELEMENTS ETCHED ON A STRIPLINE BOARD

Figure 5.30

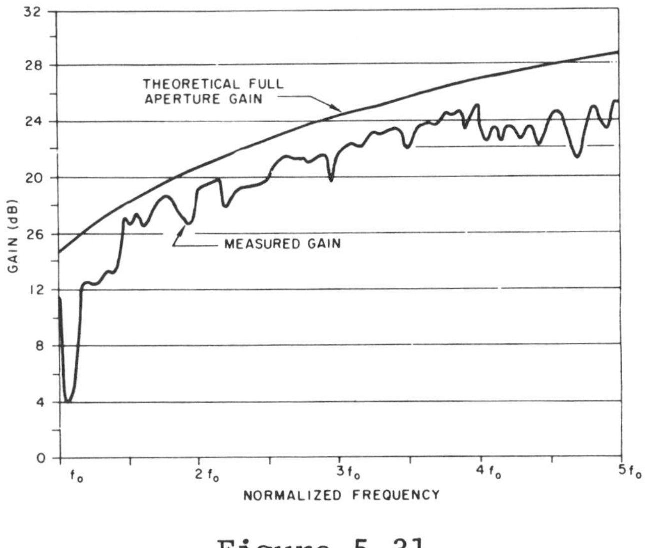

Figure 5.31

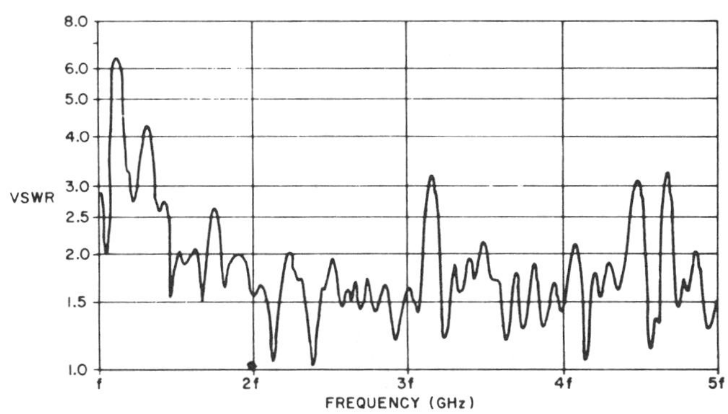

Figure 5.32

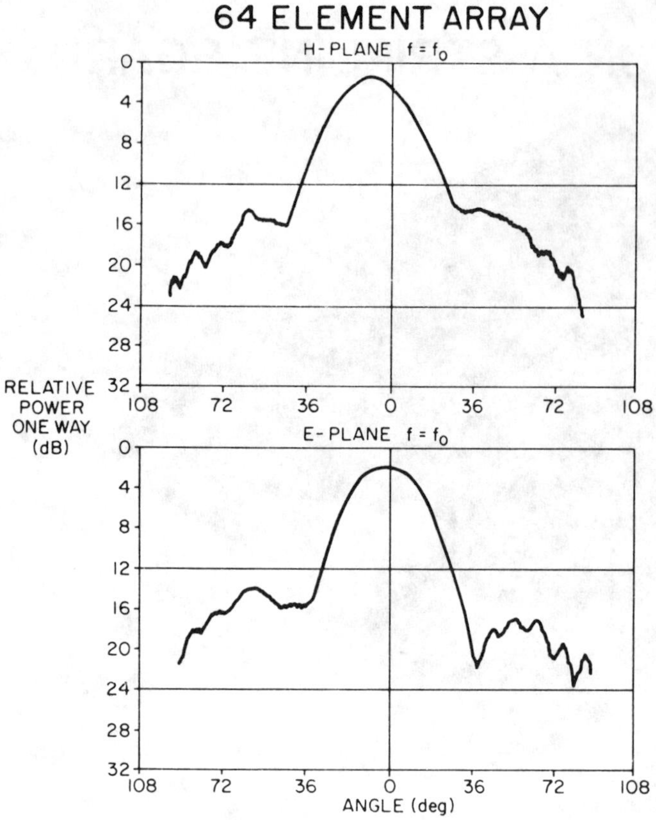

Figure 5.33a

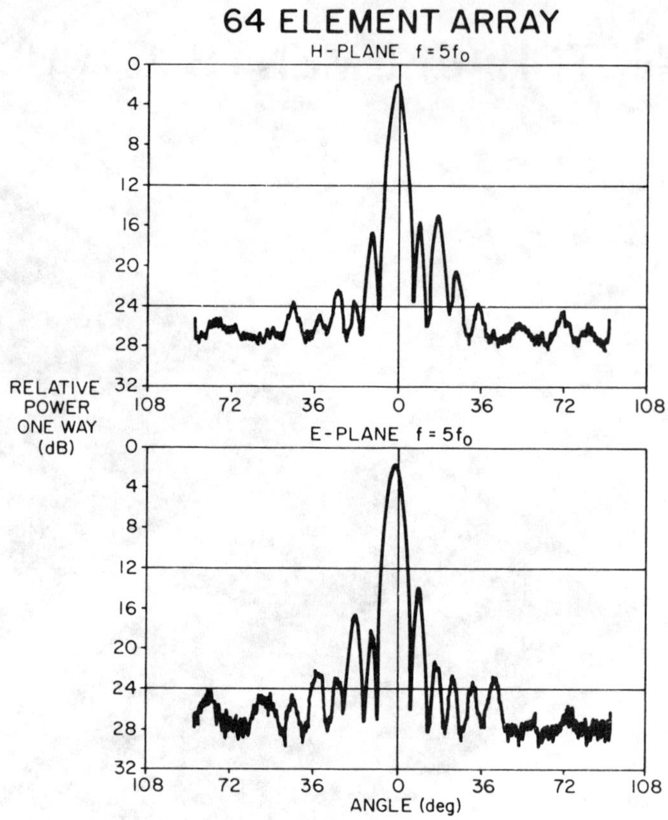

Figure 5.33c

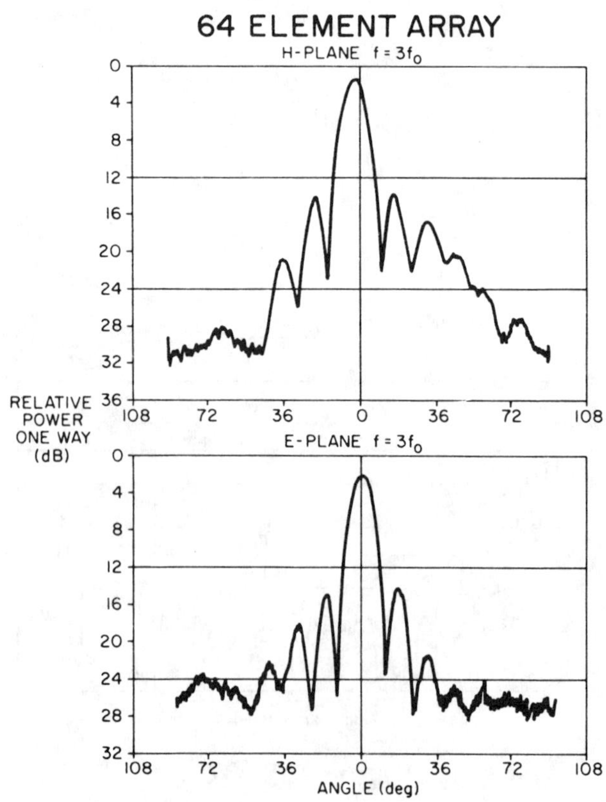

Figure 5.33b

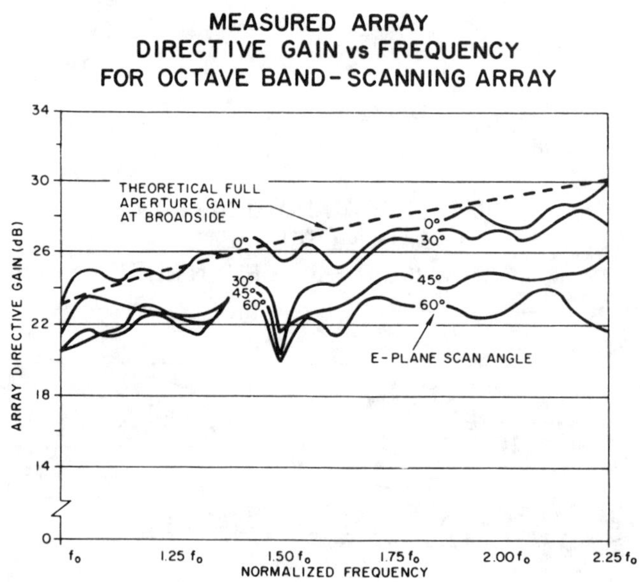

Figure 5.34

INDIVIDUAL RADIATORS (a) AND (b) ARE ASSEMBLED TO FORM POLARIZATION DIVERSE RADIATOR (c)

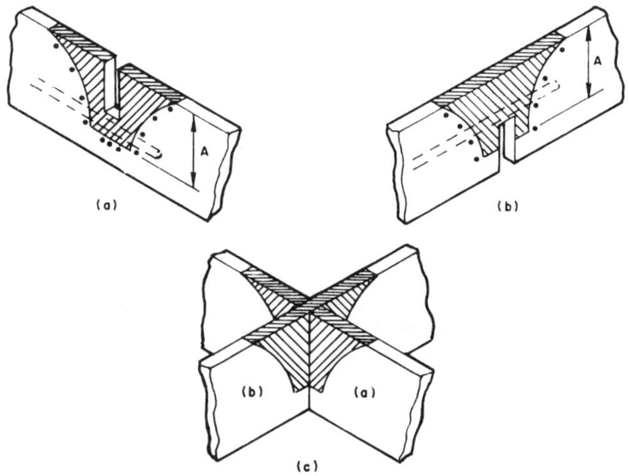

Figure 5.35

BANDWIDTH CRITERIA-APERTURE EFFECTS

CW

A CRITERIA:

PERMIT SCAN 1/4 BW -- 0.7 dB LOSS AND 60° WIDE ANGLE

BANDWIDTH (%) ~ BEAMWIDTH (deg)

PULSE

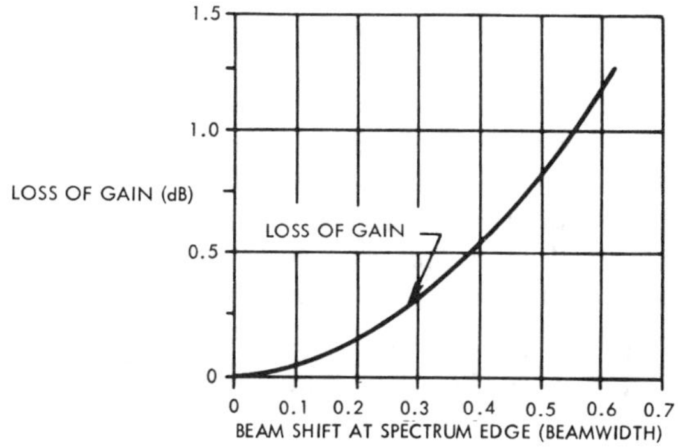

LOSS OF ENERGY DUE TO APERTURE SCANNING; (RECTANGULAR SPECTRUM)

BANDWIDTH (%) = 2 BW (deg) OR APERTURE FILL TIME T = PULSE WIDTH τ

Figure 5.38

APERTURE OR PHASE SHIFTER EFFECTS

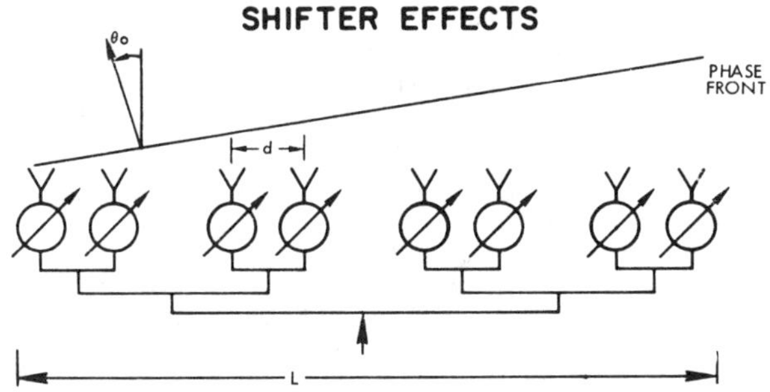

PARALLEL FEED (EQUAL LINE LENGTHS) SCANNED TO θ_o

PHASE OF ith ELEMENT - $\psi_i = \frac{2\pi i d}{\lambda} \sin\theta_o$ (RADIANS)

FIX d, L, AND ψ.. A CHANGE IN FREQUENCY f BY Δ f PRODUCES A CHANGE IN SCAN $\Delta\theta_o$

$$\Delta\theta_o \simeq -\left(\frac{\Delta f}{f}\right) \tan\theta_o$$

- NO SCAN AT $\theta_o = 0°$ (BROADSIDE)
- INDEPENDENT OF ARRAY SIZE AND BEAMWIDTH
- INCREASE IN FREQUENCY, DECREASES SCAN

Figure 5.36

INSTANTANEOUS BANDWIDTH-FEED EFFECTS

END FED SERIES FEED

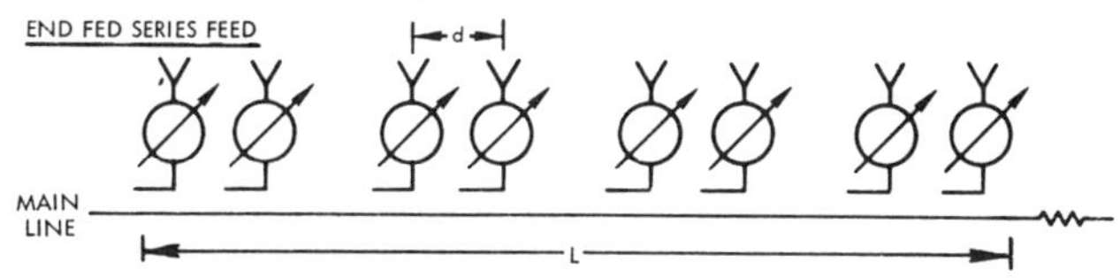

CHANGE IN PHASE $\Delta\psi = \frac{2\pi n d}{\lambda}\,\frac{\Delta f}{f}$ (LINEAR)(NO DISTORTION, JUST LIKE APERTURE EFFECT)

CHANGE IN SCAN ANGLE WITH FREQUENCY CHANGE Δf, IS,

$$\Delta\theta_o \simeq \frac{\Delta f}{f}\,\frac{1}{\cos\theta_o}$$

CENTER FED SERIES

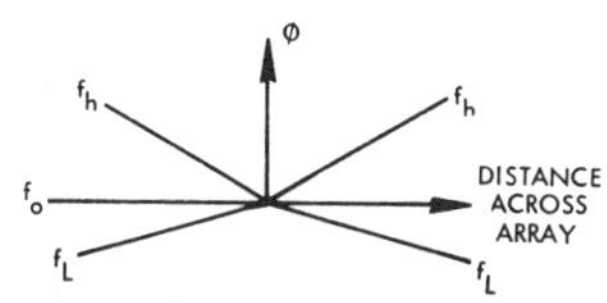

Figure 5.39

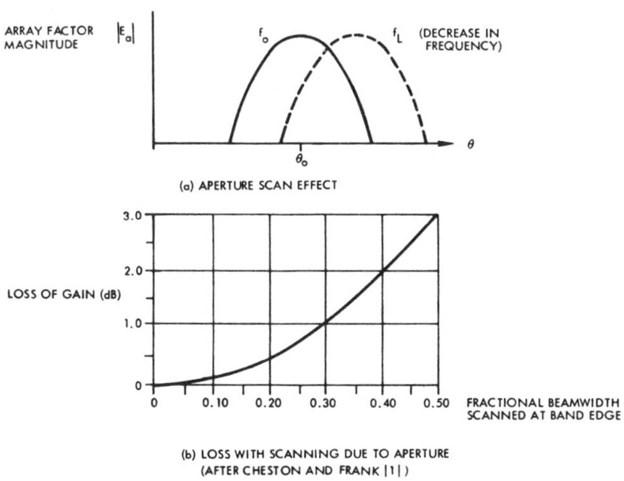

Figure 5.37

LOSS IN S/N

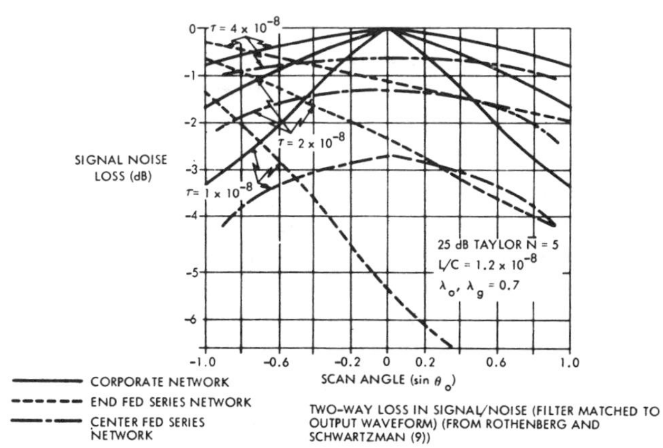

TWO-WAY LOSS IN SIGNAL/NOISE (FILTER MATCHED TO OUTPUT WAVEFORM) (FROM ROTHENBERG AND SCHWARTZMAN (9))

Figure 5.40

For pulse signals one has about twice the bandwidth as the CW case using this same criteria. The loss of gain for a rectangular spectrum pulse is shown in Fig. 5.38 versus beam shift at the spectrum edge. A small increase in loss is associated with a sinx/x spectral distribution as compared to an assumed rectangular frequency pulse spectrum.

5.5.2 Feed Effects. The parallel feed generally requires more weight and volume than a series feed alternative, however separate additive bandlimiting effects are introduced with the series feed. An end fed series feed is shown in Fig. 5.39. Energy is launched down a main line and the element excitations are coupled at precise locations along the line. Amplitude illumination is set by these coupling values, and phase shifters in each element perform steering.

For an assumed TEM feed line the phase across the aperture is $2\pi L/\lambda$ (electrical length of the feed line). The phase change with frequency at the edge element is thus,

$$\Delta\phi = \frac{2\pi L}{\lambda} \frac{\Delta f}{f} \qquad (5.6)$$

This change in phase scans the beam with a linear phase shift across the array as if phase shifters were used. There is no distortion of the beam and the band limiting is similar to that of the aperture effects.

The resulting change in scan angle from the change in frequency Δf is

$$\Delta\theta_o = \frac{\Delta f}{f} \frac{1}{\cos\theta_o} \qquad (5.7)$$

The center fed series array produces a linear phase shift across each half of the array just like the end feeding, however each half is scanned in the opposite direction. With this even phase function there is no scanning of the array with frequency. The beam will distort; for very large frequency changes the beam could split forming two beams.

The center series feed effects when added to the aperture band limiting characteristic will tend to cancel for one half of the array and add to the other. The overall effect will be about the same as the parallel feed case.

In a practical radar system the bandwidth impact on losses is manifest as loss of S/N ratio.

The antenna can be thought of as a filter, within the two way radar path. The antenna characteristic will alter the transmit and receive waveform and one must design a matched filter knowing this change in spectral shape.

Fig. 5.40 is a family of curves of S/N loss as a function of scan angle for the series and parallel (corporate) feeds with transmit pulse length as a parameter. In this case the receiver filter is matched to the receive spectrum which is the product of the transmit waveform and the two way antenna frequency and angle transfer function. This curve shows the increased loss for short pulses (wider frequency spectrum) and trends expected for each of the feeds with scan angle. It is also seen that the parallel feed performs best over most of the scan regions.

REFERENCES

[1] *Microwave Scanning Antennas* – Vol. II – R.C. Hansen Ed; Academic Press, New York 1966.

[2] *Phased Array Antennas* – Oliner and Knittel, Ed; Artech House, Inc., Dedham, Mass. 1972.

[3] *Proceedings of IEEE,* Special Issue on Electronic Scanning, Vol. 56, No. 11, Nov. 1968.

[4] Edelberg and A.A. Oliner, "Mutual Coupling Effect in Large Antenna Arrays: Part I – Slot Arrays", *IRE Trans. Antennas and Propagation AP8,* p 286.

[5] W.H. VonAulock, "Properties of Phased Arrays", *Proc. IRE,* Oct. 1960.

[6] B.L. Diamond, (see Ref. [1] pg. 303)

[7] A.A. Oliner, (see Ref. [2] pg. 107).

[8] N. Amitay and V. Galindo, "Analysis of Circular Waveguide Phased Arrays", *URSI Digest,* Washington D.C., p. 103, April 1968.

[9] B.L. Diamond, "Resonance Phenomena in Waveguide Arrays", *IEE G-AP Int'l Symp. Digest,* pp. 110-115, Oct. 1967.

[10] G.I. Farrell Jr. and D.H. Kuhn, "Muaual Coupling in Infinite Planar Array of Rectangular Waveguide Horns", *IEEE Trans.* Vol. AP-16 pp 405-414, July 1968.

[11] G.H. Knittel, A. Hessel and A.A. Oliner, (see Ref. [3] pp 1822-1836).

[12] G.H. Knittel, "The Realtion of Blindness in Phased Arrays to Higher-Mode Cut-off Conditions", *1971 IEEE G-AP International Symp. Digest,* pp 69-72, Sept. 1971.

[13] L.R. Lewis, A. Hessel and G.H. Knittel, "Performance of a Protruding-Dielectric Waveguide Element in a Phased Array", *IEEE Trans.* Vol. AP-20, pp 712-722, Nov. 1972.

[14] L.R. Lewis, L.J. Kaplan and J.D. Hanfling, "Synthesis of a a Waveguide Phased Array Element", *IEEE Trans.* Vol. AP-22 Nov. 4, pp 536-450, July 1974.

[15] G. Matthei, L. Young and E.M.T. Jones, *Microwave Impedance Matching Networks and Coupling Structures,* New York, McGraw Hill, 1964, pp 434-440.

[16] G.H. Knittel, (see Ref. [2], pg. 157).

[17] E.G. Magill and H.A. Wheeler, "Wide-Angle Impedance Matching of a Planar Array Antenna by a Dielectric Sheet", *IEEE Trans.* Vol. AP-14, No. 1, Jan. 1966.

[18] G.A. Arredondo, P.E. Butzien and R.W. Humes, (see Ref. [2] pg. 173)

[19] P.W. Hannan, P.J. Meier and M.A. Balfour, "Simulation of a Phased Array Antenna Impedance in Waveguide", *IEEE Trans.* AP-11, 1963, p 715.

[20] P.W. Hannan and M.A. Balfour, "Simulation of a Phased Array Antenna in Waveguide," *IEEE Trans.* AP-13, 1965, p 342.

[21] B.L. Diamond (see Ref. [2] pg. 127-131)

[22] W.S. Gregorwich, A. Hessel, G.H. Knittel and A.A. Oliner, "A Waveguide Simulator for the Determination of a Phased Array Resonance", *IEEE, International Symp. Digest G-AP,* 1968, Sept. Boston, Mass.

[23] B.L. Diamond (See Ref. [2] pg 127-131).

[24] L.R. Lewis and A. Hessel, "A Novel Small Array Technique for Active Phased Array Impedance Measurements" (Comm) *IEEE T-AP,* Nov. 1972, pp 107-110.

[25] Q. Balzano, L.R. Lewis and K. Siwiak, "Patterns and Gain of Finite Arrays of Apertures", *IEEE International Symp. Digest,* Aug. 1973, Boulder, Colorado, pp 136-139.

[26] L.R. Lewis, M. Fassett and J. Hunt, "A Broadband Stripline Array Element.", *IEEE Intern'l Symp. Digest,* Atlanta, Georgia 1974.

[27] G.H. Knittel, "Relation of Radar Range Resolution and Signal-to-Noise Ratio to Phased-Array Bandwidth", *IEEE T-AP* 22 May 1974.

[28] J. Frank (see Ref. [2] pgs 243-253.

Array Feeds

W.T. Patton
RCA, Moorestown, New Jersey

6.1 Selecting a Feed System. The first step in selecting a feed is identifying the critical requirements that will influence the choice of feed technique. Some of these are listed in Fig. 6.1.

6.1.1 Electrical Performance Requirements. Most of the material presented here will treat single channel feeds that can support a single polarization in the array or dual sequential polarizations where polarization control is accomplished at the output terminals of the feed system. Two channel systems required for simultaneous dual polarization can in principle be handled in the same way with the added complexity of providing two sets of feed hardware within the same inter-element spacing constraint.

The complexity of the feed itself is of course a function of the number of simultaneous beam ports that must be provided. Multiple beam systems in effect consist of multiple antennas sharing a common aperture. There is a cost saving to be realized in this way but there are also additional constraints due to aperture or beam coupling with which the designer must contend. We will consider some of these later when we examine the types of feed systems that are available to solve a particular set of requirements.

The number of transmitters to be used with the array can range from a single large transmitter to a transmitter or power amplifier at each element. The principal impact of these two situations on the feed system design is that in the first case the feed must be capable of handling the full power of the system and must have very low dissipative loss while in the latter case low power components and relatively high loss can usually be tolerated. In intermediate situations, where many transmitters are required but fewer than the total number of array elements, the choice between various feed techniques may be limited.

The type of steering control to be used must be considered along with the selection of a feed system. With phase control at the output, generally the full range of feed system types can be considered. Frequency steering is more easily handled with one type of feed. Steering by beam switching may eliminate other choices.

The sidelobe requirements address the fundamental function of a feed system which is to efficiently distribute energy to or collect energy from the many array elements with the proper amplitude and coherency relationship as has been discussed in the previous lectures. The required sidelobe level will determine the allowable amplitude quantization in the feed system; i.e., how many elements can be fed with the same amplitude? This requirement also determines the way in which the energy is to be distributed across the aperture, generally at much lower levels near the edge of the aperture than near the center. This required tapering of the illumination can be most efficiently accomplished by desi design of the relative power divisions in the feed. It can also be accomplished at the output of the feed by appropriate attenuation of the energy with some saving in feed complexity but at the cost of efficiency.

Two measures of array bandwidth must be considered in the design of a feed system. Instantaneous bandwidth is that bandwidth over which satisfactory operation is achieved with a single setting of the control devices. This bandwidth must generally be greater than or equal to the signal bandwidth. Tunable bandwidth is that band over which satisfactory operation can be obtained with multiple settings of the control devices. Although most of the feed systems which will be described can be made to have very large instantaneous bandwidths, as a practical matter they will generally be designed with some dispersion either for economy of construction or to achieve some other desired characteristic which must be compensated by the setting of the control devices. The degree to which this can be done is determined by by the instantaneous bandwidth requirement for the feed.

The efficiency required of the feed system depends upon the location of amplification on the array system. It is of relatively small importance when amplification is at the

REQUIREMENTS DETERMINING FEED SELECTION

POLARIZATION DIVERSITY REQUIREMENTS
- SINGLE OR SEQUENTIAL DUAL POLARIZATIONS (1 CHANNEL)
- SIMULTANEOUS DUAL POLARIZATION (2 CHANNELS)

NUMBER OF SIMULTANEOUS BEAMS
- SINGLE BEAM
- MONOPULSE CLUSTER
- MULTIPLE BEAMS

NUMBER OF TRANSMITTERS
- SINGLE HIGH POWER TRANSMITTER
- MULTIPLE LOWER POWER TRANSMITTERS

ARRAY BANDWIDTH
- INSTANTANEOUS BANDWIDTH
- TUNABLE BANDWIDTH

EFFICIENCY

ARRAY SIZE
- APERTURE DIMENSIONS
- DEPTH CONSTRAINT
- VOLUME CONSTRAINT

ENVIRONMENTAL REQUIREMENTS
- BLAST OVERPRESSURE
- SHOCK
- VIBRATION
- THERMAL

COST

TYPE OF STEERING CONTROL
- CONTROL AT THE OUTPUT
- DISTRIBUTED STEERING CONTROL
- BEAM SWITCHING

SIDELOBE REQUIREMENTS
- UNIFORM ILLUMINATION
- APERTURE ILLUMINATION TAPER
- QUANTIZED ILLUMINATION

ARRAY BANDWIDTH
- INSTANTANEOUS BANDWIDTH
- TUNABLE BANDWIDTH

Figure 6.1 - Requirements Determining Feed Selection

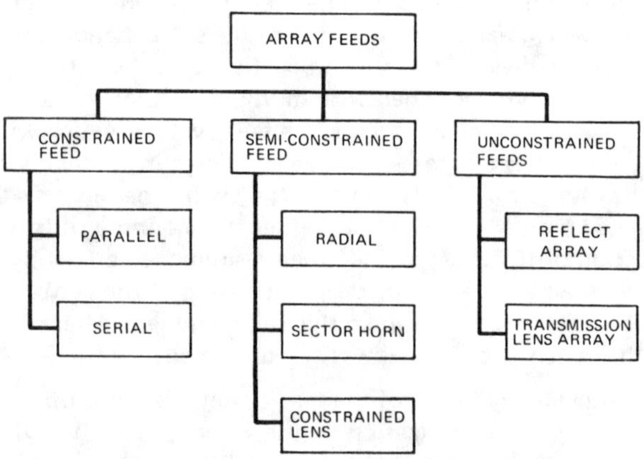

Figure 6.2 - Available Types of Array Feeds

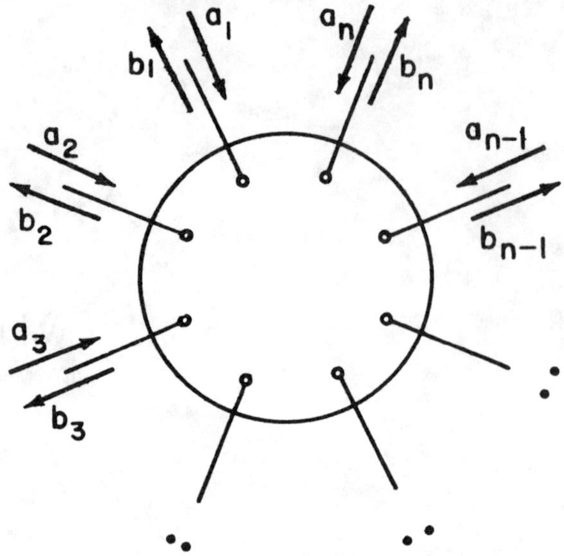

ARBITRARY NETWORK

$$b_i\big] = \big[S_{ij}\big]\; a'_j\big]$$

PROPERTIES:

RECIPROCITY $\Longrightarrow$ $S = \tilde{S}$ (Transpose)

LOSS LESS $\Longrightarrow$ $\tilde{S}^{*} \equiv S^{\dagger} = S^{-1}$ (Unitary)

Figure 6.3 - Scatter Matrix Description of a Network

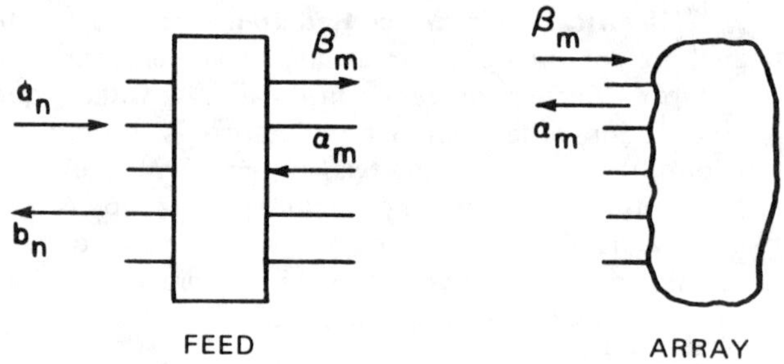

NETWORK COUPLING n SOURCES TO m ANTENNA PORTS

FEED MATRIX

$$\begin{bmatrix} b_n \\ \beta_m \end{bmatrix} = \begin{bmatrix} S_{nn} & S_{nm} \\ S_{mn} & S_{mm} \end{bmatrix} \begin{bmatrix} a_n \\ \alpha_m \end{bmatrix}$$

ANTENNA MATRIX

$$\alpha_m\big] = \big[\mathcal{S}_{mm}\big]\, \beta_m\big]$$

FEED COEFFICIENTS = β_m

$$\beta_m\big] = \big[1 - S_{mm}\mathcal{S}_{mm}\big]^{-1} \big[S_{mn}\big]\, a_n\big]$$

$$\alpha_m\big] = \Big[\mathcal{S}_{mm}\big[1 - S_{mm}\mathcal{S}_{mm}\big]^{-1} S_{mn}\Big]\, a_n\big]$$

$$b_n\big] = \Big[S_{nn} + S_{nm}\mathcal{S}_{mm}\big[1 - S_{mm}\mathcal{S}_{mm}\big]^{-1} S_{mn}\Big]\, a_n\big]$$

Figure 6.4 - Antenna Feed Interaction

array element level, but is of critical importance when amplification is at the beam port of the feed. Even in the case of amplification at the array element however, there is a trade that must be considered between loss in the feed and the required amplification or drive power. The contributors to feed efficiency will be considered later as the various types of feed systems are discussed.

Array size requirements may well be decisive in selecting a feed system. Certainly the feed system of choice for a small aperture may not be appropriate for a very large aperture, and conversely feed systems that are quite appropriate for apertures in the order of hundreds of wavelengths may not work at all well for apertures less than 10 wavelengths in extent.

6.1.2 Environmental Requirements. Quite apart from considerations of aperture size, there may be a depth constraint on the feed system that will limit the range of feed choice.

Environmental requirements can have a profound affect on the choice of feed system particularly where there is need to co-locate a ridged structural support for the array face with the feed system. The magnitude of this problem depends upon the frequency, size of the aperture and number of beams as well as shock, vibration, over pressure or hardening requirements. Thermal considerations range from removing waste heat to stabilize component temperature and minimize thermal gradients to supplying heat to prevent ice formation or to remove ice and snow accumulation.

The cost of the system of which the array feed is a part is a design requirement at least equal to performance requirements in significance. The absolute cost of the feed is a function of the number of connections between the components as well as the number of components themselves. The factors determining the relative cost will become obvious as we discuss the various feed techniques.

6.2 Types of Array Feed Techniques.

There are almost as many ways to feed arrays as there are arrays in existence. There is not a one-to-one correspondence because in some cases the same feed technique has been applied in more than one application and at the same time there are array feed techniques that have been proposed but not implemented. For the most part however the feed system tends to be unique to the particular application.

6.2.1 Array Feed Categories. Because of the great variety of implemented and proposed feed systems, it is difficult to classify the techniques. It is necessary to classify feed techniques however if we are to discuss them intelligently. One way to classify them is given in Fig. 6.2.

In this figure the term constrained feed has the usual meaning; that is, it refers to feeds which constrain the energy to flow only along lines defined by the conductors of the transmission medium. In terms of the types of propagating modes employed these are one dimensional feeds even though by repeated branching they can interconnect elements distributed over a spherical surface.

The term unconstrained feed is used here for what is more commonly called optical feed techniques or space feeds. It refers to those feed techniques where energy is radiated from a feed horn or feed horn cluster and radiates in a manner unconstrained by conducting boundaries to a collecting array of elements each of which serves as an output port for the feed. Again in terms of the propagating modes employed, these are three dimensional feeds.

Intermediate between these two types, then, is the semi-constrained feed where the energy is constrained in one dimension but is allowed to expand in another. In terms of the propagating mode use, these are two dimensional feeds. Both constrained and semi-constrained feeds can be used for line arrays as well as planar arrays while the unconstrained feed is only appropriate for planar arrays.

The constrained feed classification can be further divided into parallel and serial types. Parallel feed systems are characterized by repeated branching junctions such that the path from each output port to the input port goes through substantially the same number of junctions. Serial feeds on the other hand are composed of a cascade arrangement of junctions such that the signal to the first output port traverses only one junction, the signal to the second output port traverses two junctions and so on to the last output port whose signal has traversed all junctions. Both of these feed types are available in "lossless" reactive forms as well as matched forms using hybrid junctions. In addition, single beam, monopulse, and multiple beam versions have been built using both parallel and serial feed organizations. Although much can be learned by considering these various alternative configurations separately, many feed systems use a combination of serial and parallel combining techniques.

The unconstrained feed classification has been further divided into Reflect Array and Transmission Lens Array Subclasses. The Reflect Array uses the same radiating element as both a collecting terminal for the feed and as the output radiator for the array. The Transmission Lens Array separates these function in two different elements. Both of these are substantially parallel feed configurations.

The subdivision of the Semi-Constrained Feed classification is less well defined at this point in the development of the art. Radial types include parallel plate or stripline designs where the output ports are distributed over 360 degrees about the input port. They may all be at the same radius in a parallel configuration or at several radii in a serial parallel configuration. The sector horn classification is similar to the radial except that the output ports subtend substantially less than 360° about the feed. The constrained lens classification contains all others such as the "tin hat" Luneberg configuration, R-2R lens and those which use other methods to very the effective index of refraction through the transmission medium for beam collimation.

6.2.2 General Characteristics of Array Feeds. Some general observations about the characteristics of these feed types can be made without further detailed description which might aid in narrowing the range of consideration for a feed choice.

(1) The constrained feed can provide very precise amplitude control at each element of the array. This is required for applications where sidelobe level is particularly critical. Less precise control of illumination amplitude is afforded by the unconstrained feed due to practical feed horn design constraints and greater interelement coupling at the output terminals.

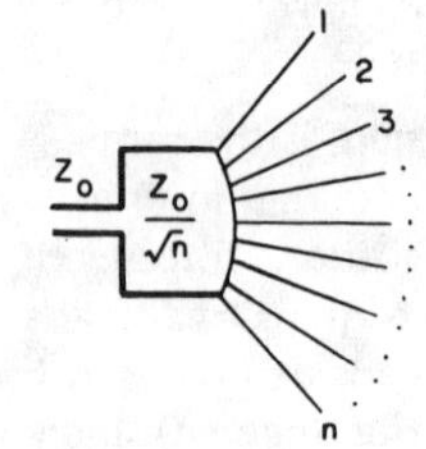

$$S=\begin{bmatrix} 0 & \frac{1}{\sqrt{n}} & \frac{1}{\sqrt{n}} & \frac{1}{\sqrt{n}} & \frac{1}{\sqrt{n}} & \cdots \\ \frac{1}{\sqrt{n}} & \frac{1}{n}-1 & \frac{1}{n} & \frac{1}{n} & \frac{1}{n} & \cdots \\ \frac{1}{\sqrt{n}} & \frac{1}{n} & \frac{1}{n}-1 & \frac{1}{n} & \frac{1}{n} & \cdots \\ \frac{1}{\sqrt{n}} & \frac{1}{n} & \frac{1}{n} & \frac{1}{n}-1 & \frac{1}{n} & \cdots \\ \vdots & \vdots & \vdots & \vdots & \vdots & \end{bmatrix}$$

MATCHED INPUT $\Longrightarrow S_{oo} = 0$

$P_{Load} = \frac{P_o}{n} \Longrightarrow S_{on} = \frac{1}{\sqrt{n}}$

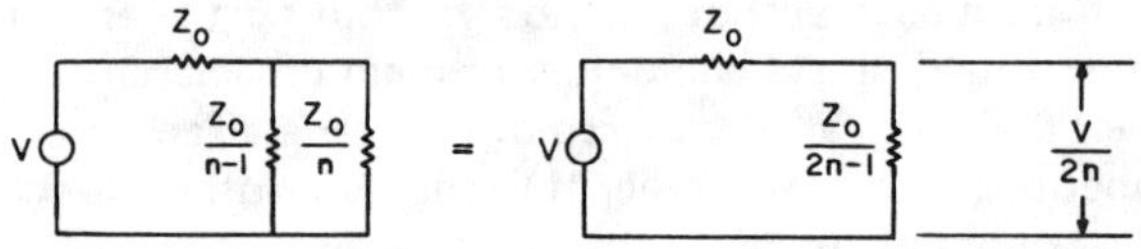

$$S_{nn} = \frac{\frac{1}{2n-1}-1}{\frac{1}{2n-1}+1} = \frac{2-2n}{2n} = \frac{1}{n}-1$$

$$S_{ij} = \frac{1}{n}$$

Figure 6.5 - Scatter Matrix for a Reactive N-Port Junction

$$\begin{bmatrix} 0 & \frac{1}{\sqrt{n}} & \frac{1}{\sqrt{n}} & \frac{1}{\sqrt{n}} & \cdots \\ \frac{1}{\sqrt{n}} & \frac{1}{n}-1 & \frac{1}{n} & \frac{1}{n} & \cdots \\ \frac{1}{\sqrt{n}} & \frac{1}{n} & \frac{1}{n}-1 & \frac{1}{n} & \cdots \\ \frac{1}{\sqrt{n}} & \frac{1}{n} & \frac{1}{n} & \frac{1}{n}-1 & \cdots \\ \vdots & \vdots & \vdots & \vdots & \end{bmatrix} \begin{bmatrix} 0 & \frac{1}{\sqrt{n}} & \frac{1}{\sqrt{n}} & \frac{1}{\sqrt{n}} & \cdots \\ \frac{1}{\sqrt{n}} & \frac{1}{n}-1 & \frac{1}{n} & \frac{1}{n} & \cdots \\ \frac{1}{\sqrt{n}} & \frac{1}{n} & \frac{1}{n}-1 & \frac{1}{n} & \cdots \\ \frac{1}{\sqrt{n}} & \frac{1}{n} & \frac{1}{n} & \frac{1}{n}-1 & \cdots \\ \vdots & \vdots & \vdots & \vdots & \end{bmatrix}$$

(FIRST ROW) X (FIRST COLUMN)

$$\sum_{i=1}^{n} \frac{1}{n} = 1$$

(SECOND ROW) X (FIRST COLUMN)

$$\frac{1}{\sqrt{n}}\left(\sum_{i=1}^{n} \frac{1}{n} - 1\right) = 0$$

(SECOND ROW) X (SECOND COLUMN)

$$\frac{1}{n} + \left(\frac{1}{n}-1\right)^2 + \sum_{i=1}^{n-1} \frac{1}{n^2}$$

$$\frac{1}{n} + \frac{1}{n^2} - \frac{2}{n} + 1 + \frac{n-1}{n^2} = 1$$

(THIRD ROW) X (SECOND COLUMN)

$$\frac{1}{n} + \frac{2}{n}\left(\frac{1}{n}-1\right) + \sum_{i=1}^{n-2} \frac{1}{n^2}$$

$$\frac{1}{n} + \frac{2}{n^2} - \frac{2}{n} + \frac{n-2}{n^2} = 0$$

$$\begin{bmatrix} 1 & 0 & 0 & 0 & \cdots \\ 0 & 1 & 0 & 0 & \cdots \\ 0 & 0 & 1 & 0 & \cdots \\ 0 & 0 & 0 & 1 & \cdots \\ \vdots & \vdots & \vdots & \vdots & \end{bmatrix}$$

Figure 6.6 - Matrix Multiplication

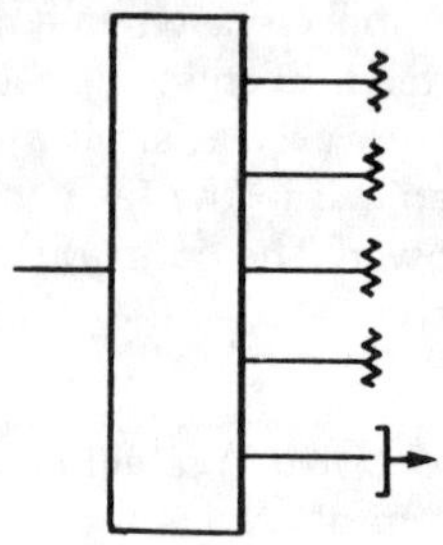

$$b = \begin{bmatrix} S \end{bmatrix} \begin{bmatrix} a_o \\ 0 \\ 0 \\ \rho_{bk} \\ 0 \\ 0 \end{bmatrix}$$

$$b_k = S_{ok}\, a_o + S_{kk}\, \rho_{bk}$$

$$b_k = \frac{S_{ok}\, a_o}{1-\rho S_{kk}}$$

$$b_k = \frac{\frac{1}{\sqrt{n}}}{1+\frac{1}{n}-1} = \sqrt{n}\; a_o \qquad b_o = a_o$$

Figure 6.7 - Feed Resonance

$$\text{IF} \quad b_i = b \qquad a_i = \rho e^{j\phi_i}\, b$$

$$b_o = \frac{\rho b}{\sqrt{n}} \sum_{i=1}^{n} e^{j\phi_i}$$

$$b_i = \frac{a_o}{\sqrt{n}} + \frac{\rho b}{n} \sum_{i-1}^{n} e^{j\phi_i} - \rho b e^{j\phi_i}$$

$$b = \frac{a_o}{\sqrt{n}\left(1+\rho e^{j\phi} - \frac{\rho}{n}\sum_{i=1}^{n} e^{j\phi_i}\right)}$$

$$b = \frac{a_o}{\sqrt{n}} \;;\; b_o = \rho\, a_o \qquad \text{IF} \quad \sum_{i=1}^{n} e^{j\phi_i} = n$$

$$b_i = \frac{a_o}{\sqrt{n}\left(1+\rho e^{j\phi_i}\right)} \qquad b_o = 0 \quad \text{IF} \quad \sum_{i=1}^{n} e^{j\phi_i} = 0$$

Figure 6.8 - Spurious Lobes

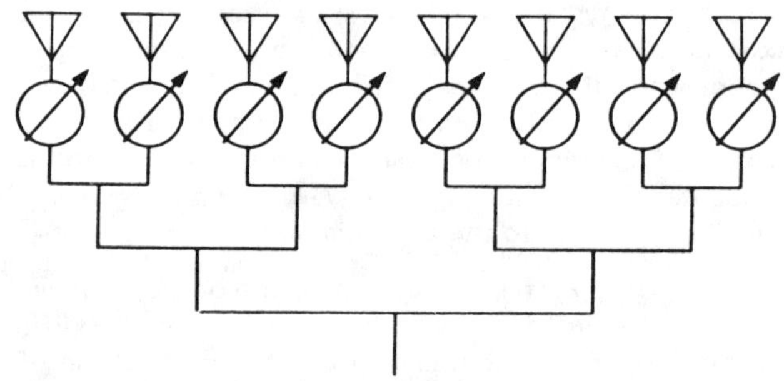

Figure 6.9 - Reactive Corporate Feed Network

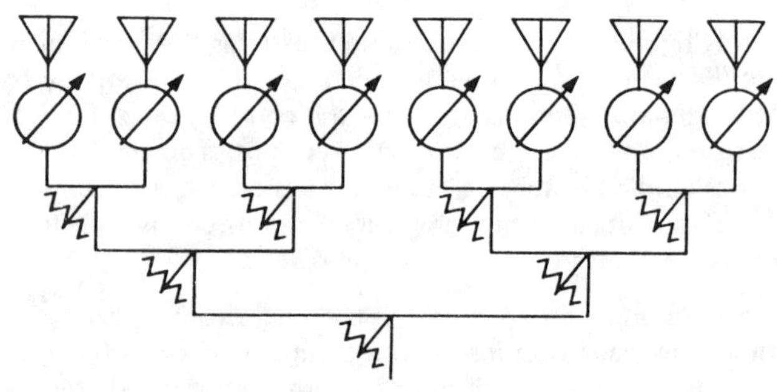

Figure 6.10 - Matched Corporate Feed Network

(2) When the number of array elements is very large, the cost of the many junctions and transmission lines of the constrained can be avoided by use of an unconstrained feed. The cost per element is generally lower in the Reflect Array configuration than in the Transmission Lens configuration but its efficiency and illumination control capability is poorer.

(3) The depth of a feed system implemented with constrained feed techniques is on the order of one tenth of the array diameter or less. The depth of an unconstrained feed is on the order of half the array diameter or larger. This may well be the decisive consideration.

(4) The constrained feed generally has advantage for arrays operating in a hostile environment since all the feed system is behind the array and room can generally be found to place the required structural support members through the feed system. The Transmission Lens on the other hand must be totally edge supported since the region between the feed horn and the back face of the lens must be clear of reflecting obstructions. The Reflect Array is ideal as far as support of the array face is concerned but the tripod and feed assembly is mounted in front of the face where it is relatively vulnerable.

6.3 Analysis of Feed Network Performance.

One way to analyze a feed network is by means of its Scatter Matrix. The definition and some properties of this matrix are shown in Fig. 6.3. This matrix describes the reflectance at several ports of the network and the coupling between them. Two important properties of this matrix follow from the properties of the feed system it described.[1]

(1) If the network is reciprocal then the matrix is symmetric. It equals its own transpose.

(2) If the network is lossless the matrix is unitary. The transposed conjugate is equal to the inverse of the matrix.

The way the scatter matrices may be used in the analysis of feed systems is shown in Fig. 6.4. The example here considers the connection of n transmitters or alternatively n beam ports to m antenna ports. The technique involves proper ordering of the feed matrix and partitioning it to feed port and antenna port groups. In this way we can express the relation between the incident and reflected wave amplitudes at the antenna ports by two equations and characterizing the feed and one characterizing the antenna. These may be solved simultaneously to give the relationship between the excitation at the antenna ports (β_m) and the excitation at the feed ports (a_n). A little more matrix algebra will also yield the scatter matrix at the feed terminals which represents the behavior of the feed system there when loaded by the antenna.

One of the beauties of matrix notation is that one can use it to represent calculations that may be too extensive even for modern computing machines. Certainly the inversion of 5000 by 5000 complex matrices is not yet a routine task. But by applying these relationships to some simple cases and by making certain simplifying assumptions they can provide useful guidance for design.

6.3.1 Reactive Divider Network.

One of the simplest feed systems is the reactive N-way divider. Let us consider N equal impedance (Z_0) lines at a common junction with an impedance transformer matching the junction to an input line of the same impedance. A schematic representation of such a divider is shown in Fig. 6.5.

Since the junction impedance of Z_0/n can be matched at one frequency by a quarter wave transformer of impedance $Z_0/\sqrt{n}$, the input reflectance S_{00} is taken to be zero. The input power to the junction through the matching transformer is equally divided between the n outputs so that the coupling from the input to any of the output ports S_{0n} is equal to $1/\sqrt{n}$ in magnitude.

Looking at the junction through one of the output lines (all others terminated in their characteristic impedance) we see the n−1 other output lines in parallel shunted by the Z_0/n load of the input circuit. This gives an equivalent load of $Z_0/2n-1$ at the junction. The reflectance at the output line due to this equivalent load is (1/n)−1. For very large n this can be a very high mismatch indeed. The voltage at the junction is 1/n times that which would be obtained with a matched load. Thus the coupling to the other output ports is 1/n. The coupling to the input port load which is only 1/nth as large as the line impedance is $1/\sqrt{n}$ as expected by symmetry.
Since the resulting scatter matrix is symmetrical, real, and represents a lossless network its square should be the identity matrix. This is often a useful test for the validity of a scatter matrix. This calculation is given in Fig. 6.6.

One potential problem with such reactive feeds is the very high mismatch seen looking back into an output port and the resulting possibility of feed resonance. For example (Fig. 6.7) if all output terminals are connected to matching loads except the Kth port which is connected to a load with reflectance ρ we find that the amplitude of the output wave is increased by the factor $1/1-\rho S_{KK}$. When port K is shorted the effective incident power on that short is n times the input power to the entire network, which would doubtless overstress the feed.

Another technique for analyzing feed system-antenna interaction is based upon the assumption that all the antenna ports have nearly the same reflectance. Making this assumption in Fig. 6.8 applied to an N-way power divider, we can see that when all of the reflectances are in-phase the output amplitudes are not disturbed and the reflectance at the antenna element appears at the input port. If the phase differences are such that the antenna reflectances add to zero at the input port, the output phase and amplitude are unbalanced. This disturbance takes the form of spurious lobes of level ρ, ρ^2, etc., relative to the main beam of the antenna.

6.3.2 Matched Divider Network.

The feed resonance problem and the spurious lobe problem can be eliminated by using hybrid junctions with the fourth port terminated or any equivalent technique which sets the reflectance and coupling between the output ports to zero. Note that such a network is not lossless.

The cost of a matched feed system can be avoided when the relative phases of the antenna reflectances do not vary as may happen when non-reciprocal (ferrite) phasers are

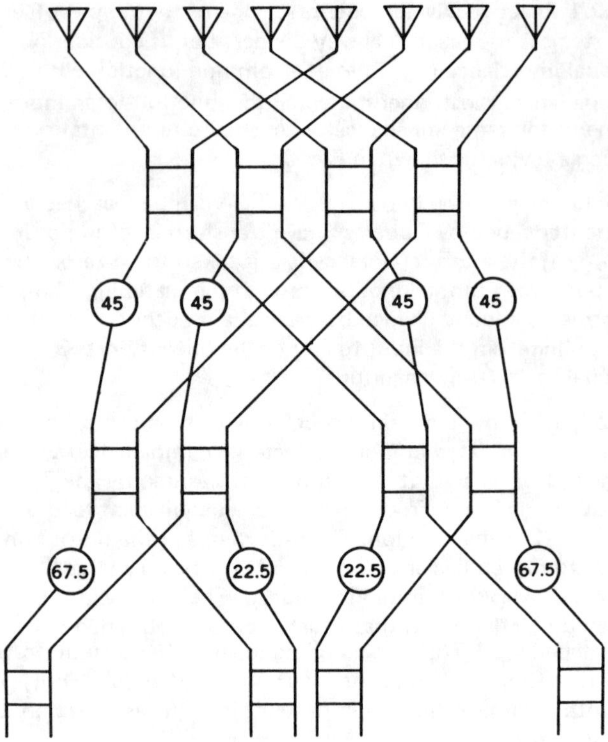

Figure 6.11 - Parallel-Multiple Beam Forming Matrix

UNBALANCED POWER DIVIDERS

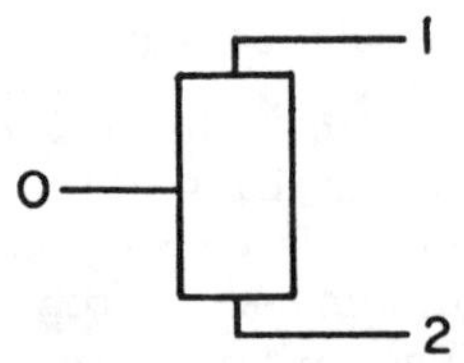

MATCHED (DIRECTIONAL COUPLER)

$$\begin{bmatrix} b_0 \\ b_1 \\ b_2 \end{bmatrix} = \begin{bmatrix} 0 & C & \sqrt{1-C_2} \\ C & 0 & 0 \\ \sqrt{1-C^2} & 0 & 0 \end{bmatrix} \begin{bmatrix} a_0 \\ a_1 \\ a_2 \end{bmatrix}$$

REACTIVE

$$\begin{bmatrix} b_0 \\ b_1 \\ b_2 \end{bmatrix} = \begin{bmatrix} 0 & C & \sqrt{1-C_2} \\ C & 1-C^2 & -C\sqrt{1-C^2} \\ \sqrt{1-C^2} & -C\sqrt{1-C^2} & C^2 \end{bmatrix} \begin{bmatrix} a_0 \\ a_1 \\ a_2 \end{bmatrix}$$

IF $\quad a_1 = \rho_1 b_1 \;;\quad a_2 = \rho_2 b^2$

$$\frac{b_1}{b_2} = \frac{C}{\sqrt{1-C^2}} \quad \frac{1-\rho_2}{1-\rho_1}$$

Figure 6.12 - Unbalanced Power Dividers

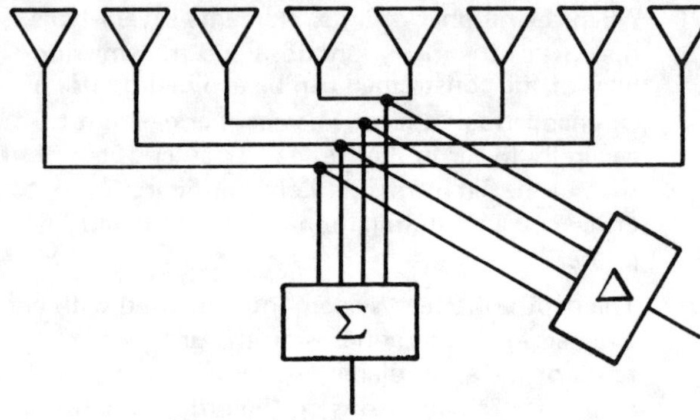

Figure 6.13 - Monopulse Combiner

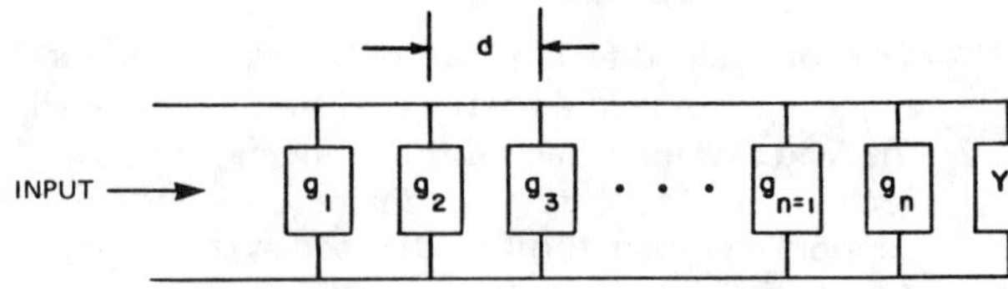

$$\rho = \frac{-e^{jNkd}\sum_{n=1}^{N}\frac{1}{2}g_n e^{-2jnkd} + \rho_L e^{-jNkd}\left(1-\sum_{n=1}^{N}\frac{1}{2}g_n\right)}{e^{jNkd}\left(1+\sum_{n=1}^{N}\frac{1}{2}g_n\right) + \rho_L e^{-jNkd}\sum_{n=1}^{N}\frac{1}{2}g_n e^{2jnkd}}$$

STANDING WAVE: $d = \frac{\lambda_g}{2}\;;\; \rho_L = 1$

$$\rho = \frac{1-\sum_{n=1}^{N} g_n}{1+\sum_{n=1}^{N} g_n}$$

TRAVELING WAVE: $\rho_L = 0$

$$\rho = \frac{-\frac{1}{2}\sum_{n=1}^{N} g_n e^{-2jnkd}}{1+\frac{1}{2}\sum_{n=1}^{N} g_n}$$

Figure 6.14 - Serial Feed Analysis

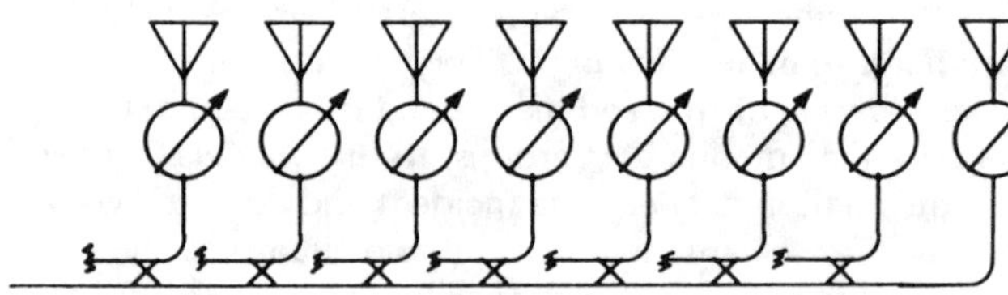

Figure 6.15 - Serial Feed Network

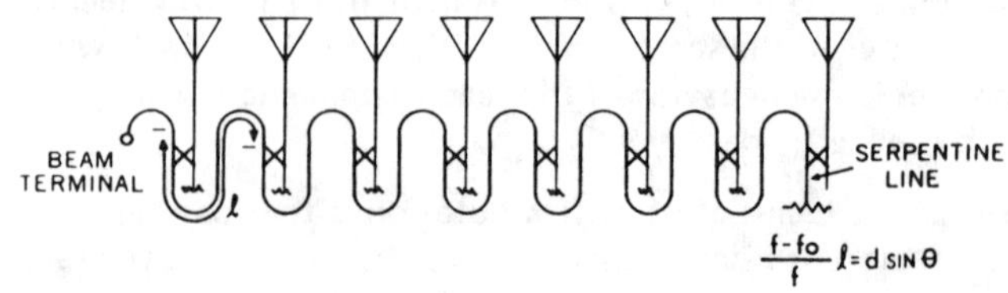

Figure 6.16 - Frequency Scanning

used for steering control. In addition, ferrite phaser shifters display a peak power limiting characteristic which will control feed resonance at an improperly phased output port to an acceptable level.

All reactive feeds need not be identical to the N-way reactive divider. It has been used to exaggerate some of the problems that can be found in feed systems if they are not carefully designed. For example, a binary branching feed with a half wavelength separation between successive junctions acts very much like an N-way power divider. If the spacing between junctions is an odd multiple of quarter wavelengths however the output mismatch will be much lower and shorting only one output port will not short out the entire feed.

6.4 Constrained Feeds.

6.4.1 Constrained Feed-Parallel.

(A) Single Beam Feed. Fig. 6.9 shows a reactive corporate feed constructed using cascaded binary junctions. This type of feed is similar in performance to the reactive N-Way divider considered earlier. It favors the use of arrays with a binary number (2^n) of elements. The number of levels of power division is n and the number of junctions is $2^n - 1$.

If each junction is replaced by a terminated "Magic T" as shown in Fig. 6.10 it becomes a matched corporate feed. This feed has excellent match at the output ports and substantially zero coupling between the output ports, assuming ideal junctions. If all the loads have the same reflectance, this reflectance will appear at the input port just as with the reactive corporate feed. If the reflectances add to zero at the input port then they are absorbed by the various terminating loads instead of being reflected back to the loads causing spurious lobes observed with reactive feeds.

This points out a fundamental difference between reactive and matched feed structures when used as power combiners with a receive array. When the incoming phase wave is aligned with a null in the antenna pattern it is reflected by the reactive feed but absorbed by the matched feed. It is just this energy absorbed in the terminating loads of the feed when the incident wave corresponds to a pattern null that is recovered by the multiple beam matrix.

(B) Multiple Beam Feed. A parallel multiple beam forming matrix,[2] shown in Fig. 6.11, like other corporate feeds favors a binary number of elements. It will produce as many independent (uncoupled) beams as there are elements in the array. It uses 2^{n-1} couplers at each of its n levels or $n(2^{n-1})$ couplers in total.

The feeds discussed thus far provide substantially uniform amplitude at the output terminals. For most array systems, however, some taper of illumination is required for sidelobe level control (Lecture 2). This can be accomplished by use of unbalanced power dividers of either the reactive type or the matched type (directional couplers) whose scattering matrices are given in Fig. 6.12. Note that an unbalance in the loads on the matched power divider does not unbalance the output amplitude ratios with respect to that designed into the junction. However, the reactive junction can be unbalanced if the loads it feeds are not equal.

Any desired illumination function can now be provided by proper design of the junction power division ratio. Given the desired output amplitude ratios we can determine the power division ratio of each first tier coupler by comparing the amplitude ratios of the adjacent elements fed at that level. The square root of the ratio of powers required by adjacent pairs of elements determines the coupling ratio for second tier couplers. This process continues until the highest-level power divider is reached.

(C) Monopulse-Beam. A monopulse-beam configuration can most easily be achieved by making use of the sum and difference ports of the highest-level hybrid power divider. The sum port combining the two halves of the array in phase addition and the difference port combining the same two halves of the array in phase opposition. Unfortunately the patterns one achieves in this way are a compromise between the desirable low sidelobe characteristics of a sum beam and high error slope sensitivity for a difference beam. The more taper that is applied to improve sum beam sidelobes the wider and less sensitive becomes the difference beam. To achieve desirable characteristics in both the sum and difference beams, they must be independently optimized.

One way of achieving independent control of the sum and difference beam illuminations is shown in Fig. 6.13. This Feed System makes use of the natural isolation in the "Magic T" when employed in a symmetrical geometry to provide independent control of the sum and difference illuminations. A significant characteristic of monopulse illumination functions is that the sum illumination is symmetric about the center line or mid point of the aperture while the difference illuminator is antisymmetric. With each "Magic T" connected to elements symmetrically located with respect to the array center we can make use of the sum port of the junction to control the amplitude of the symmetric illumination while the amplitude of the antisymmetric illumination is independently controlled at the difference port of the junction. The appropriate amplitude ratios between the several sum ports, or between the several difference ports, can be controlled by the techniques described earlier.

6.4.2 Constrained Feed – Serial.

(A) Standing Wave Serial Feeds. Two distinct types of serial feed organization are generally recognized; standing wave and traveling wave. The standing-wave serial feed is generally designed to produce a beam at a right angle to the feed which does not scan with frequency. The traveling-wave feed is designed to avoid the internal phasing conditions associated with the normal beam and the beam will move over the operating band. The standing-wave serial feed is more efficient but has a narrower bandwidth for a given size while the traveling-wave feed has substantially larger bandwidth in long feed structures.

An analysis of the serial feed due to Watson [3] is shown in Fig. 6.14. This equation assumes that the loading of each output port on the main line is so small that multiple reflections can be neglected. When the spacing between the serial connections is half the guide wavelength and the guide is terminated in an open circuit, the load conductances are summed at the input to the feed. This is the same effect that is obtained with the N-Way power divider. At that one frequency and standing-wave serial feed is equivalent to the N-Way power divider and will be represented by the same scatter matrix. The impedance of the serial feed, however changes rapidly with frequency. A useful rule of thumb is that for a maximum VSWR of 3 the bandwidth is approximately ±50/N%.

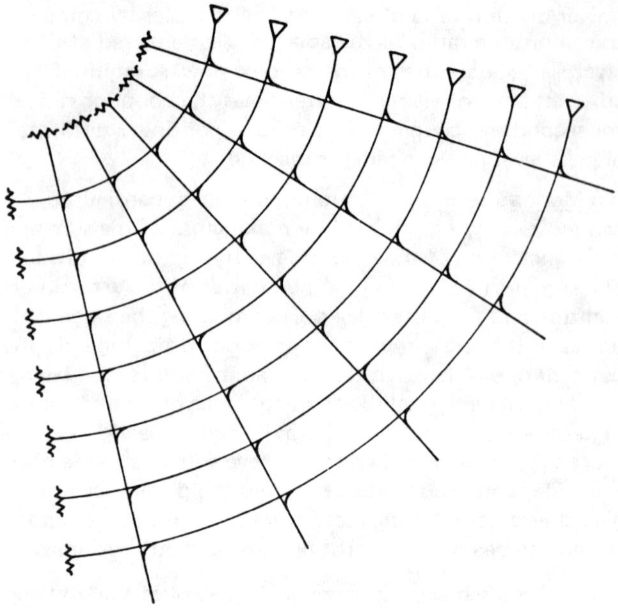

Figure 6.17 - Serial - Multiple Beam Forming Network

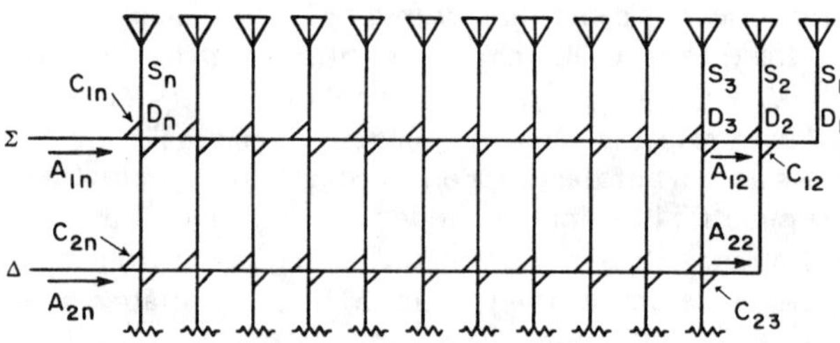

Figure 6.18 - End Fed Serial Monopulse Ladder Network

Let B_1 be a Complex column vector representing the output wave amplitudes of a lossless network when Port 1 is excited with A_1. Then by conservation of energy

$$|A_1|^2 = \langle B_1^* B_1 \rangle = \sum_{i=1}^{N} B_{1i}^* B_{1i}$$

Similarly when Port 2 is excited with A_2

$$|A_2|^2 = \langle B_2^* B_2 \rangle = \sum_{i=1}^{N} B_{2i}^* B_{2i}$$

and when both are excited simultaneously

$$|A_1|^2 + |A_2|^2 = \langle (B_1 + B_2)^* (B_1 + B_2) \rangle$$

$$= \langle B_1^* B_1 \rangle + 2 \langle B_1^* B_2 \rangle + \langle B_2^* B_2 \rangle$$

Excitations $A_1, B_1\rangle$ and $A_2, B_2\rangle$ are orthogonal, non-interacting, if and only if

$$\langle B_1^* B_2 \rangle = 0$$

Figure 6.19 - Orthogonality Condition

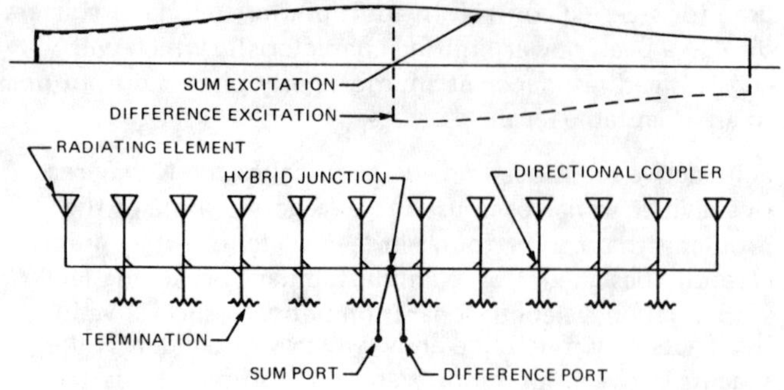

Figure 6.20 - Magic "T"

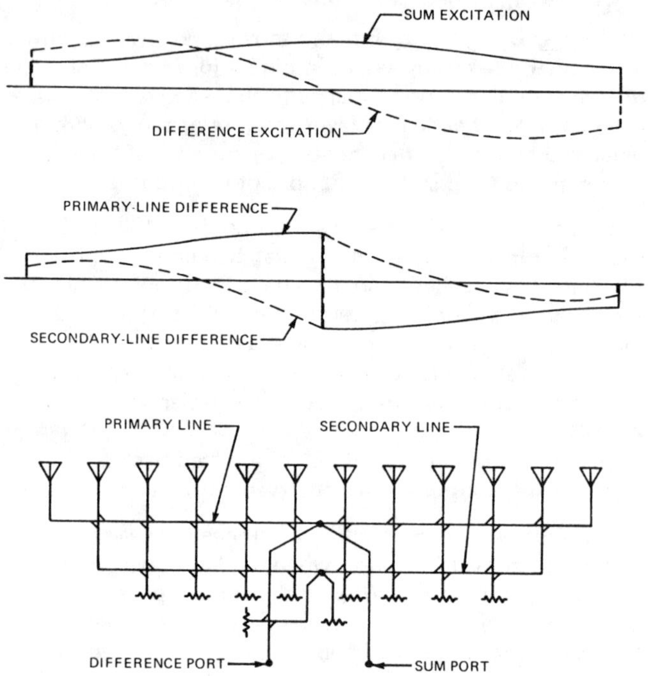

Figure 6.21 - "Center-fed" Ladder Network

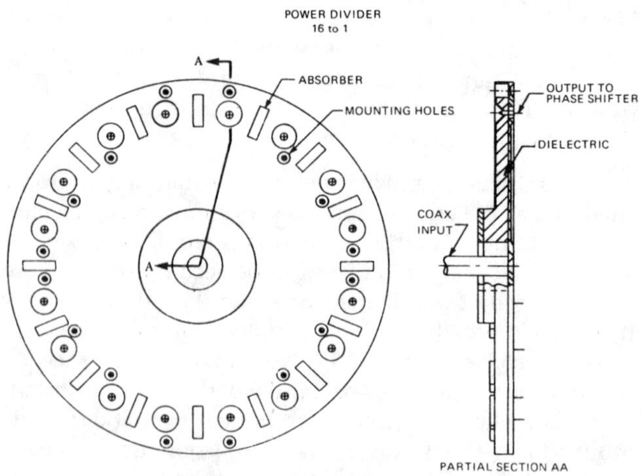

Figure 6.22 - Radial Feed

(B) Traveling Wave Serial Feeds. While the standing-wave serial feed is designed so that the reflectances from the various output ports add constructively at the input port, the traveling-wave serial feed is designed so that these reflectances add destructively. The expression for the input reflectance of a traveling wave serial feed shown in Fig. 6.14 has the appearance of an antenna pattern. The traveling-wave serial feed operates in the sidelobe region of the input impedance and must avoid the main beam region where large input mismatch would obtain. Since the sidelobe region is much broader than the main beam region, the impedance bandwidth of the traveling-wave serial feed is large. The penalty is not being able to steer the beam to broadside and in having the beam scan with frequency. This later characteristic can be amplified and utilized in frequency scanning feeds which derive from the traveling-wave serial feed.

(C) Matched Serial Feeds. The two forms of serial feed discussed thus far are reactive feeds. They have the same high mismatch at the output junctions as reactive parallel feeds Matched forms of the serial feed are implemented with terminated directional couplers as shown in Fig. 6.15 and 6.16. When an incoming wave is aligned with a null of the antenna pattern it is totally absorbed in the loads of the directional couplers. Again, as described in the case of the parallel multiple beamforming matrix, this energy can be combined to form beams aligned with these null positions. One such serial multiple beam forming matrix is shown in Fig. 6.17 where the added increments of transmission line length between the first and second lines collect the energy from the first null of the pattern formed by the first line in the second line with energy from the second null going to the third line and so on. A significant difference between the parallel and the serial multiple beamforming matrices is that the former separates the beams by phase difference which allows the beams to shift with frequency to remain always aligned peak to null, that is, always orthogonal. The serial multiple beam matrix separates the beams with differential line lengths, time delay, such that the absolute difference in pointing angle of each beam tends to remain fixed with frequency. Thus the serial multiple beam forming matrix matrix will produce orthogonal beams at only one frequency and increasing coupling between the beams is expected at other frequencies.

(D) Multiple Beam Serial Feeds. The beams of the parallel multiple beam matrix remain orthogonal and well matched with frequency but the beams move as the beamwidth increases or decreases. The beams of the serial multiple beam matrix [4] remain fixed in space but become coupled and poorly matched over wide frequency ranges.

Monopulse operation can be obtained with serial feeds by taking the sum and difference of two adjecent beams from a serial multiple beamforming matrix. However, if only a monopulse beam is required (and not more simultaneous beams), the feed can be modified to provide more optimum monopulse performance. For example in Fig. 6.18 the coupling values on the line nearest the aperture can be optimized for low sidelobe sum beam performance while the second feed line is optimized for the difference pattern. The necessary and sufficient condition on these two excitations to prevent coupling between them is the orthogonality condition shown in Fig. 6.19.

An alternate way to achieve a monopulse beam pair with a serial feed is to feed at the center through a sum and difference hybrid coupler, "Magic T" Fig. 6.20. This approach suffers the same problem of not allowing separate optimization of the Sum and Difference illumination distributions discussed earlier with parallel feeds. Lopez [5] described a method of overcoming this limitation with a center-fed ladder network such as that shown in Fig. 6.21. With this network complete independent between the sum and difference illuminations can be achieved. Its derivatives can provide satisfactory performance in the sum and difference beams with totally independent illumination control with some savings in hardware cost.

6.5 Semi-Constrained Feed.

6.5.1 Radial Feeds. A parallel plate radial waveguide with output ports on a circumference about the center feed can be an economical N-way power divider, Fig. 6.22. Although reactive, its mismatch and cross-coupling are somewhat less than obtained with a reactive constrained feed because the coupling is shared with more elements. Coupling tends to be strongest to elements immediately adjacent on the rim. Coupling to elements one quarter around the circumference is least and coupling to elements diametrically across the feed somewhat stronger. Techniques are available for controlling the coupling between adjacent output ports and improving the match of the radial feed that make it nearly equivalent to a matched constrained feed.

Multiple-beam formation has been achieved with a radial feed by exciting it with the TE_{01}, TE_{11}, and higher modes. Each of these modes has n-cycles of 2π phase shift around the circumference. When connected by equal length transmission line to a linear array, each mode will produce a beam orthogonal to each other beam similar to those formed by the parallel multiple beamforming matrix. Monopulse beams can be synthesized from any combination of these orthogonal excitations. A combination of the TE_{01}, TE_{11}, and TE_{-11} will produce a very good monopulse beam pair with a sinusoidal difference excitation and a cosine squared on a pedestal sum illumination. Derivatives of this feed have also been used for cylindrical arrays. In this application a simple phase shift between the modes can cause the illumination functions to shift around the circumference.

It is not necessary to collect all of the energy in the feed at the same radius. By placing directional couplers at different radii we have a combination serial-parallel feed which is truly two dimensional. Feeds of this type are still in the experimental stage. One problem with this feed geometry is in justifying its circular geometry with the rectangular geometry of an array.

6.5.2 Sector Horn Feeds. The sector horn might be considered a segment of a parallel plate feed except for the polarization of the electric field in the feed. It is perpendicular to the parallel plates in the radial feed but is usually parallel to the broad wall in the sector horn. This type of feed has been used for both subarray feeds and for full line array feeds with monopulse capability.

Experimental sector horn feeds for subarrays have been built at both APL and MIT Lincoln Labs. The objective of these feeds is to provide equal excitation to a few elements

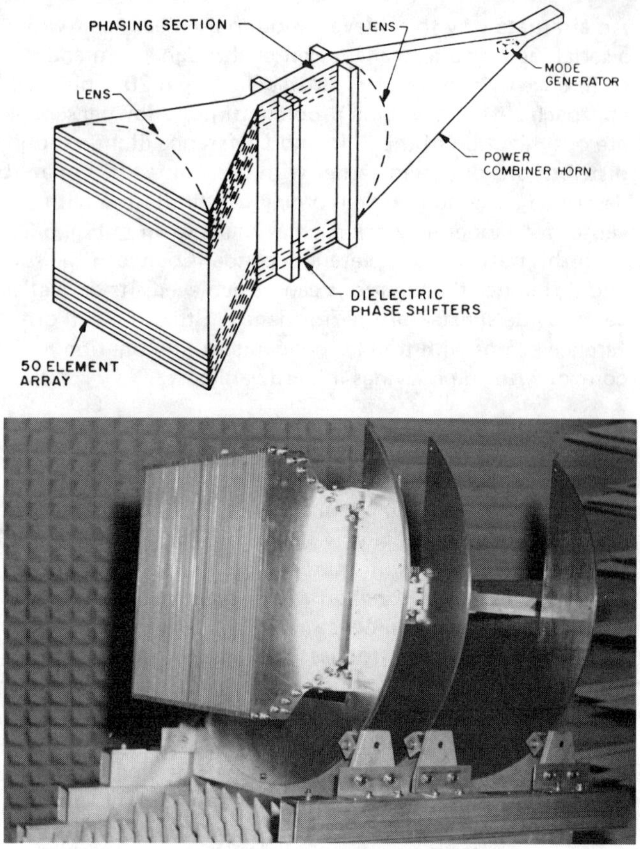

Figure 6.23 - Sector Horn Array Feed

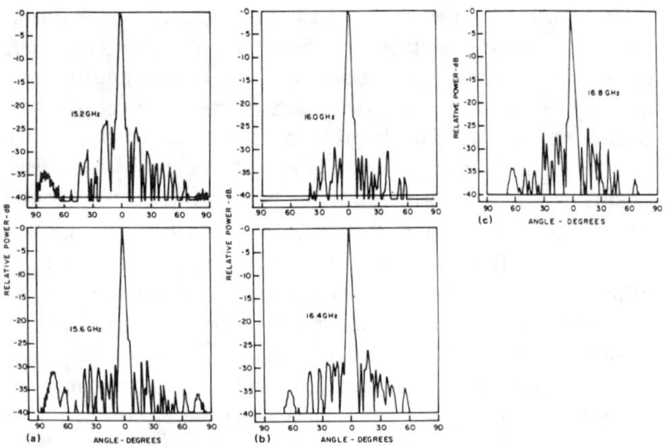

Figure 6.24 - 50-way Divider Patterns

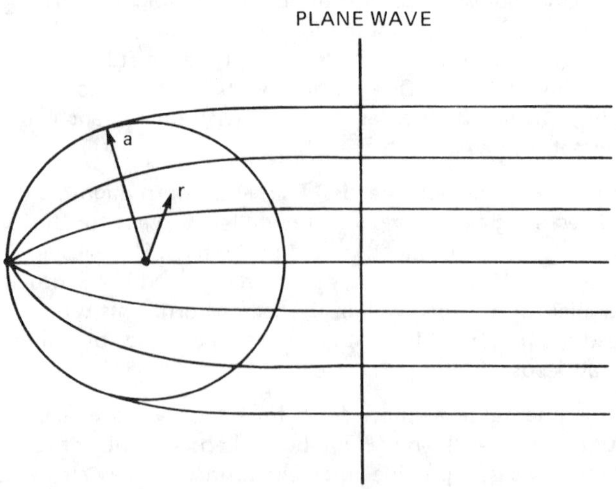

Figure 6.25 - Luneburg Lens

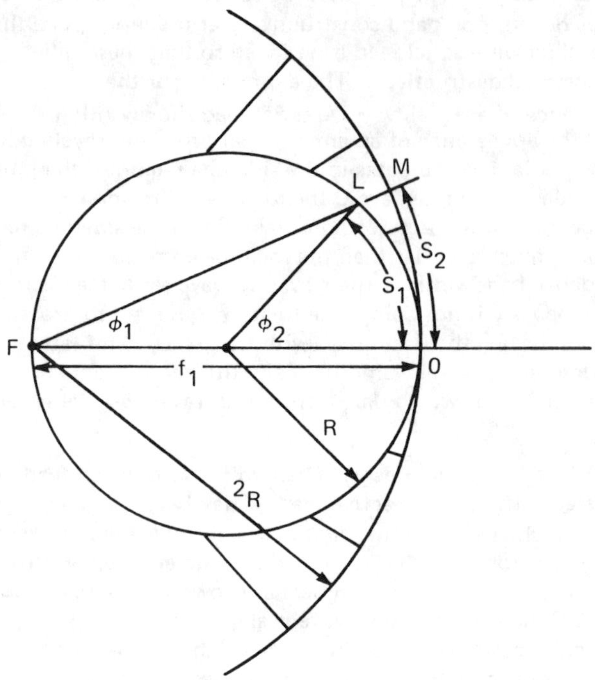

Figure 6.26 - R-2R Lens Geometry

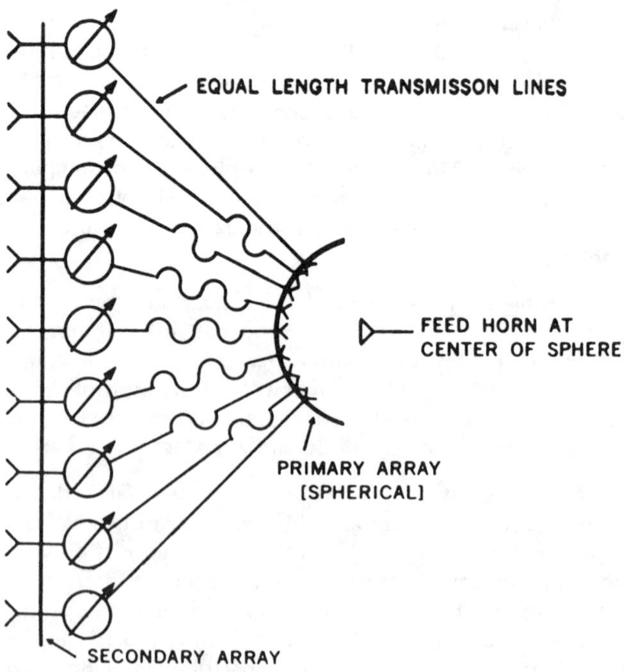

Figure 6.27 - Spherical-Planar Unconstrained Feed System

of the array. These were both reactive feeds with high mismatch at the output ports. They were intended for use with non-reciprocal phase shifters where the high output mismatch can be tolerated because the loads tend to remain balanced at all scan angles.

A much larger E-plane sector horn was built by RCA as a feed for a 15 GHz phased array radiometer, Fig. 6.23. This feed used a dielectric lens to correct the phase at the output and to ease the transition from a circular to a planar wavefront. This feed is reactive but it has better (lower) output mismatch than the smaller sector horn subarray feeds feeds because of distributed coupling to the other elements. Such a large sector horn can be fed by the equivalent of a multimode horn at its input for good mode control to form sum and difference monopulse or as was the case for the PAMS feed, for exceedingly low sidelobes, Fig. 6.24.

6.5.3 Constrained Lens Feeds. In this category of feed systems we include those semi-constrained feed systems which use variable index of refraction or a bootlace connection between an input and output surface to accomplish collimation.

(A) Luneberg Lens. A two dimensional Luneberg lens [6] is designed with a circular cylindrical section whose diffraction index varies with radius in accordance with

$$\eta = \sqrt{2 - (r/a)^2}$$

The dielectric constant of the material between the parallel plates should vary as

$$\epsilon_{(r)} = \eta^2{}_{(r)} = 2 - (r/a)^2$$

between a value of 2 at the center and 1 at the edge.

The Luneberg lens has the property of collimating the energy emanating from any point on its circumference in a direction containing that point and the center of the lens as shown berg lens.

The variable index of refraction can also be obtained by changing the geometry of the parallel plate region so as to increase the path length through the center region. This leads to the "tin-hat" or geodesic equivalent of the Luneberg Lens.

These lens feeds establish the proper phase on a circular arc to radiate a linear phase front. They work directly as feeds for cylindrical arrays or, with additional correction to translate from circular to linear geometry, as linear array feeds.

Since collimation is achieved from any point on the circumference the feed can achieve scanning by moving or switching the feed position around the circumference. It can also be used as a multiple beam feed by exciting more than one feed position.

(B) R-2R Lens Feed. The R-2R lens [6] shown in Fig. 6.26 behaves much like the Luneberg lens in forming a collimated (linear wave front) beam from any point along its feed circle (Radius R). This lens is formed by connecting probes along the input circle of radius R at central angle ϕ_1 to corresponding points on the output circle at central angle $\phi_2 = \phi_1/2$ with equal length cables. This lens system is not circularly symmetric and a shift in the amplitude distribution will occur as the feed point is moved from the axis of the system.

6.6 Unconstrained Feeds.

6.6.1 Transmission Lens Array. A transmission lens array uses two antenna surfaces connected by transmission lines and phase shifters. One surface is the primary or input face which collects energy whose distribution is controlled by a feed horn assembly similar to those used for advanced Cassegrain reflector antennas. A second surface is the secondary or output face which radiates the energy in a formed beam in a direction determined by the phase shifter settings.

(A) Transmission Lens Geometries. Two geometries can be considered for the input surface. A spherical input surface, Fig. 6.27, is analogous to the radial waveguide semi-constrained feed. Collimation can be accomplished by connecting elements on the primary surface to elements on the planar secondary surface by equal length transmission lines. This geometry is potentially very broad band since the path lengths from the phase center of the feed cluster to the secondary radiating elements are all equal. This geometry is also optimum from the point of view of spillover efficiency since the illumination at the edge of the secondary aperture is the same as that at the intercepted edge of the feed pattern. The incident wave front is substantially normal for all elements on the primary face so that little variation in primary element gain with position can be expected. There is also no excess path loss between the center and edge of the array to be considered in the illumination taper design.

A planar input surface, Fig. 6.28, is analogous to the sector horn semi-constrained feed. With this geometry collimation is usually accomplished by appropriate setting of the phase shifters although it could also be accomplished by varying the delay length through the lens to compensate for the excess delay in the path from the feed to the edge elements relative to the center elements. Although the use of phase shifters for collimation results in some small loss in bandwidth, it does decrease the peak value of phase quantization sidelobes as described in Lecture 2, and results in lower cost for the lens structure. In determining the secondary illumination taper for the planar-planar geometry, the difference in path length between central and edge elements on the primary surface must be accounted for as well as the differences in element gain with angle to the feed point. These factors will provide additional loss with position from the axis of the lens system and require higher feed illumination at the edge and hence more spillover loss for the planar-planar geometry than that obtained with the spherical-planar geometry.

(B) Transmission Lens Reflectance. Internal reflections or scattering within the feed structure can be considered in two categories: local reflections at the individual element level, and global reflections describing the interaction between the lens and the feed horn. Local reflections at the primary face are calculated in the same way that reflectances of radiating elements in a large array are calculated. To parallel the analysis of constrained feeds, we would use the active element impedance of the element when calculating the efficiency of the element in collecting energy from the

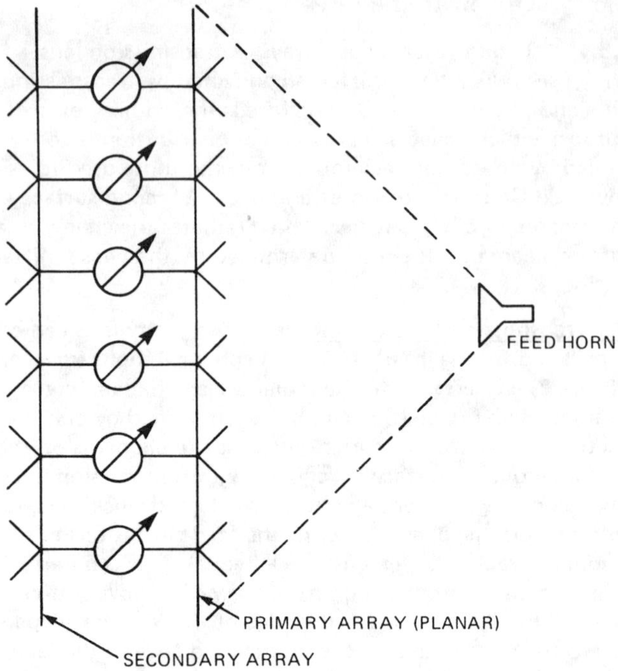

Figure 6.28 - Planar-Planar Unconstrained Feed System

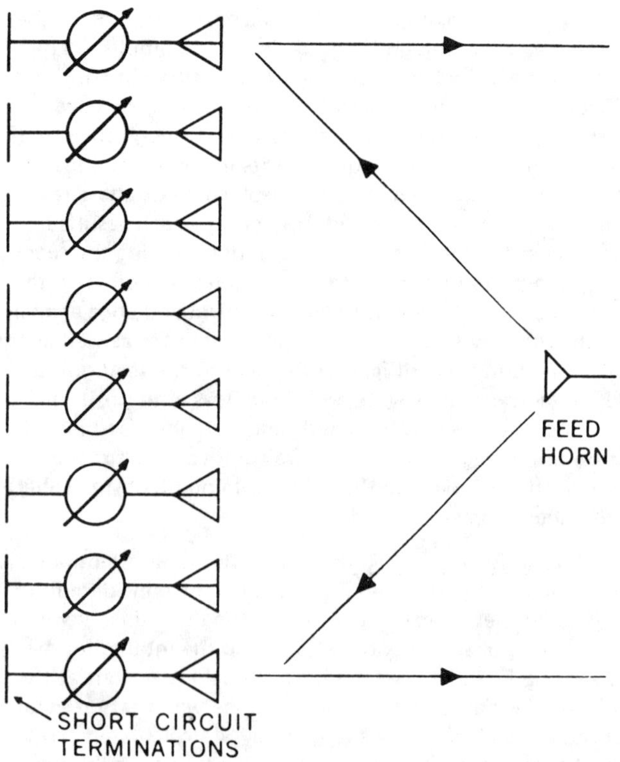

Figure 6.29 - Reflection Array Unconstrained Feed

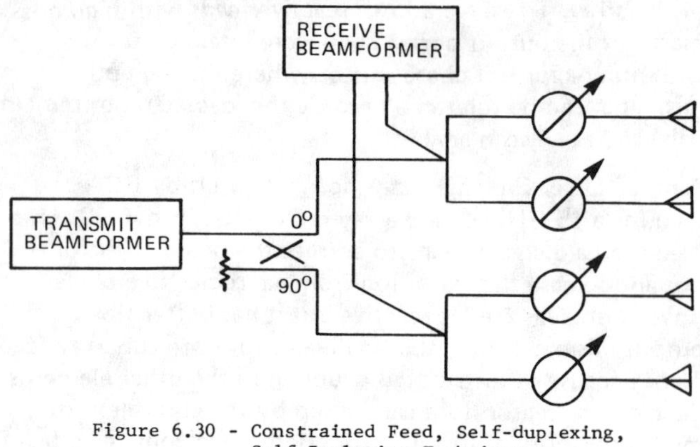

Figure 6.30 - Constrained Feed, Self-duplexing, Self-Isolation Technique

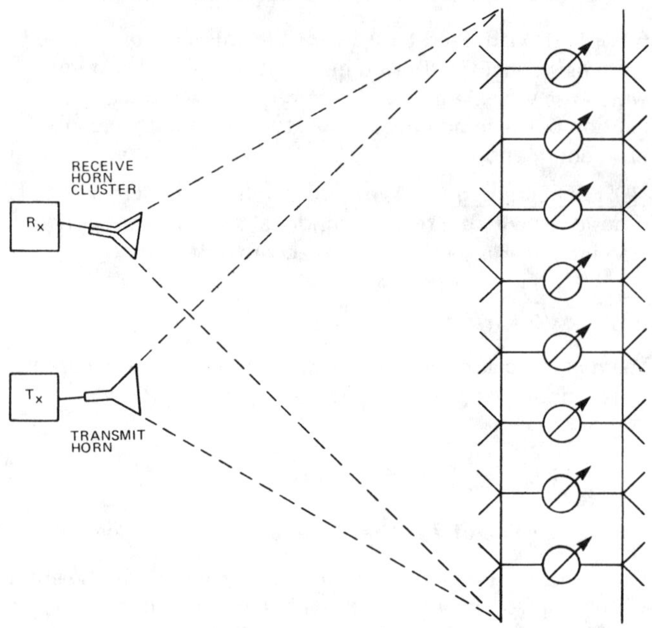

Figure 6.31 - Unconstrained Feed, Self-duplexing Technique

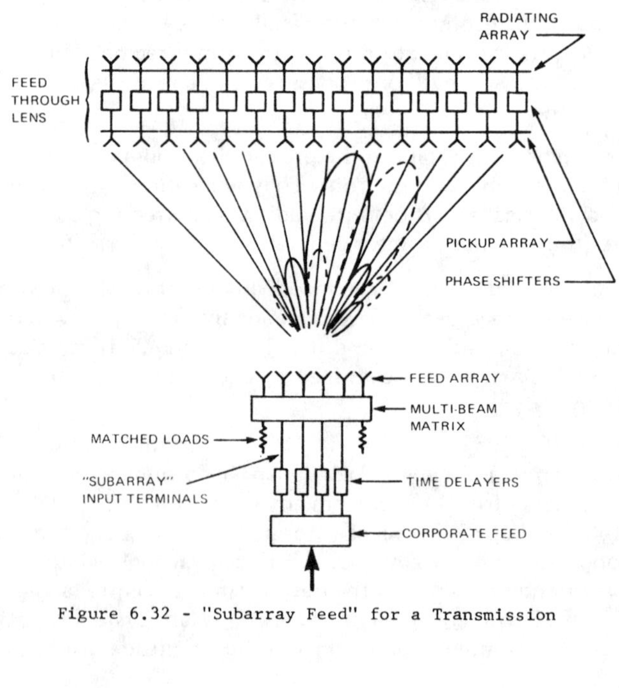

Figure 6.32 - "Subarray Feed" for a Transmission

feed horn, that is in computing the coupling coefficient between the feed horn and the primary element. When calculating the reflectance seen looking into a primary element from a secondary element however the reflectance of that element with all other elements terminated must be used. This reflectance will generally be higher than the active element reflectance with all adjacent elements fed in the proper phase. The difference between these two reflectances depends upon the spacing between the elements as well as other factors.

Global reflections depend on the type of geometry and phase shifters used in the array. With the spherical primary geometry, reflection from the primary face tends to collimate at the feed horn. With planar primary geometry reflections from the primary face tend to be specular and scatter spherically in the backlobe which may be terminated in an absorbing feed cone. Reflections from the secondary face will be collimated at the feed horn for a spherical primary when non-reciprocal phasers are used or for broadside steering using reciprocal phasers. Reflections from the secondary face are specular for a planar primary when a non-reciprocal phaser is used but are collimated at the feed horn when reciprocal phasers are set for broad-side steering. With both types of lens geometry, reciprocal phasers will shift the focal point of the reflected energy as the output beam direction is changed. This has been a source of annoyance in some very high power transmission lens arrays where the energy density at the reflectance focal point has been sufficient to cause destruction of structure or absorbing material in the vicinity of this point.

Transmission lens feed systems are not "lossless" distribution systems in the sense of constrained feeds since some of the reflection from the primary face is either radiated as a back lobe or absorbed. They are also not matched feeds in the sense of constrained feeds since they do not present a substantially zero reflectance looking back into a primary element from the secondary side. They are intermediate between these two extremes. The principal advantage of the unconstrained feed is its ability to efficiently distribute energy from a single input port to several thousand output ports without the cost, weight and loss of a constrained distribution system. Some degrees of freedom in the control of the aperture illumination function have been lost for this but, with modern feed horn technology, very satisfactory monopulse patterns can be achieved.

6.6.2 Reflect Arrays. The reflect array, Fig. 6.29, makes use of a single array face for both the primary function, collecting energy from the feed horn, and the secondary function, radiating the formed beam. Phase shift is accomplished during two passes through the phase shifters so that non-reciprocal phase shifters (those whose round trip phase is invariant with phase setting) cannot be used.* Collimation is usually accomplished using the phase shifters. It could, however, by achieved by adding delay lines between the elements and the reflecting short location to compensate for the variation in path length between the phase center of the feed cluster and the radiating elements.

The most attractive feature of the reflect array is that the cost per radiating element is least with this approach. However, it suffers from aperture blockage due to the presence of the feed structure in the radiating field of the array in addition to the spillover losses it shares with the transmission lens array. This tends to make the reflect array least efficient and places a constraint on the size of the feed structure and hence on the degree of control of aperture illumination that can be achieved.

First reflections from the input face are reflected specularly in the forward direction. When steered to broad side, energy, which enters the elements and is reflected from the shorting phase and returns to the element, is collimated to equal phase by the phase shifters and forms the beam of the array. Energy which is then reflected from the array element and makes a second round trip through the phase shifters is focused on the feed assembly. Higher order reflections attempt to focus between the feed and the array. When the array is steered off broad side the focal points move off the axis of the feed horn.

* Faraday rotation phase shifters, in conjunction with a polarization reversing reflection, can be used in this application.

6.7 Special Feed Techniques.

6.7.1 Self-Duplexing. Some means is required to connect the array to the transmitter during the transmit pulse interval and then to the receiver during the interpulse interval. For large reflector antenna systems this is accomplished using special T-R devices consisting of circulators, gas-tube and the like. In an array feed system this function can be achieved using the phase shifters as the control devices.

In a constrained feed this might be accomplished by a magic-T junction, Fig. 6.30, with the transmitter connected to the difference port. The receiver is isolated from the reflection of the transmitted pulse by the isolation of the hybrid junction. By inserting a 180 degree phase difference between the sections of the array connected to the two output ports of the junction, the array is connected to the receiver at the end of the transmit pulse.

In an unconstrained feed, Fig. 6.31, the transmit feed horn can be located at the small distance from the receive feed cluster with the phaser shifters readjusted after the transmit pulse to refocus the system on the receive feed cluster. Either of these systems require resetting the phaser shifters at the beginning and ending of the interpulse period and limit the minimum range of the radar to the phase shifter switching time. This technique does not require non-reciprocal phaser shifters but if reciprocal phaser shifters are used, there may be some beam positions for which substantial reflected power from the transmit pulse enters the receiver.

6.7.2 Self-Isolation. In order to maintain a good stable load impedance for the transmitter it is also desirable to prevent the reflection from the array face from returning to the transmitter. This can also be accomplished when non-reciprocal phasers are used. With a transmission lens array of the planar-planar geometry using non-reciprocal phasers, self-isolation occurs naturally since reflections from both primary and secondary faces are specularly reflected and do not focus at the feed. With a constrained feed, a quadrature hybrid may be used with the non-reciprocal phase shifters compensating for the extra 90 degree phase shift. Reflections from the face arrive at this hybrid in the same phase relation as the output of the hybrid and they sum in the ter-

Bandwidth

Frequency

Phase

Time Delay

Power Requirements

Loss

Power Handling Capability

Switching Speed/Rate

Minimum Range

High Data Rate

Driver Power Consumption

Reciprocity

Cost

Figure 6.33 - Requirements of Beam Steering Control Devices

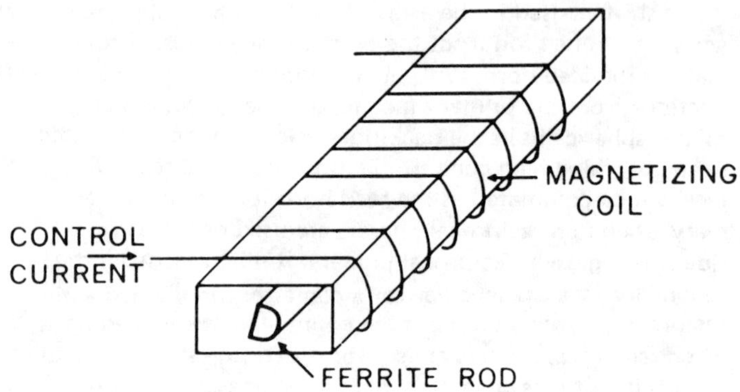

Figure 6.36 - Variable Permeability Ferrite Phaser

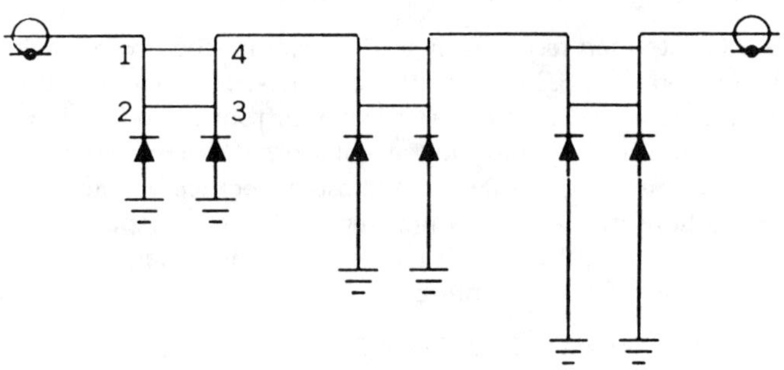

Figure 6.34 - Reflection Diode Phaser

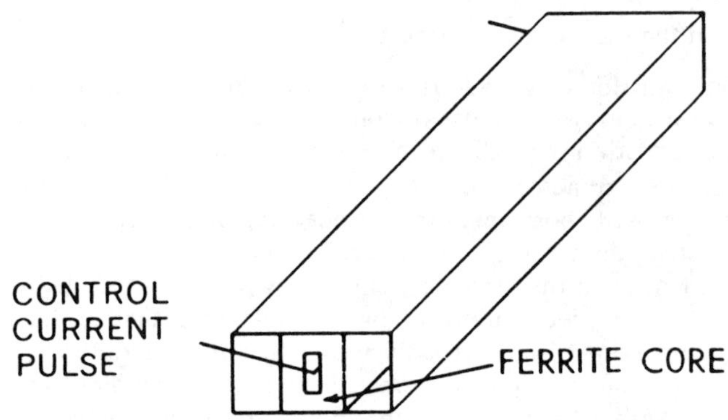

Figure 6.37 - Non Reciprocal Latching Ferrite Phaser

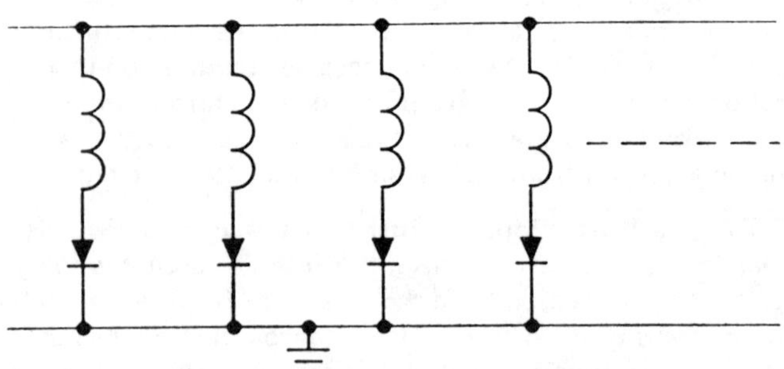

Figure 6.35 - Transmission Diode Phaser

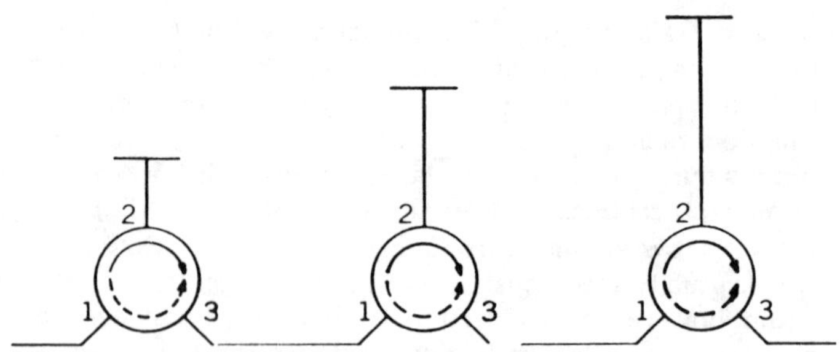

Figure 6.38 - Circulator Phaser

minated port on the hybrid. With these or similar techniques the reflection from the array face is absorbed in a load so that the input reflectance of the array remains low.

6.7.3 Sub-Arrays. Some means of organizing an array by sub-units is necessary where multiple transmitters are to be used in the array but fewer than the total number of elements. A design for wide bandwidth might also lead to subarrays since bandwidth depends upon the size of the array and time delay units are expensive.

Subdividing a constrained feed into subarray units is relatively straightforward. Either rectangular subarrays or line subarrays may be used depending on the objective of the design. For example, when a combination of phase and frequency steering is employed, the subarray takes the form of a frequency-steered line array. When time-delay steering is required at a higher level than the element level substantially square subarrays are required. Either geometry can be used when the subarray organization is employed to accommodate multiple transmitters.

Subarray organization in constrained-feed systems is also used to achieve the economy associated with modular construction. For this purpose the subarrays should be identical to each other. This dictates uniformly illuminated subarray modules and gives rise to amplitude quantization sidelobes in the antenna patterns. Thus sidelobe requirements place a constraint on the number and size of the subarrays that can be employed to give a required sidelobe level.

The benefits of subarray organization are less easily achieved with unconstrained feeds. The use of multiple feed horns with a transmission lens array does allow for multiple transmitter operation but the area in the feed horn aperture per transmitter must be less than one square wavelength to avoid loss due to grating lobes in the feed pattern and, as the feed aperture becomes larger, a larger f/D ratio must be used to insure full illumination of the primary face of the lens array.

A technique for achieving some of the broadband performance of time delay sub-array steering has been developed by Hughes under the acronym HIPSAF. This configuration Fig. 6.32, makes use of a multiple beam array feed. Each beam illuminates a small segment of the lens and defines the "subarray". Each beam port of the feed controls the excitation of the corresponding subarray so that time delay elements can be used between a constrained feed network and the unconstrained subarray feed ports to compensate for the dispersion in the lens.

6.8 Conclusions.

The techniques for feeding an array are many and varied. In this lecture some of these techniques have been catalogued in terms of the degree of constraint in the feed paths. We have also seen feeds that are highly reactive, that are well matched, and some that fall between these extremes. Many array feeds use a combination of these techniques, some combining constrained and unconstrained or semi-constrained feeding techniques, some combining serial and parallel feeding techniques. In each case the choice of feeding techniques tends to be responsive to a particular set of requirements.

We have dealt here primarily with the RF distribution system for an array. There are other distribution systems of equal importance to the ultimate performance of the array antenna system. These include the array harness, which distributes control commands and switching power to the phase shifters and their drivers; the temperature control system, which distributes coolant to the array components and controls its environment; and the array frame or structure which distributes the loads and stresses on the array to its foundation. All of these must share the available area behind the array face with the RF distribution system. It is the interaction and implementation of all of these distribution systems with their separate functional requirements which determine the ultimate form of the array needed for a given set of operating requirements.

[1] *Reference Data for Radio Engineers,* Fourth Edition, Chapter 22, pp 697, 1956.

[2] R.C. Hansen, *Microwave Scanning Antennas,* Vol. III, pp258-263, Academic Press, New York, 1966.

[3] R.C. Hansen, *Microwave Scanning Antennas,* Vol. III, p 27, Academic Press, New York, 1966.

[4] R.C. Hansen, *Microwave Scanning Antennas,* Vol. III, pp248-257, Academic Press, New York, 1966.

[5] A.R. Lopez, "Monopulse Networks for Series Feeding on Array Antenna", *IEEE Transactions*, Vol AP-16, pp 436-440, July 1968.

[6] Collin & Zucker, *Antenna Theory,* Part II, McGraw-Hill, New York, 1969, pp 131-138.

Selection of Phase Control Devices

W. T. Patton
RCA, Moorestown, New Jersey

6.9 Phase Control Device Selection Requirements.

The parameters of array design that determine the selection of the beam steering control device, Fig. 6.33 include: bandwidth, power requirements, loss, switching speed/rate, reciprocity/non-reciprocity, and cost. The bandwidth requirement is most fundamental since it determines whether time delay or phase control can be used. Power/loss requirements of course depends upon whether amplification is used between the element and the control device. The switching speed/rate requirements derives from consideration of minimum range, data rate, waveform and similar factors related to the mission and management of the radar system. The selection of a reciprocal or non-reciprocal device depends on the switching speed requirement and on the type of feed structure that will be used. The cost requirement is always there to insist that we do a good job of matching the performance of the control device to the requirements.

6.10 Selection of Phasers.

An important decision in transmitter array design concerns the use of one amplifier for each radiating element or one amplifier feeding many elements. This decision is usually made on the basis of the power to be radiated by each element, the availability of phase shifters at that power level, and the relative cost of providing many small amplifiers vs. fewer larger amplifiers. The range of choice between suitable phase shifting techniques is largest when one amplifier per element is used. However, this approach is usually more costly than one in which high-level RF phasers are used and one amplifier drives many elements.

The phasers used in electronic scanning arrays may be divided into two classes, continuous and digital. The digital device is composed of sections, each of which has two discrete phase shift states determined by its physical parameters and selectable by a control signal. The differential phase shift (difference between phases at the two states) of these devices is relatively independent of small variations in the control signal. By contrast, the insertion phase of a continuous phaser is directly dependent on the magnitude of the control signal where the various phase settings are obtained by different levels of the control signal. Phase accuracy in the the digital phaser is dependent on its construction, whereas phase accuracy in the continuous phaser depends in addition on the accuracy of the control signal. Digital phasers are most appropriate for electronic scanning arrays since they are usually controlled by a special-purpose digital computer (the beam steering controller). Continuous phasers, however, can be used as digital phasers by quantizing the control signal.

6.10.1 High-Power Phasers. High-power RF phasers can be further classified as diode phasers or ferrite phasers depending upon the active "ingredient" in the device. Diode phasers generally fall within one of two categories employing either reflection in a 90-degree hybrid circuit or propagation through a periodically loaded transmission circuit. Ferrite phasers can be either reciprocal (phase shift independent of direction through the device) or non-reciprocal (phase shift different for the two direction of transmission) and can be of either the reflection or transmission types. Diode phasers are generally reciprocal in character.

The operation of reflection diode phasers (Fig. 6.34) depends on the fact that when equal impedances are placed at ports 2 and 3 of a 90-degree 3 dB hybrid coupler, the reflected energy appears at port 4 with a phase dependent on the phase of the reflection coefficient at ports 2 and 3. In practice the reflection at ports 2 and 3 are obtained by a length of transmission line and a diode switch used to approximate an open circuit or short circuit on this line in a discrete step. Thus each hybrid junction and diode-pair circuit acts as one bit of a digital phaser. Alternatively, a pair of varactor diodes could be employed to vary the phase of the reflected wave as a function of applied bias voltage to obtain a continuous phaser.

In diode phasers of the transmission type (Fig. 6.35), the diode switch is part of a lumped periodic loading circuit in the transmission line. In digital operation, the diode is switched from its forward bias to its reverse bias condition to vary the reactive loading and therefore the phase delay through the section. Here again varactor diodes could be employed for continuous phasing.

Ferrite phasers can be designed on the variable permeability of ferrite materials as a scalar quantity, or can utilize the tensor nature of ferrite permeability. In the first type, shown in Fig. 6.36, the ferrite is generally a rod located in the center of a waveguide and magnetized longitudinally. The phase shift is dependent on the magnitude of the applied field. Phasers of this type [1] are generally continuously variable and reciprocal. A dominant disadvantage of this type of ferrite phaser is the high inductance of the magnetizing coil and the requirement for maintaining rather large currents over the operating period of the device.

The disadvantage of the variable permeability phaser is overcome by the closed-loop twin-slab phaser [2] (Fig. 6.37) which operates on the remanent magnetization of a ferrite material with a square hysteresis loop. It is essential for this type of operation that a low-reluctance closed magnetic path be available for the remanent flux. Such devices utilize the tensor character of ferrite permeability and are nonreciprocal. The differential phase shift is obtained by switching the remanent magnetization from one direction to the other. These phasers are called "latching" because they remain in one state with no further expenditure of energy until a switching pulse reverses the magnetization to the second state where it remains until subsequent switching pulses are applied. An interesting feature of this phaser is that the round-trip phase shift is the same for either state. The disadvantage of a nonreciprocal phaser is evident in its use in a duplexed array, where it must be reversed after transmission to receive from the same direction as the transmitted beam. If the switching pulse does not drive the ferrite to saturation, the differential phase shift will depend on the strength of the switching pulse. In this way it can be operated as a continuously by variable phaser

Nonreciprocal ferrite phasers are also built using a switching three-port ferrite circulator as an active element. This device, shown in Fig. 6.38, is a reflection type phaser with the differential phase adjusted by the phase of the reflection at the second port of the circulator. The transmitted energy then either takes the 1-3 reference path through the circulator or the 1-2-3 delayed path, depending on the sense of circulation. Such devices are built in a latching configuration, operating on the remanent magnetization in a closed magnetic loop.

The nonreciprocal ferrite phaser tends to be more efficient than a reciprocal phaser because off-diagonal components of the permeability tensor are much stronger functions of magnetization than are the diagonal components. A technique for obtaining the benefits of a reciprocal phaser with the efficiency of the nonreciprocal design is used in the Dual Mode [3] phase shifter. This device employs a nonreciprocal linear to circular polarization converter at each end of a ferrite section operating in the efficient nonreciprocal Faraday rotation mode. This combination of nonreciprocal effects results in a reciprocal device with the fields polarized for maximum interaction with the magnetization for both directions through the device without needing to reverse the direction of magnetization.

6.10.2 Low-Level Phasers. When phasing is to be accomplished at low levels (one amplifier per element) all of the RF phasing techniques available at high power can be used, and in addition heterodyne phasing techniques can be em ployed. It may also be desirable to accomplish phasing and beam forming at an intermediate frequency.

A heterodyne phasing technique [4] developed by W. H. Huggins operates by mixing a control frequency f_c with the operating frequency f_o and sending the difference fre-

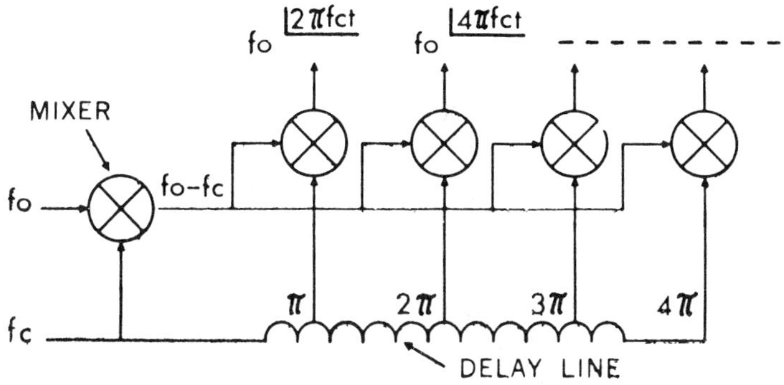

Figure 6.39 - Heterodyne Phaser

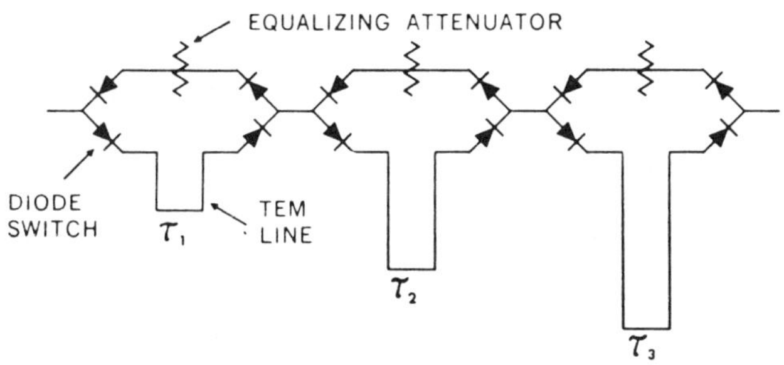

Figure 6.40 - Diode Time Delayer

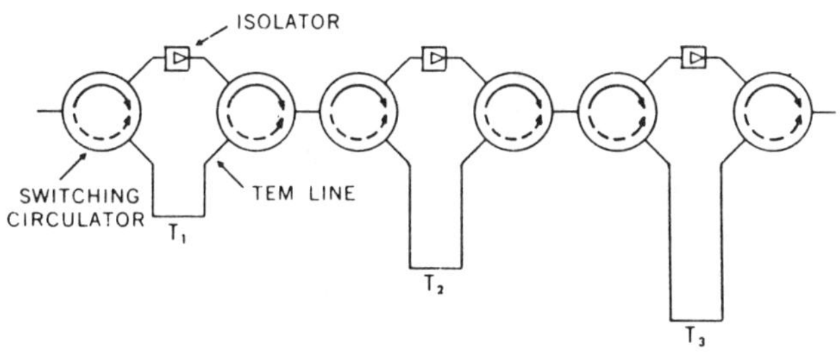

Figure 6.41 - Switching Circulator Time Delayer

quency ($f_o - f_c$) to a second mixer where it is combined with the control frequency delayed by an amount τ (see Fig. 6.39). The phase of the sum frequency at the second mixer f_o is $2 \pi f_c \tau$, proportional to the control frequency and the time delay. In an array, with the control frequency delayed by an amount $\eta\tau$, the output phase at the operating frequency is then continuously variable in proportion to the control frequency.

Phasing and beam forming at an intermediate frequency open up many additional possibilities, such as conventional lumped-constant phasers, phase-locked variable-frequency oscillators, and others. This approach also opens up the field of so called adaptive or self-phasing arrays. A discussion of this area, however, is beyond the present scope.

6.11 Selection of Time Delayers. Since phase is the product of frequency and time delay, those techniques used to produce phasers can, in principal, be used to produce time delayers. The critical difference between phase and time delay, as used in arrays, is the fact that phase is computed on the basis of molulo 2π; that is only the fractional part of the total number of cycles of phase is used. In fact, phasers employ time delay in lengths less than one wavelength, while time delayers require transmission paths nearly as long as the longest dimension of the array. Short delay lengths allow dispersive media to be employed in phasers and provide low loss. Large delay lengths require non-dispersive media and lead to higher losses and weight.

Time delay steering is required in wide instantaneous bandwidth arrays. Thus the time delayer must be designed to function over large bandwidths to be useful. It is essential that the time delayer have a linear phase-frequency characteristic over its bandwidth, since this determines the range accuracy and resolution of the radar. The many switches separated by relatively long delay lengths require an exceptionally good match at each switch junction. For example, just two connectors of 1.2 VSWR can produce a phase error of about 0.5 degree and two sets of these connectors can produce a maximum phase error of about 2 degrees. Thus care must be taken in the design of a time delayer, not only to reduce junction mismatches but also to space them so that their phase errors do not accumulate.

The insertion loss through a time delayer can vary several decibels from its minimum-delay to its maximum-delay state. This variation in the output power level can exceed the tolerance of the power amplifier driven by the time delayer, or can lead to an undesirable amplitude taper across the array. It is therefore necessary to compensate for the insertion loss to reduce its variation to a fraction of a decibel.

Because of the requirement for a linear phase-frequency relationship in the device, RF time delayers generally operate by selecting alternate TEM transmission lines of different lengths, with either a diode switch (Fig. 6.40) or a switchable circulator (Fig. 6.41). Reflection type devices can also be used, but these tend to be more sensitive to junction-match conditions and are more difficult to equalize with respect to insertion loss. Two-path time delayers can be equalized by making the loss in the short path equal to that of the long path. Some of the errors induced by junction match can be reduced by the use of isolators in the two paths. This is particularly important when circulator switches are used, since the two paths act as a resonant ring with respect to leakage through the input circulator.

The choice between diode-switched and ferrite-switched time delayers depends largely on the frequency and the mission of the array. Diode delayers tend to have the shortest switching time, whereas circulator delayers tend to have less loss and higher power-carrying capacity. Diode switches are particularly attractive at the lower microwave frequencies where their loss is least.

6.12 Conclusion. The choice between diode and ferrite devices ultimately depends upon a detailed examination of the unique characteristics of each and their suitability to the given requirements. A summary of some of the relative advantages of these techniques is given below.

(1) **High Power Handling Capability** — Ferrite devices have the advantage here since they are a bulk control medium. Diode phasers have been built, however, with power capacity adequate for most phased array applications.

(2) **Insertion Loss** — Ferrite phase shifters have typical insertion loss in the range of 0.5 to 1 dB. Diode phasers have typical insertion loss in the range of 1 to 2 dB.

(3) **VSWR** — Ferrite devices are a more nearly uniform propagating medium with fewer interacting mismatches than the diode device and can be expected to have better match over operating state and frequency.

(4) **Switching Speed** — Diode devices have smaller bulk and tend to be more easily switched than ferrite devices. Diode switching speeds are generally less than 2 us while ferrite switching speed is typically in the range from 5 to 10 us.

(5) **Driver Complexity** — The driver for ferrite phase shifters that are built as analogue devices with digitized drive are somewhat more complicated than diode device drivers but fewer are required (one per device instead of one per bit). The need for high reverse bias voltage in high power diode phasers also leads to more expensive components in the driver.

(6) **Temperature Effects** — The diode switch elements in the diode phase shifter contribute negligible change in phaser performance with temperature. The remanent magnetization in a ferrite phase shifter and corresponding differential phase are dependent on both temperature and mechanical stress. This dependence can be controlled over reasonable operating temperatures by proper formulation of the material and by driver techniques. Ferrite phase shifters tend to survive thermal overstress better than diode devices.

(7) **Reciprocity** — All diode control devices are reciprocal. Both reciprocal and nonreciprocal ferrite devices are available. The advantage of reciprocal control devices is that they need not be reset between transmit and receive intervals. The advantage of non-reciprocal phase shifters is that they can be used with reactiave feeds without the formation of spurious reflection sidelobes and can provide a self-isolation function in conjunction with the feed.

(8) **Size and Weight** — Diode phase shifters tend to be smaller and lighter than ferrite phase shifters for equivalent performance. This is particularly important in airborne systems and in solid state array applications where loss is less significant.

(9) **Reliability** — Great strides have been taken in the last decade in improvement of diode reliability so that today diode phase shifters under normal operating or design operating conditions are competitive with ferrite phase shifters in reliability. Ferrite phasers, however, are more resistant to overstress.

(10) **Frequency** — Diode phase shifters dominate the frequency range in the UHF band and below because of the much greater size of ferrite devices in this band. In the SHF frequency band both devices have a role to play.

Case Study--Design of Air Defense ESRS

Peter J. Kahrilas

Raytheon Company Bedford, Massachusetts 01730

7.1 Threat Model/Performance Requirements.

The task is to design an ESRS to cope with the threat model described in Fig. 7.1.

100 AIRCRAFT - PRIMARY MISSION: AIR SUPERIORITY

OR

50 AIRCRAFT - PRIMARY MISSION: AIR-TO-GROUND ATTACK

+

50 MISSILES - AIRCRAFT WEAPONRY

CHARACTERISTICS:

	CROSS SECTION	V_{MAX}	a_{MAX}	R_{MAX}
AIRCRAFT	1.0 m^2	MACH 2	3 g	200 nmi
MISSILES	0.05 m^2	MACH 1	10 g	100 nmi

TRACKING COVERAGE: ±45° AZIMUTH, +60° ELEVATION

MAXIMUM ALTITUDE: 50,000 FEET

MINIMUM RANGE: 10 nmi

FLIGHT FORMATION: 30 METER RADIAL SEPARATION

TERRAIN: FOREST (HEAVILY WOODED HILLS)

REACTION TIME: 8 MINUTES TO SEND OWN AIRCRAFT OFF THE GROUND; 2 MINUTES TO FIRE COUNTER WEAPONS OF 70 nmi RANGE

ACCURACY: 4 MILLISINES (1σ); 9 METERS (1σ)

BASED ON SINGLE HIT REQUIREMENTS TO DETERMINE TRAJECTORY FOR COUNTER WEAPONS FIRE CONTROL AGAINST AIRCRAFT AND MISSILES

Fig 7-1 Threat model/performance requirements

7.2 Search.

The transformation of the threat model/performance requirements to radar parameters is initiated with the search mode. The determination of detection range in terms of reaction time for the aircraft and missile is shown in Fig. 7.2.

7.2.1 Detection Range — Aircraft.

TO PROVIDE THE 8 MINUTES OWN AIRCRAFT REACTION TIME, THE ENEMY AIRCRAFT MUST BE DETECTED 8 MINUTES BEFORE THEY REACH ZERO RANGE, ASSUMING OWN AIRCRAFT ARE LOCATED AT AN AIRPORT NEAR THE RADAR.

$$\therefore R_{A/C} = 8 \text{ min} \times 22.1 \text{ nmi/min} = 176.8 \text{ nmi}$$

WHERE MACH 2 = 22.1 nmi/min

AIRCRAFT DETECTION RANGE $\geq$ 180 nmi

7.2.2 Detection Range — Missile.

THE AIR-TO-GROUND MISSILES HAVE AN EFFECTIVE RANGE OF 100 nmi AND TRAVEL AT MACH 3 (AIRCRAFT PLUS MISSILE VELOCITY RELATIVE TO GROUND) TOWARDS THE TARGET. THEY MUST BE DETECTED CLOSE TO THEIR MAXIMUM LAUNCH RANGE BUT AT LEAST A SUFFICIENT DISTANCE TO ALLOW THE 2 MINUTES REACTION TIME REQUIRED FOR THE ANTIMISSILE MISSILES.

$$\therefore R_{MSL} = 2 \text{ min} \times 33.3 \text{ nmi/min} = 66.6 \text{ nmi}$$

WHERE MACH 3 = 33.3 nmi/min

MISSILE DETECTION RANGE $\geq$ 70 nmi

Fig 7-2 Maximum detection range

The resolution and false alarm number determination are shown in Fig. 7.3.

7.2.3 Resolution.

THE RADIAL RESOLUTION REQUIREMENT OF 30 METERS IS EQUIVALENT TO A 0.2 MICROSECOND PULSE WIDTH. THIS VALUE IS DECREASED BY A FACTOR OF TWO TO ACCOUNT FOR ADVERSE RF PHASE RELATIONSHIP BETWEEN TWO ADJACENT RETURNS.

EQUIVALENT PULSE WIDTH = 0.1 μsec

NO ANGULAR RESOLUTION IS SPECIFIED.

7.2.4 False Alarm Number.

ALLOW ONE FALSE ALARM PER SECOND AS A REASONABLE VALUE FOR THE SYSTEM TO HANDLE WITHOUT OVERTAXING THE COMPUTER AND RADAR ON POWER AND TIME SPENT ON VERIFICATION OF FALSE ALARMS.

$$\therefore P_N = \tau \text{ FAR} = 10^{-7} \text{sec} \times 1/\text{sec} = 10^{-7}$$

$$P_N = 10^{-7}$$

Fig 7-3 Resolution and false alarm

7.2.5 Search Pattern.

Initially, select a beamwidth of 1° as suitable for the search and track modes. The eventual beamwidth selected will be based on obtaining acceptable values for search frame time, power-aperture product, aperture size and tracking performance. Several iterations may be required to arrive at a final design value, It may even be necessary to retain a 1° beamwidth for track and defocus the beam in search to obtain an acceptable frame time at a sacrifice in power.

The search sector is scanned with columned or interlaced beams. For a circular beam the interlaced pattern requires 13.4% (1 - .866) more elevation beam positions compared to columned beam but it is still preferred and more commonly employed because the interlace pattern decreases the depths of the holes between beam positions (Fig. 7.4).

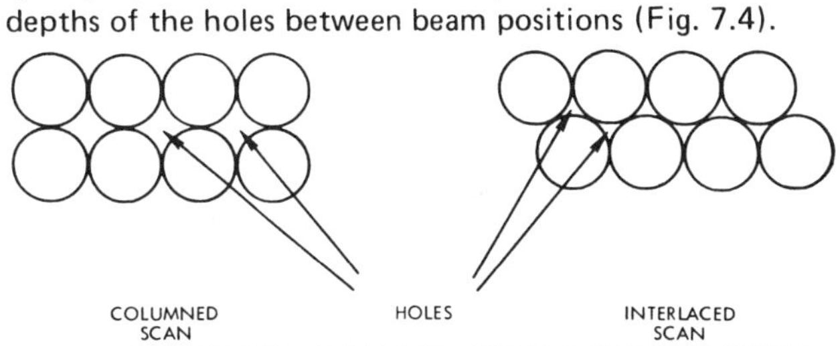

THE SEARCH SECTOR WILL BE SCANNED WITH INTERLACED INSTEAD OF COLUMNED BEAMS TO DECREASE THE DEPTH OF THE HOLES BETWEEN BEAM POSITIONS.

Fig 7-4 Search pattern

A second consideration is the size of the beam position steps. The pattern loss decreases as the beam stepping increment decreases but more beam positions, and, there fore, more time are required to cover the search sector (Fig. 7.5). The most common preference is the selection of the 3dB beamwidth steps. However, there are situations where a very high probability of detection is required that smaller increments are employed; especially if the coverage sector is small.

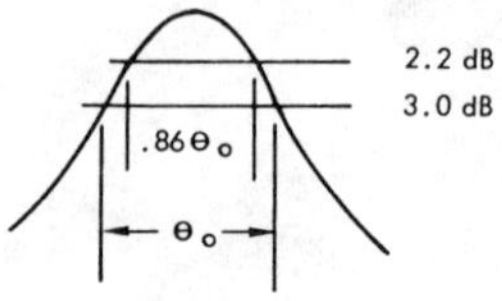

(GAUSSIAN PATTERN)

CROSS OVER	STEP SIZE	PATTERN LOSS
3.0 dB	θ_o	3.2 dB
2.2 dB	$.86\,\theta_o$	2.6 dB (TYPICAL)

THE COMPROMISE SELECTED IN THIS DESIGN IS TO STEP THE BEAM IN FULL BEAMWIDTH (3 dB POINT) STEPS.

∴ SCAN PATTERN: 17.45 MILLISINE STEPS IN AZIMUTH
15.11 MILLISINE STEPS IN ELEVATION

THE ARRAY IS DESIGNED TO OBTAIN 98° CONICAL COVERAGE (ELEMENTS LOCATED ON EQUILATERAL TRIANGULAR GRID STRUCTURE) AND FACES TRUE NORTH.

∴ N = 0

T = 22.2° FOR ±45° AZIMUTH AND 0° TO 60° ELEVATION

THE RELATION BETWEEN THE ARRAY AND SPACE COORDINATES IS SHOWN IN FIG 7-6 AND THE FORMULAS ARE CONTAINED IN FIG 7-7.

Fig 7-5 Beam step increments

7.2.7 Coordinate Transformation Summary.

$A_z, E \rightarrow \alpha, \beta$

$$\alpha = \cos E \sin A_z \cos N - \cos E \cos A_z \sin N \quad (7\text{-}1)$$

$$\beta = -\cos E \sin A_z \sin N \sin T + \sin E \cos T - \cos E \cos A_z \cos N \sin T \quad (7\text{-}2)$$

$\alpha, \beta \rightarrow A_z, E$

$$E = \sin^{-1}\left[\beta \cos T + \sqrt{1-\alpha^2-\beta^2}\,\sin T\right] \quad (7\text{-}3)$$

$$A_z = \tan^{-1}\left[\frac{\alpha \cos N - \beta \sin N \sin T + \sqrt{1-\alpha^2-\beta^2}\,\sin N \cos T}{-\alpha \sin N - \beta \cos N \sin T + \sqrt{1-\alpha^2-\beta^2}\,\cos N \cos T}\right] \quad (7\text{-}4)$$

$\theta, \phi \rightarrow \alpha, \beta$

$$\alpha = \sin \theta \cos \phi \quad (7\text{-}5)$$

$$\beta = \sin \theta \sin \phi \quad (7\text{-}6)$$

$\alpha, \beta \rightarrow \theta, \phi$

$$\sin^2 \theta = \alpha^2 + \beta^2 = \sin^2 A + \sin^2 B \quad (7\text{-}7)$$

$$\tan \phi = \frac{\alpha}{\beta} = \frac{\sin A}{\sin B} \quad (7\text{-}8)$$

Fig 7-7 Coordinate transformation summary

7.2.6 Array-Space Coordinates.

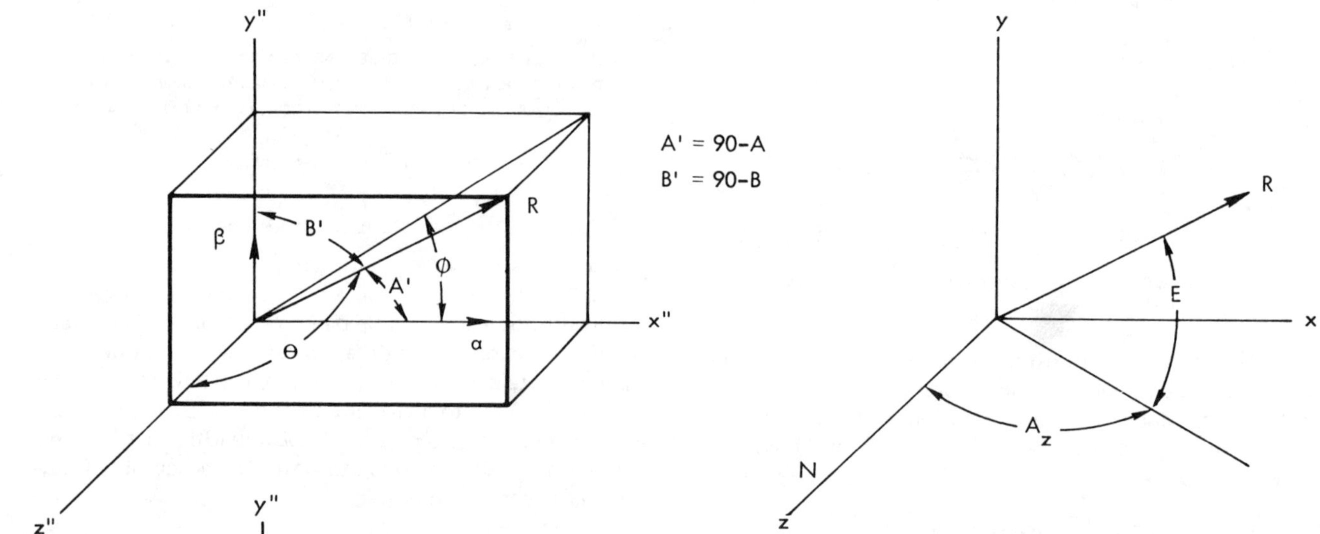

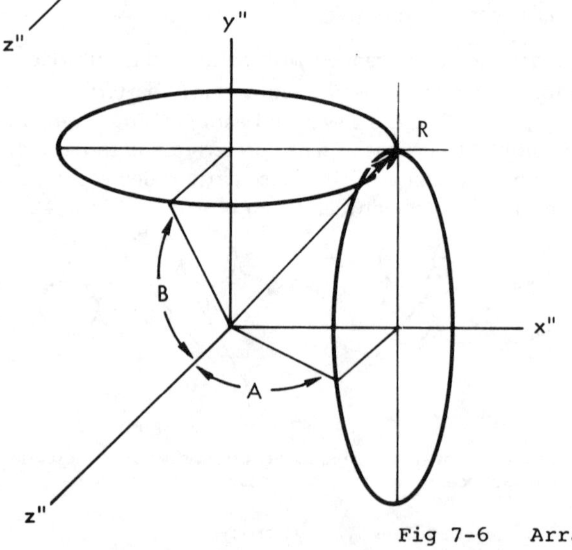

x", y" = PLANE OF ARRAY

$\alpha = \sin A = \frac{x''}{R}$

$\beta = \sin B = \frac{y''}{R}$

θ, Φ = ANTENNA SPHERICAL COORDINATES

A_z = AZIMUTH ANGLE

E = ELEVATION ANGLE

T = TILT ANGLE OF ARRAY FROM VERTICAL

N = ANGLE BETWEEN TRUE NORTH AND PLANE ⊥ TO FACE OF ARRAY

Fig 7-6 Array-space coordinates

The search scan pattern coverage is determined by the procedure described in Fig. 7.8 and equations (7.1) and (7.2) in the form given in Fig. 7.9.

1. BEGIN SEARCH PATTERN WITH ELEVATION BEAM RAISED 0.5 BEAMWIDTH ABOVE THE HORIZON.

   ∴ E = 0.5° FOR ELEVATION BEAM POSITION NO. 1

2. AT EACH ELEVATION BEAM POSITION DETERMINE (1) THE MAXIMUM RANGE COVERAGE FOR THE <u>BOTTOM</u> OF THE BEAM AS LIMITED BY AN ALTITUDE OF 50,000 FEET FROM RANGE VERSUS ALTITUDE CHART AND (2) THE NUMBER OF BEAMS REQUIRED TO SCAN ±45° IN AZIMUTH.

3. TERMINATE SEARCH IN ELEVATION AT APPROXIMATELY 70 nm WHICH IS THE MINIMUM MISSILE DETECTION RANGE.

Fig 7-8 Search coverage

FOR $N = 0,\ T = 22°,\ A_z = 45°$

$$\left.\begin{aligned} \sin A &= \alpha = .707 \cos E \\ \sin B &= \beta = .926 \sin E - .267 \cos E \end{aligned}\right\} A_z = 45°$$

$$\left.\begin{aligned} E &= T + B_o \\ \sin B_o &= \sin (E-T) \end{aligned}\right\} A_z = 0°$$

THE CALCULATIONS FOR ELEVATION BEAM POSITIONS NO. 1, 2 AND 8 (LAST) ARE SHOWN IN FIG 7-10, 11 AND 12, RESPECTIVELY.

Fig 7-9 Search pattern equations

$E = 0.5°$

$\sin B_o = \sin (0.5° - 22.2°) = -\sin 21.7° = -.3697$

$\sin B_{oL} = -.3697 - .008725 = -.3784$

$B_{oL} = -22.2°$

$E_L = 0°$

$R_{max} = 200$ nm

$\alpha = .707 \cos 0.5° = .707$

$A = \sin^{-1} .707 = 45°$

NO. OF AZ BEAM POSITIONS = 2 x .707/.01745 = 81

$\beta = .926 \sin 0.5° - .267 \cos 0.5°$

$\beta = .00808 - .267 = -.2589$

$B = \sin^{-1} - .2589 = -15.00°$

Fig 7-10 Elevation beam position #1

$\sin B_o = -.3697 + .01511 = -.3546$

$B_o = -20.77°$

$E = 1.43°$

$\sin B_{oL} = -.3546 - .008725 = .3633$

$B_{oL} = -21.30°$

$E_L = .90°$

$R_{max} = 200$ nm

$\alpha = .707 \cos 1.43° = .707$

$A = 44.97°$

NO. OF AZ BEAM POSITIONS = 81

$\beta = .926 \sin 1.43° - .267 \cos 1.43°$

$\beta = .02311 - .26692 = -.2438$

$B = -14.11°$

Fig 7-11 Elevation beam position #2

$\sin B_o = -.2791 + .01511 = -.2640$

$B_o = -15.31°$

$E = 6.89°$

$\sin B_{oL} = -.2640 - .008725 = -.2727$

$B_{oL} = -15.83°$

$E_L = 6.37°$

$R_{max} = 72$ nm - LAST ELEVATION BEAM POSITION

$\alpha = .707 \cos 6.89° = .7019$

$A = 44.58°$

NO. OF AZ BEAM POSITIONS = 2 x .7019/.01745 = 80.45 = 81

$\beta = .926 \sin 6.89° - .267 \cos 6.89°$

$\beta = .1111 - .2651 = -.1540$

$B = -8.86°$

THE MAXIMUM SCAN ANGLE, ξ_{max}, OFF BROADSIDE OCCURS AT ELEVATION BEAM POSITION NO. 1. FROM EQUATION (7-7)

$$\xi_{max} = \sin^{-1} \sqrt{\alpha_1^2 + \beta_1^2}$$

$$\xi_{max} = \sin^{-1} \sqrt{.7071^2 + .2589^2} = 49°$$

THE SINE SPACE COVERAGE PATTERN IS PLOTTED IN FIG 7-13 AND SUMMARIZED IN THE CHART OF FIG 7-25.

Fig 7-12 Elevation beam position #8 (last)

THE TOTAL NUMBER OF SEARCH BEAMS IS 8 EL x 81 A_z 648

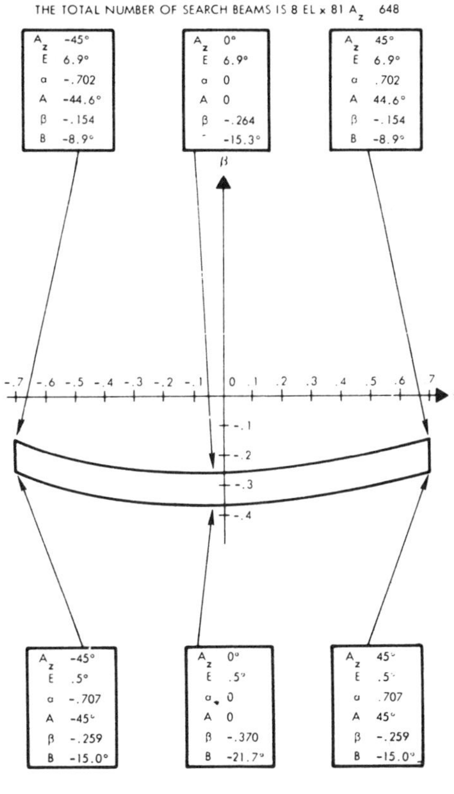

Fig 7-13 Search coverage pattern

7.2.8 Probability of Detection. The next step is to determine the detectability factor (signal-to-noise ratio required to meet the probability of detection at maximum range for the selected false alarm rate and target fluctuation model) for both aircraft and missile targets. The design conditions for obtaining the cumulative probability of detection and the tabulated results are contained in Figs. 7.14, 7.15 for the aircraft and Figs. 7.16, 7.17 for the missile.

AIRCRAFT - SW I

DESIGN FOR $P_{CUMULATIVE} = 99\%$ $P_N = 10^{-7}$

FRAME TIME = 2.42 sec (REFER TO FIG 7-25)

V = MACH 2 = 2 x .184 nm/sec = .368 nm/sec

RADIAL RANGE INCREMENT PER FRAME TIME = .368 x 2.42 = .89 nm

$$P_c = 1 - \prod_{i=1}^{M} (1 - P_{D_i})$$

WHERE:

P_c = CUMULATIVE PROBABILITY OF DETECTION FOR M SCANS

P_{D_i} = PROBABILITY OF DETECTION ON i^{th} SCAN

M = TOTAL NUMBER OF SCANS THAT THE TARGET CAN BE DETECTED

LET S/N_I = 10.0 dB AT MAXIMUM RANGE OF 200 nm
(THIS CALCULATION MAY REQUIRE SEVERAL REITERATIONS)

$\therefore P_D$ = .275 (FROM PROBABILITY CHARTS FOR SW I $P_N = 10^{-7}$)

AT OTHER RANGES $S/N_N = S/N_I + \left(\frac{R_I}{R_N}\right)^4_{dB}$

Fig 7-14 P_{CUM} for aircraft (1)

AIRCRAFT - SW I

LET S/N = 10.0 dB AT R = 200 nm

N	R nm	S/N dB	P_D	$1-P_D$	$\Pi(1-P_D)$	P_{CUM}	
1.	200.00	10.00	.240	.760	.760	.240	
2.	199.11	10.08	.245	.755	.574	.426	
3.	198.22	10.16	.250	.750	.430	.570	
4.	197.33	10.23	.255	.745	.320	.680	
5.	196.44	10.31	.260	.740	.237	.763	
6.	195.55	10.39	.270	.730	.173	.827	
7.	194.66	10.47	.275	.725	.125	.875	
8.	193.77	10.55	.278	.722	.090	.910	
9.	192.88	10.63	.285	.715	.058	.942	
10.	191.99	10.71	.290	.710	.041	.959	
11.	191.10	10.79	.295	.705	.021	.979	
12.	190.20	10.87	.300	.700	.020	.980	
13.	189.32	10.95	.310	.690	.014	.986	
14.	188.43	11.04	.312	.688	.010	.990	OK

$\therefore R_{A/C}$ = 188 nm (> 180 nm)

Fig 7-15 P_{CUM} for aircraft (2)

MISSILE - SW III

$V_{CLOSING}$ = MACH 3 = 3 x .184 = .552 nm/sec

RADIAL RANGE INCREMENT PER FRAME TIME = .552 x 2.42 = 1.34 nm

AT 100 nm 3 dB RANGE IMPROVEMENT $\left(\frac{200}{100}\right) \times 4 = 12.0$ dB

CROSS SECTION LOSS $\sigma_{A/C}/\sigma_{MSL} = \frac{1.0}{.05} = 20 = 13.0$ dB

$\therefore$ NET LOSS = 1.0 dB

LET S/N_I = 9.0 dB AT MAXIMUM RANGE OF 100 nm

Fig 7-16 P_{CUM} for missile (1)

MISSILE - SW III

LET S/N = 9.0 dB AT R = 100 nm

N	R nm	S/N dB	P_D	$1-P_D$	$\Pi(1-P_D)$	P_{CUM}	
1.	100.00	9.00	.150	.850	.850	.150	
2.	98.66	9.23	.165	.835	.710	.290	
3.	97.32	9.47	.180	.820	.582	.418	
4.	95.98	9.71	.205	.795	.463	.537	
5.	94.64	9.96	.225	.775	.359	.641	
6.	93.30	10.20	.245	.755	.271	.729	
7.	91.96	10.46	.270	.730	.198	.802	
8.	90.62	10.71	.285	.715	.142	.858	
9.	89.28	10.97	.315	.685	.097	.903	
10.	87.94	11.23	.340	.660	.064	.936	
11.	86.60	11.50	.370	.630	.040	.960	
12.	85.26	11.77	.400	.600	.024	.976	
13.	83.92	12.05	.430	.570	.014	.986	
14.	82.58	12.33	.460	.540	.008	.992	OK

$\therefore R_{MSL}$ = 82 nm (> 70 nm)

Fig 7-17 P_{CUM} for missile (2)

7.2.9 Energy Requirements. The system parameters are determined for various ESRS and a cost-effective study is then performed to select the most suitable design for this application. The method will be demonstrated for an optically-fed thinned array.

$$E = \frac{S/N\ R^4\ L\ NF}{\alpha\ G^2\ \sigma\ \lambda^2}$$

S/N = 10.0 dB
R^4 = 200^4 nm = 92.04 dBnm } from A/C P_{cum} = 99% calculations.

L = Total losses not included in system noise temperature.

$L_{atmospheric}$ = 3.3dB (R.T. @ 0.5°E, 3 GHz, 200 nmi)

Antenna

Radome	0.5 dB	
Collector Illumination taper	0.2	
Spillover	1.2	
Feed	0.2	
3 bit Phase Shifter Insertion Loss	1.2	
3 bit Quantization	0.2	
Element mismatch	0.5	
Phase tolerance -10° RMS	0.1	
Amplitude tolerance	0.3	
Manufacturing tolerance	0.3	4.7dB

Duplexer	0.5 dB	
Waveguide	1.1	
Isolator	0.5	
Harmonic filter	0.25	
Matched filter	1.5	3.85 dB
Beam pattern (3dB steps)		3.2dB
Beam Scanning $\sec^3 49° = 3.54$	=	5.5dB
L = 3.3 + 4.7 + 3.85 + 3.2 + 5.5	=	20.55dB

System Noise Temperature

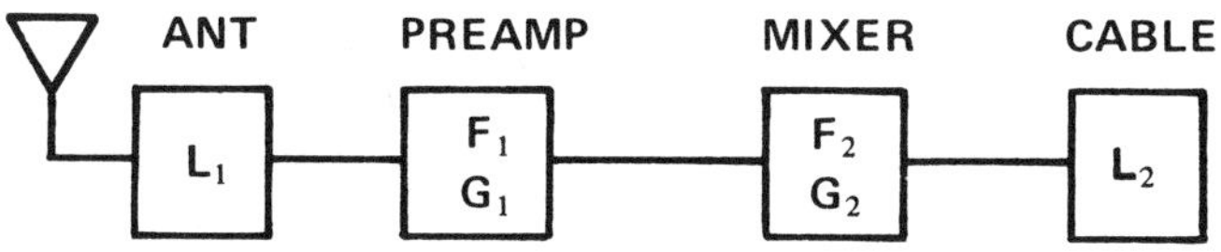

$$T_S = T_A + T_O\left[(L_1 - 1) + L_1(F_1 - 1) + L_1\frac{(F_2 - 1)}{G_1} + \frac{L_1(L_2 - 1)}{G_1 G_2}\right]$$

T_A = 40°K (from charts)

T_O = 290°K

L_1 =	L_{ANT}	4.7 dB
	Duplexer	0.5
	TR Switch	1.0
		6.2 dB = 4.17

F_1 = 2.0 dB = 1.58 uncooled parametric amplifier

G_1 = 30 dB = 1000

F_2 = 8.0 dB = 6.31 mixer

G_2 = -6.0 dB = .25

L_2 = 3.0 dB = 2.0 cable

$$\therefore T_S = 40 + 290\left[\underbrace{4.17-1}_{3.17} + \underbrace{4.17 \times (1.58-1)}_{2.42} + \underbrace{\frac{4.17\ (6.31-1)}{1000}}_{.022} + \underbrace{\frac{4.17\ (2.0-1)}{1000 \times .25}}_{.0167}\right]$$

$$T_S = 40 + 290 \times 5.629 = 40 + 1632.32 = 1672.32$$

$$NF = \frac{T_S}{T_O} = \frac{1672.32}{290} = 5.77 = 7.61 \text{ dB}$$

α = 0.28 dB

$$G = \frac{4\pi A}{\lambda^2}$$

$$D = \frac{74\lambda}{BW} \quad \text{(Taylor taper 34dB max sidelobe } \bar{n} = 4)$$

$$\lambda = \frac{c}{f} = \frac{3\times10^{10}\text{ cm}}{3\times10^{9}\text{ Hz}} = 10 \text{ cm}$$

$$D = \frac{74\times10}{1^\circ} = 740 \text{ cm} = 7.4\text{m} = 24.3'$$

A = $3.47R^2 = 3.47 \times 370^2 = 4.75 \times 10^5$ cm^2 (See Fig. 7.18 and 7.19).

Hexagonal Array

Maximum scan angle ξ_{max} = 49°

Allowing 10° margin, design for scan of 59°

$$d = \frac{\lambda}{1 + \sin\xi} = \frac{\lambda}{1+\sin 59^\circ} = \frac{\lambda}{1.86} = .539\lambda = 5.39 \text{ cm}$$

For element spacing on an equilateral triangle grid

$$a = \frac{2d^2}{\sqrt{3}} = \frac{2\times5.39^2}{\sqrt{3}} = 33.5 \text{ cm}^2$$

$$\text{Number of elements} = \frac{A}{a} = \frac{4.75\times10^5}{33.5} = 14{,}180$$

Use 50% thinning

∴ No. of active elements = 7090

No. of dummy elements = 7090

$$G = \frac{4\pi \times 4.75 \times 10^5}{100} = 5.97 \times 10^4 = 47.76 \text{ dB}$$

	47.76 dB
-50% thinning	−3.00
	44.76 dB

G^2 = 89.52 dB

σ = 1 m^2 = 0 dB m^2

λ^2 = 100 cm^2 = 20 dB cm^2

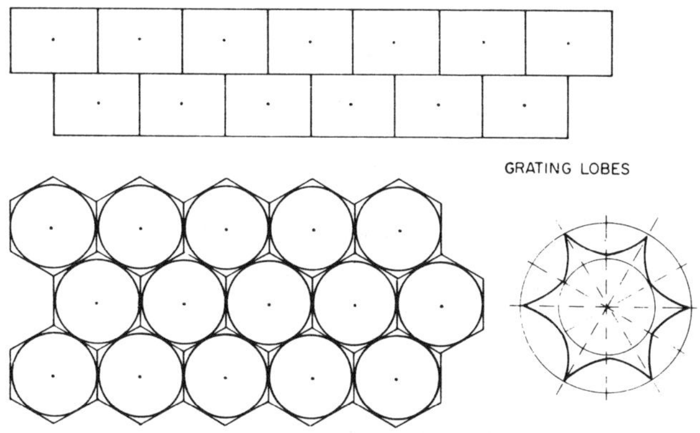

Fig 7-18 Element area - equilateral spacing (1)

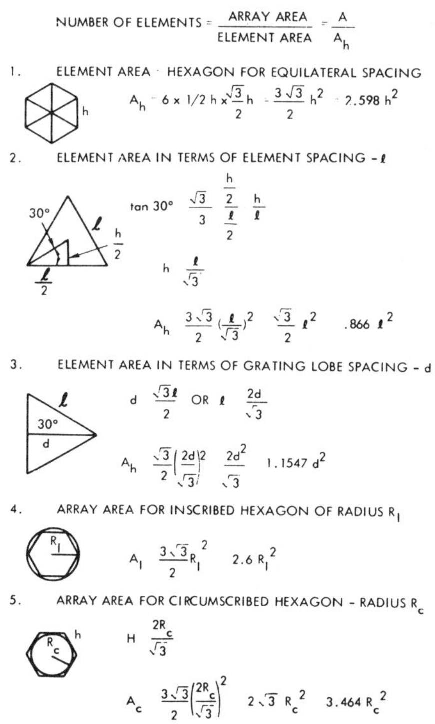

Fig 7-19 Element area - equilateral spacing (2)

The solution for the energy form of the range equation is contained in Fig. 7.20.

$$E = \frac{S/N\ R^4\ L\ NF}{\alpha\ G^2\ \sigma\ \lambda^2}$$

S/N	=	10.00	α	=	0.28
R^4	=	92.04	G^2	=	89.52
L	=	20.55	σ	=	0.00
NF	=	7.61	λ^2	=	20.00
		130.20			109.80

$$-109.80$$

$$E = 20.40\text{ dB} = 110\text{ JOULES}$$

$$R_{min} = 10\text{ nm} = 123.6\ \mu\text{sec}$$

$$\therefore\ \tau_{uncompressed} = 110\ \mu\text{sec}\ (< 123.6\ \mu\text{sec})$$

$$\therefore\ P_P = \frac{E}{\tau} = \frac{110}{110 \times 10^{-6}} = 1\text{ MW}$$

Fig 7-20 Range equation - energy form

7.3 Track.

7.3.1 Waveform. Let the waveform be the same for search and track as a first iteration. It makes for a simpler system.

τ = 100 μs/0.1 μs (1100:1 linear fm pulse compression)

Signal-to-noise ratio

S/N = 11.0 dB @ R = 188 nm

The signal-to-noise ratio is approximately the same for search and track. The range gate loss is offset by the improvement in beam shape loss. In track the target is closer to the peak of the beam position.

7.3.2 Accuracy — One Sigma Values

Angle

1. Antenna pointing = .09 ms
2. Quantization (16 bits) = .03 ms
3. A/D conversion = .05 ms
4. Thermal

$$\sigma_T = \frac{.53\theta}{\sqrt{S/N}} = \frac{.53 \times 17.452\text{ ms}}{\underset{3.55}{\sqrt{12.59}}} = 2.61\text{ ms}$$

where

$\theta = 1° = 17.45$ ms

S/N = 11.0 dB = 12.59

5. Multipath

$$\sigma_m = \frac{\theta\rho}{\sqrt{8G}}$$

θ = beamwidth

ρ = surface reflectivity coefficient

G = main lobe to side lobe ratio in direction of reflected ray

$\theta = 1° = 17.452$ ms

$\rho = -20$ dB = .1

G = 28 dB = 630.96

$$\sigma_m = \frac{17.452 \times .1}{\underbrace{\sqrt{8 \times 630.96}}_{71.05}} = 0.2\text{ ms}$$

6. Instrumentation

4 horn multimode feed with 3 channel monopulse limiting (IAGC for single hit operation) receiver

	Errors	
	Amplitude	Phase
Precomparator	0.5 dB	2.5°
Post comparator	0.5 dB	5.0°
Post post comparator (limiter)	—	1.0°

σ_I = 0.20 ms

7. Propagation (and calibration)

σ_D (no calibration) = 0.25 ms

8. Dynamic Lag

The tracking loop design formulas are contained in Fig. 7.21.

DESIGN FOR A TYPE II SAMPLED DATA SERVO LOOP TO THE COMPUTER FOR ALL 3 SINE SPACE COORDINATES FOR BOTH AIRCRAFT AND MISSILE TARGETS.

USE ONE SAMPLE PER SECOND DATA RATE FOR AIRCRAFT TARGETS AND TWO SAMPLES PER SECOND DATA RATE FOR MISSILE TARGETS.

DETERMINE IF THE SMOOTHING CONSTANTS OF A CRITICALLY DAMPED SYSTEM ARE SATISFACTORY AND THEN COMPUTE THE DYNAMIC LAG ERROR.

$$\epsilon = \frac{a\,\tau^2}{\beta}$$

$(2\alpha + \beta) \leq 4$ FOR STABILITY

$0 < \alpha < 1$

$\beta > 0$

$\alpha = 2\sqrt{\beta} - \beta$

$\epsilon_D = \epsilon(1 - \alpha)$

WHERE:

ϵ_D DYNAMIC LAG ERROR

ϵ ERROR IN PREDICTED POSITION

a = MAXIMUM ACCELERATION OF TRACKED TARGET

T = SAMPLE PERIOD

β = VELOCITY SMOOTHING CONSTANT

α = POSITION SMOOTHING CONSTANT

Fig 7-21 Tracking loop design

Aircraft

Maximum angular acceleration occurs at minimum range.

a = 3 gs = 3 x 32.2 = 96.6 fps

V = M2 = 2 x .184 = .368 nm/s

R_{min} = 10 nm

$$\ddot{\phi}_I = .65 \frac{V^2}{R_{min}} = \frac{.65 \times .368^2}{10^2} \times 10^3 = .88\text{ mils/s}^2$$

$\ddot{\phi}_2 \; R_{min} = a$

$$\therefore \ddot{\phi}_2 = \frac{96.6}{10\text{x}6080} \text{ x } 10^3 = 1.59 \text{ mils/s}^2$$

$$\epsilon = \frac{17.45}{2} \text{ mils} = 8.73 \text{ mils (½ beamwidth)}$$

T = 1.0 second

a = 1.59 mils/s^2

$$\beta = \frac{a\,T^2}{\epsilon} = \frac{1.59\text{x}1^2}{8.73} = .182 \; (> 0 \text{ OK})$$

$$\alpha = \underbrace{2\sqrt{.182}}_{.427} - .182 = .672 \; (< 1 \text{ OK})$$

ϵ_D = 8.73 (1-.672) = 2.87 ms

Missile

a = 10 gs = 10 x 32.2 = 322 fps

V = M3 (relative to ground) = 3 x .184 = .522 nm/s

R_{min} = 10 nm

$$\ddot{\phi}_1 = \frac{.65\text{x}.552^2}{100} \text{ x } 10^3 = 1.98 \text{ mils/s}^2$$

$$\ddot{\phi}_2 = \frac{322}{10\text{x}6080} \text{ x } 10^3 = 52.96 \text{ mils/s}^2$$

ϵ = 8.73 mils

T = 0.5 seconds

a = 52.96 mils/s^2

$$\beta = \frac{52.96\text{x}0.5^2}{8.73} = 1.517 \; (> 0 \text{ OK})$$

$$\alpha = \underbrace{2\sqrt{1.517}}_{1.232} - 1.517 = 0.946 \; (< 1 \text{ OK})$$

ϵ_D = 8.73 (1-.946) = .47 ms

The angle errors are summarized in Fig. 7.22.

	AIRCRAFT (ms)	MISSILE (ms)
1. ANTENNA POINTING	0.09	0.09
2. QUANTIZATION (16 BITS)	0.03	0.03
3. A/D CONVERSION	0.05	0.05
4. THERMAL	2.61	2.61
5. MULTIPATH	.02	.02
6. INSTRUMENTATION	0.20	0.20
7. PROPAGATION (AND CALIBRATION)	0.25	0.25
8. DYNAMIC LAG	2.87	0.47
TOTAL rss	3.89	2.67

Fig 7-22 Angle errors

Range - one sigma values

1.	Quantization (LSB = 1m)	= 0.29m
2.	A/D conversion	= 0.20m
3.	Main time synchronization pulse jitter for 10 nano-second sinusoid	= 0.50m
4.	Non-synchronous time measurement for 80 MHz crystal clock	= 0.54m
5.	Doppler shifts gives rms range error dependent on velocity uncertainty for linear fm pulse compression	= 2.50m
6.	Propagation (½ for missile targets)	= 3.00m
7.	Multipath	= .02m

8. Thermal

$$\sigma_T = \frac{c\,\tau_c}{2\sqrt{2\,S/N}}$$

τ_c = 0.1 μs compressed pulse width

$$\therefore \frac{c\,\tau_c}{2} = 3\text{x}10^8 \text{ x } 10^{-7} = 15\text{m}$$

S/N = 11.0 dB = 12.59

$$\sigma_T = \frac{15}{\underbrace{\sqrt{2\text{x}12.59}}_{5.01}} = 3.00\text{m}$$

9. Dynamic Lag

Aircraft

T = 1.0 seconds

ϵ = 7.5m (½ compressed pulse width)

$$a = 3 \text{ gs} = \frac{3\text{x}32.2}{3.281} = 29.44 \text{ m/s}^2$$

$$\beta = \frac{29.44\text{x}1^2}{7.5} = 3.925 \; (< 4 \text{ OK})$$

$$\alpha = \underbrace{2\sqrt{3.925}}_{1.981} - 3.925 = .0375$$

ϵ_D = 7.5 (1-.0375) = 7.22m

Missile

T = 0.5 seconds

ϵ = 7.5m

$$a = 10 \text{ gs} = \frac{10\text{x}32.2}{3.281} = 98{,}14 \text{ m/s}^2$$

$$\beta = \frac{98.14\text{x}.5^2}{7.5} = 3.271$$

$$\alpha = \underbrace{2\sqrt{3.271}}_{1.809} - 3.271 = .346$$

$\epsilon_D = 7.5\ (1 - .346) \quad = \quad 4.91\text{m}$

The range errors are summarized in Fig. 7.23.

	AIRCRAFT (METERS)	MISSILE (METERS)
1. QUANTIZATION (LSB = 1 m)	0.29	0.29
2. A/D CONVERSION	0.20	0.20
3. MAIN TIME SYNCHRONIZATION PULSE JITTER	0.50	0.50
4. NON-SYNCHRONOUS TIME MEASUREMENT	0.54	0.54
5. DOPPLER SHIFT-VELOCITY UNCERTAINTY	2.50	2.50
6. PROPAGATION	3.00	1.50
7. MULTIPATH	.02	.02
8. THERMAL	3.00	3.00
9. DYNAMIC LAG	7.22	4.91
TOTAL rss	8.78	6.50

Fig 7-23 Range errors

7.4 MTI — Moving Target Indication.

The MTI Processor is designed to provide an adequate improvement factor on distributed clutter for the full MTI coverage range for all the azimuth beams at the lowest two elevation beam positions.

The MTI design considerations are indicated in Fig. 7.24.

The maximum range of clutter return is 100 Km

where

σ_o = reflection coefficient = -20dB (from charts, λ = 10 cm)

$\frac{c\,\tau_c}{2}$ = 15m

R_m = 100 Km

θ = 17.45 mils

$\therefore\ \sigma_c = 10^{-2} \times 15 \times 10^5 \times 1.745 \times 10^{-2}$

$= 261.75 = 24.18$ dB

Required Improvement Factor

$IF_{dB} = S/N + \frac{\sigma_c}{\sigma_t} + 6$ dB

$S/N = 9.0$dB

$\frac{\sigma_c}{\sigma_t} = 24.18 + 13\ (\sigma_{T_{msl}} = .05\ \text{m}^2)$

$= 37.18$ dB

$IF_{dB} = 9.0 + 37.18 + 6 = 52.18$ dB

REQUIRED IMPROVEMENT FACTOR

$$IF = S/N + \frac{\sigma_c}{\sigma_T} + 6\ \text{dB}$$

WHERE

$\frac{S}{N}$ = DETECTABILITY FACTOR

σ_c = REFLECTED CLUTTER AREA

σ_T = TARGET RADAR CROSS SECTION

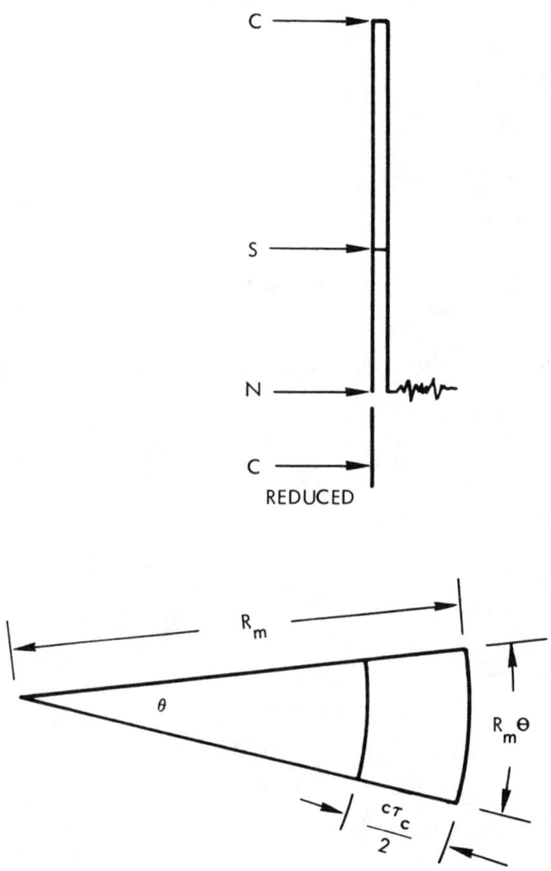

MAXIMUM REFLECTED CLUTTER AREA

$$\sigma_c = \sigma_o \frac{c\tau_c}{2} R_m \Theta$$

WHERE

σ_o = REFLECTION COEFFICIENT

c = VELOCITY OF LIGHT

τ_c = EQUIVALENT PULSE WIDTH

Θ = BEAMWIDTH

R_m = MAXIMUM RANGE OF CLUTTER RETURN

Fig 7-24 MTI design

Two Pulse Canceller

$$IF_{2p} = \frac{\omega_o^2 T^2}{4\beta}\left[1 - \frac{\omega_o^2 T^2}{8\beta}\right]$$

$$\omega_o^2 = [2\pi x 3x10^9]^2 = 3.553 \times 10^{20}$$

$$T = \begin{array}{ll} 667\ \mu s & (100\ Km) \\ +125\ \mu s & (retrace\ time) \\ \hline 792\ \mu s & \end{array}$$

$$T^2 = [7.92x10^{-4}]^2 = 6.27 \times 10^{-7}$$

β = 2.3 x 10^{17} (from charts — heavily wooded hills)

$$IF_{2p} = \underbrace{\frac{3.553x10^{20} \times 6.27x10^{-7}}{4x2.3x10^{17}}}_{2.4214x10^{-4}} \quad 1 - \underbrace{\frac{2.4214x10^{-4}}{2}}_{negligible}$$

$\therefore IF_{2p}$ = 2.4214 x 10^{-4} = 36.16 dB inadequate

Three Pulse Canceller

$$IF_{3p} = \frac{\omega_o^4 T^4}{8\beta^2} + \frac{\omega_o^2 \delta^2}{3\beta} \quad (Let\ \delta = 0.05T)(10\%\ stagger)$$

$$IF_{3p} = \tfrac{1}{2}(2.4214x10^{-4})^2 + \frac{3.553x10^{20} x 6.27x10^{-7} x 10^{-2}}{3x2.3x10^{17} x 2^2}$$

$$IF_{3p} = 2.9315 \times 10^{-8} \qquad \begin{array}{r} +0.8003 \times 10^{-6} \\ \rightarrow .0293 \times 10^{-6} \\ \hline 0.8296 \times 10^{-6} \end{array}$$

IF_{3p} = 0.8296 x 10^{-6} = 60.81 dB (> 52.18 dB OK)

Use 5 pulses per beam position ($\therefore$ 3 independent residues)

Blind Speeds.

With 10% staggering, blind speeds occur at 10 times the pulse repetition frequency.

$$\therefore f_{BS} = 10 \times prf = \frac{10}{T} = \frac{10}{7.92x10^{-4}} = 1.262 \times 10^4\ Hz$$

$$V_{BS} = \frac{f_{BS}\,\lambda}{2} = \frac{1.262x10^4 x 10^{-1}\ m}{2} = 631\ m/s = M\ 1.86$$

Rejection velocities.

Assuming the threshold level is 30dB below the $\sin^2 \phi$ function of a 3 pulse canceller combed filter response

$$\sin^2 \phi = 30\ dB$$
$$\sin \phi = 15\ dB\ (voltage) = .1778$$
$$\phi = 10.24^\circ$$

$$\frac{10.24}{180} \times 63.1\ m/s = 3.59\ m/s = M\ .0105 = 7\ Knots$$

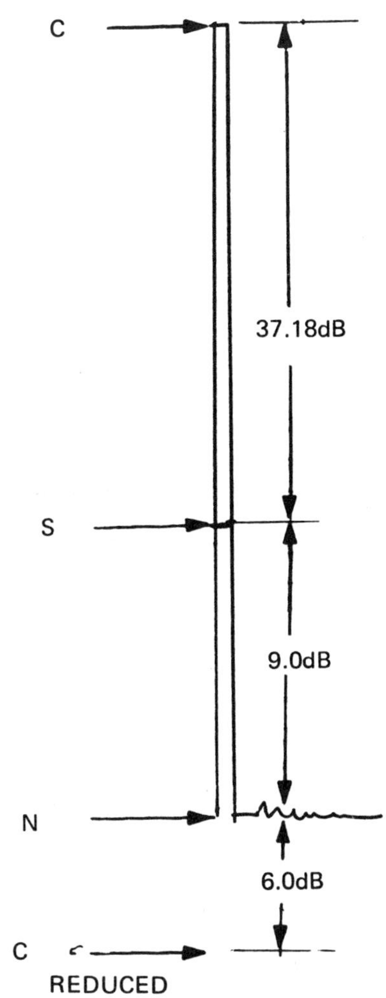

$\therefore$ Reject 0-7 Knots

7.5 Parameter Management

7.5.1 Time.

A. Search.

The maximum range as bounded by 50,000 feet altitude is a function of the earth's curvature, atmospheric refraction and elevation beam position.

The interpulse period is decreased as the beam is stepped in elevation. The time interval corresponds to the maximum range for each elevation beam position. The maximum range is determined from charts using the lower 3dB angle of the elevation beam (see Fig. 7.25).

B. Track.

The track sampling rate can be decreased as the target dynamics decrease. The error signal in range (and angle) can be monitored to control the sampling rate on each tracked target.

7.5.2 Energy.

A. Search.

The energy is decreased in 3dB increments (The reason for use of binary increments will become evident when waveform management is discussed.) as the maximum range decreases with elevation beam position as determined by the following formula:

$$\epsilon_{R_{N_{dB}}} = \left(\frac{R_I}{R_N}\right)^4_{dB} + \Delta_{ATD_{dB}}$$

where

ϵ_{R_N} = Energy reduction at Nth elevation beam position

R_I = Initial range where the process begins

R_N = Maximum range at Nth elevation beam position

Δ_{ATD} = Decrease in atmospheric loss between R_I and R_N

Energy management is usually determined relative to the lowest elevation beam position; but in the Air Defense ESRS full energy is required from 200 nm to 82 nm to obtain the required cumulative probability of detection for the missile. Consequently the energy can only be reduced in elevation beam position No. 8 (refer to Fig. 7.25).

No. 8.

$$\epsilon_{R8} = \left(\frac{82}{72}\right)^4_{dB} + (3.3\text{-}0.8)\ dB$$

$$\epsilon_{R8} = 2.26 + 2.5 = 4.76\ \ dB$$

Use 3dB = 0.5 (binary increment)

$$\therefore\ \epsilon_8 = 110\text{x}0.5 = 55\ \text{joules}$$

Without the missile detection requirement much more energy can be saved by using energy management in elevation beam positions No. 3 through No. 8 with R_I equal to 200 nm.

B. Track.

The energy is decreased in binary increments as the range, R_T, of each tracked target target decreases as determined by the formula:

$$\epsilon_{R_T} = \left(\frac{200}{R_T}\right)^4$$

C. MTI.

The energy required for MTI can be reduced because the maximum MTI range is 54 nm.

$$\therefore\ \epsilon_{R_{MTI}} = \left(\frac{82}{54}\right)^4_{dB} + (3.3 - 1.5)_{dB}$$

$$= 7.26 + 1.8 = 9.06\ dB$$

Use 9dB = 0.125

$$\therefore\ \epsilon_{R_{MTI}} = 110 \times 0.125 = 13.75\ \text{joules}$$

D. Beam Scanning.

The beam scanning loss decreases as the scan angle off broadside decreases. In the region where $\sec^3 \xi \leqslant 2.5$dB the energy can be reduced by a binary increament. This saving applies to search, track and MTI operational modes.

$$\sec^3 \xi_{max_{dB}} - \sec^3 \xi_{0_{dB}} = 3.0dB$$

$$\underbrace{\sec^3 49^\circ}_{5.5\ dB} - \sec^3 \xi_{0_{dB}} = 3.0dB$$

$$\sec^3 \xi_0 = 2.5dB$$

or

$$\sec \xi_0 = 0.83dB$$

$$\cos \xi_0 = .8254$$

$$\xi_0 = 34.37^\circ$$

$$\sin \xi_0 = \sin 34.37^\circ = .5645$$

At each elevation beam position in the search pattern use full energy for $34.38^\circ \leqslant |\theta| \leqslant 49^\circ$ and half energy for $0^\circ \leqslant |\theta| \leqslant 34.37^\circ$.

$\therefore$ Total energy reduction is: (1-.626) x 100 = 37.4%

$$\frac{.5645}{.7547}\,\frac{E}{2} = .7480 \times \frac{E}{2} = .3740E$$

$$.2520 \times E = \underline{.2520E}$$

$$.6260E$$

A simple computer algorithm can compute θ for every scan angle and determine the energy management to be used for search, track and MTI operational modes.

7.5.3 Waveform. Waveform management is feasible with either a digital or analog signal processor. The number of waveforms used is determined by a cost-effective trade-off of energy savings versus equipment complexity. Five waveforms are employed in this design. They are obtained by decreasing in binary increments both the pulse length and the time-bandwidth product of a linear fm waveform while the compressed pulse width remains unchanged. The range resolution is unaltered for all operational modes and the receiver bandwidth is the same for all five waveforms. Binary incremental control simplifies the design of the signal processor. In an analog signal processor the various waveforms are obtained by simply switching in-or-out segments of the pulse forming networks.

A. Search.

Elevation Beam Positions	Waveform	
	$0 \leqslant \lvert\theta\rvert \leqslant 34.37^\circ$	$34.38^\circ \leqslant \lvert\theta\rvert \leqslant 49^\circ$
	μs	μs
No. 1 thru No. 7	55/0.1	110/0.1
No. 8	27.5/0.1	55/0.1

B. Track.

The following waveform management is obtained by the computer as a function of range and beam scan angle:

Range nm	Waveform $0 \leqslant |\theta| \leqslant 34.37^\circ$ μs	$34.28^\circ \leqslant |\theta| \leqslant 49^\circ$ μs
200-169	55/0.1	110/0.1
168-142	27.5/0.1	55/0.1
141-120	13.75/0.1	27.5/0.1
119-101	6.88/0.1	13.75/0.1
100-10	6.88/0.1	6.88/0.1

C. MTI

The MTI waveform corresponding to the energy required at 54 nm is:

$0 \leqslant |\theta| \leqslant 34.37^\circ$ μs	$34.38^\circ \leqslant |\theta| \leqslant 49^\circ$ μs
6.88/0.1	13.75/0.1

7.5.4 Power.

A. Search.

ϵ'_{s1} = 78,292.5 joules (Refer to Fig. 7.25).

with energy management

$$\epsilon_s = .626\,\epsilon'_{s_1} \qquad \epsilon_{s_1} = .626 \times 78{,}292.5\text{J} = 49011\text{J}$$

where

.626 = reduction factor

$$\therefore\ P_{av_s} = \frac{49011}{2.417\text{s}} = 20.3\ \text{Kw}$$

B. Track.

Determine maximum average power by assuming 100 aircraft are being tracked at maximum range.

$$\therefore\ P_{avT} = \frac{100\text{x}110_\text{J}}{\text{ls}} = 11\ \text{Kw}$$

7.5.5 Time and Power Budgets (refer to Fig. 7.25).

	Time %	Power %
Search	74	65
Track	26	35

7.5.6 Duty Cycle.

A. Search.

$$\text{D.C}_S = \frac{P_{av_s}}{P_p} = \frac{20.3\ \text{Kw}}{1000\ \text{Kw}} = .02$$

B. Track.

$$\text{D.C}_T = \frac{P_{avT}}{P_p} = \frac{11\ \text{Kw}}{1000\ \text{Kw}} = .011$$

Overall D.C. = .031

C. Short-term Duty Cycle.

The short-term and long-term duty cycle transmitter specifications are met by computer controlled duty cycle management. An algorithm is developed to provide an appropriate mixture of search, track and MTI scans per major cycle to meet the duty cycle specifications. A computer major cycle is usually between 0.1 and 0.01 seconds. Another algorithm can be developed to continually monitor the long/short-term duty cycles of the transmitter and not allow them to exceed acceptable design limits.

The following are the short-term duty cycles for the various operational modes;

SEARCH

Elevation Beam Position	$0 \leqslant |\theta| \leqslant 34.37^\circ$	$34.38^\circ \leqslant |\theta| \leqslant 49^\circ$
1	.0211	.0423
2	.0211	.0423
3	.0254	.0508
4	.0309	.0617
5	.0370	.0740
6	.0423	.0846
7	.0483	.0966
8	.0271	.0542

TRACK

Range nmi		
200-169	.0212-.0298	.0423-.0496
168-142	.0125-.0148	.0249-.0296
141-120	.0074-.0085	.0147-.0170
119-101	.0048-.0050	.0086-.0100
100-10	.0026-.0253	.0051-.0506

MTI

.0087	.0174

where

$$\text{D.C.}_{ST} = \frac{\tau}{T}$$

D.C._{ST} = short-term duty cycle

τ = uncompressed pulse length

T = interpulse period

Air Defense ESRS Functional Block Diagram

Fig. 7.26 contains the major subsystems of the ESRS. The functional block diagram serves as a road map for describing the operation of the hardware and is a starting point for the development of a full scale schematic diagram.

REFERENCE MATERIAL

1. Skolnik – *Radar Handbook,* McGraw-Hill, 1970.
2. Barton & Ward – *Handbook of Radar Measurements,* Prentice Hall, 1969.
3. Nathanson – *Radar Design Principles,* McGraw-Hill, 1969.
4. DiFranco & Rubin – *Radar Detection,* Prentice Hall, 1968.
5. Cook & Bernfeld – *Radar Signals,* Academic Press, 1967.
6. Barton – *Modern Radar Analysis,* Prentice Hall, 1964.
7. Skolnik – *Introduction to Radar Systems,* McGraw-Hill, 1962.
8. Meyer & Mayer – *Radar Detection Theory,* Academic Press, 1973.
9. Berkowitz – *Modern Radar,* Wiley, 1965.

REFERENCES

[1] L.V. Blake – "Radar/Radio Tropospheric Absorption and Noise Temperature", NRL Report No. 7461, October 30, 1972.

[2] E.J. Barlow – "Doppler Radar", *Proc. IRE,* pp 340-355, April 1949.

[3] D.K. Barton – "Radar Equations for Jamming & Clutter", *IEEE Trans. Aerospace & Electronic Systems,* Vol. AES-3, No. 6, Eastcon Supplement, pp 340-355, Nov. 1967.

[4] D.K. Barton, "Simple Procedures for Radar Detection Calculations", *IEEE Trans. Aerospace & Electronic Systems,* pp 837-846, Sept. 1969.

[5] W. VonAulock – "Properties of Phased Arrays," *IRE,* pp 1-13, Oct. 1960.

[6] P.J. Kahrilas – "SAM-D Multifunction Phased Array Radar, Integration & Test Evaluation", *18th Tri-Service Symposium Record,* June 1972.

[7,8] P.J. Kahrilas – "HAPDAR-An Operational Array Radar & Design of Electronic Scanning Radar Systems", Two papers in special issue on *Electronic Scanning, Proc. IEEE,* November 1968.

[9] Mallet & Brennan – "Cumulative Probability of Detection for Targets Approaching a Uniformly Scanning Search Radar", *IRE,* pp 596-601, April 1963.

[10] J.I. Marcum – "A Statistical Theory of Target Detection by Pulsed Radar," Rand Report RM-753, July 1, 1948, also in special issue of *IRE Trans., Prof. Group on Information Theory,* Vol. IT-6, pp 145-167, April 1960.

[11] P. Swerling – "Probability of Detection for Flucutating Targets", Rand Report RM-1217, March 17, 1954, also in special issue of *IRE Trans., Prof Group on Information Theory,* Vol. IT-6, pp 269-308, April 1960.

N	E deg	E cov deg	α ms	A deg	β ms	B deg	ξ deg	R nm	T μsec	T + 125 μsec	$\left(\frac{R_I}{R_N}\right)^4$ dB	E j	τ μsec	DC
1	0.5	0–1.0	707	45.0	-258.9	-15.0	48.9	200	2472.0	2597.0	0	110 (55)	110 (55)	.0423 (.0211)
2	1.4	0.9–1.9	707	45.0	-243.8	-14.1	48.4	200	2472.0	2597.0	0	110 (55)	110 (55)	.0423 (.0211)
3	2.3	1.8–2.8	706	45.0	-228.8	-13.2	47.9	165	2039.4	2164.4	3.3	110 (55)	110 (55)	.0508 (.0254)
4	3.3	2.7–3.7	706	44.9	-213.7	-12.3	47.5	134	1656.2	1781.2	7.0	110 (55)	110 (55)	.0617 (.0309)
5	4.2	3.7–4.7	705	44.8	-198.8	-11.5	47.1	110	1359.6	1484.6	10.4	110 (55)	110 (55)	.0740 (.0370)
6	5.1	4.6–5.6	704	44.8	-183.8	-10.6	46.7	95	1174.2	1299.2	12.9	110 (55)	110 (55)	.0846 (.0423)
7	6.0	5.5–6.5	703	44.7	-168.9	- 9.7	46.3	82	1013.5	1138.5	15.5	110 (55)	110 (55)	.0966 (.0483)
8	6.9	6.4–7.4	702	44.6	-154.0	- 8.9	45.9	72	889.2	1014.2	17.8	55 (27.5)	55 (27.5)	.1084 (.0542)

MTI .0174 (.0087)

MAXIMUM ALTITUDE IS 50,000 FEET	14076.1 μsec	825j
81 BEAM POSITIONS AT EACH ELEVATION ANGLE	x 81	x81
	1.140164 sec	66825j
3 FALSE ALARMS (1 PER sec) AT 200 nmi	0.007791	330j (3 x 110)
5 MTI PULSES (54 nm) AT BEAM POSITIONS NO. 1 & No. 2 = 2x5x81x795	0.641520	11137.5j (2 x 5 x 81 x 13.75)
(NO TIME SHARING) TOTAL SEARCH TIME	1.789475 sec	78292.5j TOTAL SEARCH ENERGY

100 A/C AT 1 SAMPLE PER SECOND AT 200 nmi = 100 x 2597 μsec = .2597 s/s

TIME REMAINING FOR SEARCH = .7403 s/s

(WITH TIME SHARING) TOTAL SEARCH TIME = $\frac{1.789475}{.7403}$ = 2.417 sec

Fig 7-25 Parameter management chart

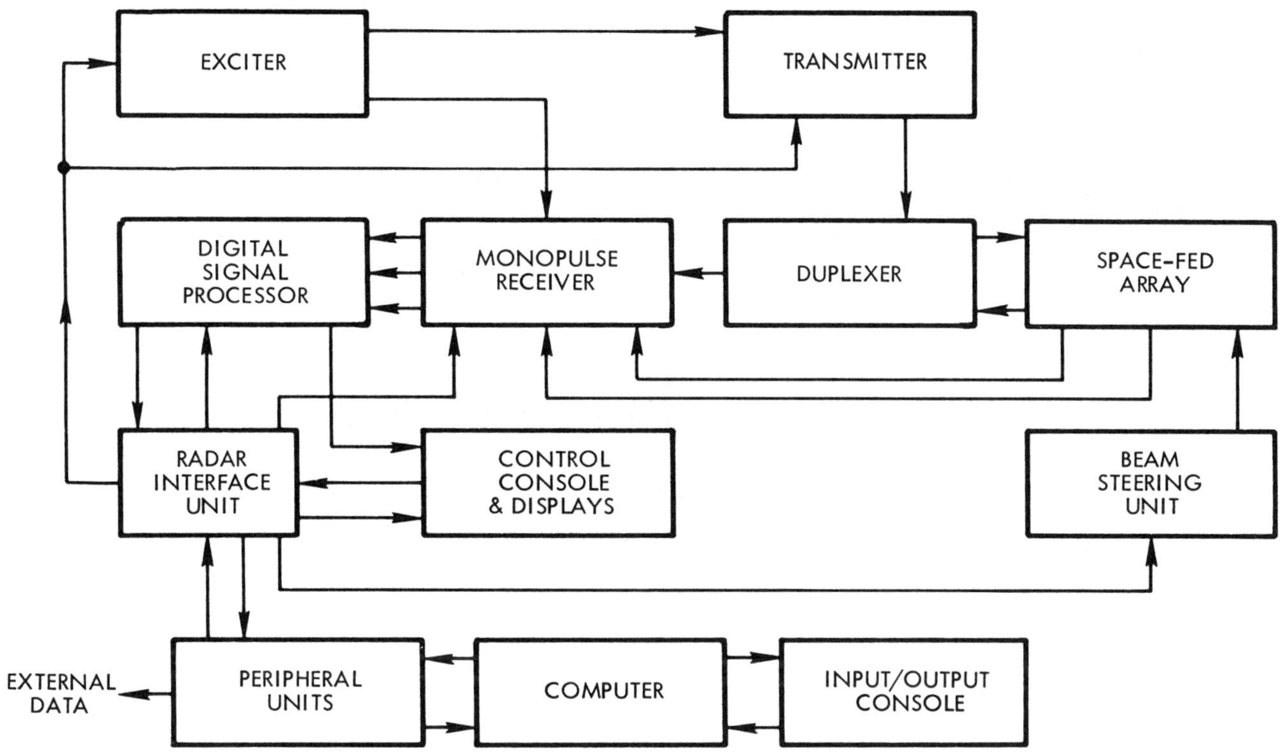

Fig 7-26 Air defense ESRS functional block diagram

RADAR TYPE	2D ES SPACE FED ARRAY 22.2° TILT 98° CONIC COVERAGE
FREQUENCY	3 GHz, 10% BW, 10 MHz INST BW
PEAK POWER	1 MW
AVERAGE POWER	31 kW (DC = .031)
SEARCH FRAME TIME	2.42 sec (100 TGT TRACK) 1.79 sec (NO TRACK)
MTI (54 nm)	IMPROVEMENT FACTOR = 55 dB (3 PULSE CANCELLER)
(BOTTOM 2 BEAM POSITIONS)	(5 PULSES PER BEAM POSITION)
TRACK CAPACITY	50 A/C (1 LOOK/sec) & 50 MSL (2 LOOK/sec) OR 100 A/C (1 LOOK/sec)
COVERAGE	(P_{cum} = 99% $P_n = 10^{-7}$) (A/C-SWI; MSL-SWIII)
($\sigma_{A/C} = 1\ m^2$ $\sigma_{MSL} = .05\ m^2$)	(1 FALSE ALARM PER SECOND)
RANGE	188 nm (A/C) 82 nm (MSL)
ELEVATION	0° - 7° SEARCH 0° - 60° TRACK
AZIMUTH	±45°
ALTITUDE	50,000 FEET
INTERPULSE PERIOD	795 - 2597 μsec VARIABLE
WAVEFORMS (10 MHz LINEAR FM)	110 μsec, 55 μsec, 27.5 μsec, 13.75 μsec, 6.88 μsec
ANTENNA	OPTICALLY FED ARRAY 4 HORN MONOPULSE FEED 50% THINNED
SIZE	7.4 m DIAMETER
BEAMWIDTH	1°
DIRECTIVE GAIN	44.8 dB
LOSSES	4.7 dB
MAXIMUM SIDELOBES	28 dB (TAYLOR TAPER SL = 34 dB MAXIMUM $\bar{n}$ = 4)
NO. OF ACTIVE ELEMENTS	7090
NO. OF DUMMY ELEMENTS	7090
rms SIDELOBES	40 dB
NULL DEPTH	33 dB (τ = 2.5°)
POLARIZATION	VERTICAL
PHASE SHIFTERS	3 BIT DIODE (1.2 dB INSERTION LOSS)
rms PHASE ERROR	10°
AMPLITUDE ERROR	0.3 dB
NOISE TEMPERATURE	1672° K (2.0 dB UNCOOLED PARAMP G = 30 dB)
RESOLUTION: RANGE	30 m
ANGLE	1°
ACCURACY: RANGE	9 METERS (1σ) SINGLE HIT
ANGLE	4 MILLISINES (1σ) SINGLE HIT
BEAM STEERING	16 BITS INPUT, 4 BIT ROW AND COLUMN
BEAM SWITCHING	0.5 μsec
BEAM COMPUTATION	125 μsec
T_x INTERPULSE NOISE	-100 dBm/MHz
COMPUTER TYPE	X
SPEED	1.5 MIPS
STORAGE	96000 WORDS
WORD LENGTH	32 BITS
INPUT/OUTPUT CHANNELS	16
MEANTIME BETWEEN FAILURE	100 hrs
MEAN DOWN TIME	1 hr
CLOCK FREQUENCY	10 MHz (RUBIDIUM BEAM STANDARD)
RECEIVER (LIMITING IAGC)	80 dB DYNAMIC RANGE ΔG_1 = 0.5 dB, ΔG_2 = 0.5 dB α = 2.5° β = 5.0° γ = 1.0°
SIGNAL PROCESSOR TYPE	DIGITAL
TRACKING ALGORITHM	TYPE II RANGE AND ANGLE

Fig 7-27 Air defense ESR system parameters

Compact, Constrained Feed Phased Array for AN/SPY-1

W.T. Patton

RCA, Moorestown, New Jersey

8.1 General Functional Requirements of the Array as a Sub-Segment of the AN/SPY-1 Radar. The phased array described in this paper is a critical item of the AN/SPY-1 radar, which in turn is a segment of the AEGIS System. It operates in conjunction with other major sub-segments including the transmitter, signal processor and the beam steering control unit, which make up the remainder of the AN/SPY-1 segment. Primary interfaces of the array antenna also include monitoring functions of the AEGIS Operational Readiness Test System (ORTS).

AN/SPY-1 radar functions have major influence on the basic design of the antenna include the following:

- Search, both active and passive
- Monopulse Angle Tracking, both active and passive, of friendly or hostile targets and the interceptor missile
- Sidelobe Blanking logic applied to both the above functions

Environmental constraints having primary impact on the basic design of the array.

- Space, and to some extent weight, parameters to be consistent with a deckhouse emplacement.
- External temperature and wind velocity conditions, particularly operation with a combined 70 knot relative wind at -20°F under icing conditions.
- Survival of nuclear blast limits.
- Vibration, including components up to 25 Hz induced by ships propellers dictating a stiff structure with small magnification to components mounted on the array structure.

8.2 Functional Description of the Array Antenna. Fig. 8.1 illustrates the organization of the array aperture, including the subarray layout for both the transmitting and receiving functions of the main radar channels, plus the auxiliary functions outlined earlier. Transmitting and receiving functions utilize subarrays of different sizes. The receiving function also uses slightly more total elements than does the transmitting, primarily to achieve a symmetrical arrangement for best monopulse pattern characteristics and peak sidelobe control. These additional elements are contained in the four "receive only" subarrays shown. The illumination function in the transmit mode is uniform; that for receive is a modified circular Taylor distribution for the monopulse reference channel, and utilizes modified Bayliss distributions for the azimuth and elevation error channels.

As shown in Fig. 8.1, the basic building blocks of the array face are termed array modules; each module includes 32 radiating elements. Two modules constitute a receiving subarray, of which there are 68 for the main array fucntion and two others which are separately utilized for auxiliary functions. Four modules (two receiving subarrays) constitute a transmitting subarray, of which there are 32 corresponding to the number of high power amplifier stages in the transmitter sub-segments.

The block diagram of Fig. 8.2 indicates the make-up of the array modules, and the relationship of the other major components of the array. The assembly of 32 phase shifters integrated with a 32:1 power divider and eight phase shifter driver boards constitutes the array module. From inside the array structure, these modules plug into assemblies of 32 rectangular horn elements which make up the array face.

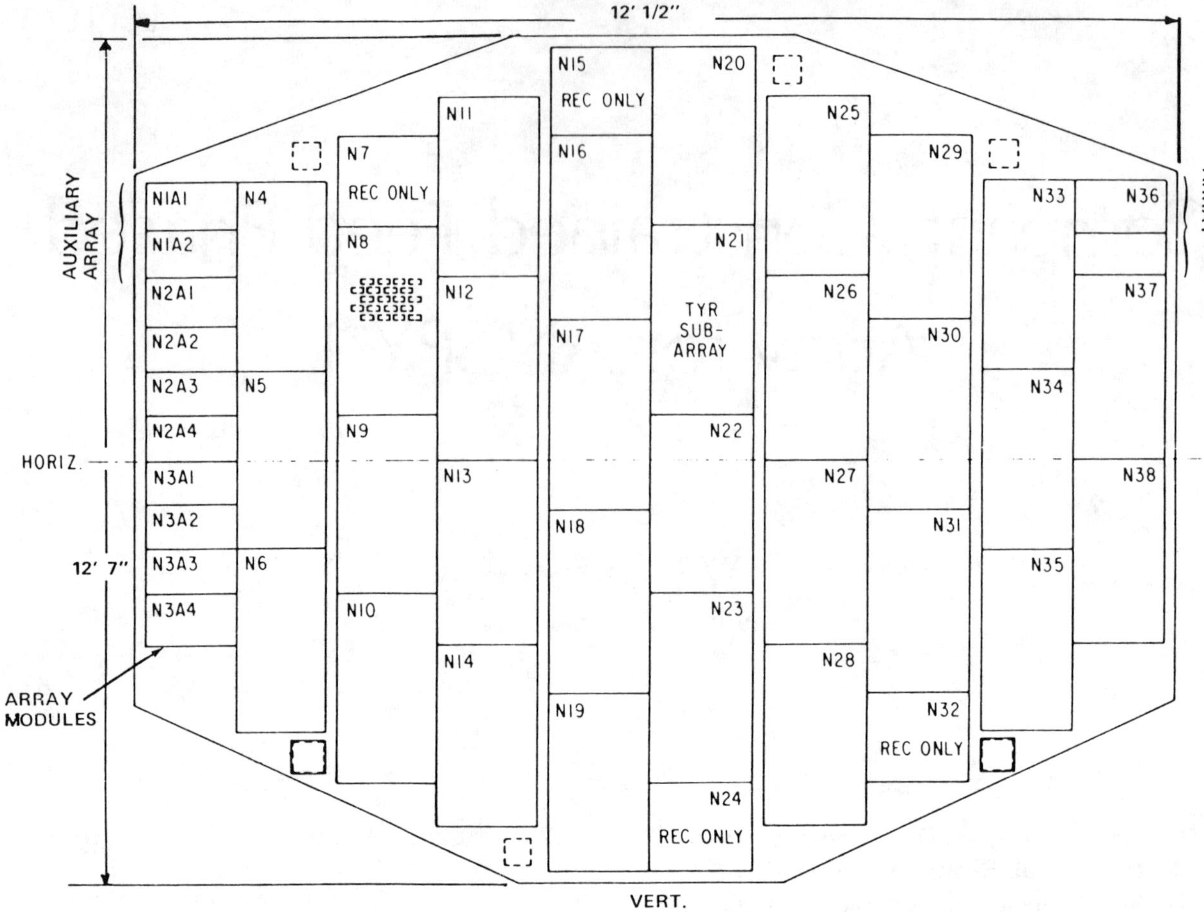

Fig. 8.1 Subarray Arrangement, Rear View

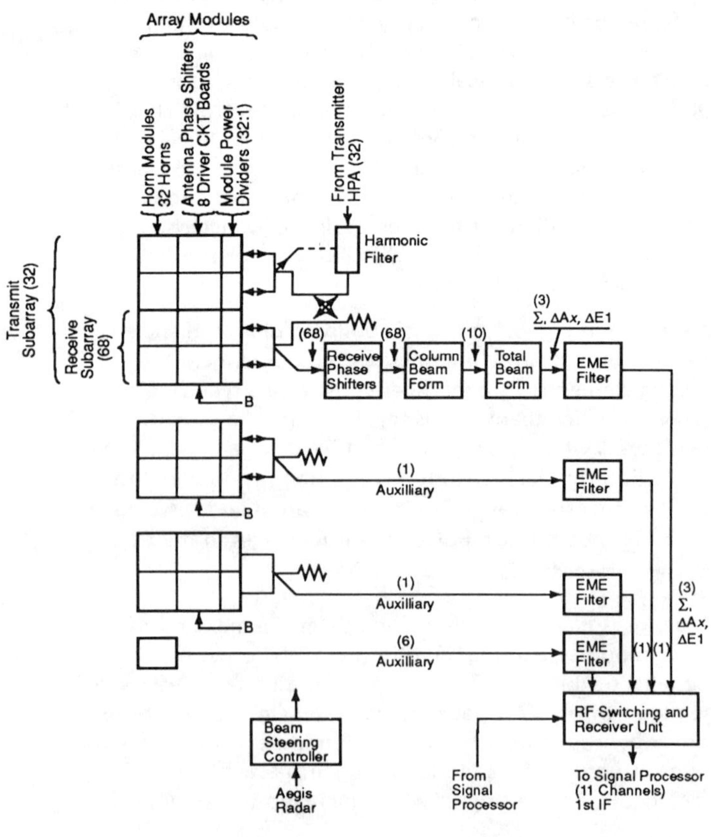

Fig. 8.2 Array Functional Block Diagram

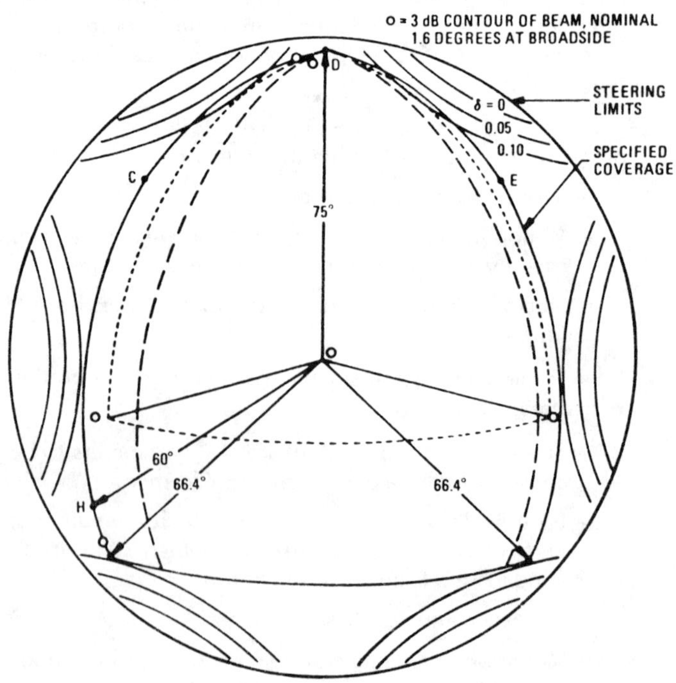

Fig. 8.3 Specified Coverage and Steering Limits Due to Grating Lobes

Four array modules are combined via a network of the three RF hybrids shown such that transmitter/receiver isolation (self duplexing) is obtained via phase control of the non-reciprocal antenna phase shifters. This hybrid network further implements receive subarrays which are half the size of transmit subarrays, and provides isolation of array face mismatch from the transmitter high power amplifiers.

In effect, all subarrays are given identical phase steering commands, thus allowing a simplified beam steering unit, and a minimum of interface cabling between that unit and the array face. Relative phasing of the subarrays in the transmit mode is accomplished within the transmitter unit at the low power input of the 32 final amplifiers. Relative phasing of the subarrays in the receiving mode utilizes 68 of the receiver phase shifters shown on the block diagram.

Receiving beam formation is accomplished in two steps. The receive subarrays in each of the 10 columns are combined to produce sum and elevation difference outputs; then in the second step, these outputs are combined to produce the complete array sum (reference) beam and the complete elevation difference beam. In addition, the azimuth difference beam is derived in the second step.

The three monopulse signals thus derived are routed off the array structure to the receiver unit that is time-shared via RF switching between two array faces within a single deckhouse. As indicated in Fig. 8.2, the auxiliary array functions are similarly routed to other RF receive channels within the same receiver unit.

The following paragraphs further describe the major components of the array, indicate their expected performance characteristics, and conclude with a summary of the predicted overall array performance.

8.3 Description of Major Array Components.

8.3.1 Radiating Elements. The array radiating elements are rectangular horns spaced on a triangular lattice with grid dimensions of 3.60 inches by 1.98 inches interlaced. This spacing was established on the basis of the required coverage for each array face as illustrated in Fig. 8.3, which makes allowances for ship motion and coverage overlap margin between the four arrays utilized in the AEGIS system. The figure indicates the maximum scan angles required of the array, based on a +15° array tilt above the deck plane, and the margins of safety for the onset of grating lobes.

Electrical design of the horn configuration utilized computer simulation [1] techniques of an infinite array which optimized the internal matching structure required. This design by simulation accounted for the beryllium oxide window necessary to provide ice inhibition (by heat conduction through the B_eO window), and the outer covering dictated by multiple environmental constraints. The design was experimentally confirmed via waveguide array simulation techniques described in the reference cited. Fig. 8.4 summarizes the design results in terms of predicted element admittance for beam steering in elevation, azimuth, and diagonal planes for the frequencies spanning the operating band. Fig. 8.5 plots the admittance characteristic of the element at maximum operating frequency for scan angles out to the limit of visible space, confirming the absence of array resonance effects or "holes" in the array coverage volume. The onset of grating lobes is observed in the H-plane cut at 60°.

Figs. 8.6 and 8.7 show the array elements as they are fabricated into horn module assemblies (the radome and beryllium oxide windows are omitted in the example pictured). The horn module is a dip brazed aluminum assembly; the beryllium oxide windows are soldered in position for heat conduction.

8.3.2 Array Modules. The rear view of Fig. 8.7 shows the openings into which the 32 antenna phase shifters of the array module, Figs. 8.8 and 8.9, make a flexible sliding joint RF connection. The major component of the array module is the high quantity antenna phase shifter and driver elements. These elements, as well as the receiver and transmitter phase shifters, are described in a later section of this paper.

The array module power divider/combiner assembly is a corporate structure of coaxial transmission line having coaxial axial interface connections with the 32 antenna phase shifters, Fig. 8.9, and a waveguide interface with the self-duplexing hybrid networks which define the subarrays. Fig. 8.10 is a side-view of the device indicating the compactness of the design. Fig. 8.11 shows the internal coaxial network utilizing round inner conductors and rectangular outer conductors permitting manufacture by layers of fine blanked aluminum plates. Each coaxial tee junction in the device is transformer matched and the dielectric inner conductor supports are compensated. The electrical length of the latter echelon divisions is such that a single shorted output terminal will not effectively short circuit the remaining 31 elements. The coaxial interface between the antenna phase shifters and the power divider netowrk is designed to permit some angular flexibility to accommodate initial manufacturing assembly tolerances and eventual maintenance procedures.

A reactive divider network, rather than a network employing loaded hybrid junctions, is used to gain the advantages of compactness and minimum cost. As such, the equality of power division has some sensitivity to variations of load impedances among the 32 output ports which can result from manufacturing tolerances of the antenna phase shifters. These these variances in phase and amplitude which may exist from element to element are accounted for in Tables III and IV. Other performance characteristics of the divider network which have been achieved in production are:

RF Loss	0.2 dB maximum
Power handling capability, peak:	60 kW peak @ 9:1 VSWR
Power handling capability, average:	500 W @ 9:1 VSWR
Phase tracking; port to port, matched:	3° rms
Phase tracking, module to module:	3° rms
VSWR:	1.35

8.3.3 Combinations of Array Modules. Fig. 8.12 is a schematic representation of the beam forming networks which are included in the array assembly.

The initial step in beam formation, applicable to both the transmitting and receiving functions, combines the output of the array modules into transmit and receive subarrays, with

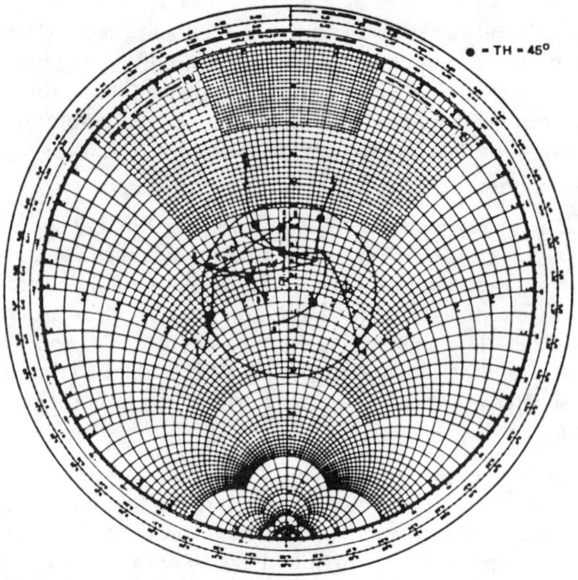

Fig. 8.4 Admittance-Step Plus Internal Matching

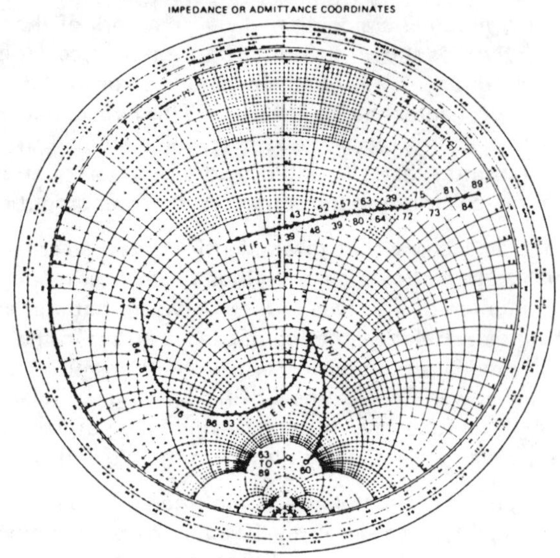

Fig. 8.5 Admittance of Horn Over Wide Steering Angle

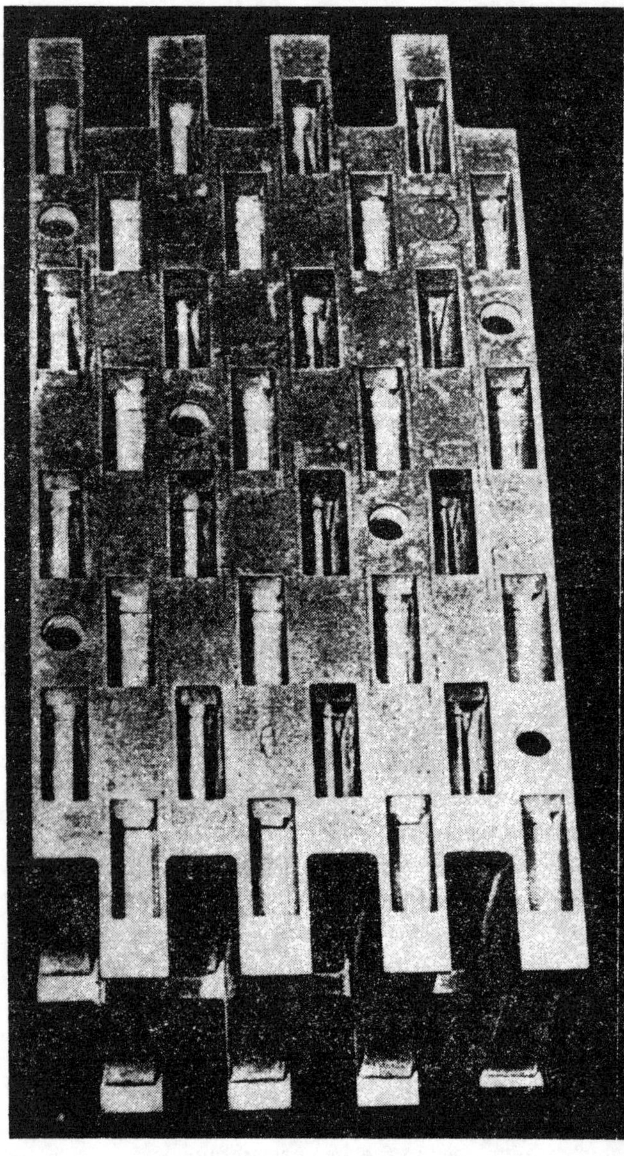

Fig. 8.6 Antenna Horn Module (Front View)

Fig. 8.7 Antenna Horn Module (Rear View)

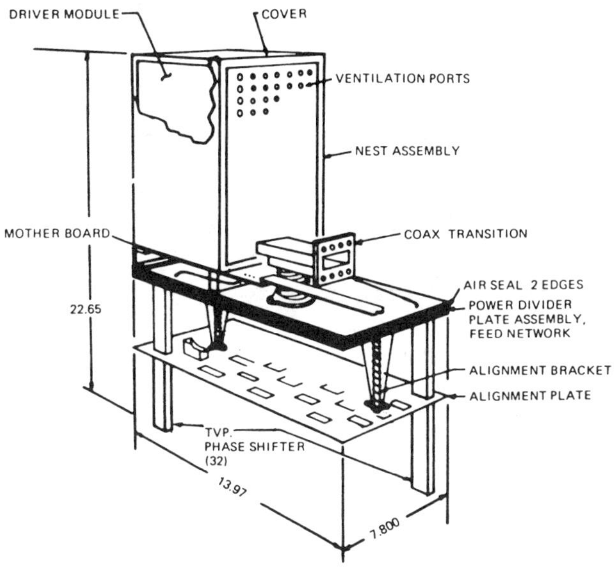

Fig. 8.8 Nomenclature of Array Module

Fig. 8.9 Photograph of Array Module

Fig. 8.10 Side View of 32:1 Power Divider

Fig. 8.11 Bottom View of 32:1 Power Divider

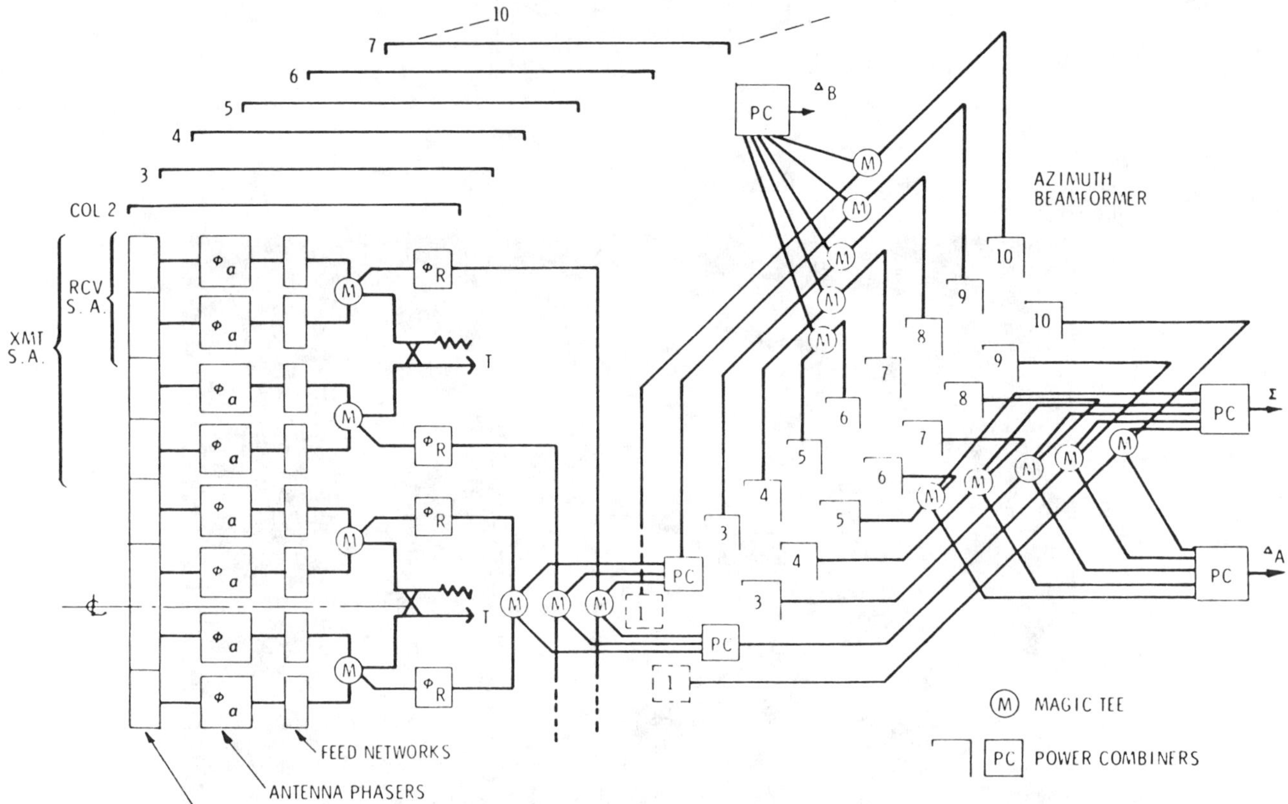

Fig. 8.12 Schematic of Microwave System

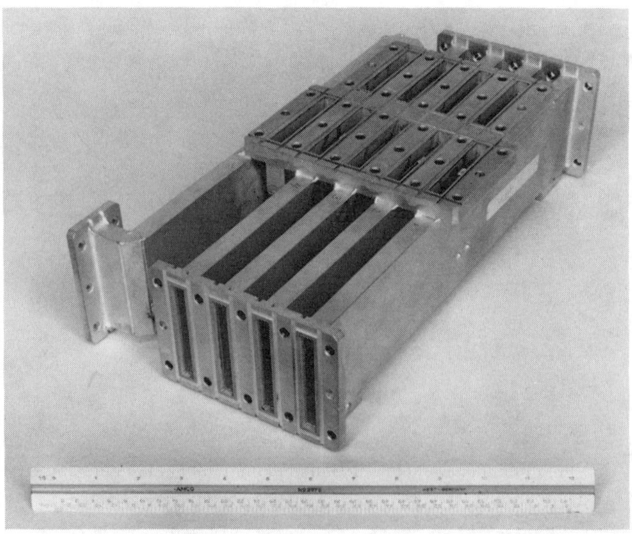

Fig. 8.13 Hybrid Combiners

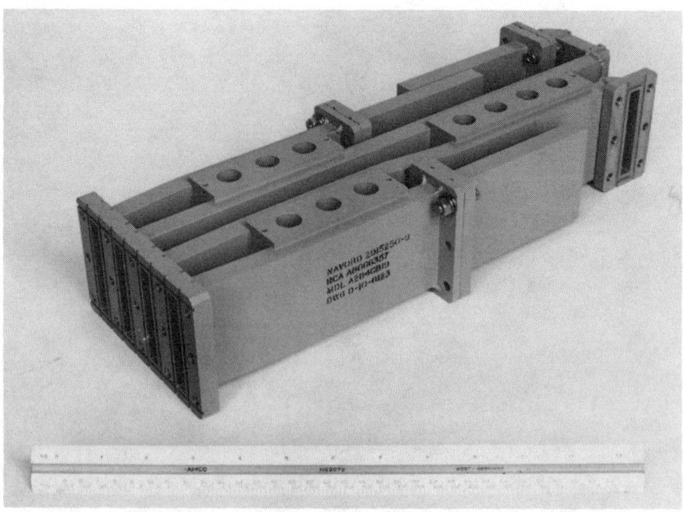

Fig. 8.14 Directional Coupler Amplitude Weighting Combiner Assembly

separate terminals for each. Hybrid tee junctions combine pairs of array modules, where the E-arm of the junction is defined as the receiver subarray encompassing four array modules. The remaining hybrid terminal is connected to an RF load capable of handling the power of the associated transmitter HPA. This arrangement basically provides the function of isolating high array face mismatch (such as that which may be caused by heavy rain) from the transmitter amplifiers, while the magic tees isolate the transmitter power from the receiver terminals. This, of course, is made possible through the use of nonreciprocal phase shifters. During the transmit mode, the antenna phase shifters in array modules must be set to accommodate the quadrature phase relationship of the transmitter hybrid. During the receive mode, the phasing between the array modules is altered to an an antiphased condition such that received signals appear at the E-arm of the magic tee hybrids, and from there proceed to the 68 receiver phase shifter components and to the next level of receiver beam formation.

8.3.4 Receiver Beamforming Networks. Fig. 8.12 indicates the two levels of beam formation which implement the monopulse reference, the azimuth and elevation difference array pattern functions.

In the beam formation sequence, columns of receive subarrays are first combined such that the sum of all in each column is obtained, as well as the difference between those subarrays in the upper and lower half of each column. For those columns having odd numbers of receive subarrays, the central subarray is omitted in the generation of the difference function.

In the second step, the 10 sum outputs from the 10 columns are summed to produce the complete reference pattern; the 10 column difference outputs generated in the first step are also summed to produce the complete array elevation difference pattern. In addition, the 5 sum outputs of one side of the array are subtracted from those of the opposite side to produce the total array azimuth difference pattern.

This beamforming network utilizes waveguide of reduced height as a compromise between space/weight considerations and RF loss characteristics. Magic tees perform the sum and difference functions; amplitude weighting utilizes networks of directional couplers to maintain adequate control of phase and amplitude excitation coefficients on the receive subarrays. Relatively tight control of the excitations of the receive subarrays by the beamformer components is required, since any errors in the beamforming networks are correlated either over complete receive subarrays (64 elements) or over a complete column of receive subarrays. In either case, the effects of errors are magnified in terms of this effect on pattern sidelobes near the main beam.

Key components of the receive beamforming network are illustrated in Figs. 8.13 and 8.14, showing respectively an assembly of combining magic tees and a coupler combiner assembly which establishes the required array amplitude weighting functions. These components are dip-brazed aluminum assemblies.

8.3.5 Array Environments. The following indicate the primary criteria for the design of the temperature control system.

Environment Exterior to Deckhouse:

1. Maximum Hot Day –
   The system is designed to maintain required component temperature limits for the following conditions:
   a) Outside air temperature of +120°F
   b) A relative wind velocity of 70 knots from any direction in the horizon plane with respect to the array.
   c) Solar flux of 90 watts per square foot from any aspect with respect to the array.
   d) Humidity up to 100%.
2. Maximum Cold Day –
   The system is designed to operate continuously within the required component temperature limits for the following conditions:
   a) Outside air temperature of -20°F
   b) A relative wind of 70 knots from any direction in the horizon plane with respect to the array.
   c) Humidity up to 100%.
3. Icing Conditions –
   The system to limit buildup to 10% or less of the total aperture area where:
   a) Icing conditions are defined as potentially existing for outside air temperatures from -20°F to +32°F.
   b) Relative wind velocity during icing conditions is defined as 70 knots from any direction within the horizon plane with respect to the array.
   c) The ice inhibiting scheme is predicated upon maintaining the radome at 32°F or greater at all times.

Environment Interior to Deckhouse:

1. Maximum hot day air temperature controlled to +80°F maximum with relative humidity controlled to 55% maximum.
2. Minimum cold day air temperature controlled to +65°F minimum with relative humidity uncontrolled.
3. For failure mode operation of the array temperature control system, the air within the deckhouse can be utilized to cool the array. In this event, the deckhouse air may be +95°F maximum (+80°F if deckhouse cooling system is not failed) and the array will remain operational with degrated performance. Permanent damage during such operation will not be incurred.

8.3.6 Performance Requirements. The thermal control system is considered to be fully operational if all of the following temperature limits are maintained for steady state conditions.

1. The radome exterior is maintained at 32°F or greater during icing conditions as defined during the extreme cold day. Supplementary air heating is required for the low temperature operation or when the RF system is not being powered.
2. Antenna phasers are maintained within the range +55°F to +160°F, but restricted to the region +80°F to +160°F except on the extreme cold day.

3. The air surrounding the antenna phaser driver circuit components within the driver nests is maintained in the range +32°F to +122°F.

4. All receive phasers and drivers are maintained between 80°F to +105°F.

5. Generally, the temperature difference between all receive waveguides and phasers within the array (from hot spot of one to the cold spot of another) is 14.5°F or less. A temperature monitoring and data correction will be used to correct for gradients (top to bottom of array) in excess of 14.5°F up to a limiting gradient of 25°F.

6. Temperature sensors are mounted on the rear face of the 1¼ inch thick structure plate (horn modules mounted on the front face) such that the average temperature of this plate is sensed to within +10°F of the true average temperature. This signal is then used for angle data correction to account for errors induced by array structure thermal expansion.

8.3.7 Internal Heat Load Sources and Air Flow Distribution. Fig. 8.15 identifies each of the separate compartments with respect to a horizontal section through the array and presents a typical illustration of an air flow path in a column of the array.

The heat loads are distributed along the height of the columns in each of the 32 transmit subarrays, in the four receive only units, the driver nests and one sidelobe blanking unit. The heat loads and required air flow distribution are shown in Fig. 8.15. As noted, the flow divides into five parallel paths from the main input duct (one to each of the five vertical columns). Within the columns parallel/series flow is utilized with air first passing through the driver compartments and then through the phasers in a down and up loop before exhausting into the top plenum. A smaller parallel flow occurs through the receive waveguide compartment before joining the driver cooling air and then passing through the phaser compartment. The de-icing air is supplied as a separate parallel loop entering at the bottom of the array and emerging at the top in the common exhaust manifold.

The overall simplified schematic of the system is given in Fig. 8.16 which shows the starboard and port arrays being controlled from a central system. The heat exchangers, ship's cold water supply and separate de-icing heaters which have a proportional silicon controlled rectifier (SCR) control are indicated. A separate monitor unit, not shown, which gives continuous readings of the array temperature profile, is also provided.

8.3.8 Ships Resources Utilized. Chilled Water – Maximum of 20 gallons per minute of water at a temperature of 45.5 ±2%F at 60 PSI gauge nominal (115 PSI guage maximum) for each array. Pressure drop through the array cooling system will not exceed 20 PSI.

Electrical Power –

1. A minimum of 4.7 kW of 440 V, 400 Hz power per array on 100% duty cycle for operation of blowers.

2. A minimum of 0.5 kW per array for temperature control auxiliary components such as sensors and control modules.

3. A minimum of 10 kW of electrical power per array for icing conditions during the extreme cold day with the transmitter operating at 100% output. This is gradually increased to 22.6 kW per array with the transmitter off (inverse function of transmitter output) for icing conditions during the extreme cold day. In either case the duty cycle for the above power will match that of the icing conditions, and is expected to be three eight-hour periods in any one calendar year.

8.3.9 Array Structure. The antenna structure must provide mechanical support for an array of 4480 radiating antenna elements and all supporting equipments. These include power distribution devices, RF-beamsteering components and cooling elements. The load carrying structure consists of a perforated 1½ inch base plate, welded to an octagonal outside frame and four inside T-beams. The frame and T-beams made of AL 5456-4321, Temper H321, are 37 inches deep and have heavy flange sections, which are connected by braces to increase the lateral stiffness. The total weight is 17,000 lbs, including all equipment. Maximum dimensions are 153 x 159.5 x 38.25 inches. The structure provides mounting for the 140 array modules which must be precisely located with respect to each other and to the connecting receive waveguide system. Consequently, locating tolerances are controlled to a maximum of ±0.020 inch by means of tooling templates at time of fabrication. Common locating dowel pins for the horn clusters and the array modules are precision mounted on the perforated front panel.

Two arrays are located in a deckhouse straddling the axis of the ship. The plane of the base plate has an azimuth angle of 43 degrees with fore and aft and an elevation angle of 15 degrees with the vertical.

The deckhouse section that supports the array contains a matching frame, two reinforcing side beams and a set of heavy brackets above and below the array. Additionally a stiff tie-off is made at the intermediate deck level. A cut through the antenna structure is shown in Fig. 8.17. A photographic reproduction of the antenna structure under construction is given in Fig. 8.18.

8.3.10 Structural Analysis. The DYNA Analysis which grew out of Navy initiated dynamics analysis programs was used to validate the structural design, and is described in [2]. The efforts of the Navy, APL, NRL and the David Taylor Model Basin all served as background leading up to the final DYNA Program produced by RCA.

Typical of the program output is the computerized plot shown in Fig. 8.19, which gives the applied time variant loading and plots the accelerations and excursions at mass nodal points in the structure.

The lowest natural frequency of the configuration was calculated to be 26.05 Hz which exceeds the 25 Hz requirement. The static load test indicates the actual value might be 1 or 2 cycles higher and that the stress levels are on the safe side.

In Fig. 8.20, the maximum g-levels and deflections are plotted with the minimum stress safety factors for the complete analysis. The values are shown at the locations at which they occur.

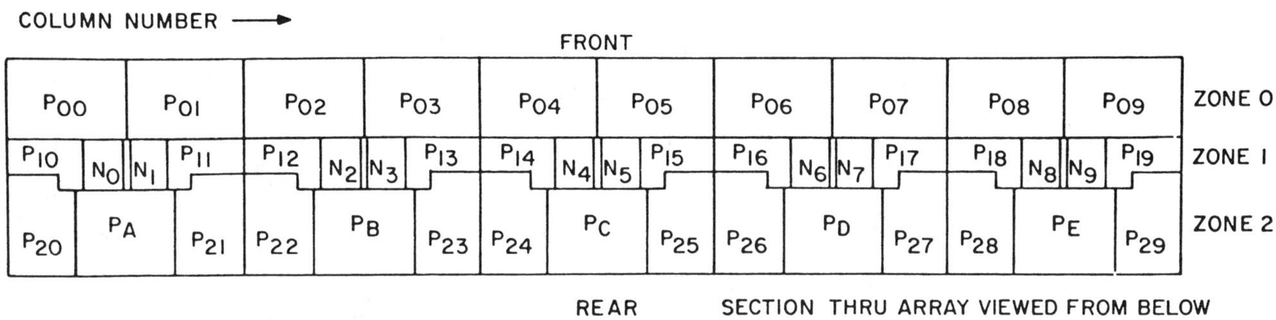

P_{XY} = COMPARTMENT NO. WHERE I_X = ZONE

X = 0 = ANTENNA PHASE SHIFTER ZONE

1 = NEST INLETS AND OUTLETS

2 = WAVEGUIDES AND RECEIVE PHASE SHIFTERS

Y = 0 THROUGH 9 SIGNIFIES COLUMN NUMBER

N_Y = NEST ASSOCIATED WITH A PARTICULAR COLUMN

P_4 THROUGH P_E = AIR CHAMBER ASSOCIATED WITH EACH OF FIVE AIR CIRCUITS

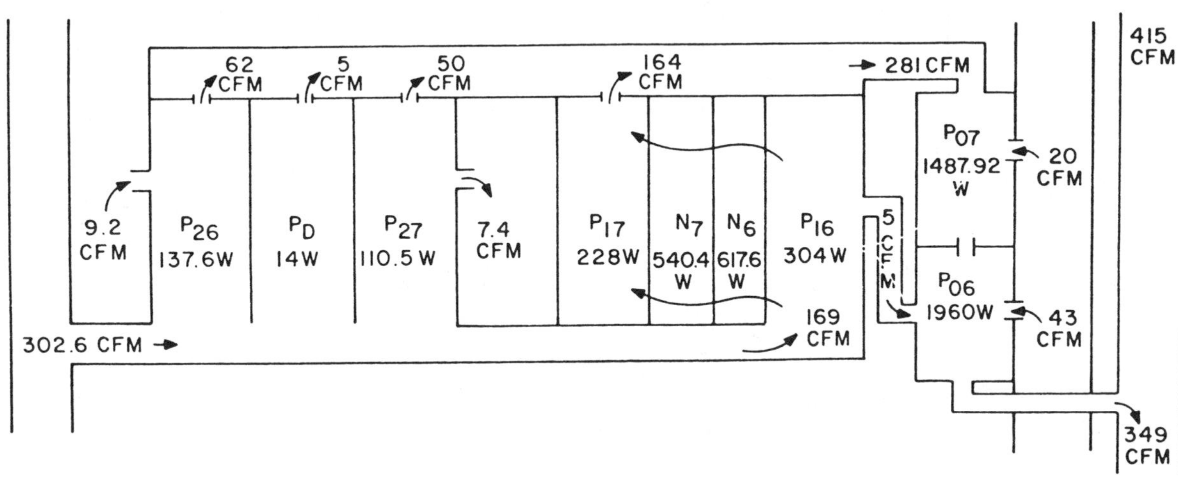

Fig. 8.15 Array Temperature Control Compartment, Nomenclature and Air Flow Diagram

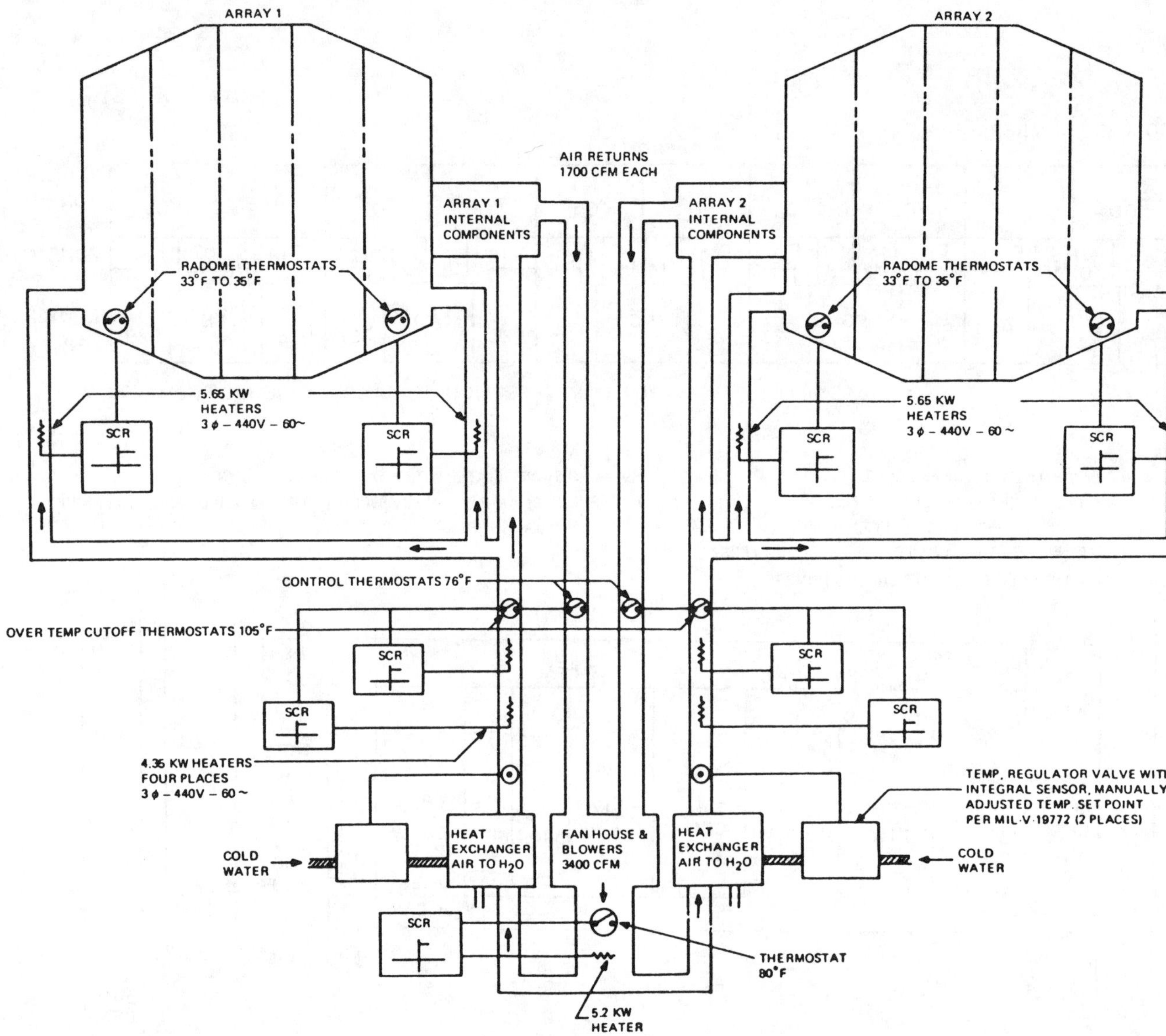

Fig. 8.16 Active Control Flow Diagram of Antenna Temperature Control System

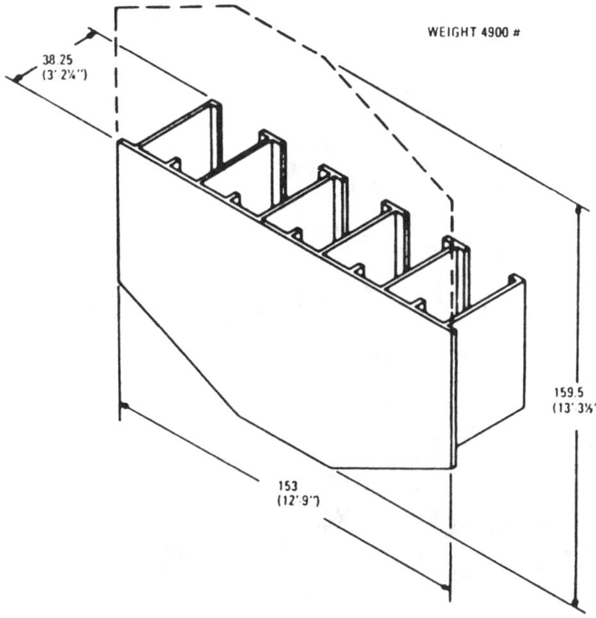

Fig. 8.17 Section Through Structural Element of Array

Fig. 8.18 Photograph of Antenna Structure During Fabrication

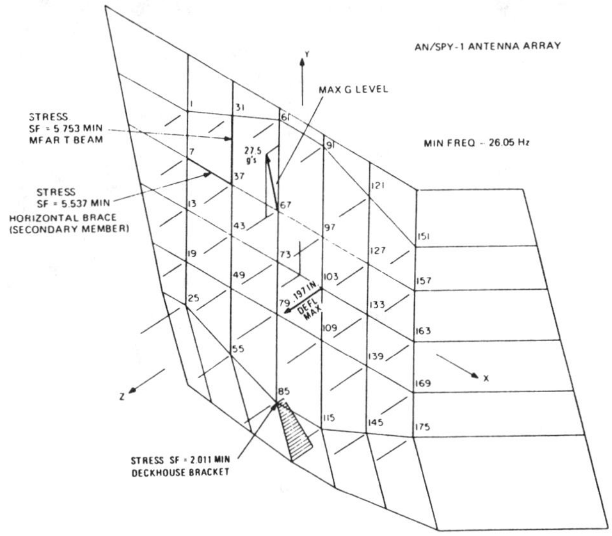

Fig. 8.20 Maximum G Levels and Deflections

MFAR – 4 IQ # VERTICAL SHOCK

TIME PLOT – 1

TRACE	DF	TYPE	SYM		SCALE	
1	74	5	1	—	1.000E–03	IN/IN
2	74	6	2	—	5.000E 00	GS/IN
3	86	5	3	—	2.000E–03	IN/IN
4	86	6	4	—	1.000E 01	GS/IN
5	104	5	5	—	1.000E–03	IN/IN
6	104	6	6	—	1.000E 01	GS/IN
7	44	6	7	—	5.000E 00	GS/IN
8	44	2	8	—	5.000E 00	GS/IN

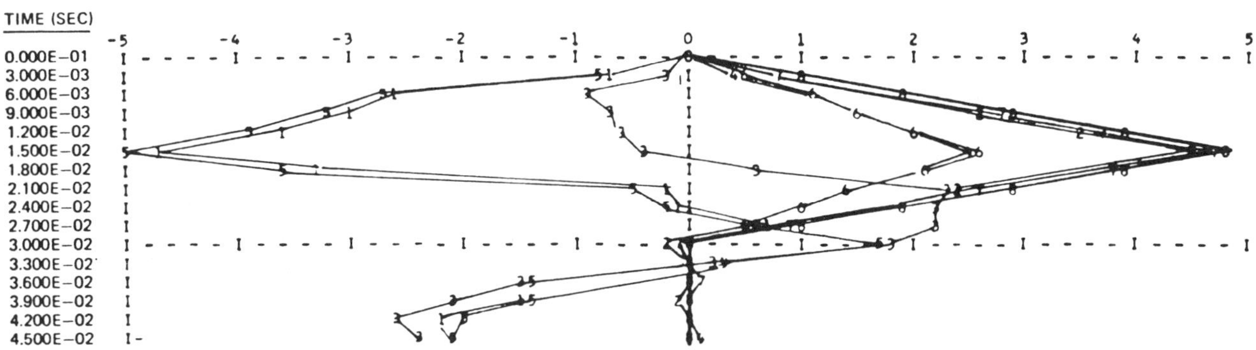

Fig. 8.19 Computer Printout of Accelerations and Excursions at Structural Nodal Points

SUCCESSFUL COMPLETION OF DYN111

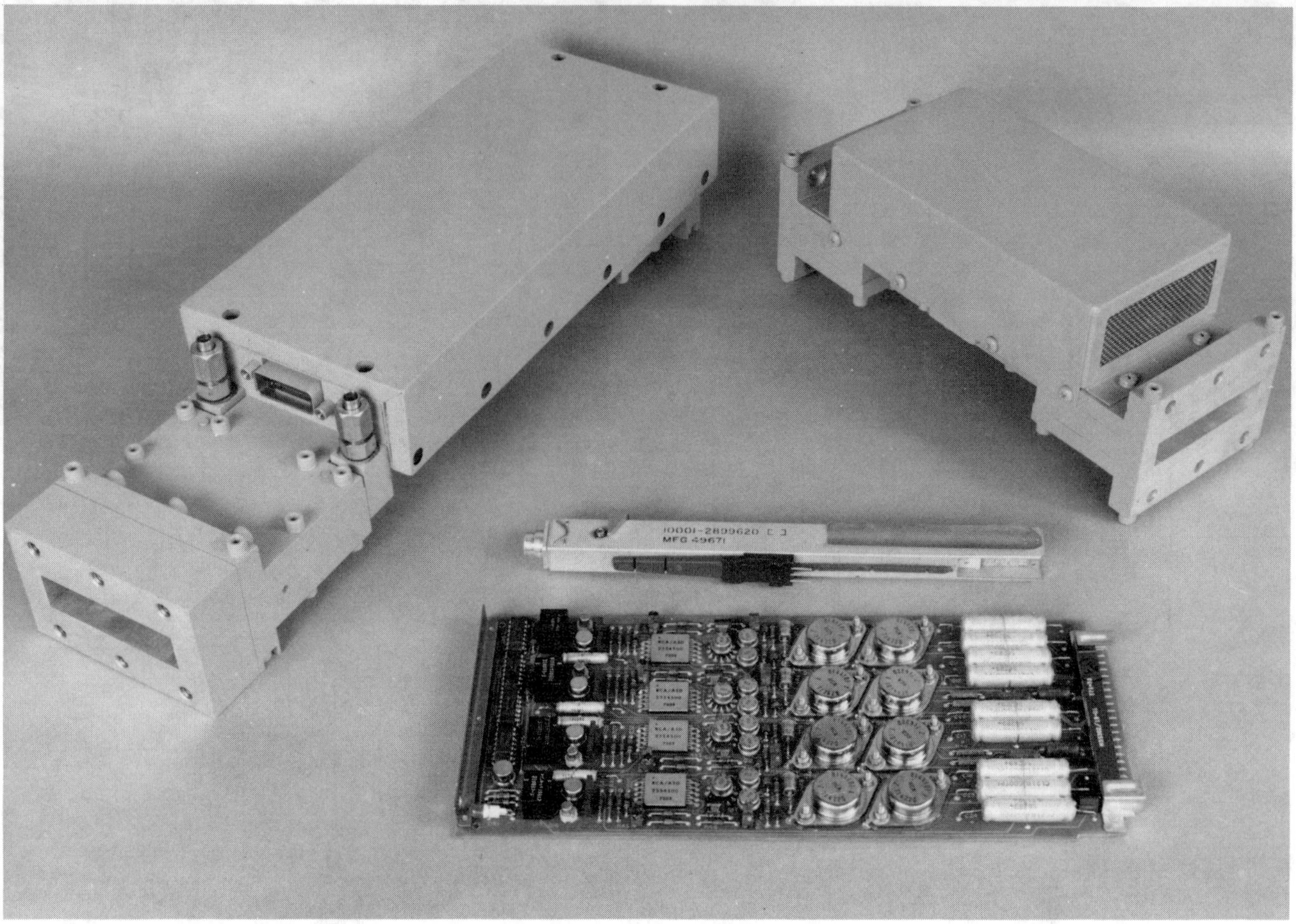

Fig. 8.21 AN/SPY-1 Phase Shifters

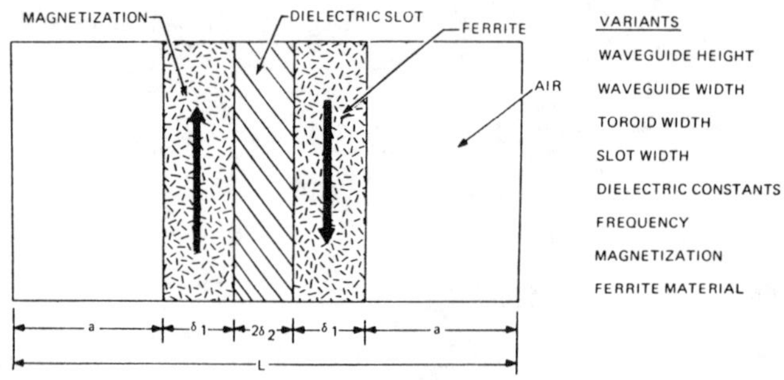

Fig. 8.22

Cross Section of Analytical Model of Phase Shifter

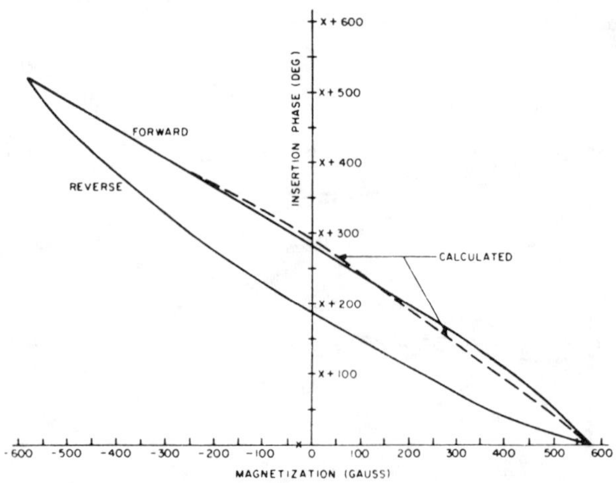

Fig. 8.23 Calculated and Measured Insertion Phase vs Magnetization of Antenna Phaser

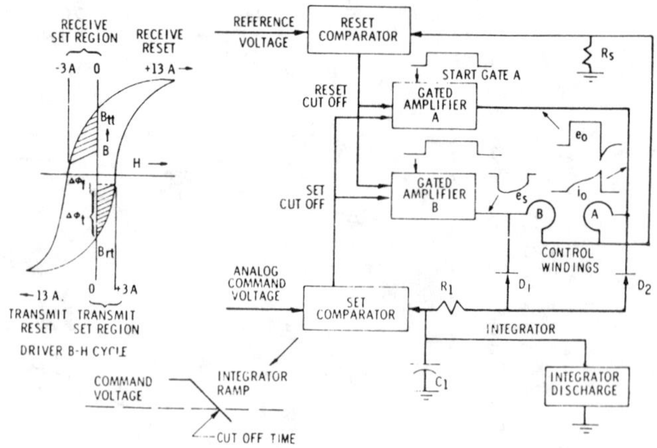

Fig. 8.24 B-H Cycle and Flux Feedback Driver of Phase Shifter

The maximum g-level of 27.5 may be used as peak input for the evaluation of electronic equipment inside the array. This value occurs only at one point; at all other locations the analysis indicates lower values.

The maximum stresses which result in minimum safety factors, occur only at one time instant; they therefore do not represent the normal operational stress levels of the structure. These are so low that fatigue is not considered.

The maximum deflection of 0.197 inch is a good indication of the clearances that are needed for the equipment behind the array. Deflections due to nuclear blast are slightly smaller.

The analysis demonstrates that the structure is able to comply with and exceed all imposed design specifications. This is proven by two analysis approaches for shock inputs, one of which is the specified NAVSHIPS-250-423-30 method. The other is presented in [2].

8.5 Summary of Array Performance Characteristics.

8.5.1 RF Loss. Table II summarizes the RF loss contributions of the components and subassemblies previously described, including both dissipation loss and losses due to phase/amplitude tolerance which degrade the antenna directivity. Note that the contribution of the phase/amplitude tracking between the 32 high power amplifiers of the transmitter sub-segment are also included in this tabulation.

The round trip losses shown in the table are within the original allocation to the antenna group, and the major contributors have been experimentally confirmed at this time.

8.5.2 Computed Pattern Characteristics. Figs. 8.26 a through 8.34 are computed major plane patterns of the array and one typical subarray. These computations do not include the aperture excitation errors outlined in the previous paragraphs, but do include the amplitude quantization effects of the uniformly illuminated subarrays, and the expected element pattern shape. Fig. 8.35 is a computed contour plot of the receiver reference pattern including excitation errors.

The amplitude quantization sidelobes noted in the single-plane, one-way patterns result from the periodic nature of the subarray layout, including both amplitude tapering from column-to-column and the steps at the periphery of the array.

The largest diffraction lobes are not in the principal plane patterns of the array, but are on diagonal cuts as indicated on the contour plot of Fig. 8.35. The largest of these, of which there are four, is -26.1 dB located approximately 100 mrad from the main beam. Including excitation errors shown in Table II, an 18% probability exists that any one of these four lobes will reach -23 dB.

8.5.3 Overall Assembly. Figs. 8.36 and 8.37 illustrate the complete assembly of the various components described in the foregoing pages, showing front and rear views respectively. The photographs were taken just prior to installation on the antenna pattern range for conduct of performance qualification testing including transmit and three channel receiving patterns, evaluation of monopulse angle tracking errors, directive gain over the coverage volume, and RF loss characteristics.

The front view illustrates the layout of the horn modules, where the front surface is made electrically continuous by EMI shielding material between the modules. The rear view shows the access pannels which enclose the array components and implement the required internal temperature control. It also shows the final combining waveguide network which evolves the three monopulse receiving channels. Interfacing waveguide connections to the 32 external transmitting high power stages are also visible on the rear of the structure.

8.6 Array Testing.

8.6.1 Test Facility. The antenna pattern data presented in this report was measured using the RCA Automatic Antenna Pattern Data Acquisition System (AAPDAS), an automated test range located at the RCA, Moorestown, N.J. plant. The facility, which is depicted in Fig. 8.38, provides automatic test capability by means of computer control of the following:

- Selection of array beam position via the AN/SPY-1 Beam Steering Controller (BSC)
- Two axis angular positioning of the AN/SPY-1 array antenna for pattern measurements.
- Selection of test frequency via remote-controlled microwave frequency synthesizer.
- Data acquisition and magnetic-tape data recording.
- Monitoring of the array and BSC self-test equipment (ORTS).

A group of computer-program modules support these functions and may be combined to compose a specific test as depicted in Fig. 8.39.

8.6.2 Test Range. The test range employs a minimum length of about $2D^2/\lambda$, elevated towers, and a shaped-beam source antenna to suppress multipath. Key dimensions of the range are shown in Fig. 8.40. The array mounted on the test pedestal is illustrated in Fig. 8.41.

8.6.3 Calibration and Verification Facilities. The facility is designed to provide convenient means for calibrating and verifying the test equipment and test range. Calibration features include boresight alignment optics, an RF field probe, and a gain standard. Several computer programs are available to aid in calibrating and verifying the test receiver. These programs create and maintain disk-based data files which are used in establishing gain and phase references for each receiver.

The AAPDAS calibrations are performed using the SYSCAL Program Routine in the computer. One of three basic system mode configurations is selected manually or automati cally depending on the calibration or check to be performed Fig. 8.42 is a basic block diagram of the AAPDAS test system which will aid in understanding the three configurations described below:

(1) Antenna Check: The AAPDAS is operated in the CAL/GAIN mode where the gain standard horn is switched into the front-end test ports to establish gain values with respect to a calibrated gain standard.

(2) System Calibration: Calibrations are performed with the AAPDAS in the MEASURE mode (switches as shown in block diagram) with special

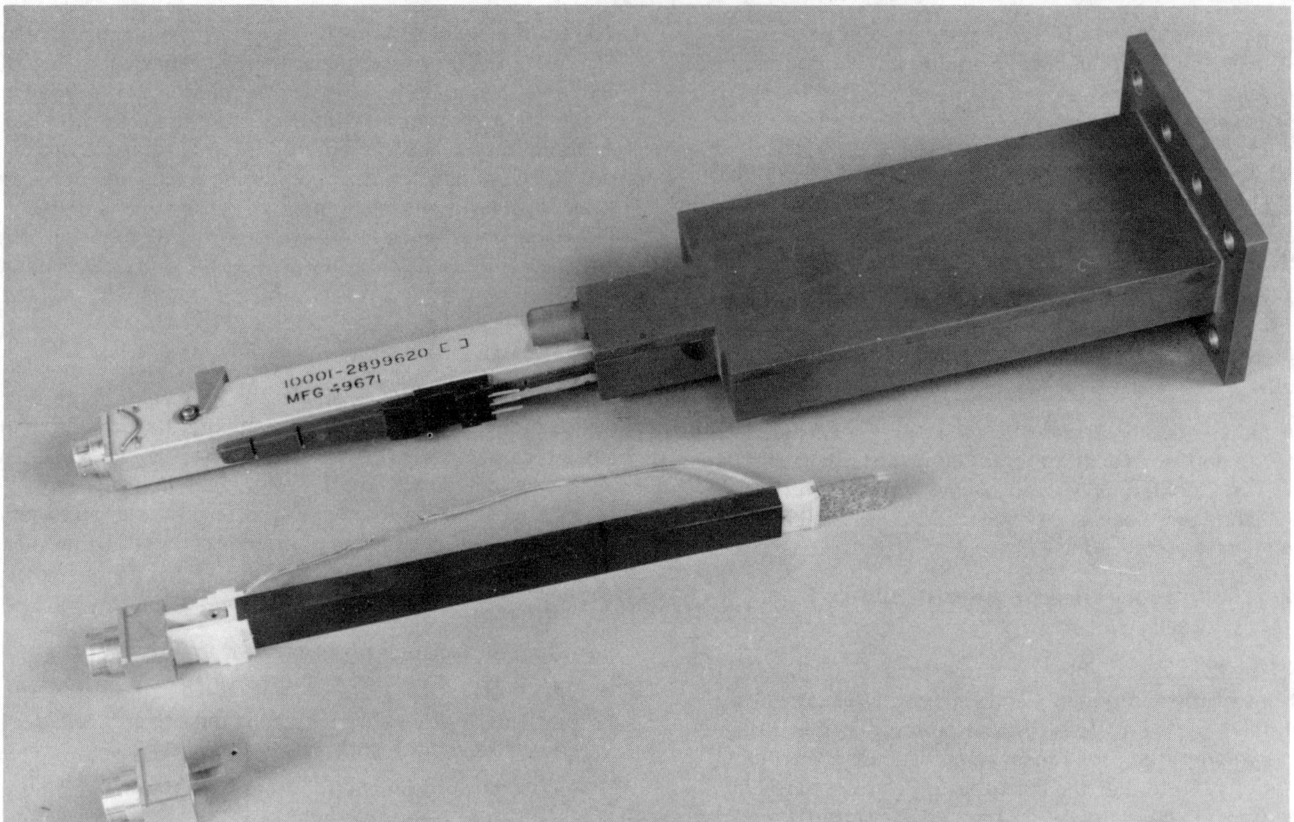

Fig. 8.25 Phaser Interfaces

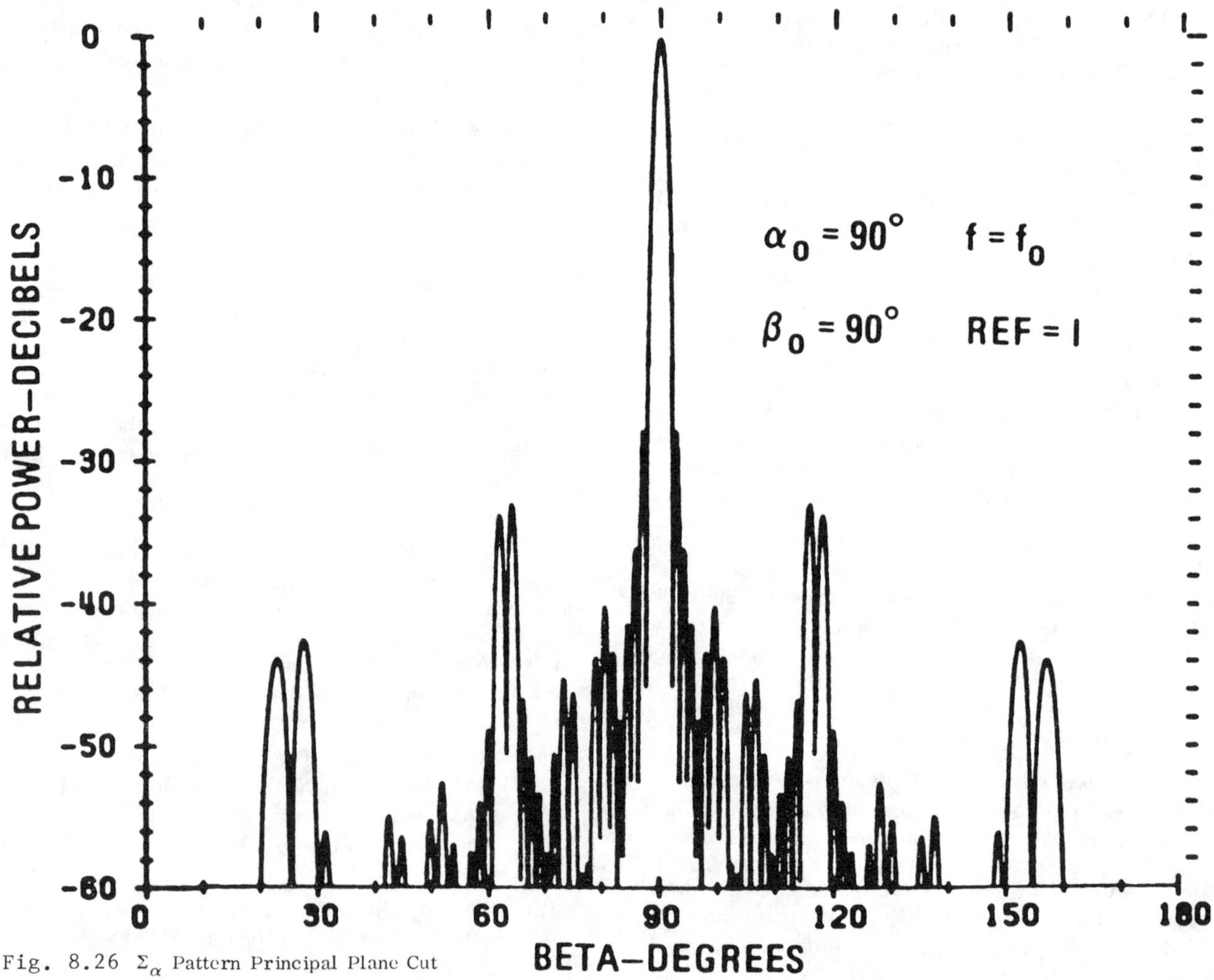

Fig. 8.26 Σ_α Pattern Principal Plane Cut

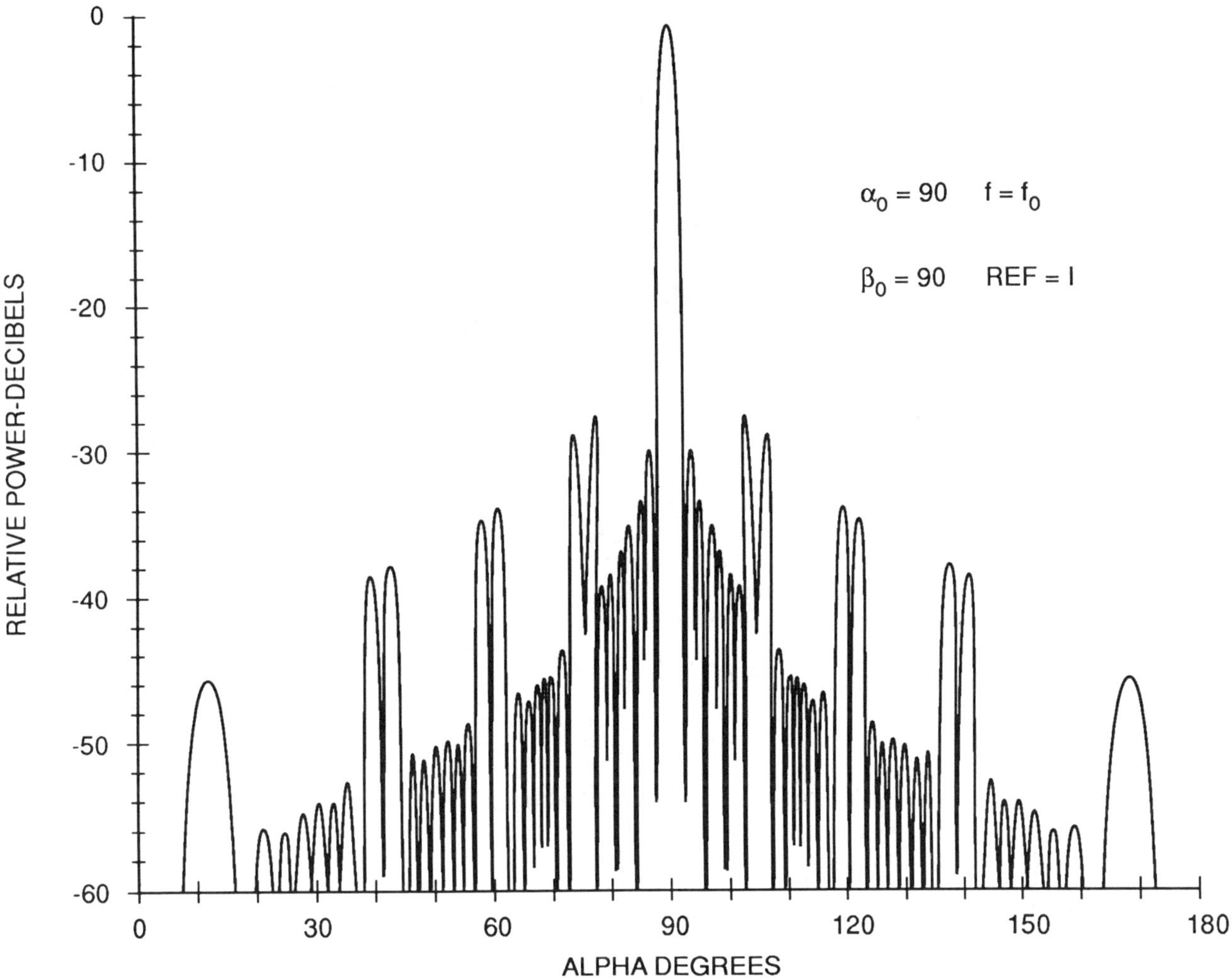

Fig. 8.27 Δ_β Pattern Principal Plane Cut

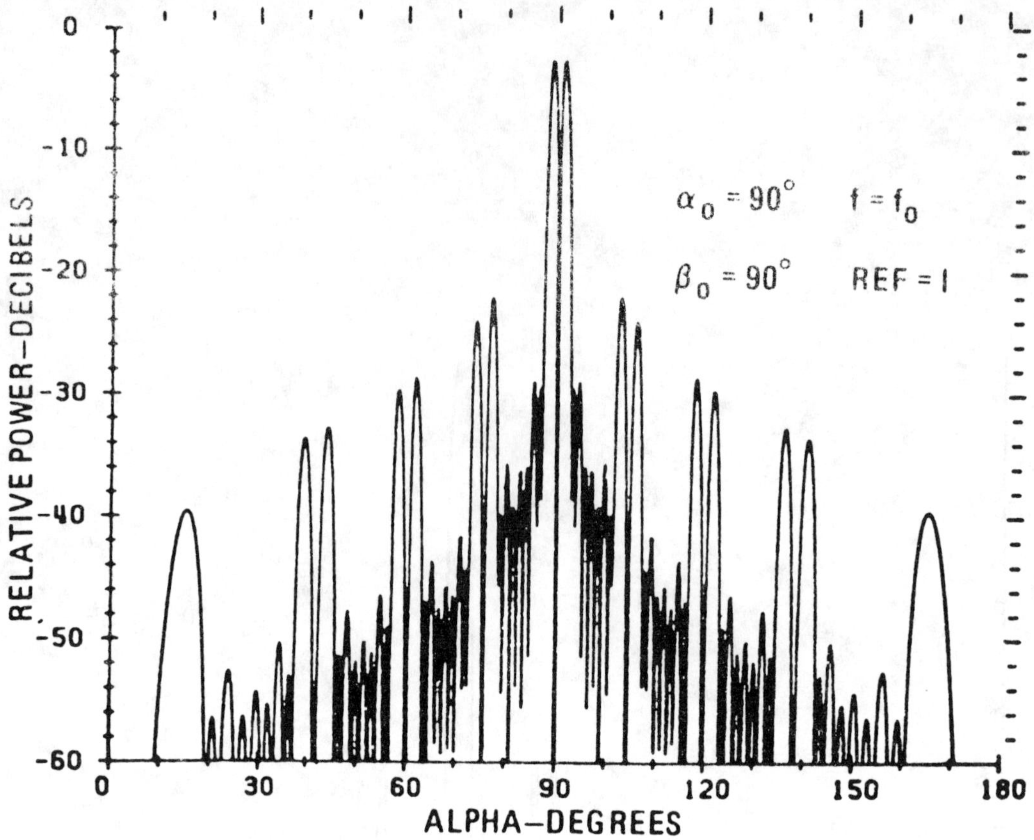

Fig. 8.28 Δ_α Pattern Principal Plane Cut

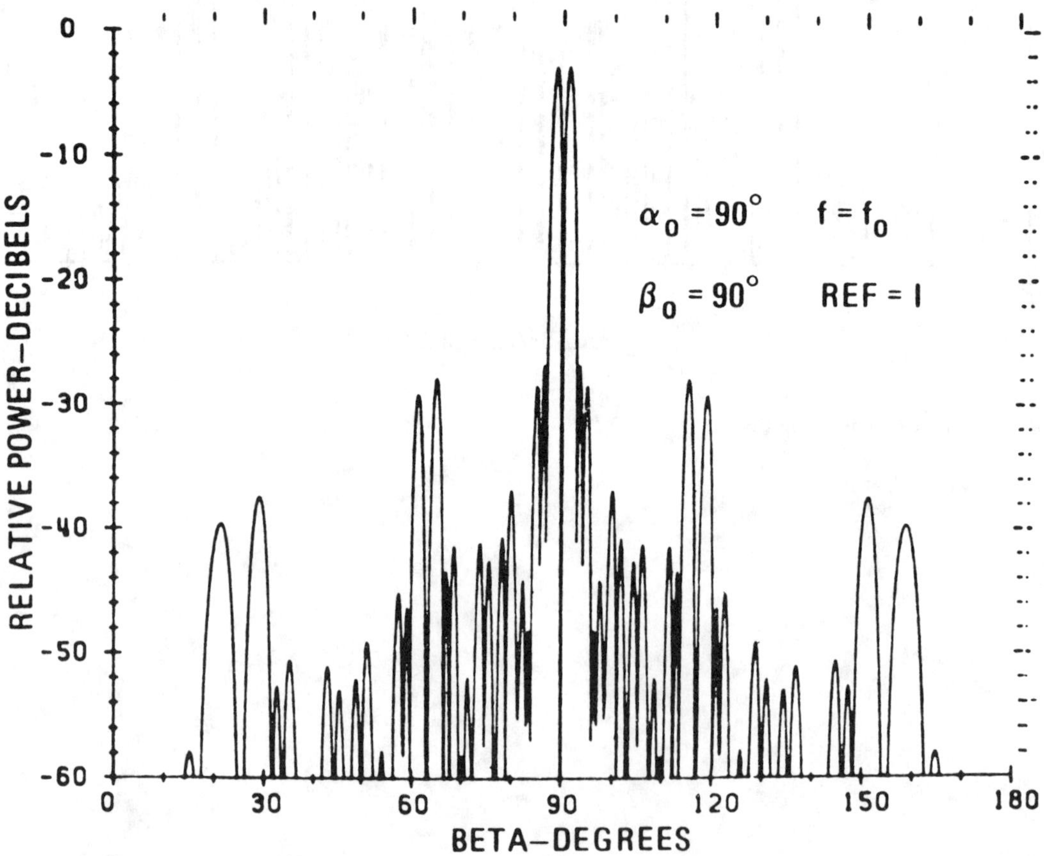

Fig. 8.29 Δ_β Pattern Principal Plane Cut

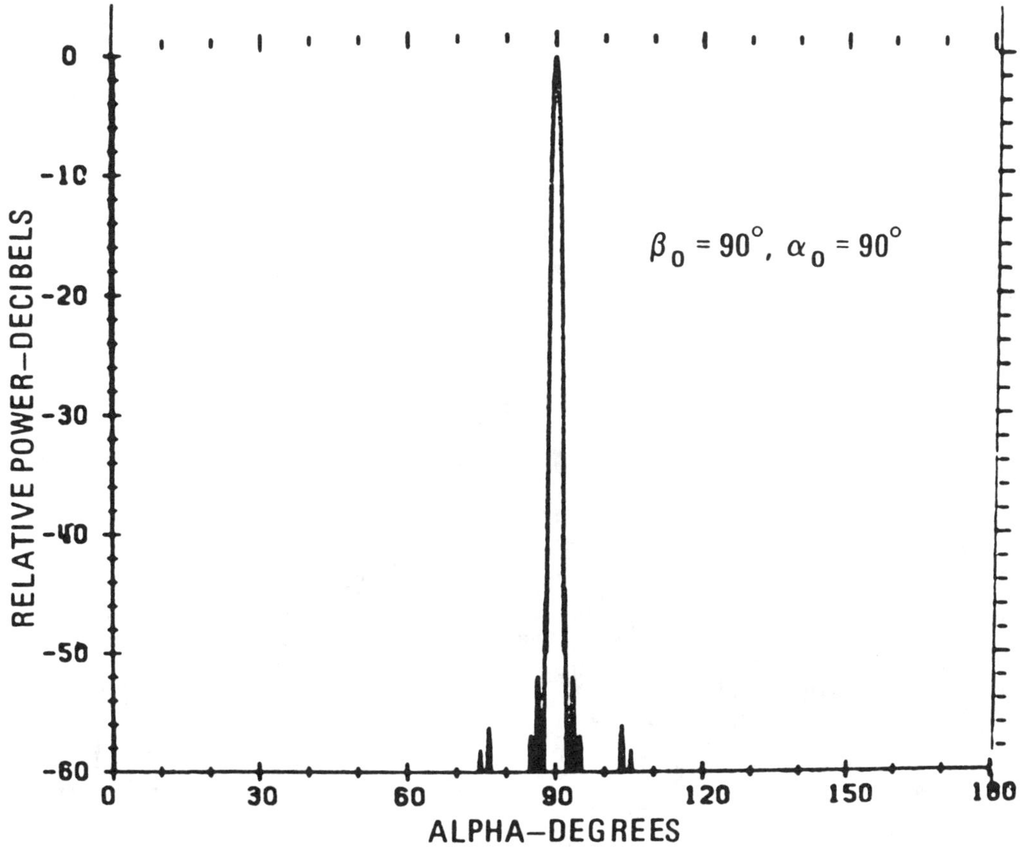

Fig. 8.30 Two-Way Reference Pattern α Cut

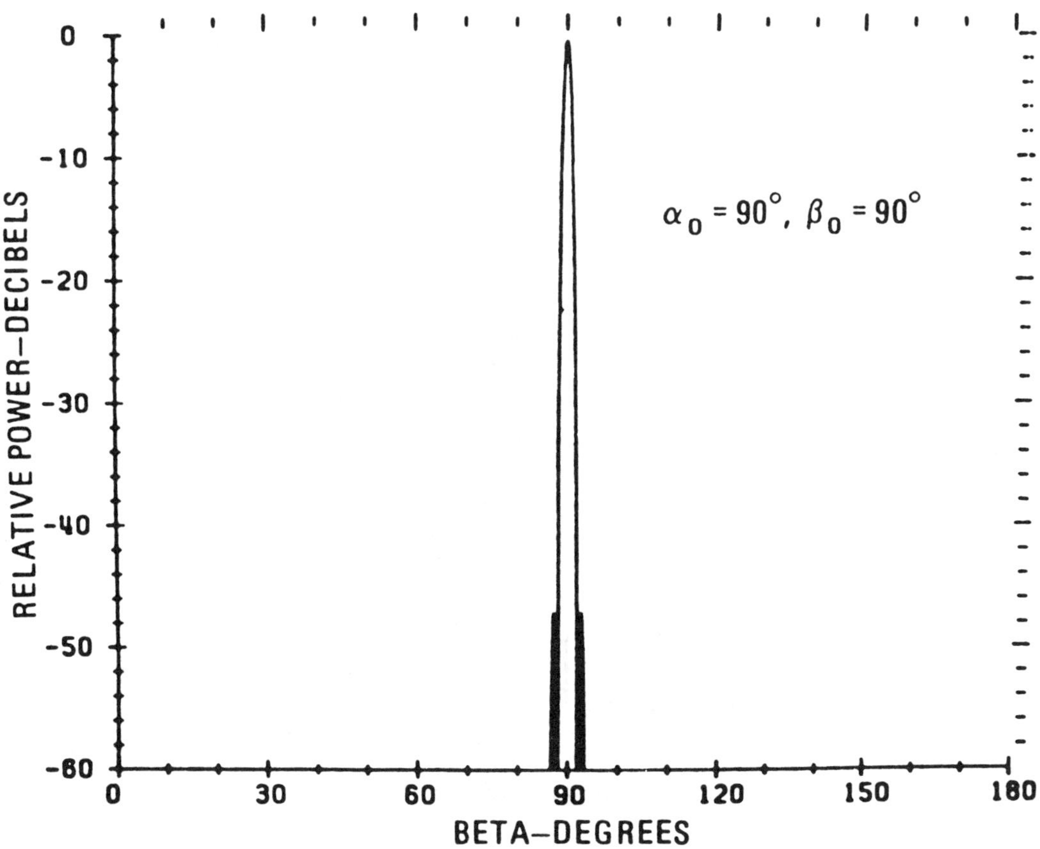

Fig. 8.31 Two-Way Reference Pattern β Cut

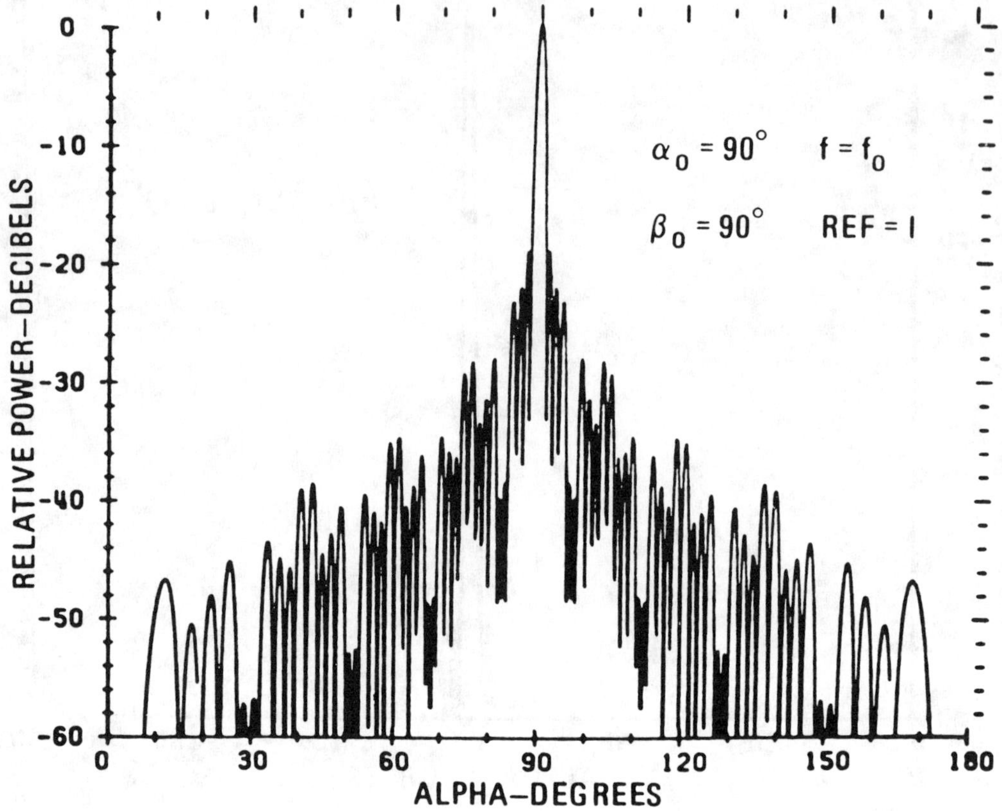

Fig. 8.32 Transmitting Pattern, Principal Plane, α Cut

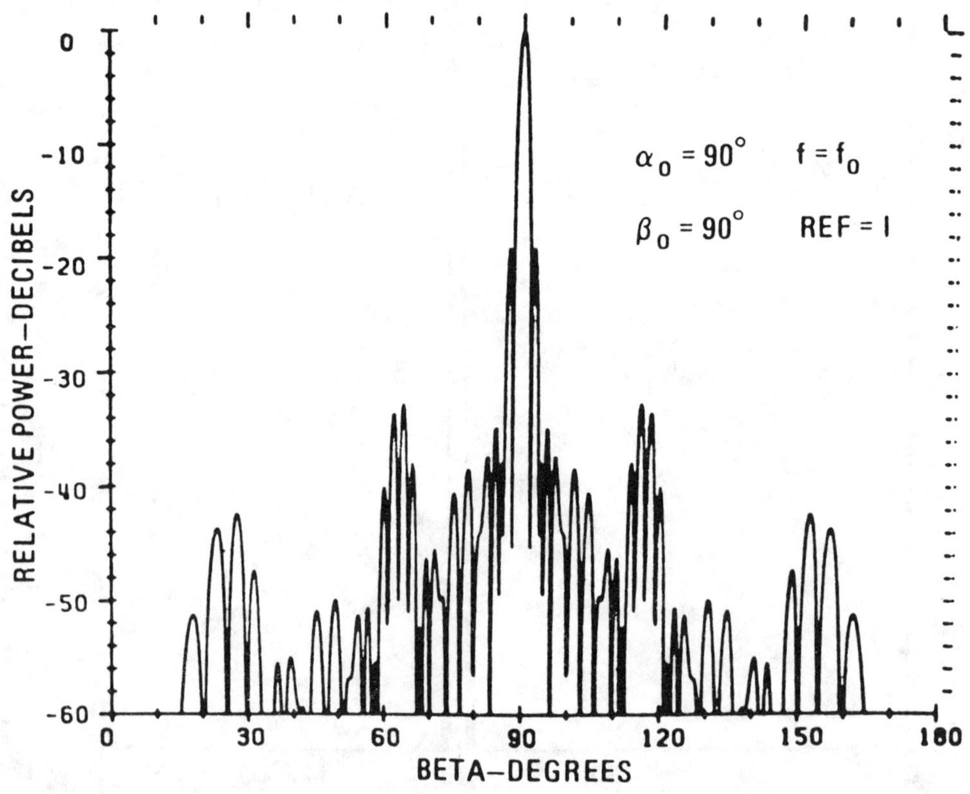

Fig. 8.33 Transmitting Pattern, Principal Plane, β Cut

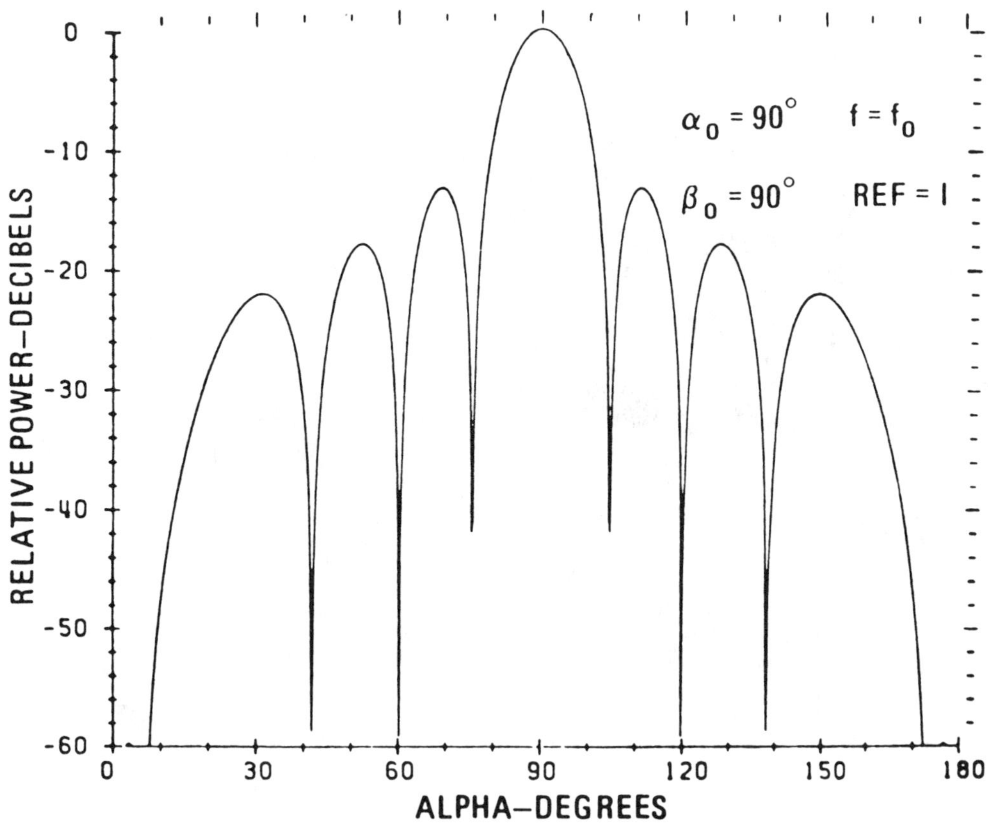

Fig. 8.34 Sub-array, Principal Plane, α Cut

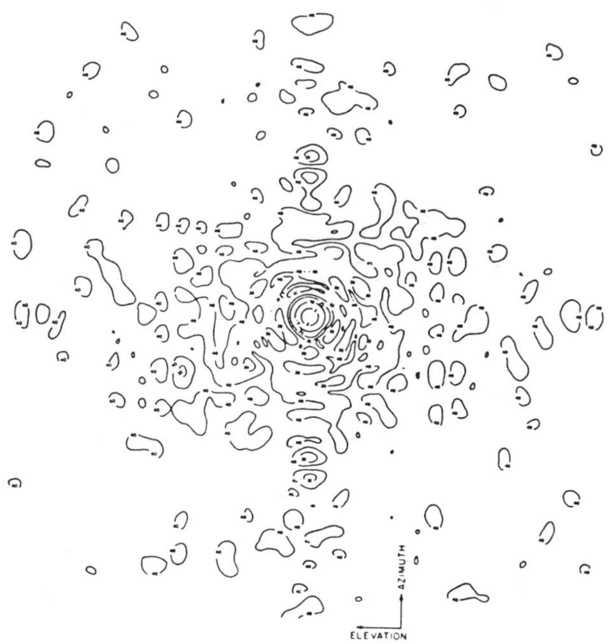

Fig. 8.35 Reference Pattern Contour with Illumination Errors

Fig. 8.36 Complete Assembly of Phased Array, Front View

Fig. 8.37 Complete Assembly of Phased Array Back View

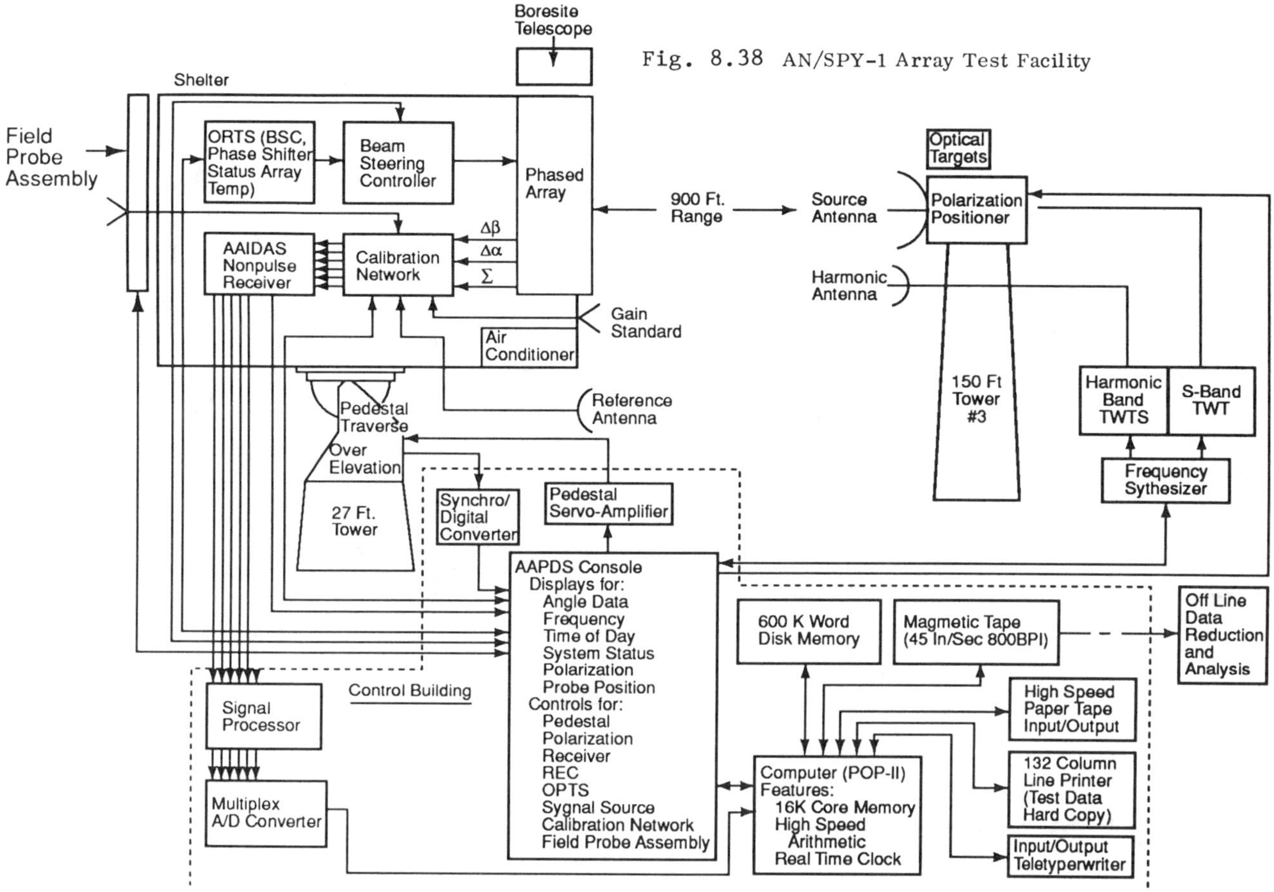

Fig. 8.38 AN/SPY-1 Array Test Facility

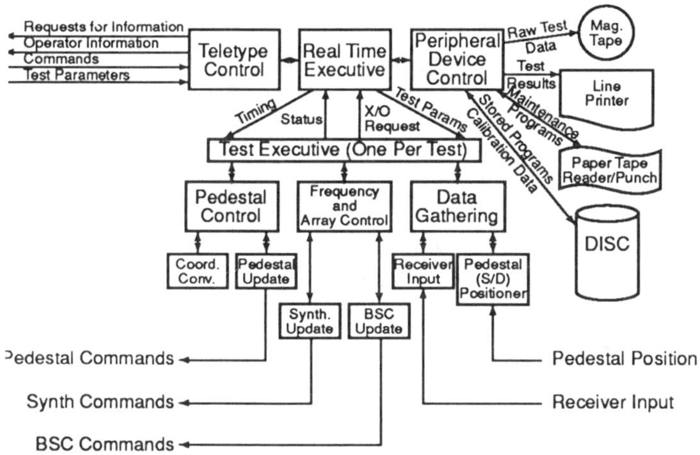

Fig. 8.39 Automatic Antenna Pattern Data Acquisition Computer Program

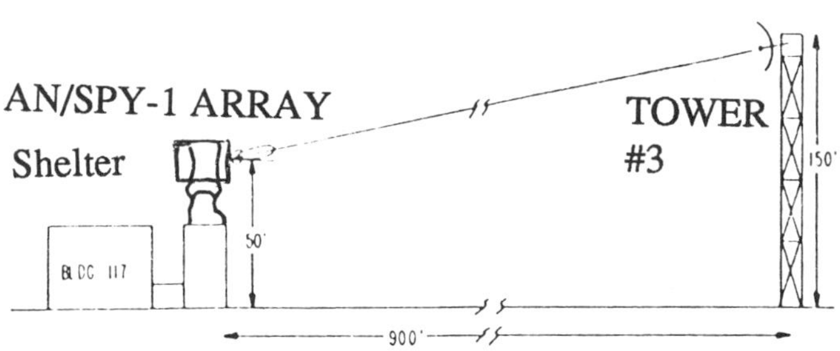

Fig. 8.40 Test Range Dimensions

Fig. 8.41 AN/SPY-1 Array Installed on Test Pedestal

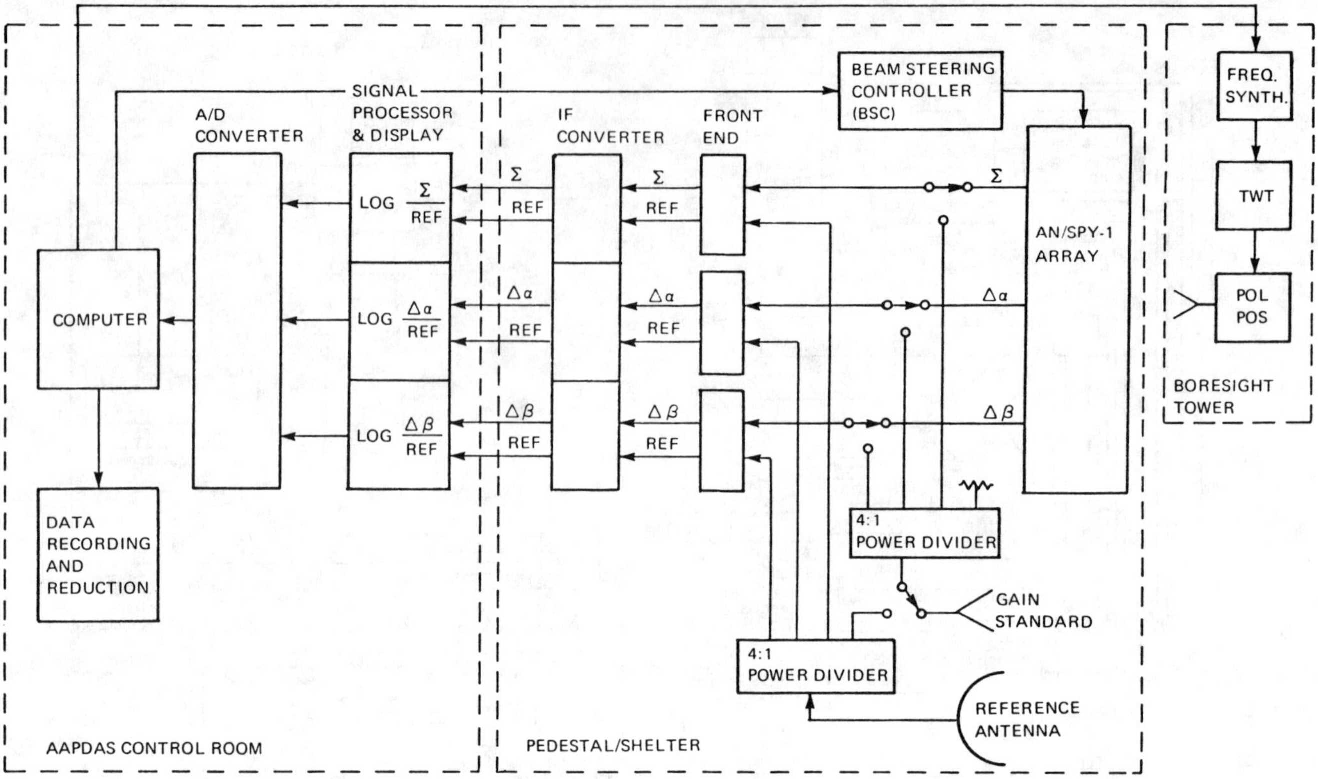

Fig. 8.42 AAPDAS 3-Channel Test Receiving System

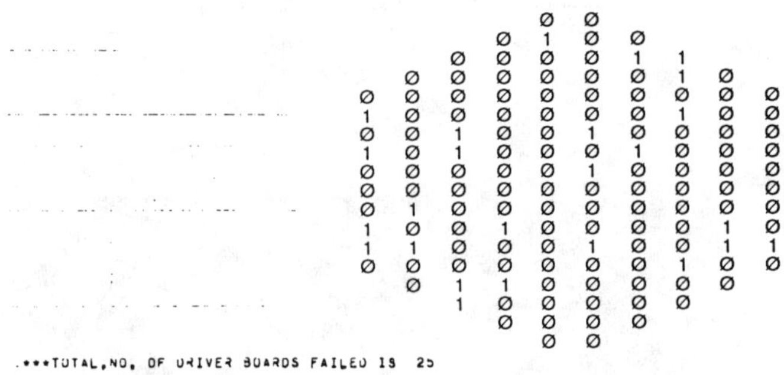

Fig. 8.43 TABOR Antenna Phase Shifter Driver Board Status Map

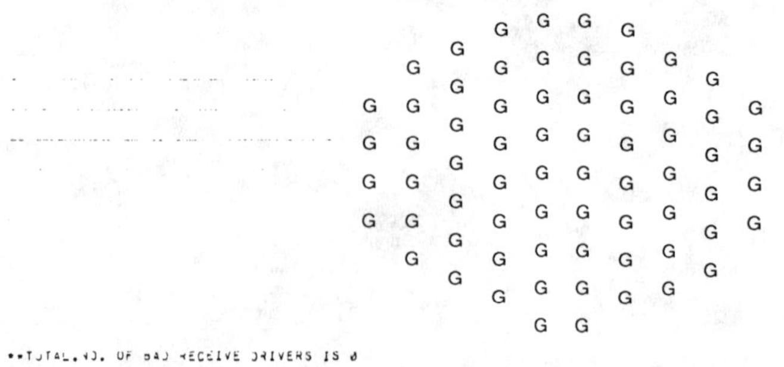

Fig. 8.44 TABOR Receive Phase Shifter Driver Status Map

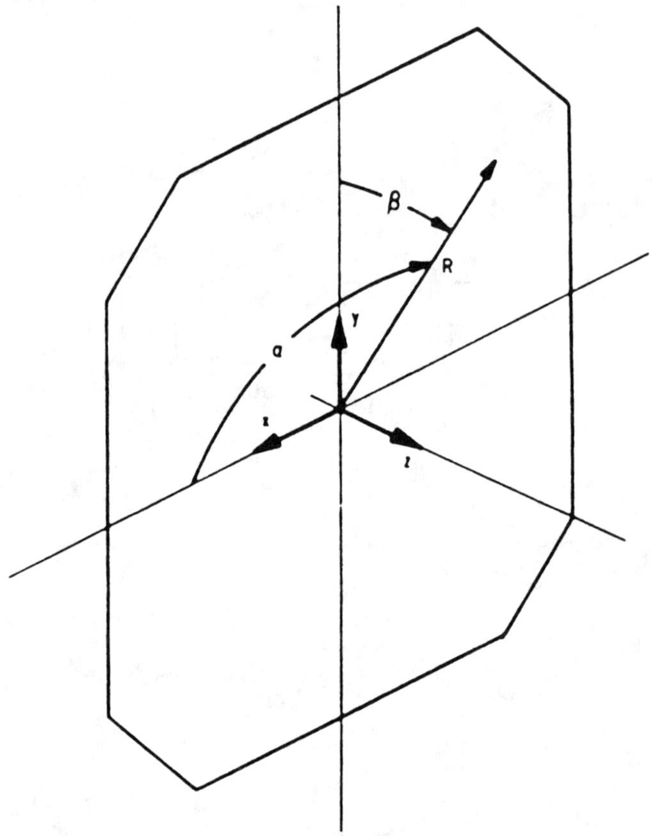

$$\cos \alpha : \frac{\hat{R} \cdot \hat{x}}{|R|}$$

$$\cos \beta : \frac{\hat{R} \cdot \hat{y}}{|R|}$$

Fig. 8.45 AN/SPY-1 Array Coordinate System

connections to input cables depending on specific calibration, i.e. Receiver Dynamic Range, Amplitude Linearity, Phase Linearity Calibrations.

(3) Receiver Check: Receiver gain checks are performed with the AAPDAS in the CAL/REF mode, where the switches are positioned such that a portion of the reference antenna output is divided equally into each of the three-channel front-end test ports.

For each configuration, measurements are made and displayed via the line printer. Each measurement is compared with a table of previous calibration data. An option is provided the test operator to store new data in each calibration table if necessary due to a component change.

8.6.4 Array and BSC Operational Status Check. The operational status of the AN/SPY-1 Array and Beamsteering Controller (BSC) is checked automatically before and after conducting each array performance test using the TABOR (Testing of Array and BSC Operational Readiness) test routine in the AAPDAS computer. Upon operator command, TABOR checks status of the ORTS (Operational Readiness Test System) test equipment contained in the array and BSC and displays to the operator via computer printout array and BSC status. The center of the TABOR system is the AAPDAS computer which is controlled by a preprogrammed disk. A typical sequence is initiated by addressing the BSC interface which translates the command data into compatible BSC data. The BSC reacts to the data by internal response or by controlling the array. The Data Aquisition Assembly (DAA) interface is then addressed to interrogate a particular address in the appropriate array or BSC DAA. The response from the DAA is transformed in the DAA interface to computer compatible data. The data is analyzed by the computer, usually by comparison to preprogrammed values, and then the results are displayed via the line printer. The tests are divided into six sequential functions as listed below:

- **RU TABOR Test,** which analyzes the general status of the ORTS associated with the array and BSC
- **BSC Status Test,** which monitors temperature sensors, power supplies, arithmetic operations, Read-only memory units, and D/A converters within the BSC unit.
- **Array Pre-conditioning Test,** to verify that a portion of the logic in both the antenna and receive phase shifters is operating and that the ORTS lines are complete.
- **Array Fault Values,** which evaluates the magnetic flux change for all phase shifters in response to a standard phase shift command.
- **Antenna Phase Shifter Status Map,** a function which displays faults identified for that group of phase shifters in graphic form, such that number and locations are readily identified by the test operators. An example is shown in Fig. 8.43.
- **Receive Phase Shifter Status Map,** a display function for these 68 units which prints out fauly locations for this group of phase shifters. Fig. 8.44 is an example of this display printout.

8.6.5 Coordinate Systems. Pattern measurements were made using a two-axis test pedestal to position the array relative to a fixed boresight tower. The pedestal, shown in Fig. 8.44, employs a traverse (azimuth) over elevation axis configuration in order to maintain vertical incident polarization with respect to the array.

Array beam-steering commands were issued under computer control via a special interface to the AN/SPY-1 Beam Steering Controller. The interface also provided manual control of the BSC. The array coordinate system is shown in Fig. 8.45.

A computer program was employed to position or raster scan the pedestal and/or array beam as required for specific tests. This program included transformation routines to convert the pedestal axis coordinate system ot the sine-space coordinate system of the array including corrections for range parallax. This feature of the system greatly simplified measurements and data analysis by allowing pedestal angles and array scan angles to be expressed in the same coordinate system.

Raster scan data could be measured using the two related array coordinate systems, α, β and cos α, cos β. In the former system, data is collected at constant $\Delta\alpha$, $\Delta\beta$ intervals, while in the latter system data is collected at constant Δ cos α, Δ cos β intervals. The latter system was used in tests such as the Beam Pointing Accuracy (BPA) test where it was necessary to normalize data to compare with data measured at broadside. Theoretically, when expressed in this coordinate system, the array beamwidths are constant for all scan angles. The unit of measurement in this normalized coordinate system, the "millisine", has a value equal to the sine of one milliradian or about 10^{-3}.

8.6.6 Angular Coverage. The angular coverage provided by the test range is depicted in Fig. 8.46. The β (elevation) boundaries shown on the figure are caused by structural limits to pedestal elevation axis motion. Because of the limitation, it was necessary to install the array upside down sot that adequate measurement capability could be obtained in the specified array coverage region. The "Real Space" boundaries result from the coordinate system definition and the constraint that $\cos^2\alpha + \cos^2\beta < 1$ in order for an α, β coordinate pair ot be physically realizable; coordinates within this boundary represent a full hemisphere of angular coverage.

8.7 Results of Performance Tests.

The following paragraphs of this section present example data measured on the AN/SPY-1 antenna during performance qualification testing of the unit. The data shown includes the results of both the mandatory procedures applicable to its function in the AN/SPY-1 radar, and other tests intended to supplement those results.

8.7.1 Self-Duplexing. The self-duplexing isolation test is a measure of the amount of transmitter power coupled into the receive beamformer sum-channel output during the time the array is operating in the transmit mode. Since the amount of isolation is dependent on operating frequency and beam steering position, partially as a result of changing aperture match to space as a function of those parameters, tests were conducted at four transmit beam positions over the full operating frequency band.

Fig. 8.47 illustrates an example of the measured results

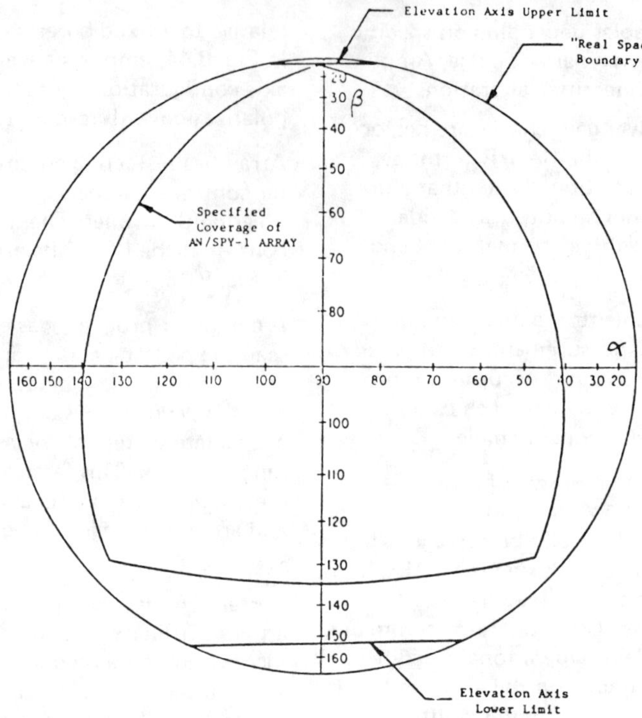

Fig. 8.46 Angular Coverage Provided by Test Facility

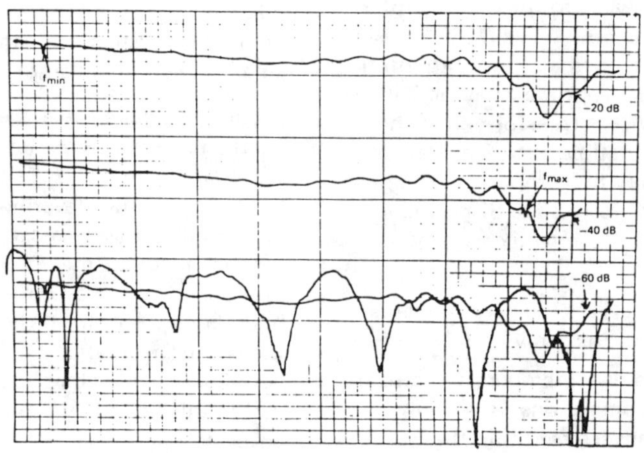

Fig. 8.47 Self-Duplexing Isolation ($\alpha = \beta = 90^o$)

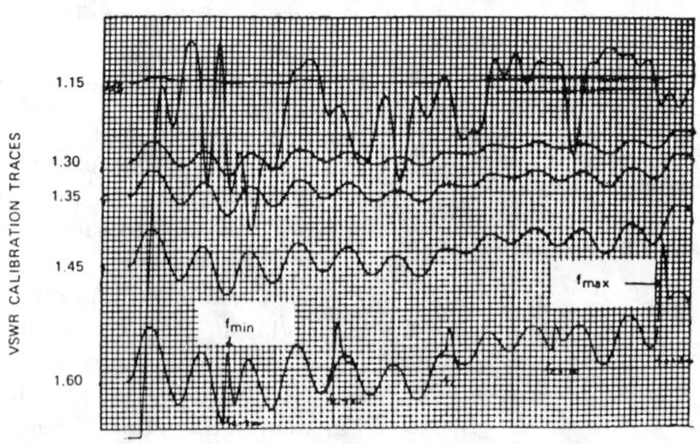

Fig. 8.48 Antenna Transmit Subarray VSWR (NR, $\alpha = 60^o$, $\beta = 63^o$)

superimposed on the isolation calibration traces. Frequency band edges are noted on the graphs as notches created by the reactive wavemeter utilized.

The measured relative power coupled into the receive beamformer output port is a combination of the cross-coupled power from all 32 transmit subarrays into all 68 receive subarrays. The design of the T-R limiters used to protect the RF receiver unit is based on a coupling value of -20 dB, thus providing adequate margin.

8.7.2 Antenna Mismatch. This test evaluates the VSWR characteristics presented ot the transmitter units as a function of operating frequency and four array beam positions. The evaluation is based on a sampling of 5 of the total of 32 transmit subarrays, having identification numbers N12, N13, N17, N18, and N19. The subarrays are located as shown in Fig. 8.48 and represent central and edge locations on the array face. The measured data included the complete transmit path from the transmitter interface through the array, including harmonic filters, waveguide runs, and internal array components. The test was conducted at the AN/SPY-1 Land Based Test Site, just prior to integration of the array and the transmitter equipment.

Fig. 8.48 gives an example of the swept-frequency VSWR data recorded. The measurements indicate that the VSWR will be under 1.30:1 for more than 86% of the operating conditions of beam positions and transmission frequencies.

8.7.3 Coverage. The coverage test was performed to demonstrate the gain of the AN/SPY-1 array beam peak as the beam is raster scanned throughout the entire specified array coverage sector. The sum-channel gain was measured relative to the peak sum-channel gain at broadside as the array beam was raster scanned in 1-degree increments throughout the limits of real space for α limits of 40 degrees to 140 degrees and β limits of 15 degrees and 140 degrees. The measurement was made at the upper operating frequency band F_H, since any array resonance phenomenon is most likely to occur there. Center band data were also measured.

The AAPDAS antenna pattern test setup (Fig. 8.38) was used with the coverage test executive computer program routine in the AAPDAS computer. In this test the AAPDAS pedestal was raster scanned in α and β as the array beam was counter scanned so that it was always pointed towards the source antenna mounted on the boresight tower. The relative gain was measured and the data stored on a magnetic tape.

The display for this test is a contour plot showing relative gain versus array-beam scan coordinates of α and β.

The objectives of this test were to confirm the absence of holes in the array coverage and demonstrate the array performance with respect to the gain taper requirement.

The results of this test show that there are no holes in the array coverage, confirming the absence of resonance effects in the array aperture.

The display of the test data measured at F_H shown in Fig. 8.49 has been marked to show the measured points slightly below the gain taper objective. These points represent 5.3% of the angular coverage area which measured below that criteria.

Two factors have a significant effect on the data as measured during the coverage test, causing the results shown in the two contour plots to be pessimistic. These factors are: (1) the array polarization variation from vertical at scan angles off broadside in both α and β and (2) taper in the received signal amplitude off the boresight axis due to the beamwidth of the source antenna.

The polarization effect is discussed in paragraph 8.7.9, showing example results of coverage tests conducted with the source polarization angle oriented to +20°, -20°, +45°, -45°, +67.5°, and 90° from vertical. From this data the relative gain at the following beam positions which are all at the $\theta = 45°$ boundary were calculated:

α	β	Relative Gain	Polar. Angle
60	60	1.9 dB	20°
60	120	1.8 dB	-20°
120	60	1.5 dB	-20°
120	120	1.4 dB	20°

When the effects of these factors plus minor corrections for test signal tapering are taken into account, the array coverage data satisfies the gain taper requirement over the required 45° off-boresight scan limits.

Fig. 8.50 is a similar contour plot measured at operating frequency band center. In this case, the gain function essentially adheres to the objective without correction for polarization effects. Again, there is no tendency noted for local areas of low gain.

8.7.4 Instantaneous Bandwidth. The instantaneous bandwidth test evaluates the amplitude transfer function of the antenna as a function of operating frequency over the primary AN/SPY-1 signal waveform bandwidth. Three sub-bands were evaluated, representing the center and each extreme of the total operating frequency range of the antenna. The amplitude characteristic was measured with the array beam steered to broadside ($\alpha = \beta = 90°$) and with the beam steered 60° off broadside in the elevation direction ($\alpha = 90°, \beta = 30°$).

The performance data was evaluated in terms of the characteristics of a best-fit quadratic curve of the measured data. Data points were measured at 2-MHz intervals over each sub-band tested in order to provide a sufficient sampling basis for the curve-fitting process. In addition, test data was extended beyond the defined band limits, typically spanning a total of more than 160-MHz for the broadside beam position tests and more than 120 MHz for the 60° scan case. Data was accumulated over these wider ranges also to provide a sufficient basis for the data reduction.

During the conduct of the instantaneous bandwidth tests, the measured amplitudes of the array were directly related to the signal amplitude output of the standard-gain horn to implement evaluation of antenna gain in the receive mode, covered in the following paragraphs.

Figs. 8.51 and 8.52 illustrate examples of the measured data and best-fit curves overlayed for both test beam positions.

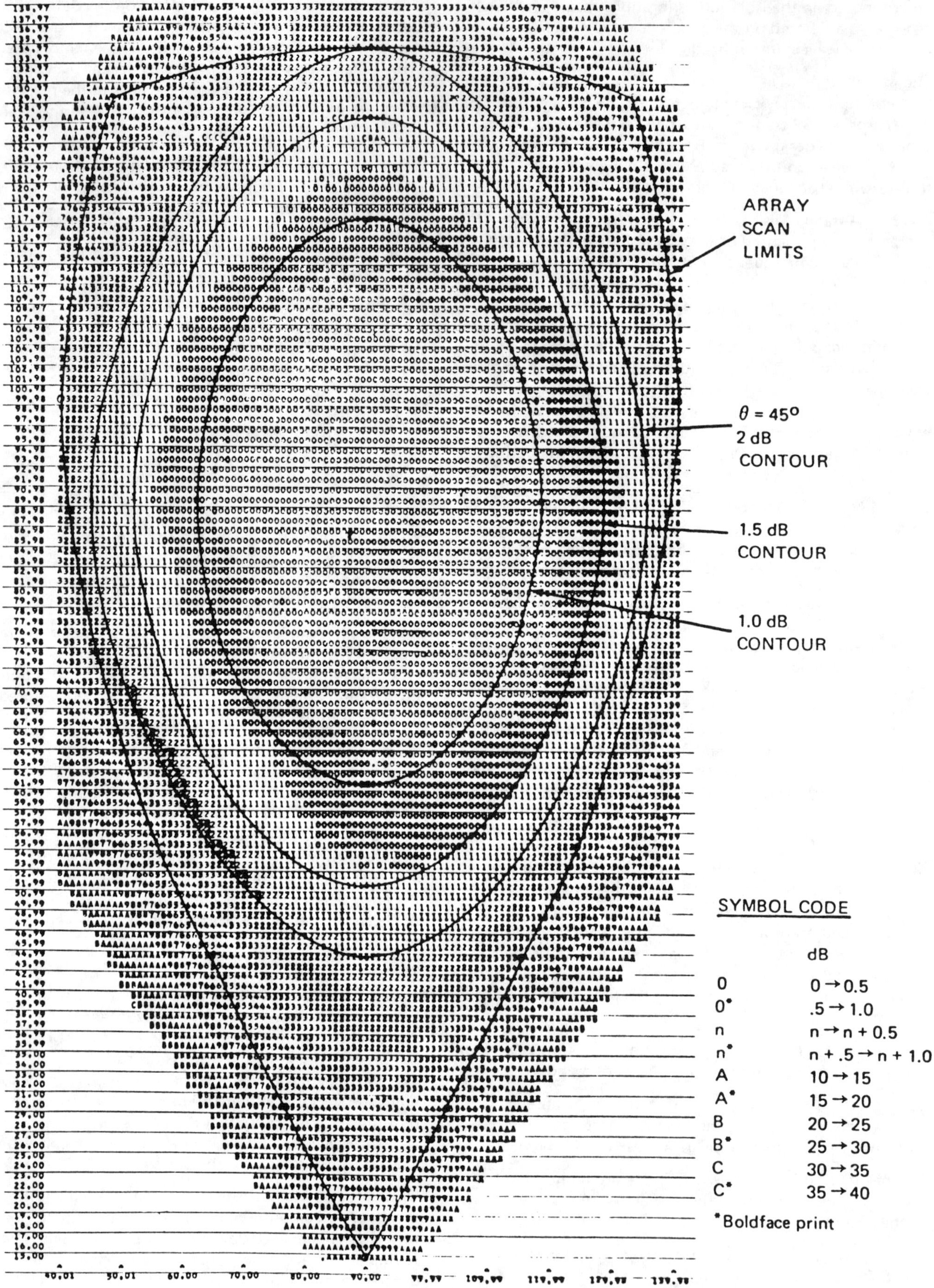

Fig. 8.49 Coverage Contour, f_H

Two primary factors control these frequency response curves. For the beam broadside case ($\alpha = \beta = 90^\circ$), the small drop in gain at band edges is a result of the use of unequal length waveguides in the column beamforming networks described in paragraph 2. For the $\alpha = 90$, $\beta = 30^\circ$ case this effect is combined with the inherent phased array frequency-beam position response to cause additional gain reduction at sub-band edges.

The Beam Steering Controller unit utilizes read-only memory units to compensate for the unequal line length beamformer waveguides as a function of the operating sub-bands of the AN/SPY-1 radar.

8.7.5 Receive Gain. Receive gain was evaluated at the output port of the azimuth beamformer, and thus included the total RF losses from the array face through the RF filter terminal at the interface to the radar RF receiver unit. Data utilized in evaluating the gain characteristics were obtained in the process of measuring the instantaneous bandwidth properties for the beam broadside case, since that measurement was made in terms of the ratio of the output of the array reference channel to the output of the standard-gain-horn output as the reference signal in the coherent test receiver unit. The relative signal level used for the gain computation at each of the test frequencies corresponds to the peak of the best-fit quadratic curves, examples of which are shown in Figs. 8.51 and 8.52.

The measured results are significantly better than the budgeted criteria. The major factor accounting for the difference is that the contributions to RF loss shown in Table II are generally maximum values. The overall RF losses in the receiving mode are therefore indicated to be 0.4 dB below the total loss value of 4.9 dB shown in that table.

8.7.6 Transmit Gain. The transmit gain, transmit RF loss combination was measured in two steps, since the RF loss includes the off-array waveguide which provides the interfacing connections between the transmitter subsegment and the array. The first step, carried out at the AAPDAS facility included the measurement of the combined directive gain and the RF losses of the array components through the self-duplexing hybrid networks and a coaxial test beamforming network utilized to drive the 32 transmit subarrays with cophased, equal-amplitude signals. This test beamformer functionally substituted for the transmitter subsegment in normal AN/SPY-1 system operation, except that variable relative phasing between transmit subarrays was omitted from the test setup.

The second step included the RF loss measurement of a sample waveguide runs as installed at the AN/SPY-1 Land based Test Site as required to complete the evaluation of the total RF loss to the defined array interface in the transmit mode.

The gain-measurement process for the transmit mode utilized methods similar to those used for the receive mode where the array signal amplitude as received from the boresight tower was directly compared to that received via the standard-gain horn from the same source. This ratio was measured over the full sub-band width at intervals of 2 MHz for each of the three sub-bands centered at F_L, F_C, and F_H. Since the test beamformer utilized equal-length transmission lines, the response across the sub-bands was essentially constant, and the relative difference in signal level between the array and the standard-gain horn was obtained by averaging the 21 data points recorded for each sub-band measured.

8.7.7 Transmit and Receive Patterns. Most of the patterns measured for this antenna were recorded and reduced in contour format in order to provide a complete evaluation. Both transmit and receive functions were actually evaluated using the antenna in a receiving mode. The Beam Steering Controller was manually modified to permit transmit patterns to be measured in this fashion through the non-reciprocal antenna phase shifters. Check patterns of the transmit measured in the transmit mode validated this approach.

The contour patterns were recorded as the test pedestal was driven by the AAPDAS computer in α cuts (constant β) over the operator selected sector. At the end of each cut, the pedestal was automatically caused to increment a preselected β step, and then moved again through the selected α range in the reverse direction. The pedestal was generally driven at 2° per second during these raster scans. The test program provides the flexibility to select the position of the raster pattern anywhere within the array real space, as well as the limits of scan in array α and β coordinates. In addition, the array beam position is operator selectable in the same coordinate system along with the test operating frequency. Data was recorded on magnetic tape, identified by machine assigned file numbers.

When the data is reduced, the computer, via the line printer, provides the data in two forms for evaluation. The first is a contour plot which displays an overall graphical presentation of the data using symbols for relative amplitude levels compared to the main beam. This presentation is used to establish the overall characteristics of the particular parameter of interest. The second form of presentation is a plot of sidelobe peaks where a preselected range of relative amplitudes is presented for precisely determining the sidelobe levels. This presentation provides those amplitude peaks that fall within the selected range, encompassing the test criteria levels. The peaks are presented so that the angular location and the relative amplitude are printed out.

In order to determine the beam width of antenna, principal-plane patterns were made using an analog recorder on an expanded scale. This provided the necessary detail to measure the beamwidth limits.

<u>A. Transmit Patterns.</u> Figs. 8.53 through 8.55 are examples of the transmit patterns for beam position $\alpha = \beta = 90^\circ$. Fig. 8.53 is the large area scan covering 100° in both α and β at center frequency F_C. This pattern is used primarily for purposes of getting an overall view of the array performance. It clearly shows that a large portion of the area scanned is free of sidelobes greater than -40dB, and, as expected, most sidelobes are concentrated around the main lobe. The symbols used for relative amplitude with respect to the peak of the main lobe are listed in Table III. The block in the center of Fig. 8.53 is that area covered by the more detailed scan shown in Fig. 8.54.

Fig. 8.54 is a detailed contour of the area around the main lobe. This pattern also uses the same symbol presentation as listed in Table III. In order to more clearly define the

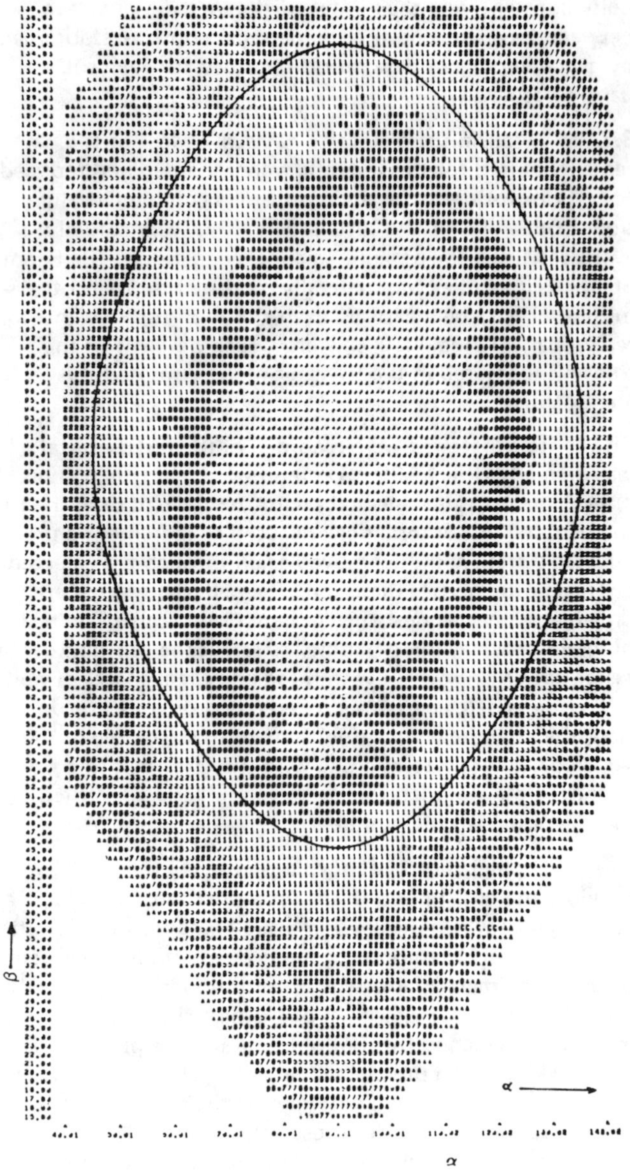

Fig. 8.50 Coverage Contour, f_C

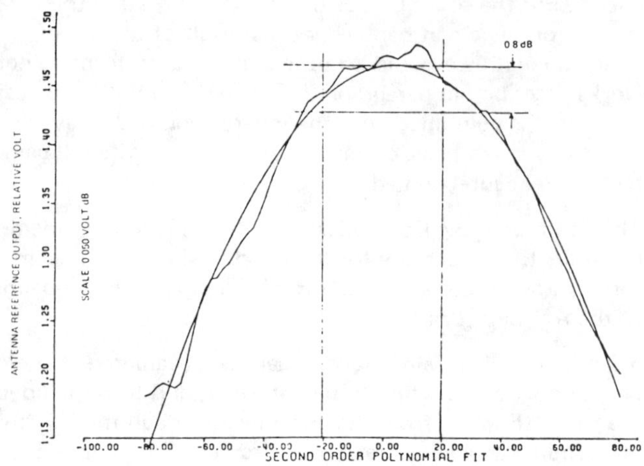

Fig. 8.52 Instantaneous Bandwidth (α = 90, β = 30, F_L)

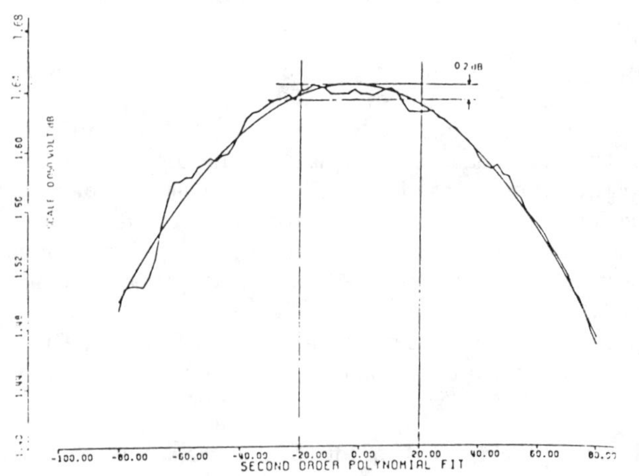

Fig. 8.51 Instantaneous Bandwidth (α = 90, β = 90, F_L)

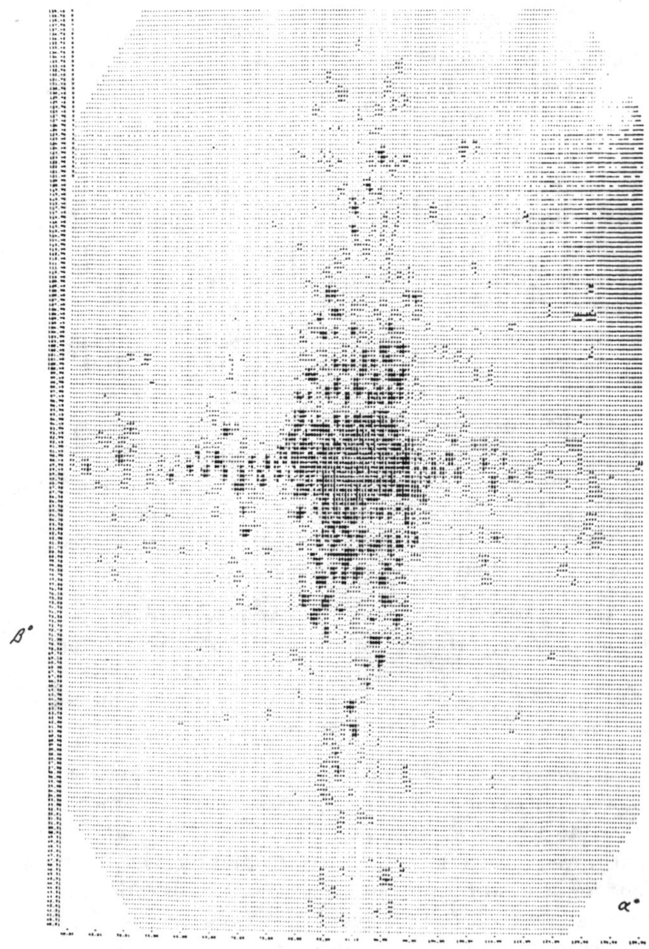

Fig. 8.53 Transmitter Contour, $100^o \times 100^o$, Broadside, F_C

various contours, every other symbol is printed in bold-face type to provide contrast only.

This pattern shows in detail the characteristics of the main lobe and all significant sidelobes. It does not, because of the granularity, accurately determine the sidelobe peaks. It does show quite clearly the location and the extent of any lobe within the scan area.

The second form of presentation is the sidelobe peaks pattern shown in Fig. 8.55. This uses the same data used to generate the pattern of Fig. 8.54 but is shown in a form that accurately determines all significant sidelobe levels and locations within the scan area. The sidelobe peaks are shown as an "X" on this presentation, located at the proper α, β coordinates. The peak amplitude of the sidelobe relative to the main lobe is printed out just below and to the right of the β coordinate. Where there is more than one peak at the particular β coordinate, the levels are printed in the same left-to-right sequence as the peaks. Because the scan area includes the main lobe, the presentation was limited to a particular range or window of levels around the specified sidelobe level. In this case, the range included all amplitude levels that were at least 10 dB, but not more than 25 dB below the peak of the main lobe. Those levels of the main lobe that fall within this window have been lined out.

RELATIVE AMPLITUDE SYMBOLS

TABLE III

Symbols	Amplitude Below Main Peak (dB)
0	0 to 2
2	2 to 4
4	4 to 6
6	6 to 8
8	8 to 10
A	10 to 15
B	15 to 20
C	20 to 25
D	25 to 30
E	30 to 35
F	35 to 40
•	Greater than 40

A histogram of all transmit sidelobes within the 10 to 25 dB range is shown in Fig. 8.56. This graph is a composite of sidelobes for F_L, F_C and F_H for the near-in sidelobes displayed on the 20° x 20° contour and local maximum plots, for $\alpha = \beta = 90°$. This histogram therefore does not include the many thousands of much lower-level sidelobes distributed over the array field of view. Each bar represents all sidelobes that fall within 0.5 dB intervals. The graph includes any duplication of sidelobes which may occur in the same angular position at different operating frequencies.

B. Receive Patterns. The patterns required to determine antenna performance in the receive mode were taken using similar methods as for the transmit patterns except that the antenna/BSC was set for receive operation. The raster scans, the sidelobe peak plots, and analog patterns are similar to those described in the section covering transmit patterns. To illustrate the results, a sample set of patterns taken at F_C for the three monopulse channels is presented. A summary of all the sidelobe data (at three test frequencies) is provided in graphical form.

C. Receive Sum Patterns. The sum (Σ) patterns and composite results are shown in Figs. 8.57 through 8.61. Fig. 8.57 is a copy of the data obtained for the 100° x 100° raster scan at F_C. The marked-off rectangular area is presented in more detail in Fig. 8.58. This pattern is included to provide an overall view of the array performance over a large coverage area. The sidelobe levels which require closer examination are found to be those contained within the rectangular area. The symbols used for the relative amplitudes with respect to the peak of the main beam are contained in Table III.

Fig. 8.58 is a detailed contour plot (20° x 20° scan) of the rectangular section shown in Fig. 8.56. The central region in this figure ($\alpha = \beta = 90°$) shows the main beam. Fig. 8.59 is a peak sidelobe presentation of all sidelobes within the 20° x 20° coverage area that fall between 15 and 30 dB below the peak of the main beam. Those levels associated with the main beam that fall within this 15 dB window have been lined out. Review of Fig. 8.59 shows that there are no sidelobes that exceed the 23 dB criteria at the test frequency F_C.

The composite result for the three test frequencies is shown graphically in Fig. 8.60 for the sum channel. As in the presentation of transmit sidelobe data, each bar represents all sidelobes that fall within a 0.5 dB window, and again is restricted to only the higher level lobes near the main beam. Upon review of this figure, there appear to be seven sidelobes that exceed the 23dB criteria for the sum channel; however, many of these are the same lobes repeated at different operating frequencies.

Fig. 8.61 is a principal plane analog pattern of the receive Σ channel covering 360 ° in the α coordinate, with $\alpha_o = \beta_o = 90°$. This pattern is shown to demonstrate the low level of back radiation from the array. Except that due to the physical obstruction noted, back radiation peaks are typically below -65 dB.

D. Receive Difference Patterns ($\Delta\alpha$ and $\Delta\beta$). Test data similar to the type presented for the sum channel was simultaneously recorded for the $\Delta\alpha$ and $\Delta\beta$ channels. Again sample data at F_C as well as composite data in graphical form for these channels are shown. Figs. 8.62 and 8.63 are the contour plots for the 100° x 100° raster scans for the $\Delta\alpha$ and $\Delta\beta$ channels.

Figs. 8.64 and 8.65 are the 20° x 20° contour plots for $\Delta\alpha$ and $\Delta\beta$, and Figs. 8.66 and 8.67 are the corresponding sidelobe peak plots for these two channels. The sidelobe peak plots show all sidelobes within the range 15 to 30 dB below the peak of the main beam.

Fig. 8.68 is a composite histogram of all sidelobes at the three test frequencies which fall within the range of 15 to 30 dB below the peak of the main beam for the $\Delta\alpha$ channel. Fig. 8.69 is a similar presentation for the $\Delta\beta$ channel. Since these figures show the total number of lobes within the -15 dB to -30 dB range for 3 test frequencies, characteristics of any one pattern would in general have 1/3 the ordinate values shown.

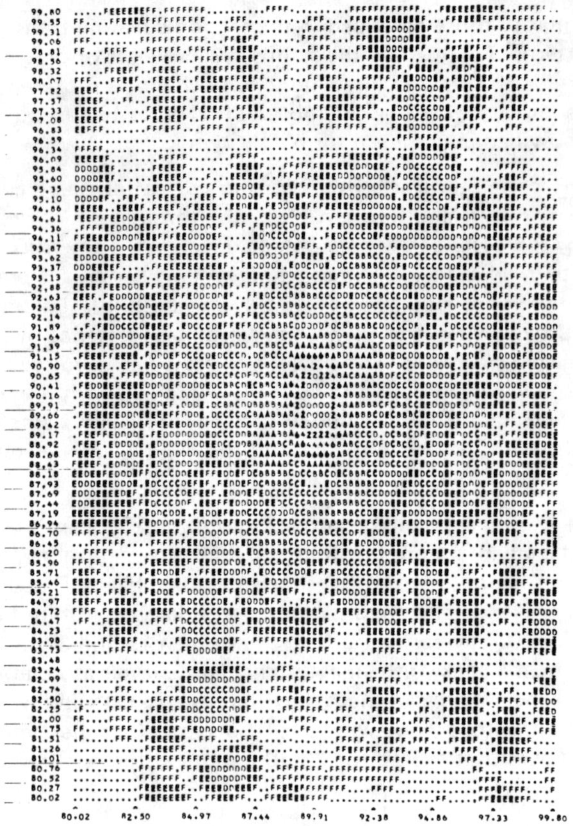

Fig. 8.54 Transmitter Contour, 20° x 20°, Broadside, F_C

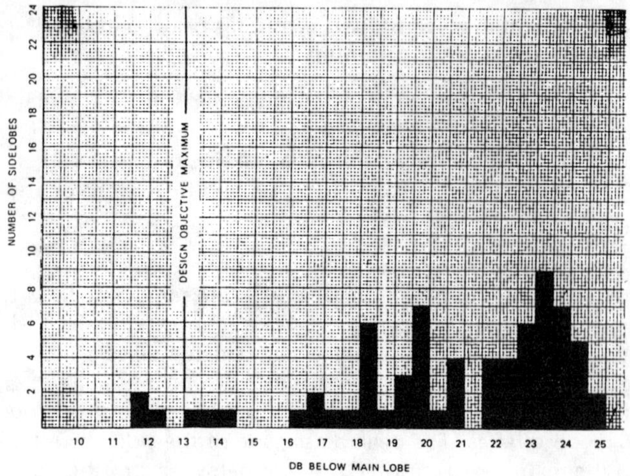

Fig. 8.56 Number of Transmitter Sidelobes Versus Sidelobe Level

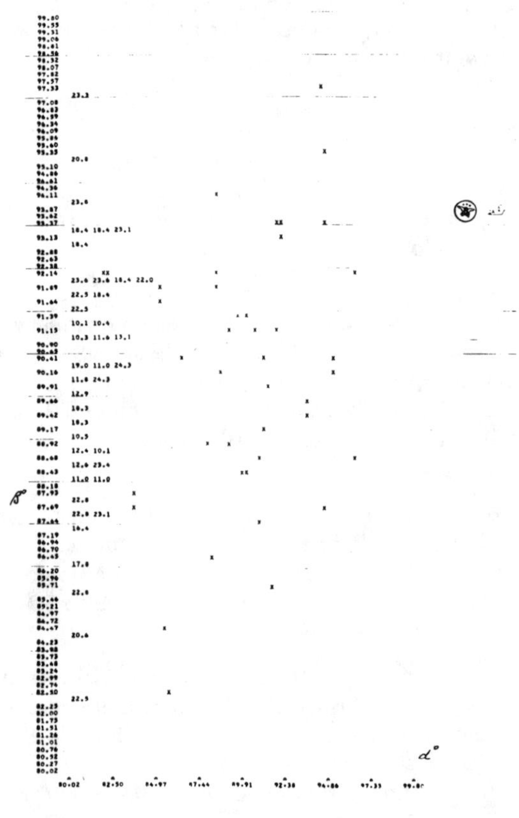

Fig. 8.55 Transmitter Sidelobe Peaks, 20° x 20°, Broadside, F_C

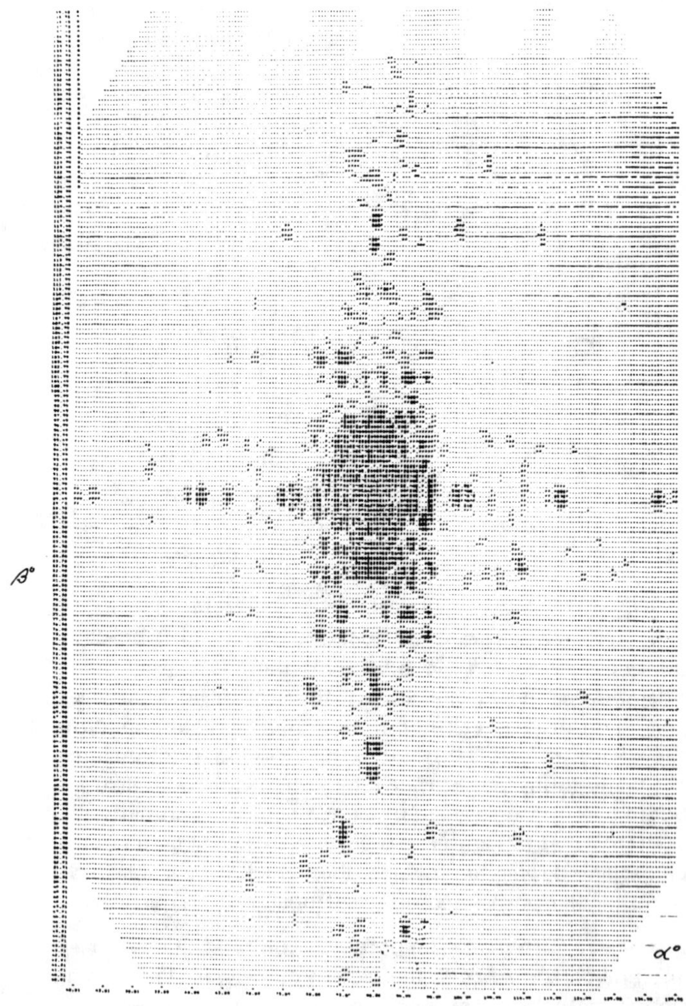

Fig. 8.57 Sum Contour, 100° x 100°, Broadside, F_C

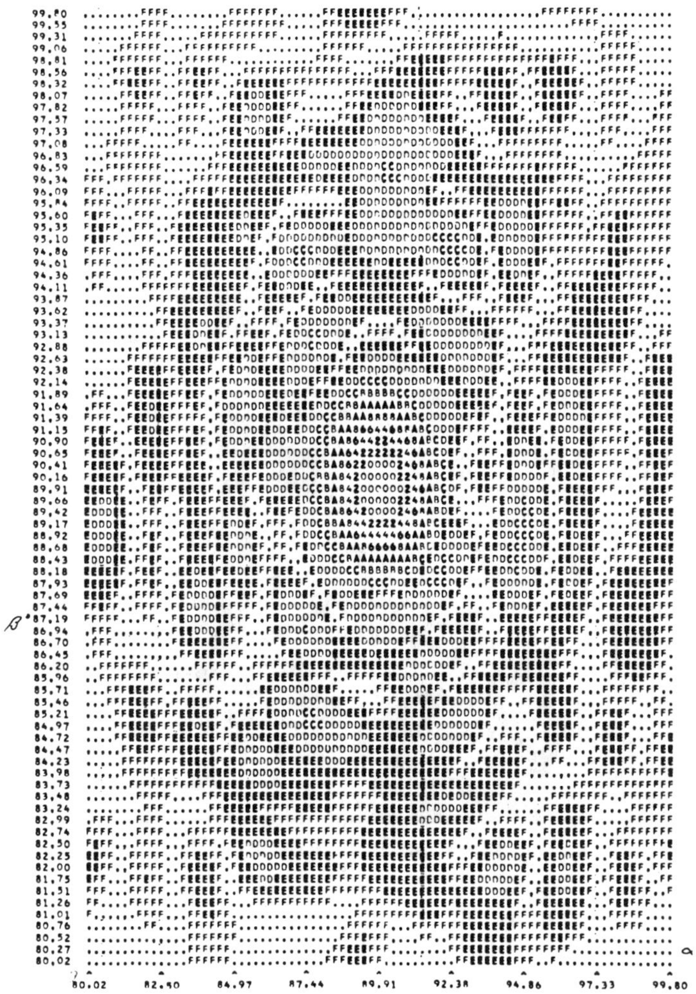

Fig. 8.58 Sum Contour, $20^{O} \times 20^{O}$, Broadside, F_C

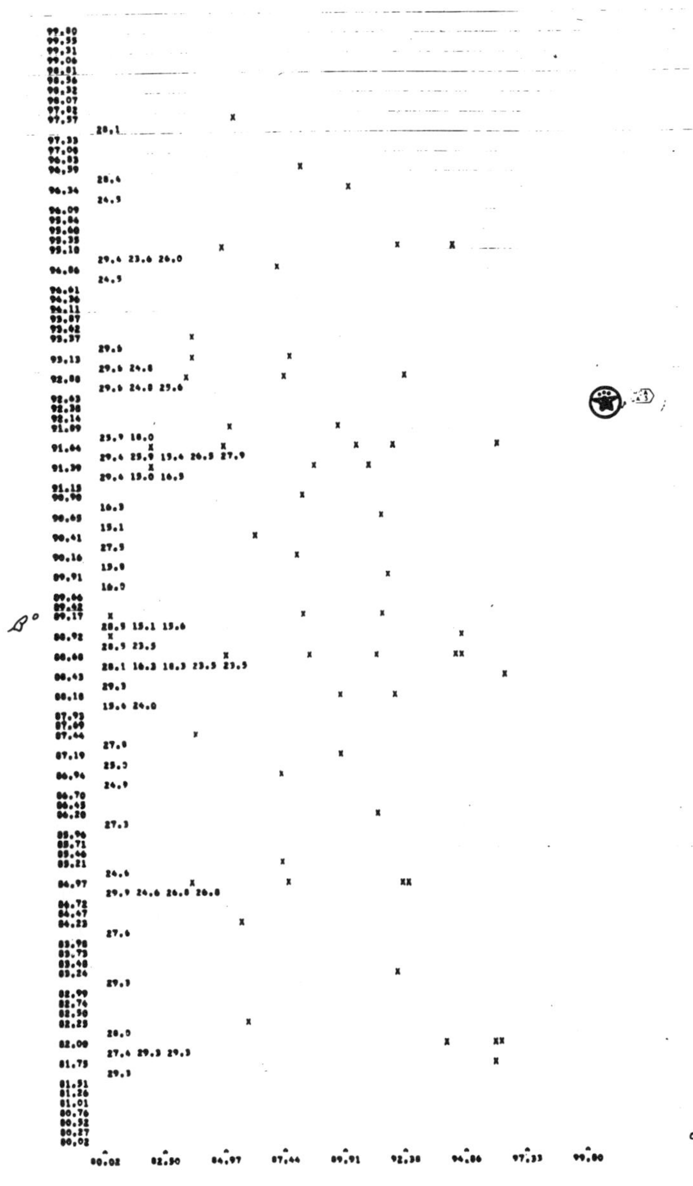

Fig. 8.59 Sum Sidelobe Peaks, $20^{O} \times 20^{O}$, Broadside, F_C

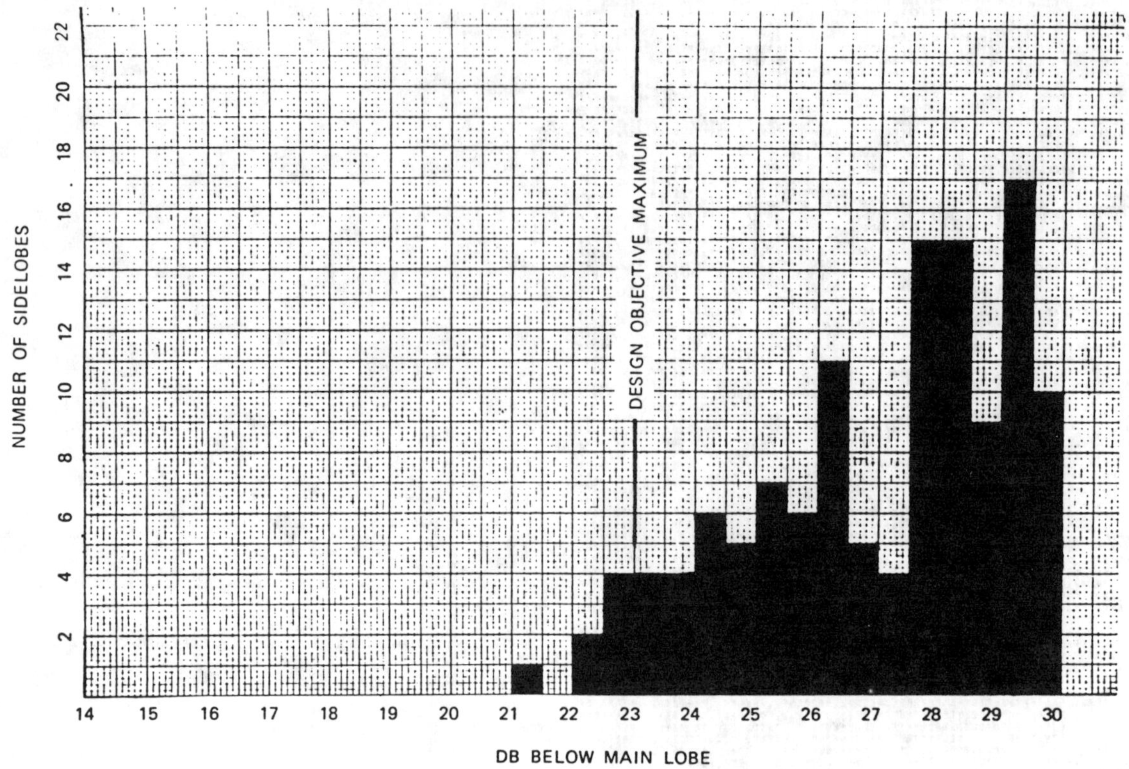

Fig. 8.60 Number of Sidelobes Versus Sidelobe Level,

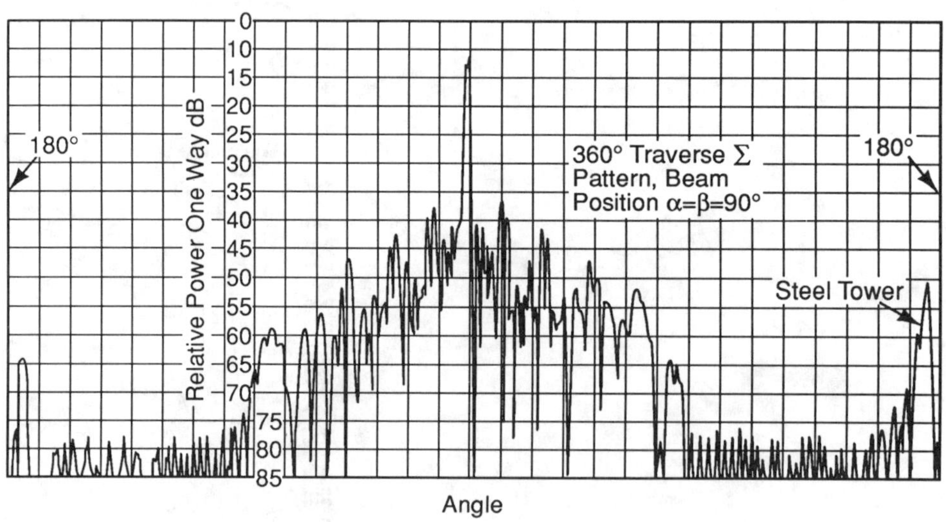

Fig. 8.61 360° Traverse Σ Pattern, Beam Position $\alpha = \beta = 90^{\circ}$

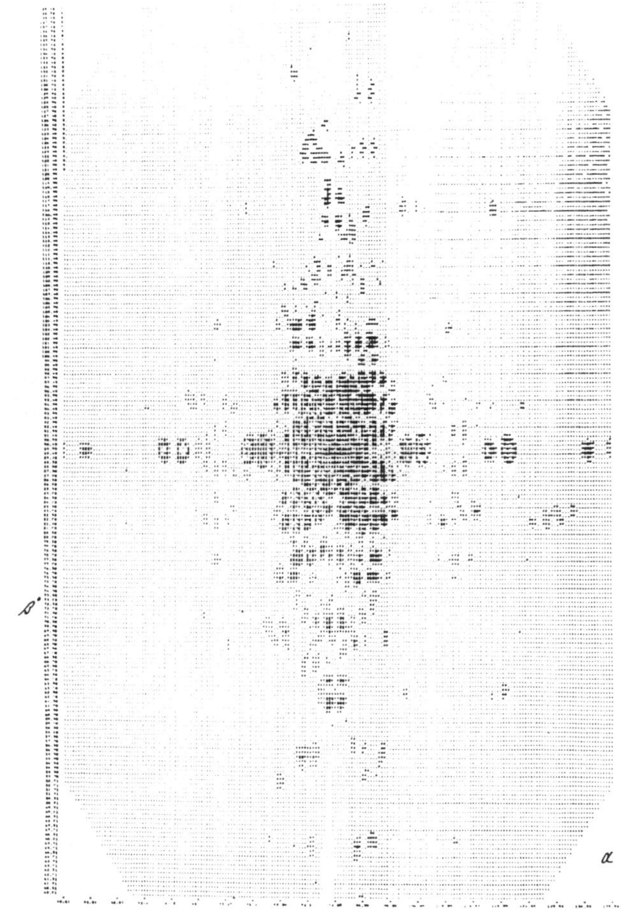

Fig. 8.62 Δα Contour, $100^{o} \times 100^{o}$, Broadside, F_C

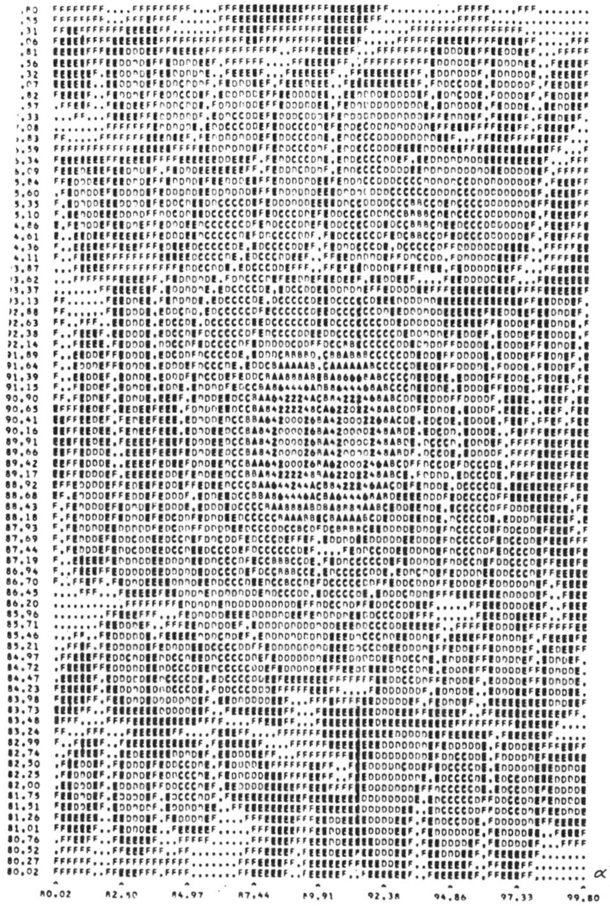

Fig. 8.64 Δα Contour, $20^{o} \times 20^{o}$, Broadside, F_C

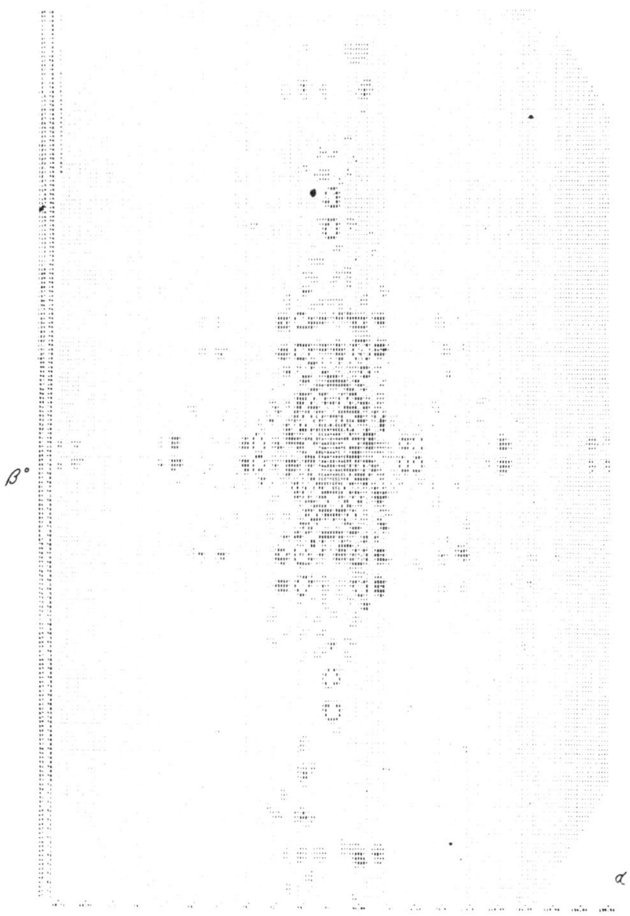

Fig. 8.63 Δβ Contour, $100^{o} \times 100^{o}$, Broadside, F_C

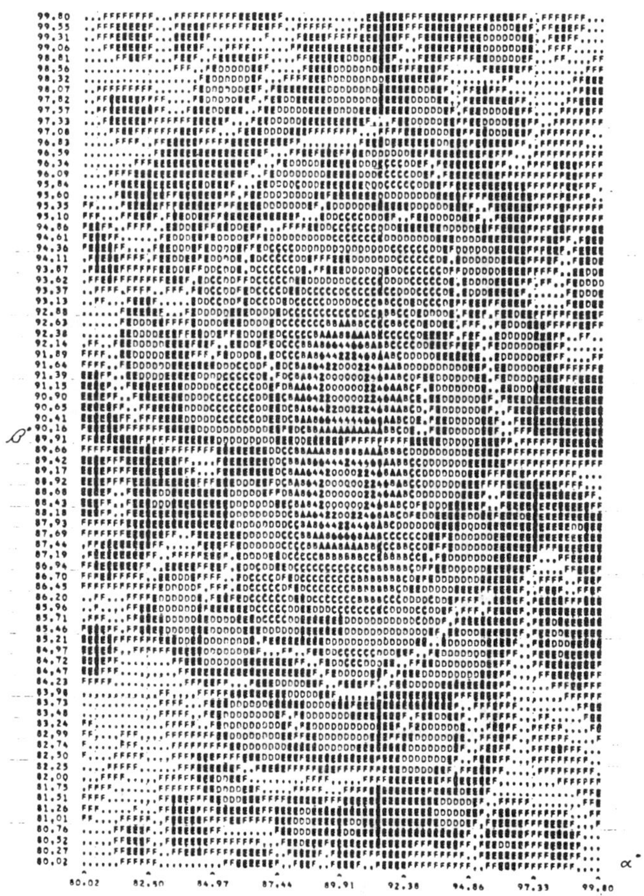

Fig. 8.65 Δβ Contour, $20^{o} \times 20^{o}$, Broadside, F_C

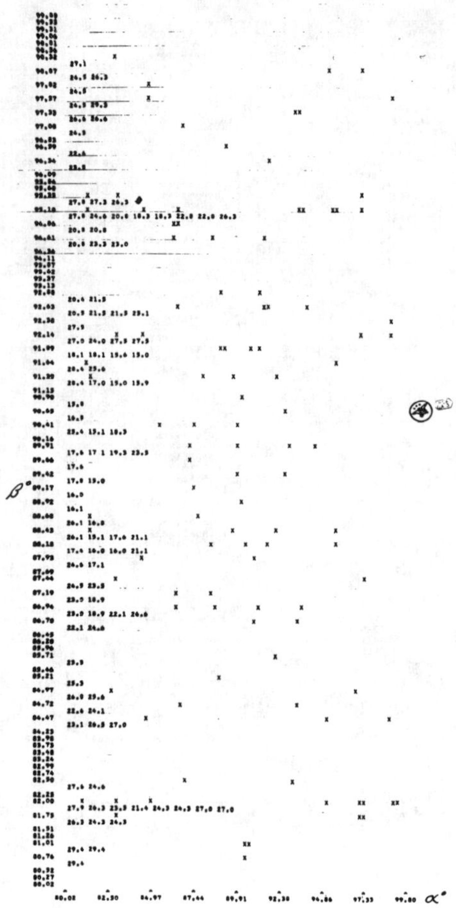

Fig. 8.66 $\Delta\alpha$ Sidelobe Peaks, $20^{\circ} \times 20^{\circ}$, Broadside, F_C

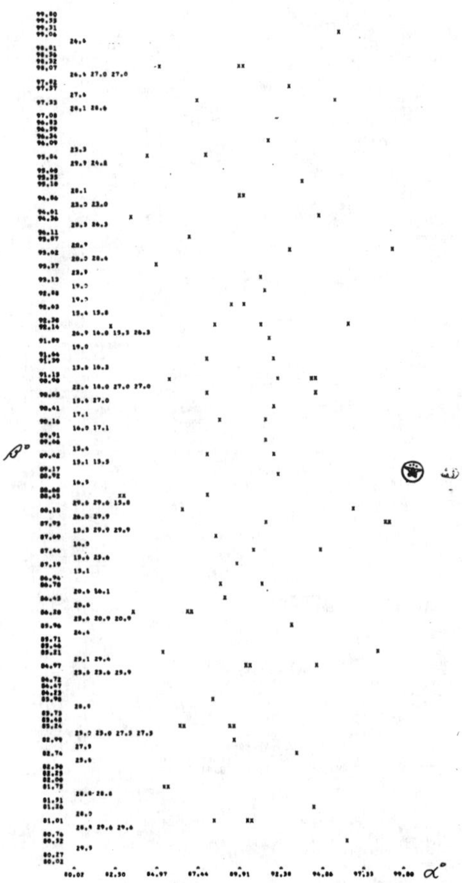

Fig. 8.67 $\Delta\beta$ Sidelobe Peaks, $20^{\circ} \times 20^{\circ}$, Broadside, F_C

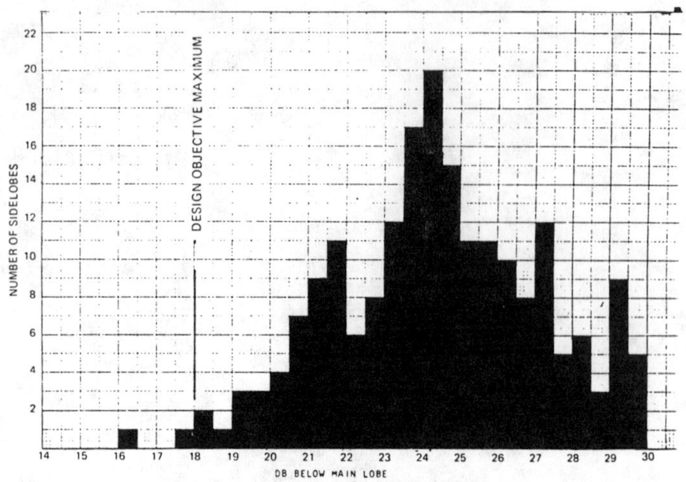

Fig. 8.68 Number of $\Delta\alpha$ Sidelobes Versus Sidelobe Level

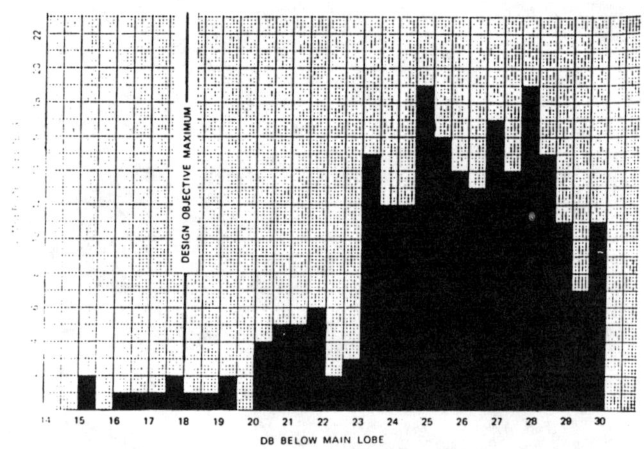

Fig. 8.69 Number of $\Delta\beta$ Sidelobes Versus Sidelobe Level

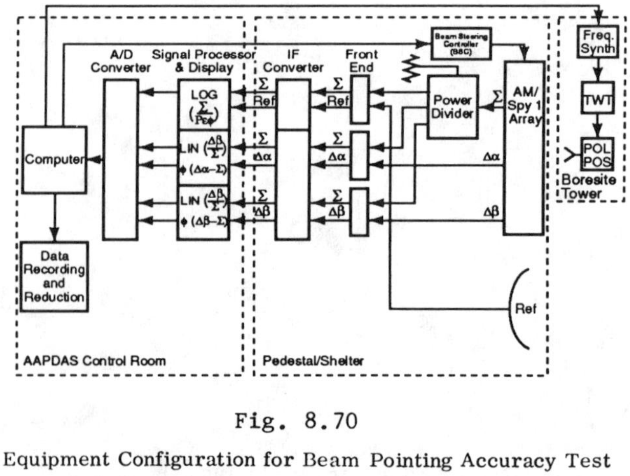

Fig. 8.70

Equipment Configuration for Beam Pointing Accuracy Test

8.7.8 Summary of Pattern Data. The importance of evaluating the antenna pattern characteristics using contour data is evident from the sample figures shown, since most of the major close-in sidelobes would not be observed using only principal plane data. The higher lobes are off the principal coordinate axes.

Measured pattern characteristics of the antenna correlate closely with earlier predictions based on computed patterns and statistical application of array component tolerances. Although the sidelobe performance objectives were exceeded in some instances, the measured results satisfy the operational needs of the AN/SPY-1 radar.

8.7.9 Antenna Monopulse Tracking Characteristics. Three parameters were evaluated in connection with the contribution of the array to AN/SPY-1 radar target tracking errors. These parameters are (a) the measurement of the departure of the actual monopulse null positions in cos α, cos β coordinates to the commanded positions; (b) the total error encompassing the variances of both monopulse null position and normalized error gradients, giving a measure of the error contributed by the antenna in providing angle data on targets up to 12 millisines off the monopulse null; and (c) the magnitude of the normalized error slopes in the null region.

A. Array Alignment. Angular data was related to three fixed optical reference points located on the array face, which define the reference plane of the antenna. The alignment process basically used an optical square on the boresight telescope to position that telescope to be perpendicular to the defined physical plane of the array face, then the optical square was removed and pedestal positioned such that the telescope was optically aligned to the boresight tower. The array test pedestal angle encoders were then set to zero.

With the measured beam position thus related to physical test points on the array face, those same points are utilized for all future emplacements of the array in system use.

B. Monopulse Test Configurations. Fig. 8.70 shows the rearrangement of the three coherent receivers used in performing this test, where the error channel signals, both amplitude and phase, are directly related to the array sum (Σ) channel.

The data obtained by this test are $\Delta\alpha/\Sigma$ and $\Delta\beta/\Sigma$ pattern cuts measured over the normalized angular region ±16 millisines from the initially commanded test beam positions.

The computer program was designed to assure that data were measured along principal planes located within ±1.0 millisine from the actual $\Delta\alpha$, $\Delta\beta$ nulls. This was achieved by using the pair of preliminary scan cuts about the initial position to obtain data to estimate the actual null position. This estimated null position was then used as the center point of a second pair of scan cuts for which data was recorded on magnetic tape.

In each data scan, receive data representing the Σ, $\Delta\alpha/\Sigma$ and β/Σ amplitudes and relative phase between the Σ and $\Delta\alpha$ and Σ and $\Delta\beta$ channels was recorded at 1.0-millisine intervals.

The data was reduced by a program which collects the raw data obtained during the actual tests, applies calibration correction factors, and then reduces the data to determine error slope and beam pointing characteristics.

The error slope parameter is defined as the linear coefficient of a cubic function which was fit to the normalized error signal versus normalized angle from the commanded beam position, where the angular data is restricted to the region ±12 mils from the commanded beam position. Beam pointing error is defined as the zero intercept of this fitted function. The normalized error signal is calculated according to the relationship

$$R = \frac{(\Delta/\Sigma)\sin\phi}{1 + (\Delta/\Sigma)\cos\phi}$$

Where Δ is the amplitude of the respective error signal, Σ is the amplitude of the reference signal, and ϕ is the relative phase between the error and reference channel. This functional relationship was used because it represents the function employed by the AN/SPY-1 radar.

Total error contributed by the array was determined by calculating apparent target directions from the measured data and comparing these to the known target direction. This analysis is based on system use of an odd cubic function for the target offset angle, which is of the form $\Delta\alpha = f_{c/f\sin\alpha}(\gamma R_\alpha + \zeta R_\alpha^3)$ where f is the frequency and γ and ζ are calculated to form a least-mean-square fit to the measured data. In this sense, total error is both a measure of the antenna performance and the adequacy of the tracking algorithm.

Test data was collected at 100 beam positions distributed over the defined array coverage volume as shown in Fig. 8.71. At each angular point indicated, patterns were run at operating band center and end points.

C. Monopulse Test Results. Figs. 8.72 and 8.73 illustrate one form of the data reduction program output, showing composite plots of the normalized error voltage in relation to commanded beam angular positions. The clusters of points represent the extent of data variation over all beam positions and operating frequencies. The measured data essentially falls within a normal distribution with typical standard deviations of 10%. For reference, the error slopes corresponding to each test frequency are superimposed on the data clusters.

8.7.10 Polarization Characteristics of the AN/SPY-1 Array. The AN/SPY-1 Array consists of an array of vertically polarized horn elements having a radiating aperture of 1.82 by .658 inches with the narrow dimension in the vertical plane. It is of interest to know what the polarization characteristics of the array are over the specified array coverage sector. Is the polarization similar to that of an electric dipole, where the electric field lines follow the longitudinals shown in Fig. 8.74a, do the electric field lines follow the small circle of the constant alpha angle, shown in Fig. 8.74b as the electric field lines do for a magnetic dipole or slot radiator, or is it somewhere between the two expected extremes?

A. Polarization Test Configuration. The polarization characteristics were measured as part of the coverage test by repeating coverage measurements with the source antenna

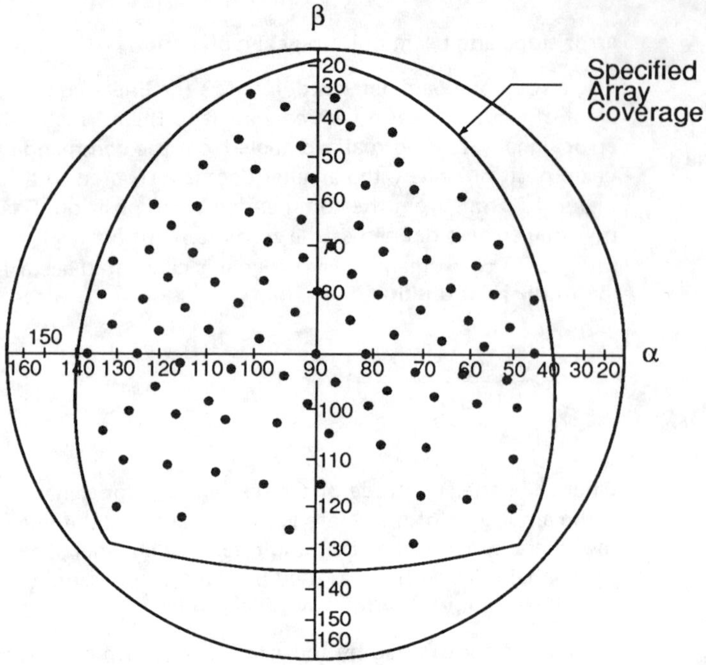

Fig. 8.71 Beam Pointing Accuracy Test Beam Directions

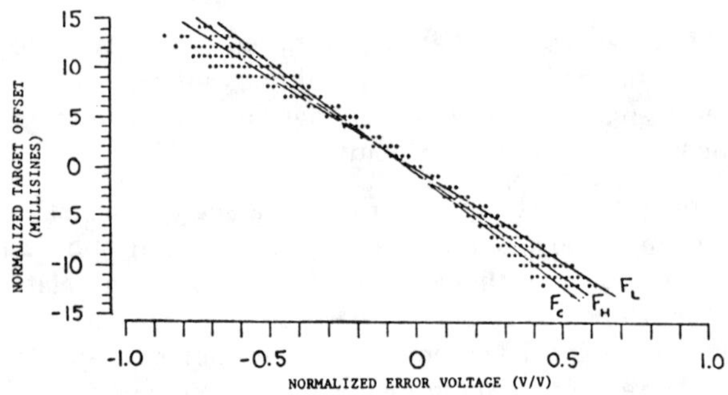

Fig. 8.72 $\frac{\Delta\alpha}{\Sigma}$ Data Cluster and RMS Error Slopes

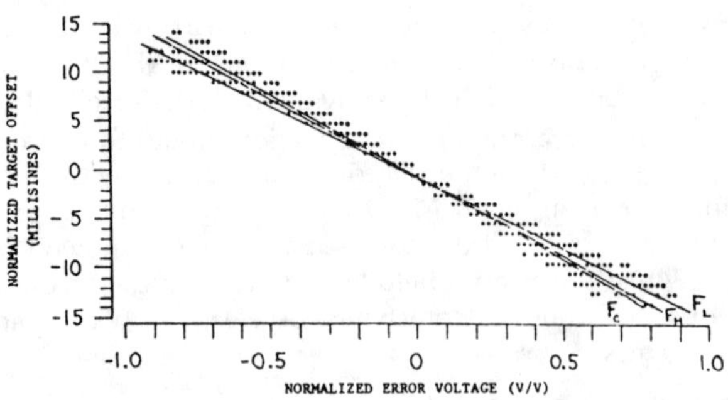

Fig. 8.73 $\frac{\Delta\beta}{\Sigma}$ Data Cluster and RMS Error Slopes

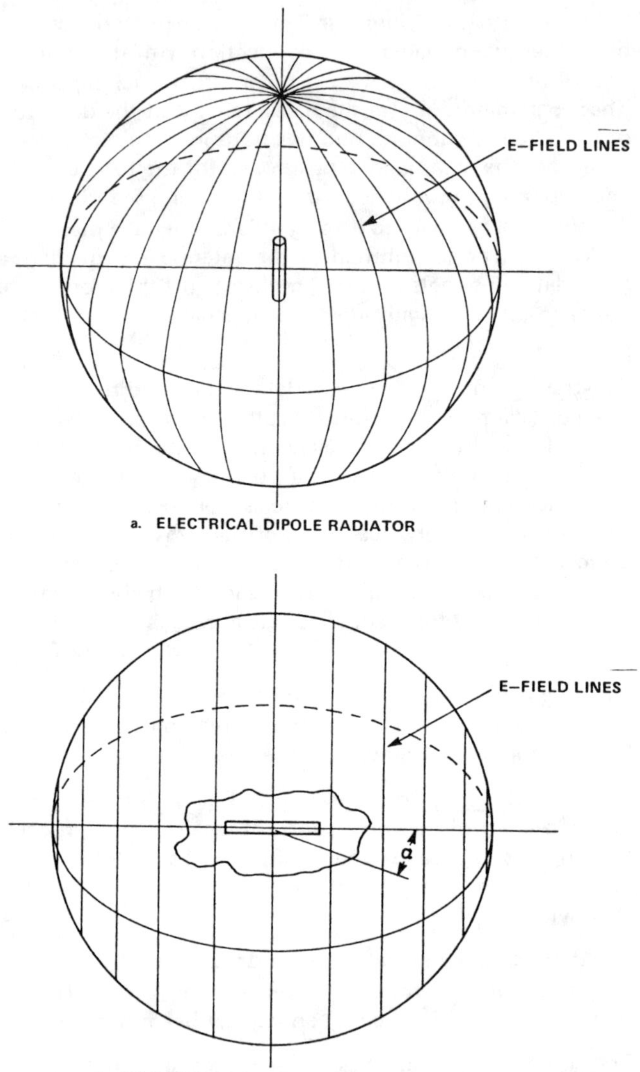

Fig. 8.74 Polarization Lines

linear polarization orientation set to the following values: 0° (vertical), -45°, +45°, +67.5° and +90°. Positive angles indicate clockwise rotation of the E-vector as observed from the array looking toward the boresight tower. An example of the coverage plot measured with the source polarization set at +67.5° from vertical is shown in Fig. 8.75. Note, from the plot, that the relative amplitudes peak up at the diagonal scan angles beam positions in the vicinity of α = 58, β = 58 and α = 122 and β = 122 while in the vicinity of α = 58, β = 122 and α = 122, β = 58 there are deep nulls. This illustrates the fact that the polarization remains very linear with just the angle of polarization deviating from the vertical longitudes.

The measured polarization characteristics were analyzed at a number of beam scan positions as follows: A vector diagram was plotted of the measured gain at their respective polarization angles as shown in Fig. 8.76. The maximum amplitude and orientation of the major polarization axis is obtained from the plot. The angle of the major axis is the polarization angle with respect to vertical for that particular beam scan angle. Since the measurements were made with the array mounted on a traverse over elevation pedestal, the orientation of the vertically polarized incident field from the source antenna would follow the constant longitudinal lines in Fig. 8.74a. Thus, the measured polarization angle indicates the array polarization orientation with respect to the lines of constant longitude.

B. Polarization Test Results. The measured polarization angles for the listed beam positions are shown in Table IV. Also listed are the angles of the constant alpha circles with respect to the intersecting longitude for the respective beam positions. These results show that the measured array polarization follows very closely the lines of constant alpha. Therefore the array exhibits polarization characteristics very similar to a magnetic dipole. Fig. 8.77 shows a plot of the polarization orientation over the array scan coverage diagram.

Beam Position α	β	Measured Pol. Angle	Angle between Longitudes and Constant α Circles
50°	49°	44°	45°
	59	32	30
	69	19	19
	79	9	9
	101	9	9
	111	19	19
	121	30	30
	131	40	45
60	49	32	30
	59	20	20
	69	13	13
	79	6	6
	101	8	6
	111	14	13
	121	20	20
	131	30	30
70	49	18	19
	59	13	13
	69	8	8
	79	4	4
	101	5	4
	111	10	8
	121	14	13
	131	23	19

[1] Campbell, J.J. and Popovich, B.V., "A Broad Band, Wide-Angle Scan Matching Technique for Large Envoronmentally Restricted Phased Arrays", *IEEE PGAP,* July 1972.

[2] Pshunder, Dr. R., *Structural and Dynamic Analysis of the MFAR Array,* RCA Report MTMR-71-TR-020, 10 May 1971.

[3] Ince, W.J. and Temme, D.H., *Phasers and Time Delay Elements,* Project Report RDT-14, Contract AF19(628)-5167, MIT Lincoln Laboratory, (July 11, 1967).

[4] Ince, W.J., Temme, D.H., Willwerth, F.G., and Huntt, R.L., "T "The Use of Manganese-Doped Iron Garnets and High Dielectric Constant Loading for Microwave Latching Ferrite Phasers", *GMTT 1970 International Microwave Symposium Digest,* pp 327-331.

[5] Ince, W.J. and Stern, E., "Nonreciprocal Remanence Phase Shifters in Rectangular Waveguide", *IEEE Transactions on Microwave Theory and Techniques,* Vol. MTT-15, No. 2 (Feb. 1967) pp 87-95.

[6] Whicker, L.R. and Jones. R.R., "Design Guide to Latching Phase Shifters", *Microwaves* (Nov. 1966) pp 31-39 and (Dec. 1966) pp 43-47.

[7] Landry, N.R., "Microwave Phase Shifters", *RCA Engineer,* Reprint RE-15-3-5, (Oct./Nov. 1969) pp 20-25.

[8] Cobine, J.D., *Gaseous Conductors, Theory and Engineering Applications,* dover Publications, Inc. New York, N.Y., (1958) pp 166.

[9] Wheeler, G.J., "Broadband Waveguide to Coax Transitions", *IRE National Convention Record,* Part 1, (1957) pp 182-185.

[10] Hackborn, R.A., "An Automatic Network Analyzer System", *Microwave Journal,* (May 1968) pp 45-52.

[11] Goodrich, H.C. and Tomsic, R.C., "Flux Monitoring Boosts Accuracy of Phased Array Radar Systems", *Electronics* (Nov. 23, 1970), pp 77-80.

[12] U.S. Patent, "Elimination of Mode Spikes in Microwave Ferrite Phase Shifters", #3,555,460, N. Landry.

[13] U.S. Patent, "Linear Flux Control Circuit", #3,510,675, H.C. Goodrich et al.

[14] U.S. Patent Application, RCA Docket Number 62089, "A Flexible and Slidable Waveguide Feed System For a Radiating Horn Antenna", N. Landry et al.

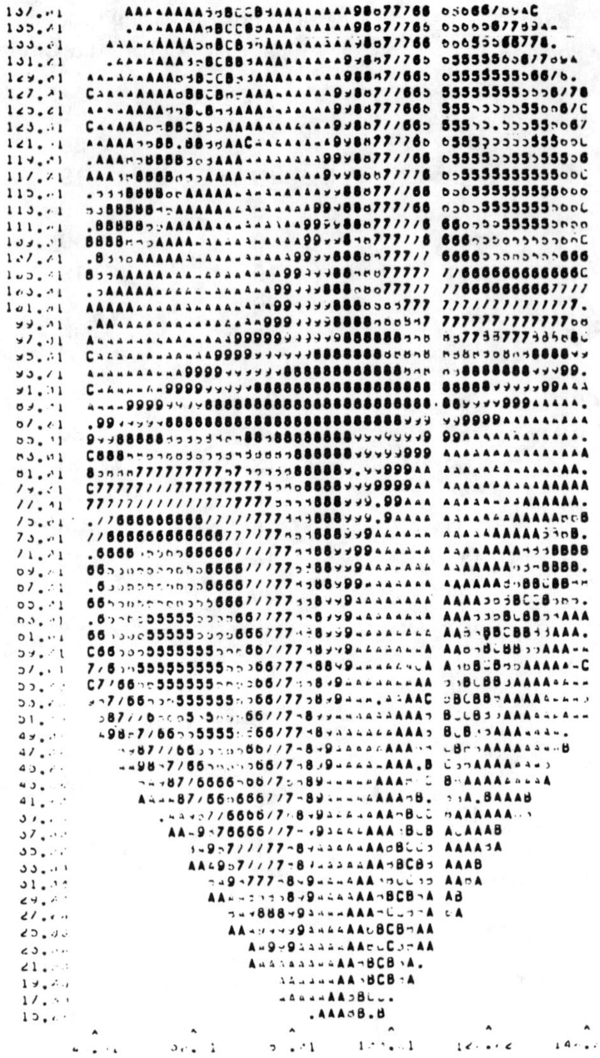

Fig. 8.75 Array Coverage with 67.5° Polarization

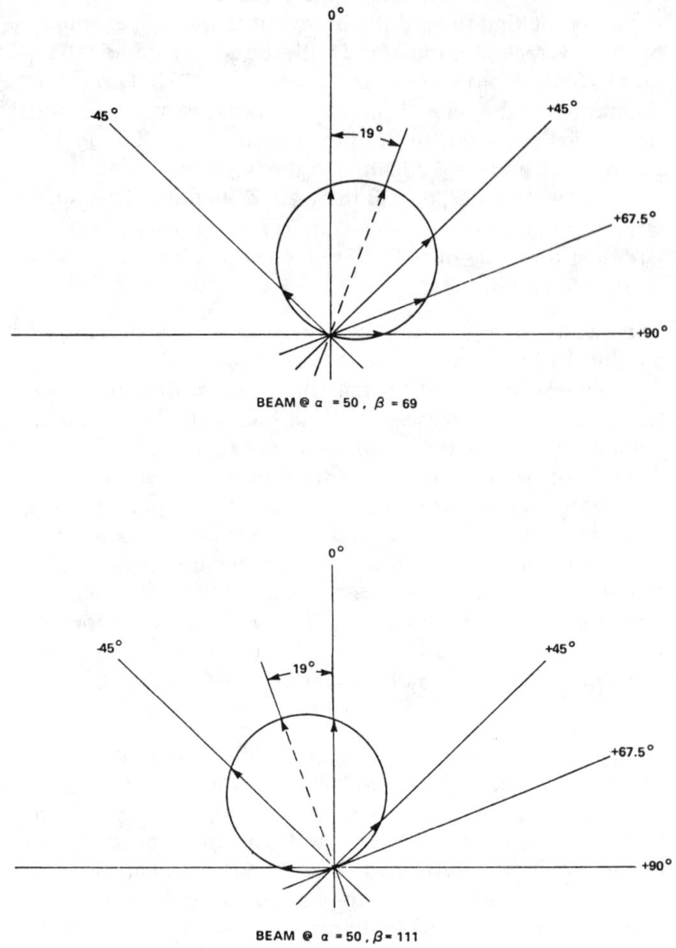

Fig. 8.76 Polarization Diagrams

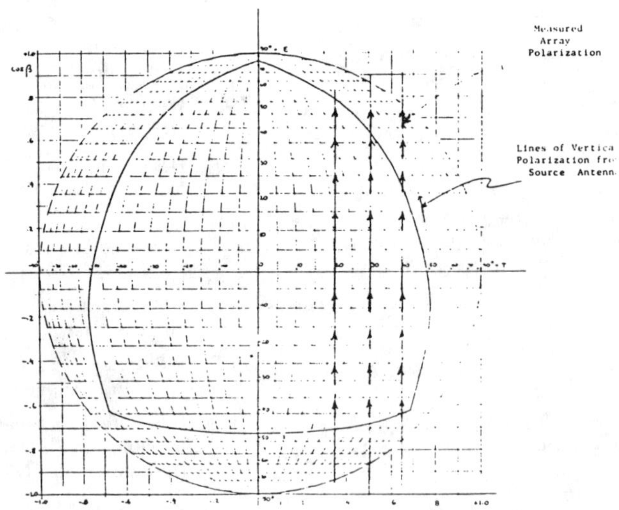

Fig. 8.77 Array Polarization

Limited Scan Arrays - Part 1

Robert J. Mailloux

Air Force Cambridge Research Laboratories, Bedford, MA.

9.1 Basic Requirements of Limited Scan Systems.

Limited scan antenna systems constitute an important area for phased array technological growth. Applications include fire control systems, weapon locators, air traffic control systems and synchronous satellite communications antennas. Each application satisfies the basic criteria for requiring a phased array: there is simply no purely mechanical way of doing the job.

The major requirement for each application is the need for gain in excess of 40dB coupled with rapid electronic scanning. A case in point; the precision approach radar (PAR) antenna of a ground control approach system. Such antennas require beamwidths of the order of a degree or less, and with 0.9λ spacing a conventional array would require approximately 3500 elements. At present day costs of $90.00 to $100.00 per phased element the antenna would cost more than $300K. Conventional array costs are excessive for this application because that technology does not take advantage of the limited scanning requirement. Array technology has not, and indeed could not evolve gradually into a form that satisfied the needs and economics of limited scan systems. It could not because regular periodic arrays have grating lobes for element spacings in excess of a wavelength, and this limits the minimum number of phase controls. To fill this gap a growing class of specialized antennas is being developed with reduced numbers of phase shifters that are more appropriate for the restricted sector coverage. Chief among these are reflector and lens type antennas which take advantage of the magnification of optical systems to illuminate large apertures with small arrays. In addition, there is a growing number of special purpose array techniques designed primarily to take advantage of the limited sector coverage by using large array elements and thus fewer phase shifters and simpler power division. These arrays use aperiodic grids, overlapped subarrays or multi-mode control of array element radiation patterns to prevent the formation of large grating lobes.

In addition to high gain, all of the above applications require (but rarely achieve) sidelobe levels of -20 to -25dB. The radar applications require sufficient monopulse null depth, and both radar and communications applications impose some minimum specification on polarization and bandwidth.

These required performance characteristics are easily satisfied by conventional arrays, but at such great cost as to make them non-competitive. The specific requirements have however caused the limited scan antenna to exist; and they will be the impetus for its continued technological growth. Antenna characteristics are indicated throughout the text as they pertain to various of these applications. In addition, there are several generalized statements that one can make about limited scan antenna specifications.

9.1.1 Instantaneous Bandwidth.

High gain array antennas usually operate over very narrow frequency ranges unless they are steered by time delay control devices. Some relaxation of these narrow-band characteristics can be obtained by limiting the scan sector, as in a limited scan array, but since these antennas usually have very high gain the instantaneous bandwidth is often still severely restrictive.

The dependence of bandwidth on scan angle is indicated by the following approximate analysis of array squint.

Consider an array of length D in one plane, scanning out to a maximum angle $\sin\theta$. For element spacing d, the inter-element phase shift ϕ is determined at the center frequency f_1 by:

$$\phi = 2\pi \frac{d}{\lambda_1} \sin\theta \tag{9.1}$$

The half power beamwidth of the array is approximately λ/D, and so at some higher frequency f_2 one can define the band edge by the condition that the peak of the beam ($\sin\theta_2$) be one half of a half power beamwidth away from the beam center at f_1.

$$\sin\theta_2 = \sin\theta_1 - 0.5\frac{\lambda_2}{D} \tag{9.2}$$

These beam positions are produced by the same phase progression ϕ, and so one can equate $(\sin\theta)/\lambda$ for the two cases to obtain

$$\frac{\sin\theta_1}{\lambda_1} = \frac{1}{\lambda_2}\left(\sin\theta_1 - \frac{\lambda_2}{2D}\right)$$

or

$$\sin\theta_1\left(\frac{\lambda_1 - \lambda_2}{\lambda_1}\right) = \frac{\lambda_2}{2D} \tag{9.3}$$

since the above defines the half bandwidth of the array, the full bandwidth can be written approximately as:

$$\frac{\Delta f}{f_1} = \frac{1}{(D/\lambda_1)\sin\theta_1} \qquad (9.4)$$

This relationship shows that restricting the scan angle ($\sin\theta_1$) directly increases bandwidth. For example, an array scanning ±10°, with one degree beamwidth (and assuming D ~ 60λ) could therefore have an instantaneous bandwidth of nearly 10%, or about five times the bandwidth of an array scanning a wide angular sector.

This definition of bandwidth obviously neglects other primary limitations. Factors like the grating lobe locations, the element or subarray patterns, the optical path lengths and the scanning circuitry all combine to place additional bounds upon the true array bandwidth, but the expression above remains the upper bound unless time delay control devices are incorporated.

9.1.2 Polarization Characteristics and Weather. Precision approach radar (PAR) antennas are often required to have low integrated cancellation ratios (ICR) in order for the system to effectively cancel the rain clutter. The ICR is defined as the ratio between the backscattered power accepted by a linearly polarized antenna from a uniform angular distribution of scatterers and the power accepted by the antenna under consideration. A perfectly circularly polarized antenna would have an infinite ICR because of the polarization reversal of the signals reflected from spherical scatterers. At present, well-designed PAR antennas have ICR's of about 20dB, and thus are expected to cancel the backscattered contribution of rain to this degree while allowing the reversesence scattered signal from the aircraft to be received unattenuated.

Raindrops however, are not spherical in shape, and so the attenuation constant and the propagation velocity are different for vertical and horizontal components of the polarized wave. In addition, the backscattered field from each drop has a cross-polarized component. In some cases the net effect of these three phenomena is to severely limit the clutter rejection capabilities of the antenna system. Recent data [1, 2] shows that these atmospheric factors play an important role for frequencies above S-band. Calculations performed by Crane [2] show circular depolarization ratio (CDR) data at K-band, X-band and S-band for two way propagation at 50 mm of rain per hour. The CDR is the ratio of cross polarization to transmitted circular polarization along a specified path, and so this ratio should be larger than the antenna ICR if that antenna specification is to be meaningful in all weather. Crane's data shows that at 2 Km range, the CDR is -29dB at S-band, -18.6dB at X-band and -14.1 dB at K-band, while at 9.25 Km these numbers are reduced to -16dB, -4.5dB and -0.8dB respectively for the conditions stated above. These data clearly show the futility of using polarization cancellation for such heavy rain rates except when the propagation path is on the order of a few kilometers, and they help to illustrate that specifying antenna ICR does not necessarily lead to a system that performs satisfactorily in all weather.

9.2 Fundamental Bounds and Concepts for Ideal Limited Scan Systems.

Although the types of limited scan systems vary widely, there are common features among them that allow for generalized comparisons. A set of phase controls apply signals at the antenna array terminals, and these signals ultimately result in a phase tilt across an antenna aperture. The final aperture may be excited directly by the steered signals, or it may be the aperture plane of a reflector or lens whose phase tilt is provided by a small array near the focal region. In either case the maximum scan angle and the size of the final aperture dictate the minimum number of active controls required for steering. Although this section deals mainly with conclusions that can be drawn from a discrete aperture point of view, one of its more important conclusions is the statement that; when an array is combined with a reflector or lens, the array scan angle should be optimized not only so that it excites the final aperture efficiently, but in addition so that it is made to scan to its own maximum scan limit consistant with the number of elements and interelement spacing.

In this section, several estimates are given that describe the minimum number of control elements to scan an arbitrary beamwidth antenna over a fixed scan sector. The fundamentals used in determining these bounds are also applied to provide design information that can simplify or improve the design of such optical devices as scanning lenses or reflectors.

9.2.1 Estimates of the Minimum Number of Control Elements. Given the maximum required scan angle and the aperture size or beamwidth, there exists a minimum number of control elements necessary to scan a beam over the coverage sector. [3]

One measure of this minimum number is shown later to be the number of orthogonal beams within the scan sector. Another is the theorem of J. Stangel [4] which states that the minimum number of elements is:

$$N = \frac{1}{4\pi} \iint G_0(\theta, \phi)\, d\Omega \qquad (9.5)$$

where $G_0(\theta, \phi)$ is the maximum gain achievable by the antenna in the (θ, ϕ) direction, and $d\Omega$ is the increment of solid angle. Figure 9.1 shows how this number varies for an array scanning a conical sector. This curve is based upon an array scanning according to the gain envelope $G_0(\theta, \phi) = G_0 \cos\theta$, and so in this general form includes beam broadening as a function of scan. The equation assumes uniform illumination for a planar array, and so does not account for beam broadening due to aperture taper.

Another measure of the minimum number of array elements is contained in the definition of a parameter introduced by Patton [5], and called the "element use factor". This parameter will be used throughout this paper in a somewhat generalized form to compare the number of phase shifters in competing systems with unequal principal plane beamwidths. The factor is N/N_{min} where N is the actual number of phase shifters in the control array, and N_{min} is a reasonable number of control elements as defined below

$$N_{min} = \left(\frac{\sin\theta^1_{max}}{\sin\frac{\theta^1_3}{2}}\right)\left(\frac{\sin\theta^2_{max}}{\sin\frac{\theta^2_3}{2}}\right) \qquad (9.6)$$

θ^1_{max} and θ^2_{max} are the maximum scan angles in the two orthogonal principal planes measured to the peak of each beam, and θ^1_3 and θ^2_3 are the half power beamwidths in these planes. Thus N - min is approximately four times the product of the number of beamwidths scanned in each principal plane, and as will be shown later, is also approximately equal to the number of orthogonal beam positions for a rectangular array with beams filling a rectangular sector in direction cosine space. The formula will be used quite loosely throughout this paper to refer to circular apertures scanning conical and elliptical sectors. This tends to give an advantage for antennas of this second type, in which the minimum number of elements is smaller than that for scanning a rectangular sector. The criterion is used in this case also, and it is used independent of array taper and sidelobe ratio because there is no truly fair way of comparing all of these cases, and it has the advantage of simplicity. Although more general, Stangel's formula reduces to Patton's in the limiting case of small scan angle and using the approximate formula

$$G_0 = \frac{4\pi}{\theta^1_3 \, \theta^2_3} \tag{9.7}$$

and integrating over a rectangular sector. This comparison further indicates that the minimum number of elements for a conical or otherwise shaped sector is given approximately by the ratio of that sector area to the area of the rectangular sector of equation (9.6). A conical sector for example, should require $\pi/4$ times as many controls as a square sector. Nevertheless, equation (9.6) will be used throughout this paper as a standard for comparison of all antennas.

The existance of a minimum number of control elements is fundamental to the nature of the scanned aperture. This, and other basic relationships can be understood from an intuitive point-of-view after a thorough understanding is gained of the array radiating properties and the concept of orthogonal beams.

The radiated field of the array of aperture elements shown In Figure 9.2 is given in direction –cosine space for a beam at (u_0, v_0) as:

$$E(u, v) = \frac{e(u, v)}{N} \sum_{m=1}^{Nx} \sum_{n=1}^{Ny} A_{mn} \, e^{j\frac{2\pi}{\lambda}(umd_x + \Delta_n u + vnd_y)}$$

where $u = \sin\theta \cos\phi$

$v = \sin\theta \sin\phi$

and

$$A_{mn} = I_{mn} \, e^{-j\frac{2\pi}{\lambda}(u_0 md_x + u_0 \Delta_n + v_0 nd_y)} \tag{9.8}$$

Here A_{mn} is the signal at the element (m, n) located at the position $x = md_x + \Delta_n, y = nd_y$ as required to form the beam at (u_0, v_0). I_{mn} is the signal amplitude and e(u, v) is the element pattern of each aperture in the array. $N = N_x N_y$ is the number of elements in the array.

The resulting array field pattern is:

$$E(u, v) = \frac{1}{N} \sum_{m=1}^{N_x} \sum_{n=1}^{N_y} I_{mn} \, e(u,v) \, e^{j\frac{2\pi}{\lambda}\left[(u - u_0)(md_x + \Delta_n) + (v - v_0) nd_y\right]} \tag{9.9}$$

Assuming that the array is excited by a separable distribution so that $I_{mn} = I^H_m I^E_n$, and assuming that the element pattern is also separable, then $e(u,v) = e_x(u) e_y(v)$ and this field distribution can be written in the following form:

$$E(u,v) = \frac{1}{N_x} \left\{ \sum_{m=1}^{N_x} I^H_m \, e_x(u) e^{j\frac{2\pi}{\lambda}(u - u_0) md_x} \right\} \cdot \frac{1}{N_y} \left\{ \sum_{n=1}^{N_y} I^E_n \, e_y(v) \, e^{j\frac{2\pi}{\lambda}\left[(v - v_0) nd_y + \Delta_n (u - u_0)\right]} \right\}$$

$$= E_x(u) \, E_y(u, v) \tag{9.10}$$

This pattern is not separable in general, but may be so for certain choices of $[\Delta_n]$. The choice of optimum $[\Delta_n]$ for grating lobe reduction will be considered later. For the present we shall set the $\Delta_n = 0$ for all n, in order to introduce the concept of grating lobes.

Under these conditions the field pattern is separable and is written:

$$E(u, v) = \frac{e_x(u)}{N_x} \sum_{m=1}^{N_x} I^H_m \, e_x(u) \, e^{j\frac{2\pi}{\lambda}(u - u_0) md_x} \cdot \frac{e_y(v)}{N_y} \sum_{n=1}^{N_y} I^e_n \, e^{j\frac{2\pi}{\lambda}(v - v_0) nd_y}$$

$$= E_x(u) \, E_y(o, v) \tag{9.11}$$

The expontial forms in the above summations have principal maxima at the grating lobe locations:

$$u_p = u_0 + \frac{p\lambda}{d_x}$$

$$v_q = v_0 + \frac{q\lambda}{d_y} \tag{9.12}$$

for all m, n bounded by the inequality

$$K_{pq} = 2\pi\sqrt{1 - u_p^2 - v_q^2} \leqslant 2\pi$$

These points are shown in (u, v) space as a regularly spaced grating lobe lattice about the main beam location (u_0, v_0) in Figure 9.3. The circle with unity radius represents the bounds of the above inequality; all grating lobes within the circle represent those radiating into real space, and those outside do not radiate.

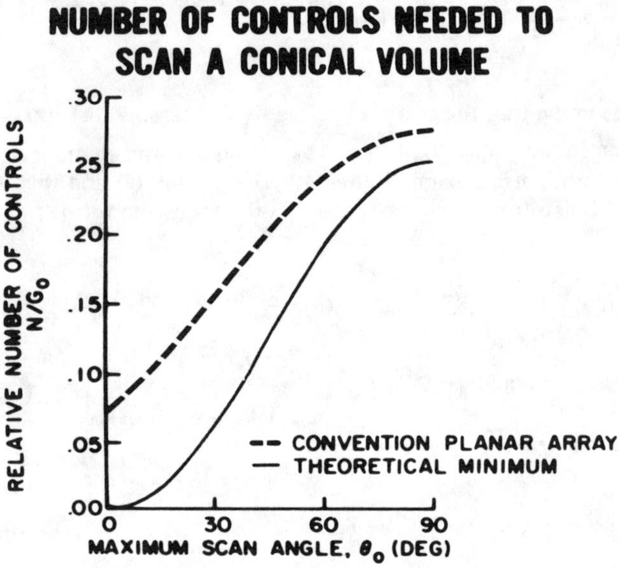

Figure 9.1 - Minimum Number of Control Elements for a Conical Sector (Courtesy of J. Stangel, Sperry Rand Corp.)

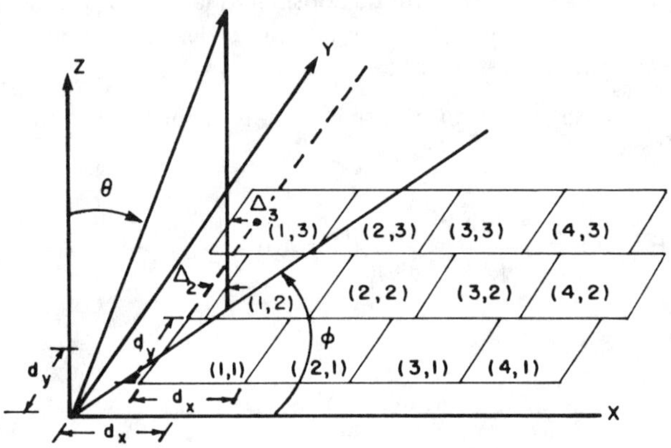

Figure 9.2 - Array Geometry, Element (m,n) center locations $x = md_x + \Delta_{ng}$ $y = nd_y$

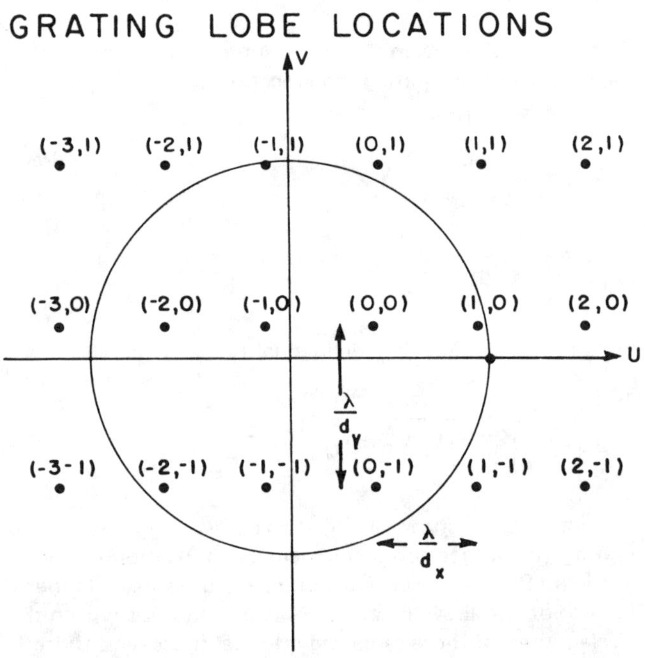

Figure 9.3 - Grating Lobe Lattice

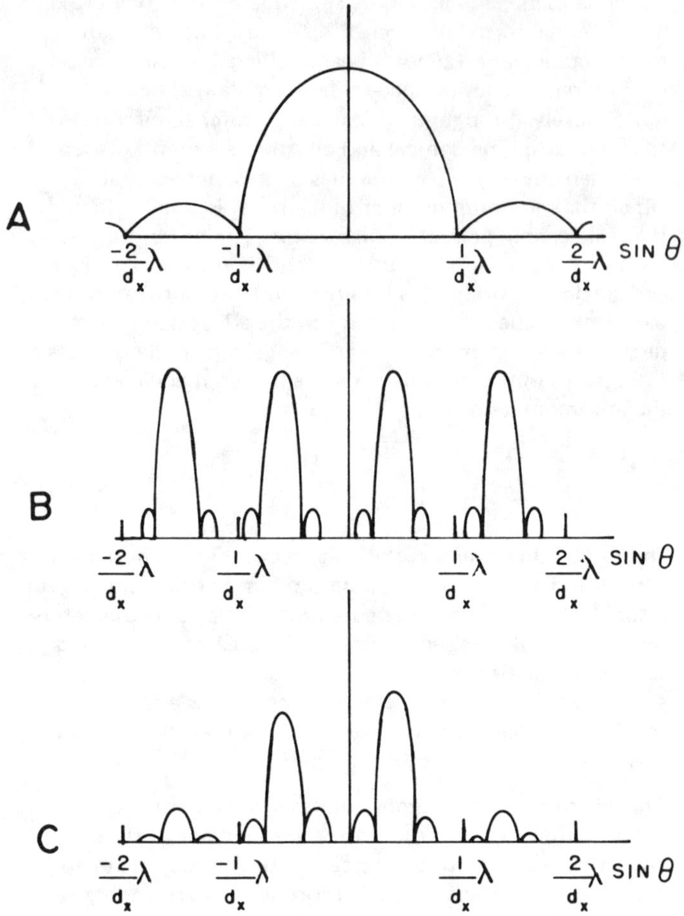

Figure 9.4 - Array Radiation Characteristics
a. Element Pattern
b. Array Factor
c. Resulting Radiation Pattern

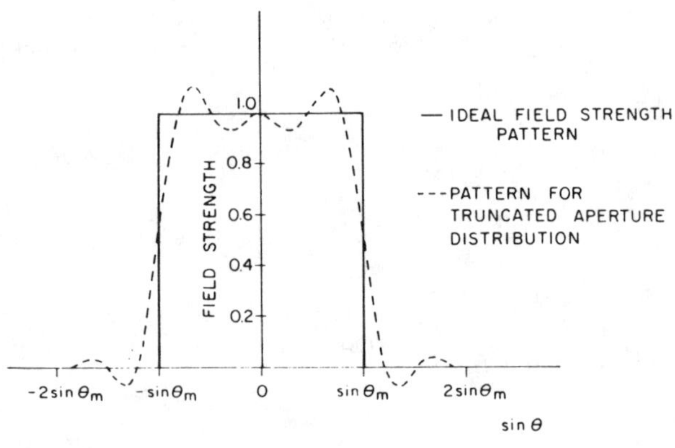

Figure 9.5 - Overlapped Subarray Distributions
a. Ideal Element Pattern

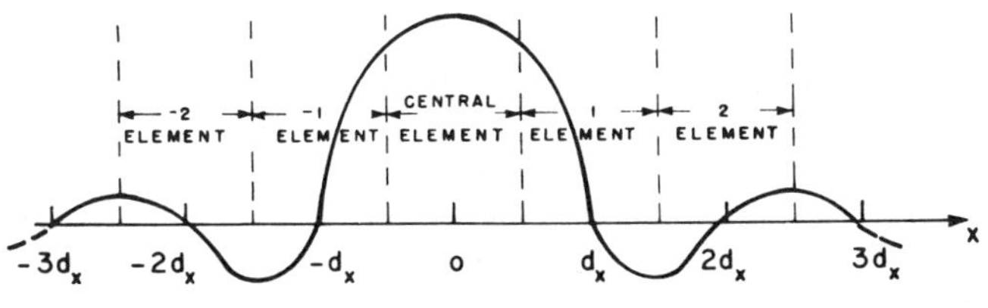

Figure 9.5 - Overlapped Subarray Distributions
b. Subarray Field Distribution for Scan to $(dx/\lambda)\ \sin\Theta_m = 0.5$

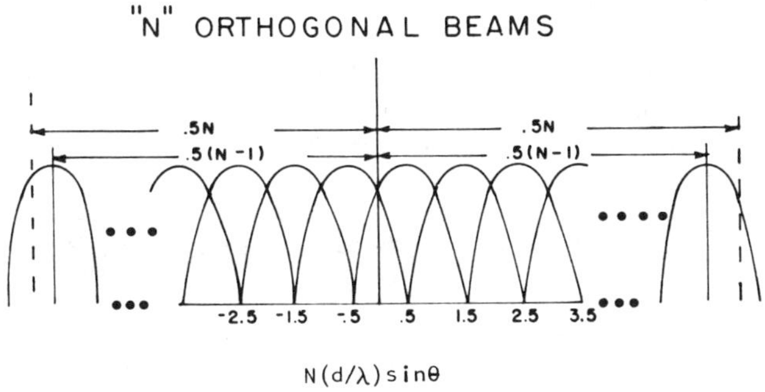

Figure 9.6 - Orthogonal Beam Positions for Array of N-Elements (N-even)

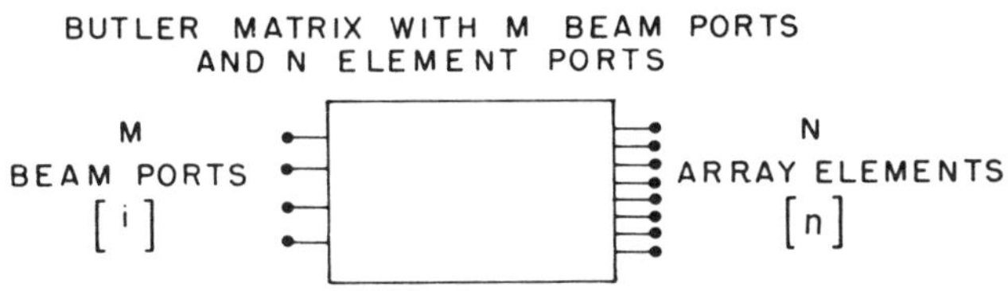

Figure 9.7 - Notation for Orthogonal Beam Matrix

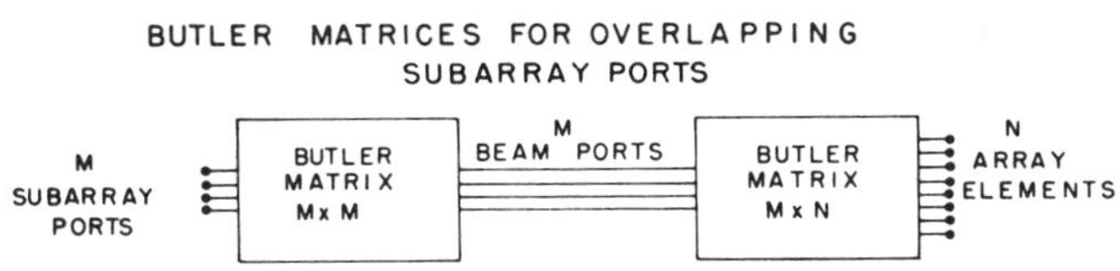

Figure 9.8 - Butler Matrices for Overlapping Subarray Ports

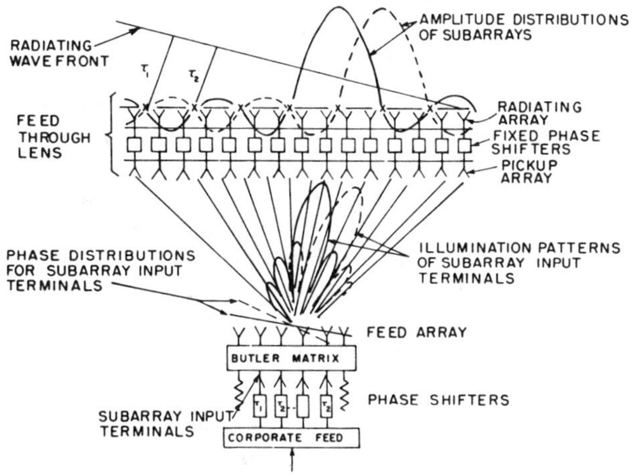

Figure 9.9 - Completely Overlapped Subarray Antenna

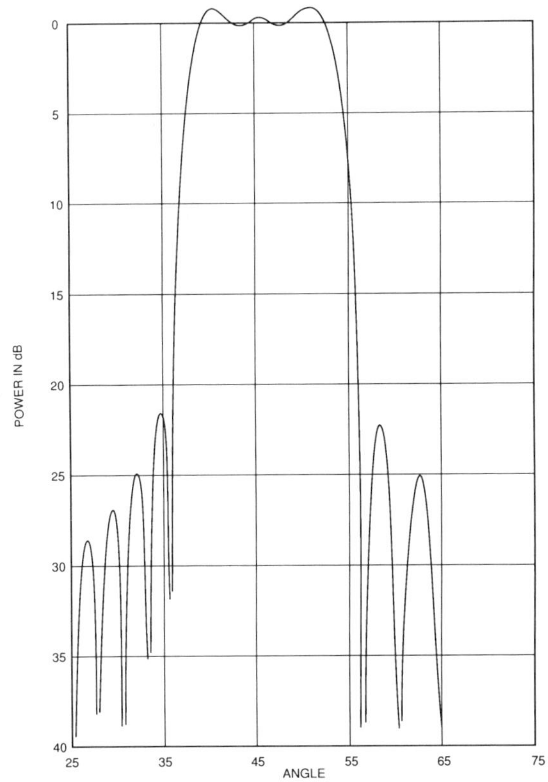

Figure 9.10 - Subarray Pattern of Completely Overlapped Subarray Antenna

Source: Skolnik, *Radar Handbook*, 1st Ed., 1970, McGraw-Hill Book Company. Reprinted with permission.

Consider an array scanning in one plane shown in Figure 9.4. Assuming that the interelement spacing in that plane is d_x, and that the array is scanned to some angle θ_{max}. The array factor, shown in Fibure 9.4b, shows the periodic repetitions of the main beam and sidelobe structure at each grating lobe position and Figure 9.4c shows the total radiated pattern as the product of the array factor and the element pattern $e_x(u)$ and emphasizes the reduction of the further grating lobes to the element pattern sidelobe levels. The first grating lobe on the opposite side of broadside from the main beam is not bound by the element pattern sidelobe level, and grows very large as the array is scanned.

The particular element pattern shown in Figure 9.4a has zeros at $\sin\theta = \pm n/d_x$ for integer n excluding zero. This pattern occurs for uniformally illuminated apertures and is characteristic of the E-plane of an element in a waveguide array. The advantage of such a distribution is that it suppresses all array grating lobes for broadside radiation. Typical H-plane element patterns for slots or horns are broader, and so have non-zero grating lobes at broadside. This subject will be discussed further in connection with techniques for achieving limited scan designs with multiple mode excitation.

9.2.2 Ideal Subarray Pattern: Concept of Overlap. The previous section describes the effect of the element pattern in reducing certain grating lobes, but element patterns of this type can offer little suppression of the grating lobe nearest broadside. An element pattern like that of Figure 9.5 would offer nearly ideal performance for a limited scan system by allowing the beam to scan out to some maximum scan angle ($\sin\theta_m$) while suppressing all grating lobes as long as they did not occur within the range $-\sin\theta_m \leqslant \sin\theta \leqslant \sin\theta_m$. Since the grating lobes occur at positions given by equation (9.12) then for a very large array one can optimize the inter-element spacing for a given maximum scan angle by choosing

$$(dx/\lambda)\sin\theta_m = 0.5 \qquad (9.13)$$

where d_x is the inter-subarray spacing. In this case a large array with main beam at $\sin\theta = (0.5\lambda/d_x) - \sigma$ for some arbitrarily small value σ will have its nearest grating lobe at $\sin\theta = -(0.5\lambda/d_x) - \sigma$, and all grating lobes will be completely suppressed. Such an array is characterized as a limited scan design because it can take advantage of limitations imposed upon the scan sector in order to increase aperture size d_x/λ. In principle an array with $\sin\theta_m = 0.1$ can use a 5λ inter-element spacing while for $\sin\theta_m = 0.05$, a 10λ spacing can be used. These size increases and associated reductions in the number of required phase controls for restricted coverage illustrate the goal of limited scan antenna designs. Later this scan condition will be shown as equivalent to the element use factor for specifying the minimum number of array controls. The aperture field corresponding to this far-field distribution is of the form:

$$i(x) = \frac{\sin\left(\frac{2\pi}{\lambda}x\sin\theta_m\right)}{\left(\frac{2\pi}{\lambda}x\sin\theta_m\right)} \qquad (9.14)$$

where x is the distance measured from the center of the subarray. If the maximum ideal spacing $d_x/\lambda = (0.5/\sin\theta_m)$ is used, then this aperture distribution has zeros at $x = \pm n d_x$ excluding $n = 0$, as shown in Figure 9.5, and one must include a number of elements in order to reproduce the i(x) distribution faithfully. Thus, each phase shifter must feed a multiplicity of sub-arrays and the sub-arrays can be said to be overlapped. Obviously, the ideal aperture field can only be approximated; it must be truncated and then approximated by realizable distributions at each element. The dashed curve shown in Figure 9.5a shows the flat-topped sub-array pattern achievable if the i(x) is truncated at $x = \pm 3d_x$. In this case, 20-dB grating lobe supression can be obtained for scan out to the angle

$$(d_x/\lambda)\sin\theta_m = 0.43. \qquad (9.15)$$

This figure also shows the illumination that must be produced at each element if the element size were to be made equal to the inter-subarray distance d_x. Clearly the distribution is quite dissimilar at each element, and so it is not possible to use such large elements to produce the required subarray illumination. In anticipation of some work that will be described in a following section, however, it should be noted that with the exception of the central element, the distribution at each element is quite asymmetric, and so perhaps could be approximated by some combination of the fundamental and a higher order odd mode.

Apart from this possible exception however, the distribution required to produce a well defined flat topped sub-array pattern requires the interconnection of a number of array elements and so is extremely difficult to fabricate in microwave circuitry. Consequently, the circuit approach has received only limited attention.

Alternatively, space feed systems can quite naturally achieve overlapped sub-array distributions [6] that have proven very practical. These systems use feed-through lenses or reflectors that can faithfully reproduce a substantial part of the f(x) distribution as compared with microwave circuit systems. A description of several circuits for achieving overlapped distributions will be given in other sections of this paper.

9.2.3 Spatial Fourier Transforms: Orthogonal Beams and Orthogonal Beam Matrices. Equation (9.8) for E(u, v) is a discrete two dimensional expontial (Fourier) transform of the array aperture distribution, with the space co-ordinates (u, v) as the transform plane. This transform pair relationship between the far field and aperture illumination is well known, and leads to several important concepts for the design of certain limited scan array types. Before investigating these properties, it is convenient first to introduce the concept of orthogonal beams and Butler Matrices.

For uniform illumination, the discrete Fourier transform represented by Equation 9.11 assumes the following form for an N_x -element array with spacing d_x in one dimension.

$$E_x(u) = \frac{e_x(u)}{N_x}\sum_{m=1}^{N_x} e^{\,j\frac{2\pi}{\lambda}(u-u_0)\left(m-\frac{N_x+1}{2}\right)d_x} =$$

$$e_x(u)\,\frac{\sin[N_x\pi/\lambda\,(u-u_0)\,d_x]}{N_x\sin[\pi/\lambda\,(u-u_0)\,d_x]} \qquad (9.16)$$

This form differs from that of equation 9.11 because $(N_x+1)/2$ is subtracted from the element index m in order to bring the zero phase point to the geometrical center of the array. The field pattern corresponding to the uniform amplitude distribution has -13 dB sidelobes and a single principal maximum for $\frac{-0.5\lambda}{d_x} \leqslant u \leqslant \frac{0.5\lambda}{d_x}$. Furthermore, it is possible to choose N_x- such beams within this sector such that the nulls of each beam occur at the maximum of every other beam. The set of beams thus chosen is orthogonal in space [7, 8] and has the maximum directive gain for each of the multiple beams. [7] The gain is maximum because of the high-gain characteristics of the array pattern and because a distant point source located at the peak of any one beam loses no power into any other beam ports. That location of this set of beams (in one dimension) is given by

$$u_i = \sin\theta_i = \frac{1}{N_x d_x}(i - ½) \qquad (9.17)$$

$$\text{for } \frac{-N_x}{2} + 1 \leqslant i \leqslant N_x/2$$

Figure 9.6 shows the location of the N beams in this set. The required input phase at the m'th input port for the i'th beam is:

$$\text{Arg}(Im) = \frac{-2\pi}{\lambda}\left(m - \frac{N_x+1}{2}\right) u_i d_x$$

$$= \frac{\frac{-2\pi}{\lambda}(i - ½)}{N_x}\left(m - \frac{N_x+1}{2}\right) \qquad (9.18)$$

$$\text{for } 1 \leqslant M \leqslant N_x$$

Networks designed to synthesize such distributions are called orthogonal beam matrices, or Butler matrices, and have been described in the literature in substantial detail for feeding linear and planar arrays. [7 — 10]

The operation of such a circuit on transmission is again that of taking a discrete Fourier transform of the aperture field to form the sum of terms that constitutes the required input signal at the i'th port:

$$A_i = \frac{1}{\sqrt{N_x}} \sum_{m=1}^{N_x} I_m\, e^{+j\frac{2\pi}{\lambda}\left(\frac{i-½}{N_x}\right)\left(m - \frac{N_x+1}{2}\right)} \cdot \qquad (9.19)$$

This expression, upon selecting the I_m corresponding to the orthogonal beam with index k gives A_i=1 for i=k and 0 for $i \neq k$. The matrix thus connects one orthogonal beam port with one orthogonal beam in space. For the sake of completeness, the radiating element signals I_m are given by the inverse transform of the multiple beam input port signals A_i.

$$I_m = \frac{1}{\sqrt{N_x}} \sum_{i=\frac{-N_x}{2}+1}^{\frac{N_x}{2}} A_i\, e^{-j\frac{2\pi}{\lambda}\left(\frac{i-½}{N_x}\right)\left(m - \frac{N_x+1}{2}\right)} \qquad (9.20)$$

Equations (9.19) and (9.20) can also be used to relate the signals at the j'th beam port when the I_m represent the aperture illumination in a receive mode, but in that case the exponential form in (9.19) must have a negative sign, and that of (9.20) has a positive sign. In this case the receive distribution $I_m{}^*$ (see Eq. 9.18) results in the received signal $A_i = 1$.

Orthogonal beam matrices with more array elements than beam positions (Fig. 9.7) are produced by terminating some of the unwanted beam positions, and thereby reducing the number of passive components by the quantity used in providing those extra beams. The mathematics of the circuit operation is unchanged (Eqs. 9.19 and 9.20) except that the indices "i" are limited, and the number of passive components is reduced.

9.2.4 Orthogonality and Some Practical Conclusions about Limited Scan Arrays.

Several very useful conclusions can be drawn based upon the concepts of independent orthogonal beams. The first concept will not be proven here but merely stated. Each of the "N" orthogonal beams, and their associated sets of signals at the element ports are independent, and taken collectively the set of N possible distributions (one corresponding to each beam) is complete, in that one can expand any given realizable distribution of signals across the array in a sum of orthogonal beam distributions. Thus the measured properties of an array excited by all of its orthogonal beams contain all of the necessary information about the properties of the associated scanning array.

The minimum number of array controls for a given scan angle can also be deduced from the above principles, as well as from Stangel's theorem or the formula of Patton. The maximum beam position angle for an even number of elements (N_x) is shown in Fig. 9.6 to be at:

$$\sin\theta_{i_{max}} = \frac{0.5\lambda}{d_x}\left[1 - \frac{1}{N_x}\right] \qquad (9.22)$$

There can be no beam with $\sin\theta_i$ larger than this because if there were, it's grating lobe at

$$\sin\theta_i = \frac{0.5\lambda}{d_x}\left[1 - \frac{1}{N_x}\right] - \frac{\lambda}{d_x} = \frac{-0.5\lambda}{d_x} - \frac{\lambda}{N_x d_x} \qquad (9.22)$$

would be of significant size and would contribute to a loss in gain and pattern ambiguity.

Since this is the maximum scan angle for a finite array, a refinement of the "ideal array concept" of Section 9.2.2 would use the above value of $\sin\theta_{i\,max}$ for a finite array instead of the value given in Equation 9.13. The flat topped element pattern should also be narrowed by the same amount.

Furthermore, since a combination of orthogonal beams can be used to synthesize any possible beam within the region $-\theta_{i_{max}} \leqslant \theta_i \leqslant \theta_{i_{max}}$, then a minimum of N_x active controls are required in order to scan an array anywhere within the stated scan sector. Thus, for an aperture of length L in one plane, the number of control elements is given by

Eq. (9.22) as:

$$N_x - 1 = \sin\theta_{i_{max}} \frac{N_x d_x}{0.5\lambda} \tag{9.23}$$

and using $L = N_x d_x$ and $\sin\frac{\theta_3}{2} \sim \frac{0.5\lambda}{N_x d_x}$, one obtains approximately

$$N_x - 1 = \frac{\sin\theta_{i_{max}}}{\sin\frac{\theta_3}{2}} \tag{9.24}$$

For large N_x, this equation reduces essentially to the single plane equivalent of the minimum number of elements as defined by Patton [5].

9.2.5 An Idealized Circuit for Spatial Overlap, and It's Implications for Design.

Shelton [11] has indicated one way in which orthogonal beam matrices can produce spatially overlapped patterns. The problem he addresses is that of producing a multiple beam feed for a reflector or lens final aperture. Since independent feed radiators cannot be placed close enough together to obtain good crossover levels, Shelton argues that by synthesizing overlapped subarrays with flat topped distributions one can obtain multiple feeds for reflector or lenses with good aperture efficiency.

The antenna system proposed by Shelton requires an additional Fourier transform provided by the large reflector or lens, and then it serves the function of connecting every multiple beam port to a beam position in space. Alternatively, by identifying each input feed port on a phased array input port, and exciting each input with a progressive phase, the overlapped subarraying features of the circuit can be used to synthesize flat topped subarray patterns for a scanning array. Figure 9.8 shows the interconnection of two Butler matrices (back to back). The input matrix has M beam ports and in elements, and the outer one has M beam ports and N elements ($N > M$). This is basically the circuit proposed by Shelton. It's operation is simply explained if one assumes that a set of M incident signals corresponding to one of the M input subarray beam positions is applied at the subarray ports. The M x M matrix will Fourier transform this distribution and the entire signal will appear at one of the M beam ports, there to be introduced into the M x N matrix whose output is then a linearly progressing uniform amplitude distribution at the N array element ports. This explanation does not describe the subarraying properties of the circuit however, and for that one must consider the transformations executed by each matrix. Consider a signal S_n applied the n'th subarray port. Eq. (9.19) using the negative sign in the exponent gives the set of resulting signals at the M beam ports of that matrix output as:

$$A_i = \frac{1}{\sqrt{M}} S_n e^{-j\frac{2\pi}{\lambda}\left(\frac{i-½}{M}\right)\left(n-\frac{M+1}{2}\right)} \tag{9.25}$$

for $\frac{-M}{2} + 1 \leqslant i \leqslant M/2$

Each coefficient A_i excites a progressive phase distribution at the element ports of the MXN matrix, and the pattern in space that corresponds to each port is given by equation (9.16) as:

$$E_x^i(u) = e_x(u) \frac{\sin\left[N_x \pi (u - u_i) d_x/\lambda\right]}{N_x \sin\left[\pi(u - u_i) d_x/\lambda\right]} \tag{9.26}$$

which corresponds to an orthogonal beam from the central subset $\frac{-M}{2} + 1 \leqslant i \leqslant M/2$. Equation (9.25) however shows that when one subarray port is excited, all of the beams in the central subset are excited, and each with unity amplitude, but with phase relationships that depend upon the subarray port index n. The composite pattern for the subarray is thus the sum of these M beams and is approximately a flat topped pattern to allow near optimum scan within the limits of it's spatial sector. Mathematically, the subarray pattern for the n'th subarray port is:

$$e_s(n) = \Sigma_i A_i E_x^i(u) \tag{9.27}$$

If a whole set of progressively phased signals are applied at the subarray ports so that:

$$S_n = e^{+j\frac{2\pi}{\lambda}\left(\frac{\delta-½}{M}\right)\left(n-\frac{M+1}{2}\right)} \tag{9.28}$$

where δ is a parameter related to scan, then the radiated pattern for the system is:

$$E_x(u) = \sum_{n=1}^{M} S_n e_s(n)$$

$$= \sum_{i=\frac{-M}{2}+1}^{M/2} E_x^i(u) \sum_{n=1}^{M} \frac{1}{\sqrt{M}} e^{-j\frac{2\pi}{\lambda}\left(\frac{i-\delta}{M}\right)\left(n-\frac{M+1}{2}\right)} \tag{9.29}$$

If the scan parameter δ corresponds to any of the orthogonal beam indices (i) then the array pattern will be that of the i'th orthogonal beam, or:

$$E_x(u) = \sqrt{M}\, E_x^i(u) \tag{9.30}$$

and if the chosen scan parameter is not one of the beam positions the $E_x(u)$ will be a combination of orthogonal beams to provide for transmission or reception of the intermediate beam positions.

Although not given explicitly in the above discussion the subarray aperture distribution for the n'th port is clearly distributed across all of the N array elements (since each A' beam excites the whole array). The aperture distribution is in fact a sin X/X type illumination and its phase center moves across the array as a function of n. This circuit thus produces a nearly ideal square topped subarray pattern that makes it possible to scan over the entire spacial sector contained within the M beam subset using exactly M control elements.

Shelton did not use this concept as a scanning array, rather he used it to provide overlapped feed elements for a reflector. The HIPSAF antenna described in a later section uses it to provide overlapped subarray ports for a multiple beam array.

The concept of orthogonal beams and of idealized transforms allows a further conclusion that is useful in designing limited scan systems. If a set of progressively phased signals is applied to the input of the circuit of Fig. 9.8 for the purpose of producing a set of progressively phased signals at M subarrays at the N array element ports, the minimum number of subarrays M is specified by Equation (9.21). Obviously, this M is also the minimum number of input ports at the subarray input terminals. This much is trivial. The phase difference between any two adjacent subarray ports is the same as that transferred to the subarrays in the final aperture, and so if the circuit of Fig. 9.8 were replaced by one in which the Fourier transforms were performed by a lens or reflector, and if the elements of the feed array were spaced a distance d_i apart, then this array would necessarily scan to $\theta_{i_{max}}$, such that

$$d_i \sin \theta_{i_{max}} = d_x \sin \theta_{max} = 0.5\lambda \left[1 - \frac{1}{N} \right] \qquad (9.31)$$

This is precisely the condition for the feed array maximum scan angle, and so one concludes that if a scanning system is made up of a scanned feed array, spatial Fourier transformers, and a final aperture, then the minimum number of control elements as specified by the maximum scan angle and the size of the final aperture can only be achieved if the feed array is made to scan to its limit. Some examples of the implications of this condition are given in Sect. 9.3.

9.2.6 The Completely Overlapped Subarray Antenna System (HIPSAF).

Tang [6] describes a completely overlapped subarray system of unique design that achieves the required overlap using a space feed system [12] that relies heavily upon the Fourier transform concepts of the last section. The principles of operation are simply explained with reference to Figs. 9.8 and 9.9. A signal applied to one of the subarray input ports of Figure 9.9 produces a uniformally illuminated linear phase front across the feed array. The feed array therefore has a sin u/u type radiation pattern in space as given by Equation (9.16) which has its peak at different points on the pickup array depending upon what subarray input terminal is used. Path length differences are compensated by the fixed phase shifters and the truncated sin u/u distribution is transferred to the front face of the array to give a close approximation to the ideal flat topped subarray pattern. This circuit is essentially an implementation of that shown in Figure 9.8, with the second Fourier transformation performed by the space feed.

Figure 9.10 shows a subarray pattern of the HIPSAF antenna, and demonstrates the effective aperture control achievable using this concept.

Access to individual subarray ports allows the use of time delay for each subarray and results in an inherent wide band capability. Access at this level also simplifies tailoring the sidelobe levels using aperture taper. The prototype model had 10% bandwidth, a 1° pencil beam and approximately -23dB sidelobes with only about one dB loss in the feed system. The orthogonal beam matrix used in the prototype model was an 8 x 8 Butler matrix constructed in waveguide.

Important in its own right, this antenna is also important because it serves as a transition between the established concepts of subarrays fed by matrices, and the world of large optical structures excited by space feeds. Some of these structures, as described in Section 9.3, are being influenced by the techniques first demonstrated in the HIPSAF program.

9.3 Arrays with Reflectors or Lenses.

The idealized orthogonal beam descriptions used throughout the last section assume a situation that may be realistic for discrete array systems with constrained feeds but which is very difficult to reproduce with reflector or lens structures. Design of this class of limited scan systems is dominated by optical considerations, and problems of spillover, aperture blockage, off axis focussing and induced cross polarization often effect the design more than the ratio of beam positions to control elements. Aperture efficiency is another parameter of importance for many applications. Many of the reflector or lens geometries require oversize apertures because the feed structure illuminates only a spot on the main aperture, and that spot moves with scan angle. For these devices, aperture efficiencies can be of the order of 25% instead of the usual 55 to 60% for non-scanning reflectors, and so scanning reflectors or lenses often require double the aperture of the more efficient fixed beam structures. The type of beam steering required is also a factor of great importance; the simplest being row and column steering with progressive phases in both planes. Certain antennas however require complex steering functions for off axis scan, and these result in slower beam steering and larger computer data storage.

Optical limited scan techniques have their origin in the development of mechanically scanned reflector and lens geometries using feed tilt [13] and displacement [14] and in the development of feeds to correct the wide angle performance of beam shaping reflectors [15]. Present devices include single and dual reflector or lens geometries in combination with a phased array, that produce an electronically scanned beam using relatively few phase controls.

Design principles for such a wide ranging collection are themselves so varied that they cannot be developed from basic principles in a text of this length. Instead, this paper will outline some of the more important contributions to the technology and to the analytical and conceptual tools that made the technology possible. Several recent sources include comprehensive surveys of developments in these areas. [16, 17]

Many of the early studies on mechanical scanners were performed using geometrical optics, and even today this method receives wide usage for computing required feed locations, focussing conditions and phase shifter controsl, and for investigating general design parameters. Recent studies [17] have emphasized the deficiencies of the geometrical optical approach for obtaining intensity information about focal region fields, and have demonstrated the use of physical optics fields for design. Analytical methods used in design of the various antennas are noted in the technique descriptions that follow.

Early studies of scanning parabolas have uncovered a number of useful design concepts. Using ray optical techniques, Sletten, et. al. [15, 16] have investigated the location of focal (or caustic) surfaces for paraboloidal reflectors receiving off axis plane waves, and have shown how these characteristics can be used to develop midpoint correctors for elevation beam shaping while maintaining a narrow focussed beam in the azimuthal plane, and ridge line correctors for forming several pencil beams in elevation without destroying azimuthal focus. Ruze [14] has used a scalar plane wave theory to analyze the scanning characteristics of a parabola with a laterally displaced feed located at the Petzval surface. This analysis was used to obtain scanning patterns, to evaluate coma lobe contributions and to derive equations for the number of beamwidths scanned by such displacement for a -10.5dB coma lobe at the scan limit. This number is given below.

$$N = 0.44 + 22\,(f/D)^2 \qquad (9.32)$$

Recent work by Rusch and Ludwig [17] has included a numerical evaluation of focal region fields for a paraboloid receiving an off axis plane wave. Results of this study show that the maximum focal field locus, and therefore the position of optimum feed location, does not coincide with the Petzval surface, but remains relatively close to it for low f/D values. Reflectors with f/D greater than 0.5 have their optimum feed location closer to the focal point than the Petzval surface, but this location tends toward that line for large scan angles.

Imbriale et. al. [18] have also considered parabolic reflectors with large lateral feed displacements, and have compared the results of Ruze's scalar theory with the complete vector theory solution and experimental data for various feed displacements. This study demonstrated that the coma-lobe level is sometimes vastly underestimated by the scalar theory when used for feeds with large displacement.

Another recent work of significant import has been reported by Rudge [19], who has demonstrated the spatial fourier transform relationship between the aperture fields of a parabolic reflector and its focal plane fields. In an extension of this work, Rudge and Withers have also shown [20] that the fields in a specified off axis focal plane bear the same transform relationship to the aperture field under the excitation of an inclined wave. This work is very suggestive of the sort of design techniques that led to the HIPSAF antenna, and Rudge has demonstrated the utility of this concept for designing mechanically scanned reflector antennas and for several other important applications. [21, 22]

9.3.1 Single Reflector and Lens Techniques. In the early 1060's, the study of White and DeSize [23] showed that an array of feeds placed on a spherical surface concentric with the focal point of a parabolic reflector can be used to obtain limited beam scanning. The technique achieved about 10 beamwidths of scan in one plane.

Winter [24] describes a flat array feed for a parabola similar to that shown in Figure 9.11 that achieved 7 beamwidths of scan with E-plane sidelobes approximately -15dB relative to the main beam in the E-plane, and approximately -10dB in the H-plane. The phase shifts required for achieving these scanned conditions were computed from geometrical optics. These were not progressive in either plane of scan and so row and column steering could not be used. The array consisted of 980 elements and so has an element use factor of approximately 5. The aperture efficiency is low because the illuminated spot is allowed to move about on the main reflector to achieve scanning. This study is of great importance because of its timeliness and because it showed that reflectors could be scanned by a phased array without the accompanying coma lobes observed for offset feeds.

A more recent study of a planar array scanning parabola is described by Tang [25], and Howell [26]. This structure, the AGILTRAC antenna, is a C-band array of 499 elements that scan an eighteen foot parabola over a 7° half angle conical sector. The antenna has sidelobes of about -18dB at broadside, and maintains a monopulse null depth of 20dB out to 5°. The gain is nearly constant at 41dB out to the scan limit, and the element use factor is about 2.5. Losses throughout the system reduce the gain about 3.8dB below the area gain for the chosen taper. System bandwidth is 9%.

Some general design principles for the development of such systems are given by Tang [25] and Howell [26]. Array size, location and approximate phasing are usually computed by geometrical optics. Tang gives an equation for the percentage of blockage as a function of reflector size, illuminated diameter scan angle and focal length, and points out that sidelobe ratios on the order of -18dB for a design of this type can be achieved with reasonable size reflectors, but if one insists on sidelobes lower than -20dB a much larger main reflector must be chosen.

These and other studies of single reflector limited scan techniques has led to the development of offset array structures with low aperture blockage and low sidelobes. Notable among these is the PAR antenna of the TPN-19 GCA system, which was developed for the Air Force by Raytheon Corporation. This antenna, shown in Figure 9.12 uses a hyperbolic main reflector and an offset phased array feed with a modified row and column steering algorithm for off axis scanning. The antenna has 1.4° azimuthal beamwidth and 0.75° elevation beamwidth, and scans ±9.5 beamwidths in elevation and ±6.65 beamwidths in the azimuthal plane. The array used 824 phase shifters, and so has an element use factor of approximately 3.25. The first sidelobes are at the -20dB level throughout most of the scan sector, although rising to -18 at the limits. The system operates over a 2% bandwidth at X-band, and has a realized gain of approximately 39dB at the scan limits. This realized gain is about 3.5dB below the directive gain on boresight due to horn feed spillover, array face matching, and phase shifter losses. The main reflector size is about 9 ft. by 11½ ft corresponding to an aperture efficiency of approximately 30%, for the illuminated spot moves across the oversize reflector as a function of scan.

The basic principle behind each of the above antennas is that of placing the array surface at a plane between the main reflector or lens, and its on-axis focal point. The array is then phased to best approximate the fields of a receiving parabola for the particular scan angle. The same is true of

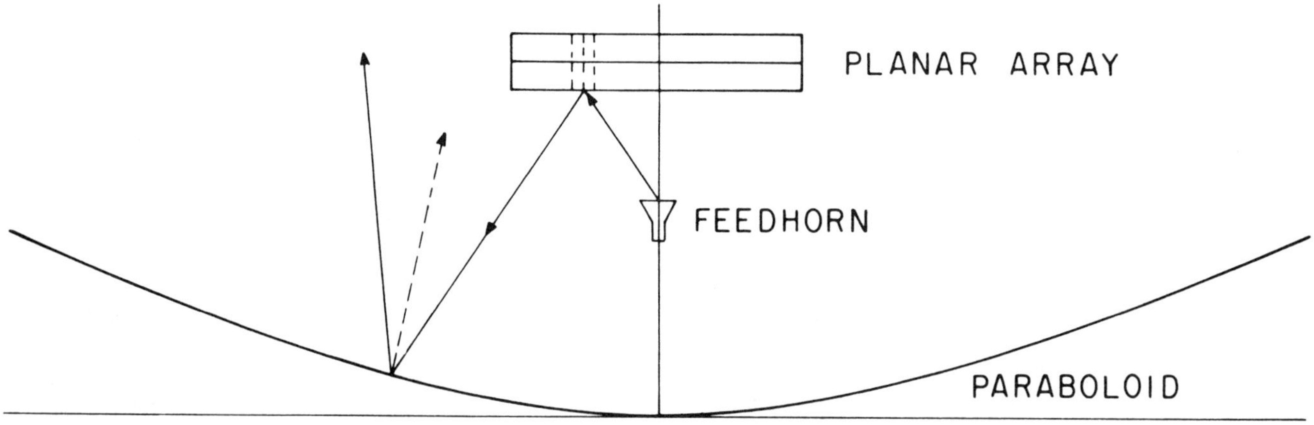

Figure 9.11 - Planar Array and Paraboloid

Figure 9.12 - Precision Approach Radar Antenna of AN/TPN-19

Source: Skolnik, *Radar Handbook*, 1st Ed., 1970, McGraw-Hill Book Company. Reprinted with permission.

the last geometry considered in this section, but with the fundamental difference that the main reflector or lens is not restricted by a focussing condition. This new concept in limited scan antennas, proposed by A.C. Schell [27] uses an array disposed around a cylinder to scan a reflector or lens surface that is contoured according to an optimum scan condition, rather than a focussing condition. The reflector is then stepped or the lens phase corrected to achieve focussing.Preliminary design results [28] show that in one plane of scan the technique achieves an element use factor of about unity, while using an oversize final aperture to again allow motion of the illuminated spot.

Figure 9.13 shows a schematic view of the array-lens combination and Figure 9.14 demonstrates its scanning properties. The array element currents are equal in amplitude and have a progressive phase given by $\beta\eta\Delta\theta$. The reflector surface (or lens back face) is chosen to transform this phase variation into a linear wavefront normal to the beam direction. The condition for determining the curvature of the reflector (or lens) is that a constant incremental phase change in θ along the circular arc tangent to the center of the back face of the lens produces a constant incremental phase change in the "y" co-ordinate along the aperture.

$$\frac{dy}{d\theta} = \text{Constant} = R \tag{9.33}$$

and, since $y = \rho \sin\theta$

$$\rho = \frac{R\theta}{\sin\theta} \tag{9.34}$$

This curvature satisfies the scan requirement, but does not guarantee that the wave will focus. Focussing is achieved for the reflector through the use of confocal parabolic sections stepped so that their centers lie along the scan surface, and for the lens by adjusting the path lengths so that they are equal at some angle. The array represented by the data of Figure 9.14 consists of 25 elements. The half angle θ_0 subtended by the array is 45°, $k_0 R = 197$, $k_0 a = 48$ and k_0 times the final aperture width is approximately 526 ($D/\lambda = 83.7$).

The far field beam angle is given by

$$\sin\theta_B = \beta/kR, \tag{9.35}$$

and for the case shown in the figure the maximum θ_B is about 11.5°. At this scan angle the gain is reduced about 2.5dB with respect to broadside and a far-sidelobe has risen to the -20dB level. The aperture illumination is nearly uniform, and the -13dB near sidelobe ratio is maintained throughout the scan sector.

These computations by McGahan [28] have been confirmed experimentally for the lens geometry scanning in one plane. An element use factor of 0.95 would result if the same economy of phase controls can be maintained for the lens scanning in two dimensions. Ultimate bandwidth limits, sidelobe levels and the development of a model for two dimensional scan remain the primary subjects for further investigation.

9.3.2 Dual Reflector or Lens Antennas. Single reflector or lens structures with a phased array feed are simple but, with the exception of the technique described by Schell [27] and McGahan [28], they require a relatively large number of phase controls (Element use factors of 2.5–3.25). All of these antennas require oversize main apertures because the illumination moves with scan (typical aperture efficiencies are 20-25%). Thus large aperture size coupled with weight and cost limitations, usually restricts the choice of final aperture to that of a reflector, and the resulting blockage can cause sidelobe problems and the need for offset feeds.

Technological Innovations. Recent technological efforts have been concerned with developing systems for scanning over wider angles with pencil beams, and for providing these scanning, capabilities with relatively small high efficiency primary apertures. Two recent studies [29, 30] at Lincoln Laboratory have dealt with the quest for larger scan multiples. One of these concerns a near field cassegrain antenna shown in Figure 9.15, which is compact and has a length to diameter ratio L/D of approximately 0.4, but the array diameter "d" required was relatively large, with typical d/D ratios considered between 0.25 for a 400λ main reflector, and 0.35 for a 250λ main reflector. These computations show up to 17 beamwidths scan with element use factors between 6 and 7. These relatively large element use factors arise because the array is only required to scan over a narrow range of angles, and so itself is a limited scan array. For this reason, element spacings greater than the 0.9λ can be used to maintain element efficiency. [29] The use of these oversized elements would reduce the element use factors for this approach to somewhere between 3 and 4. This array scheme also has the advantage of using linear phase control and since its L/D ratio is small, can be mounted and moved conveniently. It has relatively high sidelobes, on the order of -13dB due to the inherent aperture blockage of the system. The second Lincoln Laboratory study is the offset-feed Gregorian geometry of confocal coaxial parabolic sections shown in Figure 9.16. This geometry was first proposed by Dudkovsky [31] in 1962 and proposed and investigated as a scanned geometry by Skahil et. al., [32] in 1966. Fitzgerald [30] describes a system with -15 to -17dB sidelobes, ½° beamwidth which scans 14 beamwidths and, with an array 45 on a side, has an element use factor of about 2.5. The array itself scans only to about ±20°. The L/D ratio is approximately 1.5, and so the structure is awkward for pedestal mounting. More recent results [33] have shown that the scanning properties of this system can be improved (and the element use factor reduced to about 2) by optimizing the main and subreflector contours.

The reports describing these two studies [29, 30] show substantial fillin of the monopulse null from its broadside levels of -25 to -30dB to values of about −10dB for the plane of the offset reflector for the gregorian structure, and about the same for the cassegranian antenna.

Research studies sponsored by AFCRL have sought to increase aperture efficiency by allowing movement of an illuminated zone on a sub-aperture, and to decrease element use factors by maximizing the phased array scan. The work of of C. H. Tang and C. Winter of Raytheon [34] has included computer simulation of the array/lens structure shown in Figure 9.17 consisting of a planar feed of 437 elements arranged on a square grid. The array focuses a small spot on the elliptical rear face of a lens which transfers the

Source: Skolnik, *Radar Handbook*, 1st Ed., 1970, McGraw-Hill Book Company. Reprinted with permission.

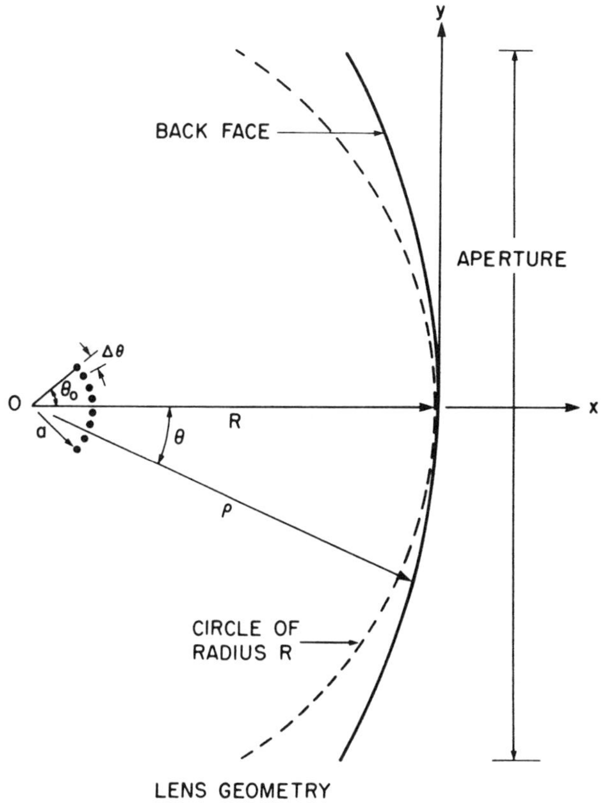

Figure 9.13 - Lens Design for Limited Scan

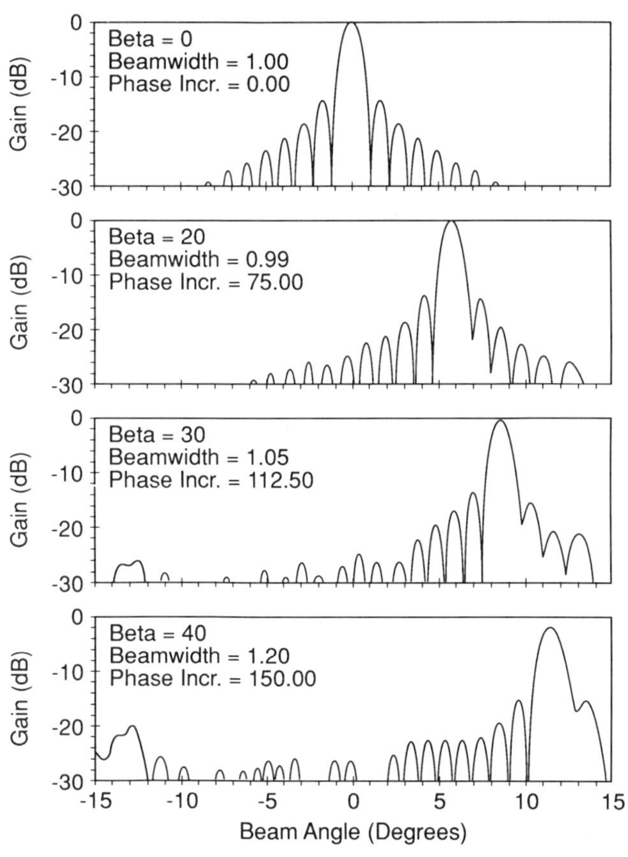

Figure 9.14 - Scanned Patterns of Limited Scan Lens (Courtesy of R. McGahan, AFCRL)

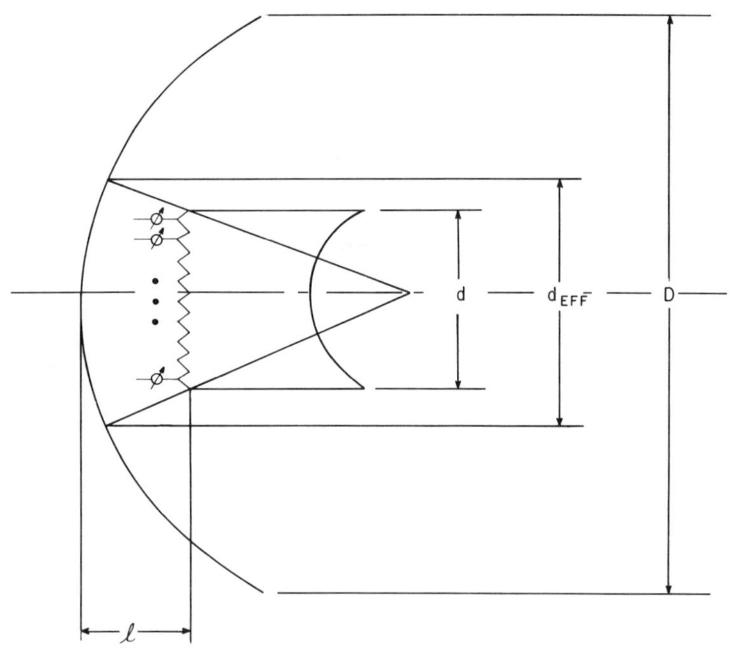

Figure 9.15 - Near Field Cassegranian Geometry

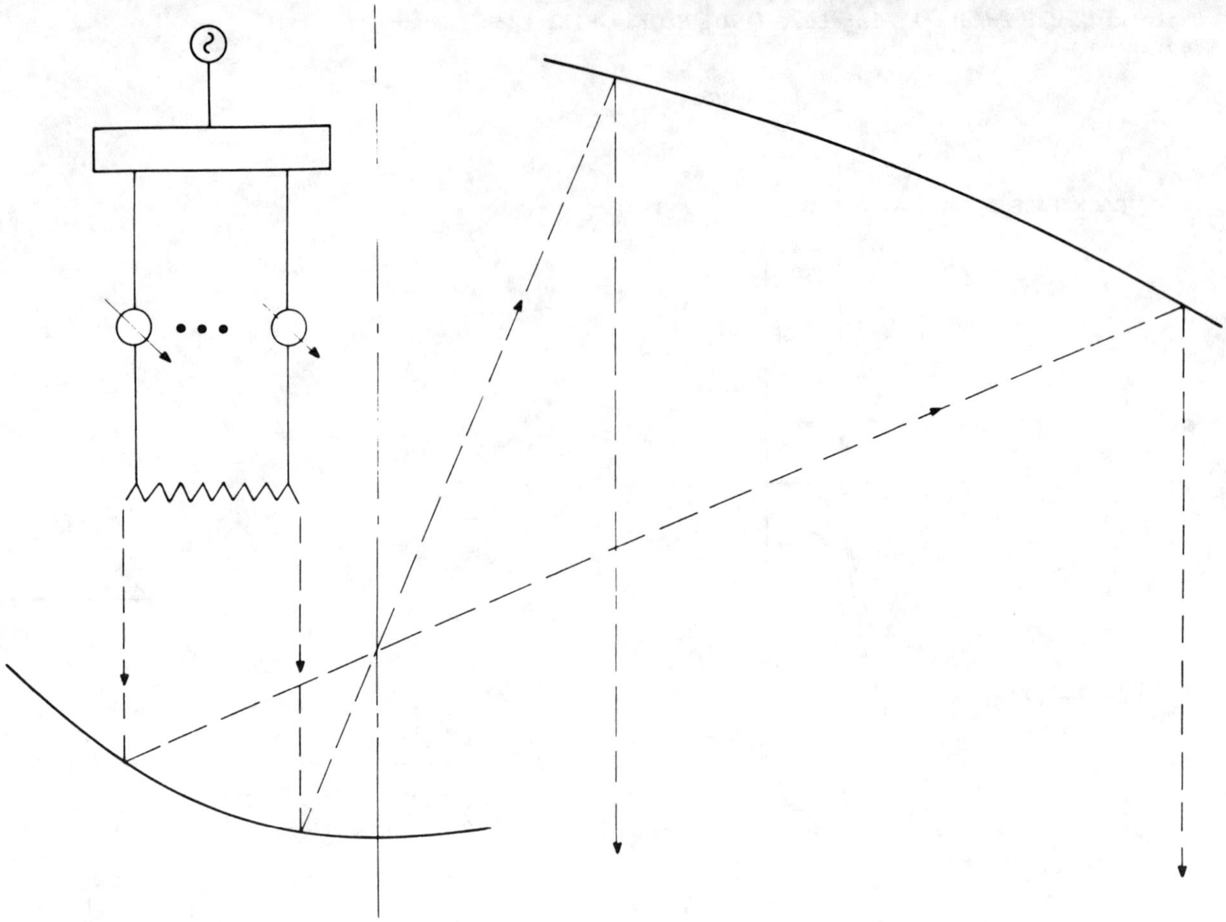

Figure 9.16 - Offset-feed Gregorian Geometry

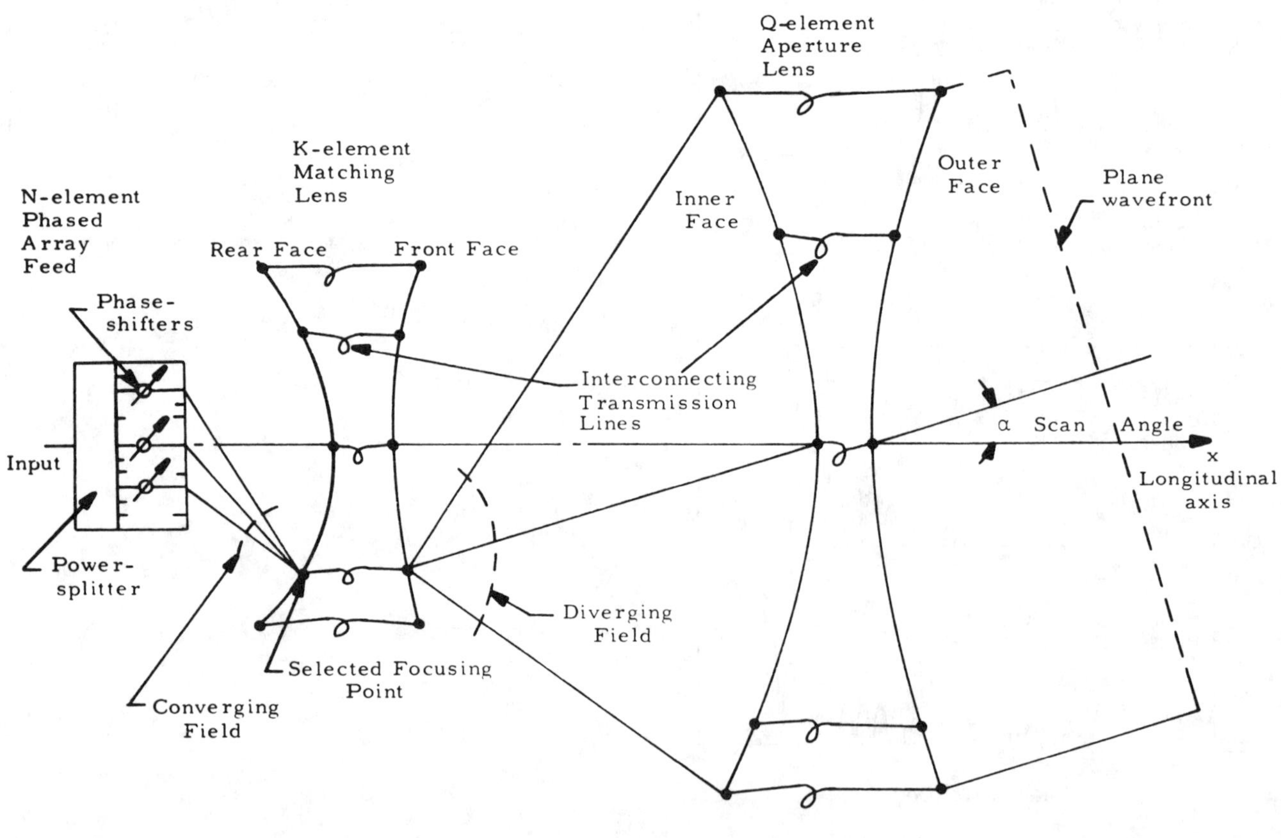

Figure 9.17 - Dual Lens Limited Scan Concept (Courtesy of C.H. Tang, Raytheon Co., Wayland, Ma.)

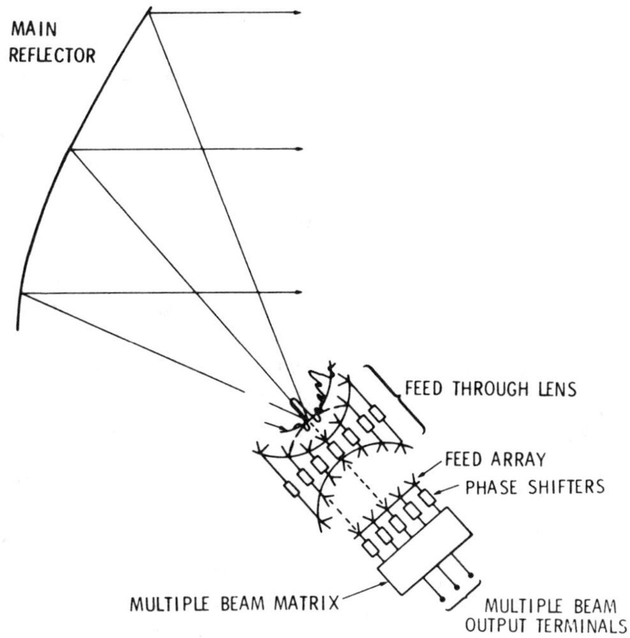

Figure 9.18 - Reflector-Lens Limited Scan Concept (Courtesy of R. Tang, Hughes Aircraft Co., Fullerton, CA)

spot to a region on the focal arc of a final lens with spherical back face. This main lens is an equal path lens with two point correction (0, ±10°). Sidelobes are at approximately the -16dB level for ±10° scan. At 65λ diameter, the antenna forms a beam of approximately 1.2° for all scan angles, and the aperture efficiency is about 60%. The element use factor is 1.5, and although the L/D ratio is at least 1.7 the main aperture diameter D is small because of its efficient illumination. The intermediate lens is about 0.7 the size of the final lens. The total system loss, including -2.1dB aperture efficiency is about -6.5dB. System bandwidth of ±20% is claimed.

R.Tang and colleagues of Hughes Aircraft Co. [35] have performed a theoretical investigation and extension of a technique described by Rudge and Withers [21], wherein a Fourier transformer (Butler Matrix) is used to match the scanned beams of a reflector. This antenna (Figure 9.18) differs substantially from the Raytheon study because the array does not focus, but provides a nearly linear phase front that is perturbed with scan to compensate for comatic phase error. The feed-through lens focuses the energy to form an approximate moving spot distribution across the reflector focal plane to scan the reflector. The initial progressive phase shifts can be produced by a multiple beam matrix or phased array. The array amplitude distribution is reproduced at the main aperture and good sidelobe control is possible. Computed patterns show -23dB sidelobes for the on-axis beam, and -19dB for the beam scanned 10 beamwidths. The array consists of 550 elements and scans the the average 1.05 pencil beam about 9.5 beamwidths corresponding to an element use factor of 1.53. The reflector diameter is 73λ and the L/D ratio is approximately 0.75, and the subreflector or lens is about 47λ, again about 0.65 the size of the final aperture.

Although not strictly an exercise in limited scan, the work of Chen and Tsandoulas [36] has demonstrated the effectiveness of the subarraying concept in dual reflector systems. These authors have used the plane wave spectrum technique of Wu and Rudduck [37] for expanding the feed radiation fields, and have shown that a small array with time delay steering can be used in combination with a dual reflector geometry as a feed for a large array with phase shifters. The space feed system creates time delayed subarrays that provide wide angle performance over bandwidth of 10-20% with main array apertures of up to 100λ on a side.

Design Principles. The design principles underlying these innovations are substantially more complex than those governing the development of conventional single reflector systems. The complexity occurs because the array is used to create a focal surface to match that of the main aperture. This focussing be done by the array alone as in the lens study by Tang [34], or through an intermediate reflector or lens as in all of the other studies. Thus, the work of Ruze [14] and Rusch [17] on focal region (caustic) surface variation with scan, and of Sletten [15] on ridge line locations serve as an aid in design.

The optimization study for Lincoln Laboratory made extensive use of computer techniques to generate main reflector and sub-reflector surfaces that optimized array scan characteristics. The lens design of Tang and Winter involved matching of focal region fields at the front and back face of the feed through lens. A major contribution to limited scan antenna concepts has been made by Rudge and Withers, [20] who have taken advantage of the spatial fourier transform properties of the scanned main reflector as observed in a generalized focal plane, to develop a spatial transformer feed for matching the fields of the reflector. Motion of this primary feed along a fixed path assures assures accurate matching of focal fields. These authors have also proposed an electronically scanned version but have not investigated this in detail. Based upon Rudge's work, the antenna design of Tang [35] results in the synthesis of moving focal spots whose location correspond to the offset feed point for each of the set of quasi-orthogonal beam positions of the main reflector. This spot location is at the point of maximum field intensity for various elevation angles, and care is taken to choose a geometry so that the average ridge line length is minimuzed for all scan angles. The focal region spot is then created by a phased array feeding an orthogonal beam matrix (a feed through lens), the feed array elements are identified with subarray terminals as done in the HIPSAF antenna and in Figure 9.8. Synthesis then proceeds at the subarray level toward an optimization of these subarray patterns, and this optimization includes the selection of the front surface of the feed through lens to obtain minimum subarray overlap, and to assure that the subarray phase centers are equally spaced across the final aperture. Since the path lengths for each subarray are not the same, phase shifters are added to each feed array element to collimate the progressively phased multiple beam output terminals. This collimation holds up over relatively wide scan angles, but a non-progressive phase distribution is required to achieve the full scanning capabilities of this technique.

Several other interesting design conclusions can be drawn from the data of these dual reflector or lens geometries. First, they can be made free of blockage, and so can potentially have very low sidelobes. Second, in the case of the Lincoln Laboratory studies the ratio of subreflector to main reflector size was kept at about 0.25 to 0.35 and the array was made to scan over only relatively small sectors; this design led to element use factors of from 2. to 2.5. The AFCRL sponsored studies however sought mainly to reduce the number of control elements and did succeed in achieving use factors of about 1.5, but this was coupled with required use of secondary lenses that were about 0.7 the size of the main reflector or lens. The distinction between these two approaches is essentially an experimental restatement of the design conclusion drawn in Section 9.2.4. The Raytheon [34] and Hughes [35] studies have minimized the element use factor by insisting in each case that the array scan nearly to its own scan limit. The resulting oversize subapertures are required to allow for motion of the illuminated region. The Lincoln Laboratory studies [29, 30] however chose to maintain relatively small subreflectors and so use arrays that scan over limited sectors. The conclusion drawn in section 9.2.4, therefore shows that low element use factors could not be achieved for the cases considered, but, as observed by Fitzgerald, further reductions in the number of phase shifters can be obtained using a array of oversize elements.

REFERENCES

[1] A.C. Schell, et. al., *The Detection Range of the AN/TPN-19 Radars in Heavy Rain (Baseline Scan Mode)*, AFCRL-72-0368, 15 June 1972, Special Report No. 141.

[2] R.C. Crane, "Propagation Through Rain", NEREM 74 Record, Part 4, *Radar Systems and Components,* Oct. 28-31, 1974.

[3] P.M. Woodward and J. D. Lawson, "The Theoretical Precision with Which an Arbitrary Radiation Pattern may be Obtained from a Source of Finite Size", *J. AIEE* Vol. 95, Pt. III, pp 362-370, Sept. 1948.

[4] J.J. Stangel, "A Basic Theorem Concerning the Electronic Scanning Capabilities of Antennas", URSI Commission VI, Spring Meeting, June 11, 1974.

[5] W.T. Patton, "Limited Scan Arrays" in *Phased Array Antennas; Proceedings of the 1970 Phased Array Antenna Symposium.* Edited by A.A. Oliner and G.H. Knittel, Artech House, Inc., Dedham, Mass., 1972, pp 332-343.

[6] R. Tang, "Survey of Time-Delay Beam Steering Techniques", in *Phased Array Antennas; Proceedings of the 1970 Phased Array Antenna Symposium,* Artech House INc., Dedham, Mass., 1972, pp 254-260.

[7] J.P. Shelton and K.S. Kelleher, "Multiple Beams from Linear Arrays", *IEEE Trans. AP-9,* March 1961, pp 154-161.

[8] J. Butler and R. Lowe, "Beam Forming Matrix Simplifies Design of Electronically Scanned Antennas", *Electronic Design,* Vol. 9, pp 170-173, April 12, 1961.

[9] J.P. Shelton, "Multibeam Planar Arrays", *IEEE Proc.* Vol. 56, No. 11, pp 1818-1821, Nov. 1968.

[10] J.L. Allen, "A Theoretical Limitation on the Formation of Lossless Multiple Beams in Linear Arrays", *IEEE Trans. AP-9,* pp 350-352, July 1961.

[11] J.P.Shelton, "Multiple-Feed Systems for Objectives", *IEEE Trans. AP-13,* pp 992-994, Nov. 1965.

[12] R.T. Hill, "Phased Array Feed Systems, A Survey", *Proceedings of the 1970 Phased Array Antenna Symposium.* Edited by A.A. Oliner and G.H. Knittel, Artech House, Inc., Dedham, Mass, 1972, pp 1970211.

[13] S. Silver and C.S. Pao, *Paraboloid Antenna Characteristics as a Function of Feed Tilt,* MIT Radiation Lab., Cambridge, Mass., Rep 479, 1944.

[14] J. Ruze "Lateral-Feed Displacement in a Paraboloid", *IEEE Trans. Antennas Propagation,* Vol. AP-13, pp 660-665, Sept. 1965.

[15] C.J. Sletten, et. al., "Corrective Line Sources for Paraboloids", *IEEE Trans. AP-6,* No. 3, pp 239-251, July 1958.

[16] R.E. Collin, F.J. Zucker, *Antenna Theory,* Part 2, McGraw-Hill Book Co., New York, 1969, Chapter 17.

[17] W.V.T. Rusch and A.C. Ludwig, "Determination of the Maximum Scan-Gain Contours of a Beam-Scanning Paraboloid and Their Relation to the Pitzval Surface", *IEEE Trans. Antennas Propagation,* Vol AP-21, pp 141-147, March 1973.

[18] W.A. Imbraile, et.al., "Large Lateral Feed Displacements in a Parabolic Reflector", *IEEE Trans. AP-22,* No. 6, pp 742-745, Nov. 1974.

[19] A.W. Rudge and M.J. Withers, "New Techniques for Beam Steering with Fixed Parabolic Reflectors", *Proc. IEEE,* Vol. 118, No. 7, July 1971, pp 857-863.

[20] A.W. Rudge "Focal Plane Field Distribution of Parabolic Reflectors", *Electronic Letters,* Vol. 5, pp 610-612, 1969.

[21] A.W. Rudge and M.J. Withers, "Beam-Scanning Primary Feed for Parabolic Reflectors", *Electronic Letters,* Vol. 5, pp 39-41, 1969.

[22] A.W. Rudge and D.E.N. Davies, "Electronically Controllable Primary Feed for Profile-Error Compensation of Large Parabolic Reflectors", *Proc. IEE,* Vol. 117 (2), pp 351-358, 1970.

[23] W.D. White and L.K. DeSize, "Scanning Characteristics of Two-Reflector Antenna Systems", *1962 IRE International Conv. Record,* Pt. 1, pp 44-70.

[24] C. Winter, "Phase Scanning Experiments with Two Reflector Antenna Systems", *Proc. IEEE,* Vol. 56, No. 11, Nov. 1968, pp 1984-1999.

[25] Tang, C.H., *Application of Limited Scan Design for the AGILTRAC-16 Antenna,* 20th Annual USAF Antenna Research and Development Symposium, Univ. of Illinois, Oct. 1970.

[26] J.M. Howell, "Limited Scan Antennas", *IEEE/AP-s International Symposium,* 1974.

[27] A.C. Schell, "A Limited Sector Scanning Antenna", *IEEE G-AP International Symposium,* Dec. 1972.

[28] R. McGahan, *A Limited Scan Antenna Using a Microwave Lens and a Phased Array Feed,* M.S. Thesis, MIT, 1975.

[29] W.D. Fitzgerald, *Limited Electronic Scanning with an Near Field Cassegrainian System,* ESD-TR-71-271, Technical Report 484, Lincoln Laboratory, 24, Sept. 1971.

[30] W.D. Fitzgerald, *Limited Electronic Scanning with an Offset-Feed Near-Field Gregorian System,* ESD-TR-71-272, Technical Report 486, Lincoln Laboratory, 24 Sept. 1971.

[31] A. Dudkovsky, "A System for Exciting Large Parabolic Antennas", Russian Patent No. 146365 (1962).

[32] G.E. Skahill, L.K. DeSize, C.J. Wilson, "Electronically Steerable Field Reflector Antenna Techniques", RADC-TR-66-354, August 1966.

[33] C.J. Miller, D. Davis, *LFOV Optimization Study,* Final Report No. 77-0231, Westinghouse Defense and Electronic Systems Center, System Development Division, Baltimore, Md., January 1972, ESD-TR-72-102.

[34] C.H. Tang, C.F. Winter, *Study of the Use of a Phased Array to Achieve Pencil Beam over Limited Sector Scan,* AFCRL-TR-73-0482, ER 73-4292, Raytheon Co. Final Report Contract F19628-72-C-0213, July 1973.

[35] R. Tang, Final Report on Contract F19628-73-C-0129, In Preparation.

[36] M.H. Chen and G.N. Tsandoulas, "A Dual-Reflector Optical Feed for Wide-Band Phased Arrays," *IEEE Trans. AP-22,* No. 4, pp 541-545, July 1974.

[37] D.C.F. Wu and R.C. Rudduck, "Application of Plane Wave Spectrum Representation to Radome Analysis", *Proc 10th Electromagnetic Windows Symposium,* pp 46-50, July 1970.

Limited Scan Arrays - Part 2

Robert J. Mailloux

Air Force Cambridge Research Laboratory, Bedford, MA

10.1 Array Techniques. Conventional phased arrays can provide excellent electrical characteristics as limited scan antennas, but are excessively expensive because the elements must be placed less than a wavelength apart to prohibit the radiation of unwanted grating lobes. A conventional phased array with moderate taper and circular aperture has an element use factor given by 0.23 $(d/\lambda \sin \theta_{max})^2$ for interelement spacing d. The element spacing 0.9λ will be used throughout this paper for purposes of comparison and leads to factors 9.3 and 37 for 10° and 5° scan respectively. Other techniques discussed have much smaller and relatively constant element use factors, and these numbers provide a direct comparison of the number of control circuits required by each system.

10.1.1 Aperiodic Arrays. Grating lobes are due to the periodic nature of conventional array latices, and can be avoided by varying the element spacings. An example is the array investigated by W. Patton. [1] This structure, shown schematically in Fig. 10.1, consists of a circular array of dipole subarrays arranged in an aperiodic fashion, and excited by an optical power divider feeding a primary spherical array. The elements have equal areas, and their size ultimately determines the maximum scan angle of the antenna at the element half power point, or approximately $\sin \theta_m = 0.44\lambda/d$ where d is the element width in any plane. This array is locally periodic, and so does have vestigal grating lobes, but these are considerably suppressed for a large array. Patton describes a 30 ft. diameter array and a 10 ft. diameter array at C-band. The 300 foot array consists of one thousand elements that scan a 0.36° beam approximately 5° with an element use factor of 1.3. The system has high average sidelobes at its maximum scan and losses which add to 5.94dB for the 10 ft. model and a projected 4.21dB for the 30 ft. array. The transmission line interconnections may also make an X-band design somewhat less practical. Peak sidelobes were measured at the -15dB level or the 10 ft. diameter array, and are projected at -20.9dB for the 30 ft. array, but the item of primary importance is the achievement of this extremely low element use factor and the low generalized f/D ratio achievable with aperiodic array technology.

A similar antenna, but using unequal size elements has recently been described by Manwarren and Minuti [2]. This antenna has been designed to provide a 1° pencil beam at 1300 MHz to scan a conical sector with 8° half angle with 20dB grating lobes. An S-band model has also been configured. The antenna consists of 412 elements and uses elements of three different sizes to make up the array surface. The elements are arranged in concentric rings to produce a pseudorandom grid as done in the Rest program, but with additional randomness introduced by the unequal size elements. The element use factor is about 1.6. Computed patterns show graceful gain degradation with scan and grating lobes at the desired levels. The program has included a substantial effort to provide monopulse difference patterns with good null control. Beam steering control is provided by using the periodicity of the concentric rings.

10.1.2 Spatially Interlaced Arrays. Tang [3] refers to several types of spatially overlapped and interlaced [4] subarrays, and points out that the distinction between the two is that the signal from each element of an interlaced array, feeds into only one subarray, but the signal from each element of an overlapped subarray feeds into several subarray ports. The principle behind the use of an interlaced subarray is that once the inter-subarray distances are chosen the subarray terminal is fed to a number of antenna elements that occupy a greater spatial extent on the array than this distance. The end result is that the subarray pattern is narrower than could be obtained for a uniformally illuminated conventional subarray. In this case the spreading is accomplished by interlacing every subarray with those adjacent to it. If this were done in an entirely periodic manner, each subarray pattern would have well defined sidelobes and one might expect large discrete lobes to emerge at the array factor grating lobe positions.

Stangel [5] has studied randomized interlaced subarray configurations that result in narrowed subarray distributions with low sidelobes because of the randomness imposed upon the various subarray patterns. Fig. 10.2 shows a square grid geometry with shadings to represent elements in common subarrays. The inter-subarray distance is chosen to give a regular lattice of sub-array centers, and the actual subarray configuration is determined by a random number generating technique. Fig. 10.3 shows the maximum phase shifter reduction using this technique and that obtained for a regular subarray as compared with a filled array of elements spaced one half wavelength apart. In comparison with the 0.9λ spacings used elsewhere in this paper the phase shifter savings are approximately 50% for conical sectors of 10° half angle. The main advantage of the technique is its requirement for

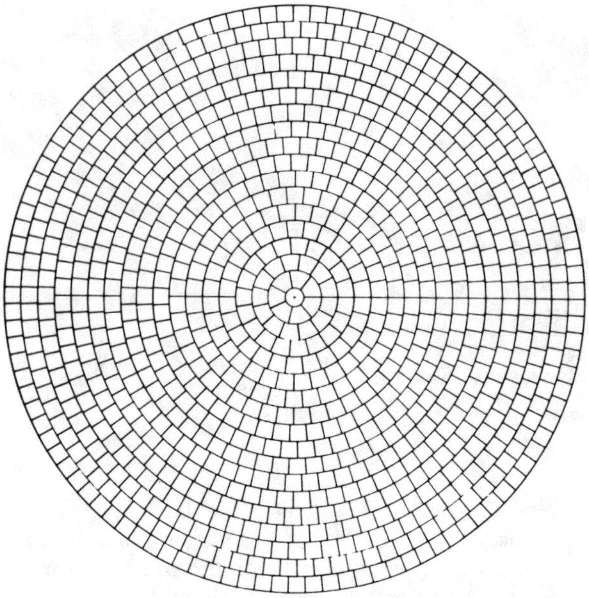

Figure 10.1 - Aperiodic Array for Limited Scan.

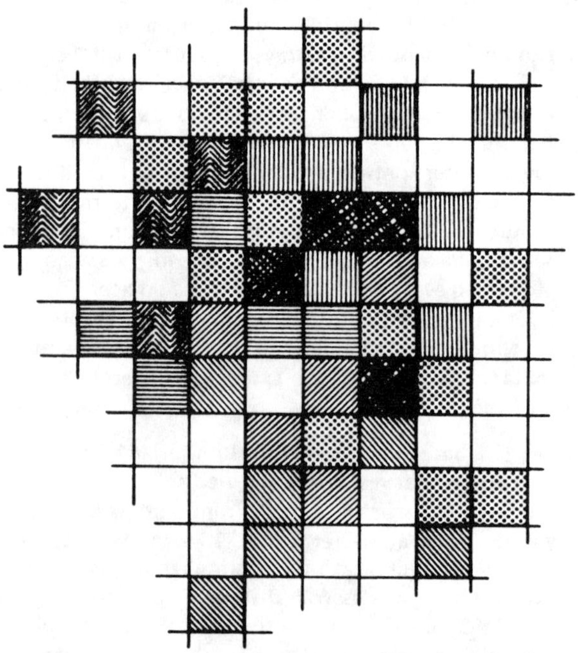

(SQUARE GRID REPRESENTS ELEMENT LATTICE.
DIFFERENT SHADINGS REPRESENT COMMON SUBARRAY ELEMENTS.)

Figure 10.2 - Interlaced Subarray Geometry
(Courtesy of J. Stangel, Sperry Rand Corp.)

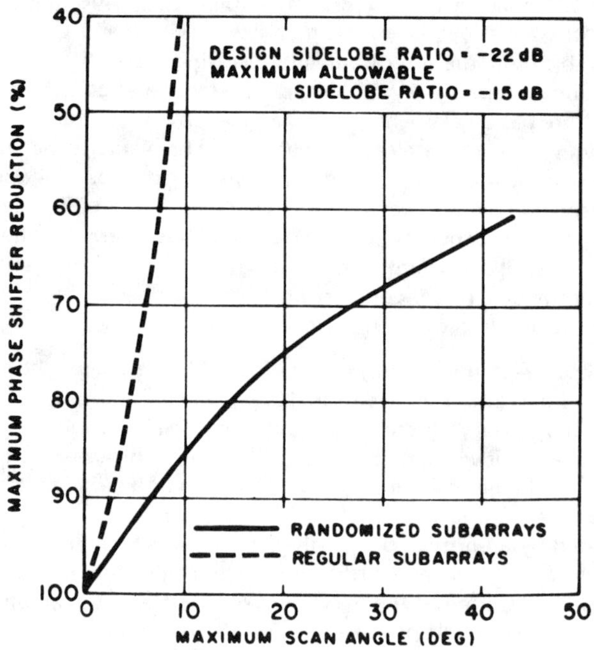

Figure 10.3 - Phase Shifter Savings for Interlaced Subarray
(Courtesy of J. Stangel, Sperry Rand Corp.)

simple progressive phase excitation; its main disadvantage is the difficulty in exciting the subarray elements, although methods of efficient implementation have been derived for optically fed and network fed arrays.

10.1.3 Multi-Mode Scanning Techniques. A recent effort at AFCRL has revealed that relatively large aperture horns can be used as elements of a limited scan array if the higher order mode amplitude is actively controlled. [6] The technique is called multi-mode scanning and consists of choosing odd mode amplitudes and phases so that the combined element radiation pattern from any horn has a zero at the angle of the grating lobe nearest to broadside.

The required odd mode amplitude and phases are obtained from a knowledge of the element patterns for even and odd modes. An array for E-plane scan has its field pattern given by (for $[\Delta_n] = 0$)

$$E(u, v) = E_x(u) \sum_n I_n^E \, e_y(v) \, e^{j \frac{2\pi}{\lambda} (v - v_o) n d_y} \qquad (10.1)$$

As indicated in section 9.2.1 and Fig. 9.4 of Part 1, the element pattern $e_y(v)$ is zero at the grating lobe positions $v_q = \pm q\lambda/d_y$ corresponding to broadwide main beam position for a uniformally illuminated element. The growth of the $q = -1$ grating lobe as a function of scan can be nearly eliminated by actively controlling $e_y(v)$ to place a zero at this grating lobe for all scan angles. In a waveguide circuit such control is accomplished by exciting the aperture with two modes (the LSE_{10} and LSE_{11}) instead of just the dominant LSE_{10} mode, so that $e_y(v)$ becomes the sum of two terms, with a zero at $v = v_o - \lambda/d_y = v_{-1}$. Choosing the ratio of odd mode to even mode as R_{11}, the combined element pattern is:

$$e_y(v) = e_{yo}(v) + R_{11} \, e_{y_1}(v) \qquad (10.2)$$

choosing $e_y(v)$ to be zero at the position of the $q = -1$ grating lobe (Eq. 9.12 of Part 1) one obtains for R_{11}

$$R_{11} = \frac{-e_{yo}(v_{-1})}{e_{y1}(v_{-1})} \, . \qquad (10.3)$$

Since the various waveguide modes have constant phase aperture distribution e_{yo} is a real function and e_{y1} is pure imaginary, so the R_{11} is pure imaginary and increases with scan in order to maintain the null position coincident with the center of the $q = -1$ grating lobe. The relative odd mode phase is thus fixed at $\pm 90^\circ$ with respect to the even mode phase, depending upon the sense of the scan angle. The simplest circuit for exciting even and odd modes to scan in one plane is shown in Fig. 10.4. This figure shows an E-plane horn and a dual mode waveguide section of length "L" excited through phase shifters. The phase shift η provides the beam control phase, and the difference excites an odd mode at the junction of the waveguide and the dual mode section with constant phase (90° away from the even mode phase) except for a 180° shift as the odd mode amplitude becomes negative. The two modes propagate with a different phase constant and so the length "L" is adjusted to give the required $\pm 90^\circ$ phase relationship at the aperture. Required phase controls are the progressive terms for beam steering plus the difference term Δ which is the same for all elements but which does vary with scan angle.

Fig. 10.5 shows a comparison of a single mode element pattern with principal effect of the superposition is to cancel the first grating lobe, in fact all of the nulls shift in the same direction and so tend to suppress other grating lobes for scan angles not too far from broadside. Another advantage of the technique displayed in these figures is that although the odd and even modes subtract at the $q = -1$ grating lobe, they add together at the main beam, thereby providing relatively constant gain with scan. This property is demonstrated in Fig. 10.6. The allowable element spacing for E-plane horns is

$$(dy/\lambda) \sin \theta_{max} = 0.6 \qquad (10.4)$$

Elements can therefore be 3.5 wavelengths on a side for $\pm 10^\circ$ scan or 4.6 wavelengths for ± 7.5 scan, corresponding to savings in phase shifters of 2 and 2.5 for one plane of scan.

The laboratory model [7] shown in Fig. 10.7, is an array designed for E-plane scan ($\pm 12^\circ$) using the power divider of Fig. 10.4. Fig. 10.8 shows the E-plane pattern for the array phased at broadside and the elements excited with the central four at uniform amplitude, the second element in from each end of the array at -3dB amplitude, and the outer elements at -6dB amplitude. This taper should have first sidelobes at about -19dB, but due to phase errors the level is approximately -17dB. The grating lobes at $\pm 19^\circ$ (-16dB) and $\pm 40^\circ$ (-26dB) can be reduced by using a dielectric lens in each horn.

Fig. 10.9 shows two cases at the maximum scan angle $\pm 12^\circ$. The dashed curve is the horn array radiation pattern without odd modes and shows that the main beam gain is reduced more than 5dB with respect to the broadside array and the grating lobe at -7° is larger than the main beam by 1.8dB. Other grating lobes are at tolerable levels. The solid curve of 10.9 shows that when element is excited by two modes the offending grating lobe is reduced to approximately the -20dB level, and the main beam increased to -1.2dB with respect to broadside because the new element pattern has its peak tilted toward the main beam. The second grating lobe ($q = -2$) is approximately the -12dB level.

Conventional aperture tapering procedures can be used to reduce near sidelobe levels to -30db or less. The nulled grating lobe is suppressed 20 to 25dB at center frequency for a small array (8 elements), but substantially more for larger arrays. Residual grating lobes at wider angles are unaffected by array tepering and remain the major limitation of the technique.

The array used for this experiment was not designed for wideband operation but does provide 18dB suppression for the $q = -1$ grating lobe over approximately a 3% bandwidth, and 16dB suppression over a 4% bandwidth. Although one can in principle design such systems with up to 10% bandwidth, such designs must include careful suppression of unwanted higher order modes that can cause null filling in the -15 to -18dB level. The array cross-polarized radiation pattern was less than -30dB for broadside radiation (even mode only) and less than -22dB at the worst case of maximum scan.

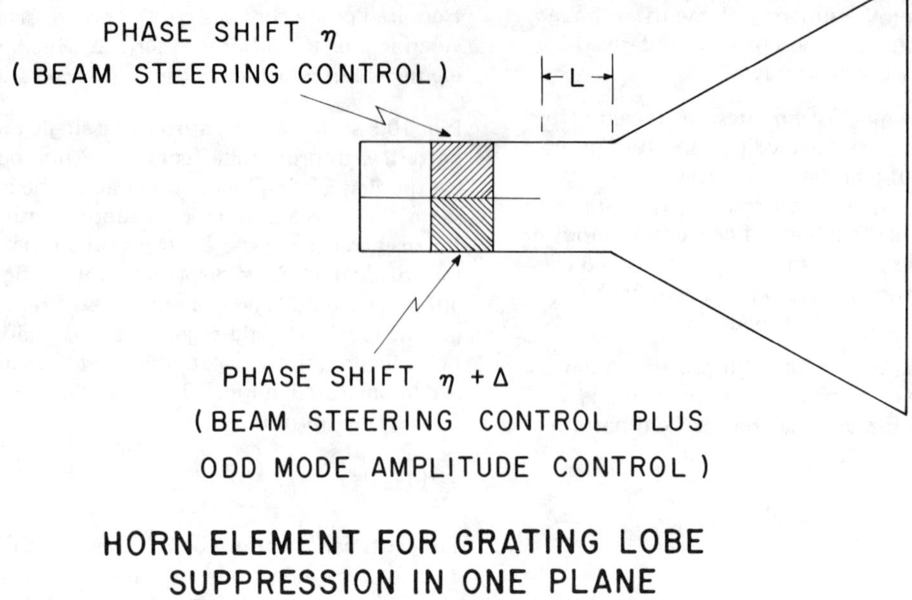

Figure 10.4 - Even-odd Mode Power Divider for E-plane Scan

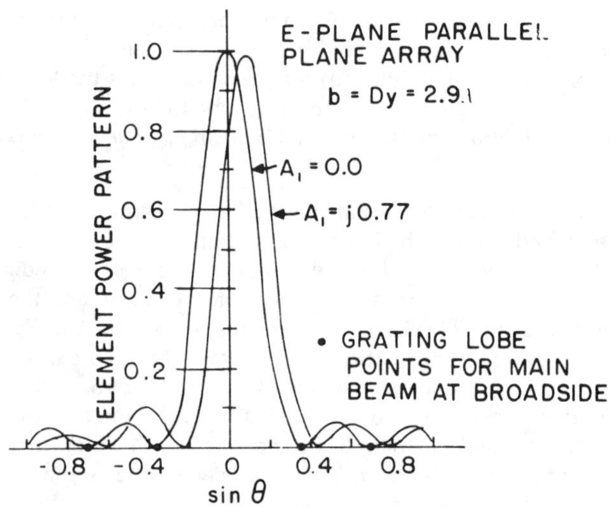

Figure 10.5 - Element Patterns of Fundamental Mode and of Fundamental Mode with Odd Mode

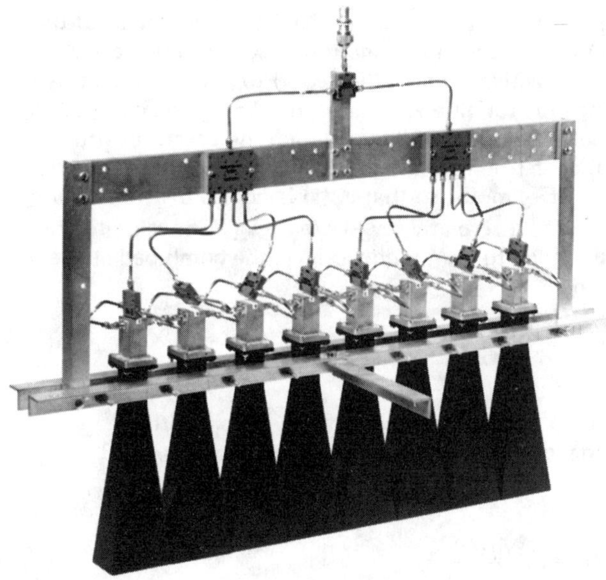

Figure 10.7 - Prototype Array for E-Plane Scan (d_y = 3.01λ)

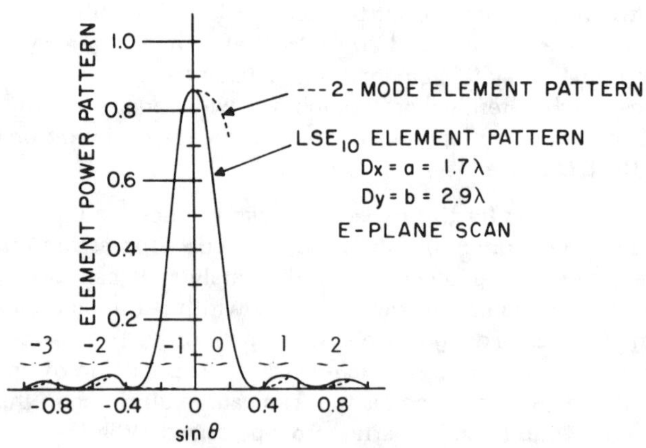

Figure 10.6 - Active Element Pattern with Odd Mode Chosen for N = -1 Mode Cancellation

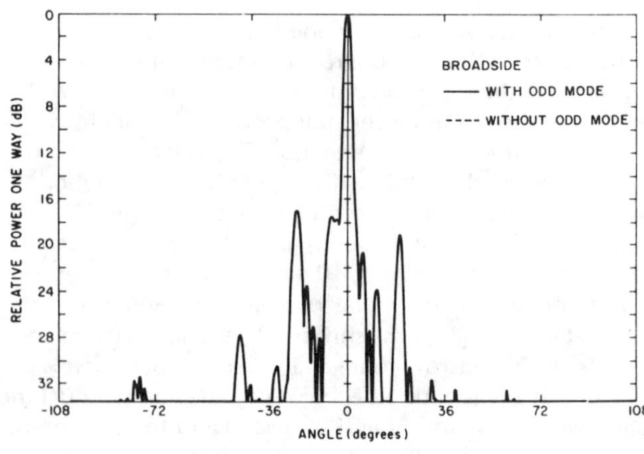

Figure 10.8 - Broadside Pattern of Prototype Array

As shown in Fig. 9.4 of Part 1 the multimode scanning technique is edeally suited to E-plane scan because the array grating lobes lie in the element pattern nulls when the array is at broadside. H-plane scan offers more difficulties because the widened element pattern has nulls beyond the grating lobe points for the array at broadside, and so energy is lost into these H-plane grating lobes. Observed grating lobe levels for a broadside main beam are indicated on Fig. 10.10 as approximately -10dB with respect to the main beam, for a 4.38λ horn described in reference [7]. Figs. 10.10 and 10.11 show that this pattern null is varied correctly by introducing the H-plane odd mode (LSE_{20}), and that in fact the null can be controlled well beyond the $(d/\lambda)\sin\theta = 0.6$ point at the expense of a high $p = -2$ grating lobe.

Unfortunately, the odd mode can do nothing to control the grating lobe level present at broadside, and so the H-plane element pattern must be modified for proper operation.

Figs. 10.12 and 10.13 show the H-plane scanned element pattern of the 4.38λ horn using 0.25 in. plexiglass dielectric slabs at the H-plane horn walls. The slabs are tapered to nearly a point to affect a smooth transition into the dielectric loaded horn section. This geometry was developed by Tsandoulas and Fitzgerald [8] for improving the aperture efficiency of large H-plane flared horns, and achieves this efficiency by making the field wave more uniform throughout the aperture. Fig. 10.12 shows that the broadside element pattern is indeed improved by this design, for the near grating lobes are reduced to approximately -18dB with respect to the main beam. Fig. 10.13 shows that the geometry provides a controlled null suitable for scanning, but that the $p = -2$ grating lobe is larger than the E-plane case and its suppression remains a problem to be solved.

The H-plane odd mode excitation for the circuit using dielectric slabs is given by direct analogy with equation (10.3) as:

$$R_{20} = \frac{-e_{x1}(u_{-1})}{e_{x2}(u_{-1})} \qquad (10.5)$$

The cross-polarized radiation for E- and H-planes was below -24dB for the even mode in either case; the odd modes had cross-polarized components of -24 and -20dB for the two cases. These low cross-polarized radiation components offer good assurance that the linearly polarized LSE modes are indeed the dominant higher order modes excited, and that cross-polarized radiation effects can be neglected for all but extremely low sidelobe requirements.

Having thus achieved scanning in both principal planes, it is not difficult to anticipate the characteristics necessary for scanning in two dimensions. Full two dimensional scan requires the suppression of three grating lobes, however, and so an added higher order mode must be controlled as a function of scan. The dominant grating lobes to be cancelled are those nearest broadside [(p, q) = (-1, 0), (-1, -1), (0, -1)] for general scan angles. Relative odd mode amplitudes as given by equations (10.3) and (10.5) provide control for principal plane scanning and by exciting the LSE_{21} mode with relative amplitude

$$R_{21} = R_{20}\,R_{11}\;; \qquad (10.6)$$

one obtains the following element pattern (using separable horn element patterns) $e_{ij}(u, v) = e_{xi}(u)\,e_{yj}(v)$

$$\begin{aligned} e(u, v) &= [e_{y0}(v) + R_{11}e_{y1}(v)]\;[e_{x1}(u) + R_{20}e_{x2}(u)] \\ &= e_y(v)\,e_x(u) \end{aligned} \qquad (10.7)$$

This separable element pattern provides excellent grating lobe control for general scan planes, and can be obtained using the circuit of Fig. 10.14. Here each horn is excited by four waveguides. The phase shift η is used for beam steering as in any conventional array. Other phase shifters are set to $\eta + \Delta_E$, $\eta + \Delta_H$, and $\eta + \Delta_E + \Delta_H$, and these serve to excite higher order modes in the horn elements. The component with even symmetry in both planes excites the LSE_{10} mode in the basic oversize waveguide feed. The signal amplitude (normalized to unity power, and dropping common phase factors) is:

$$LSE_{10} \sim \cos\frac{\Delta_E}{2}\cos\frac{\Delta_H}{2}$$

Other higher order components are:

$$\begin{aligned} LSE_{20} &\sim -j\cos\frac{\Delta_E}{2}\cdot\sin\frac{\Delta_H}{2} \\ LSE_{11} &\sim -j\sin\frac{\Delta_E}{2}\cdot\cos\frac{\Delta_H}{2} \\ LSE_{21} &\sim -\sin\frac{\Delta_E}{2}\cdot\sin\frac{\Delta_H}{2} \end{aligned} \qquad (10.8)$$

These incident modes excite the more complex horn modes with amplitude ratios as required by equations (10.3), (10.5), and (10.6) and so do provide a separable field distribution at the aperture with nulls located to cancel the three dominant grating lobes.

The circuit of Fig. 10.14 is highly schematic, and in practice one must add a waveguide section that is bifurcated in either the E-plane or H-plane and that will support two incident modes but not four. Then by varying the dimensions of the large rectangular waveguide section at the horn throat in order to adjust the relative phase of the LSE_{21} mode with either that of the LSE_{20} or LSE_{11}, and by varying the length of the bifurcated section to correct the phase of the remaining mode, one can obtain the proper null control in all planes. Such control has to date been obtained for very limited frequency ranges, and substantial development work remains to broadband the technique. Alternative means are available [6] for exciting the aperture properly over a much wider frequency range, but these require external power dividers and monopulse-type horn geometries, and thus complicate the design.

Apart from the fact that one can achieve full two-dimensional scan with the multi-mode technique, there remains the question of whether one should. Fig. 10.15 shows that although the dielectric layers aid in suppressing the H-plane grating lobes at broadside, they restrict the maximum scan about

$$(d_x/\lambda)\sin\theta_{max} = 0.45$$

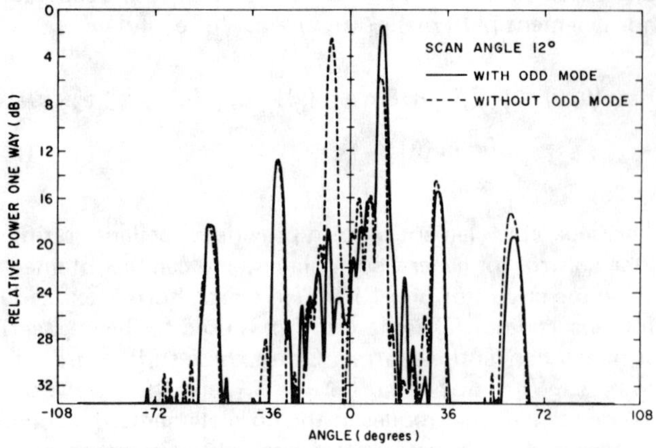

Figure 10.9 - Prototype Array Radiation Patterns (12° scan)

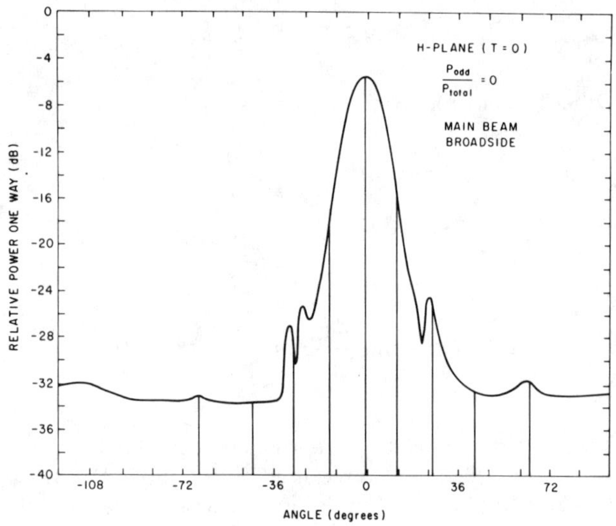

Figure 10.10 - H-Plane Broadside Pattern

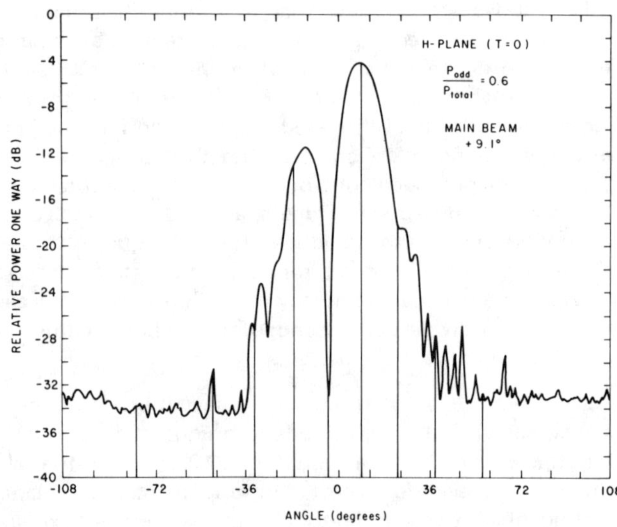

Figure 10.11 - End-of-Scan H-Plane Pattern

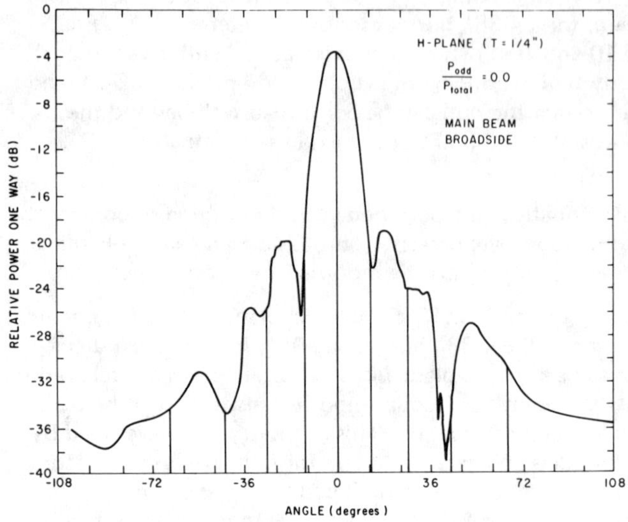

Figure 10.12 - H-Plane Broadside Pattern for Horn with Dielectric Slabs

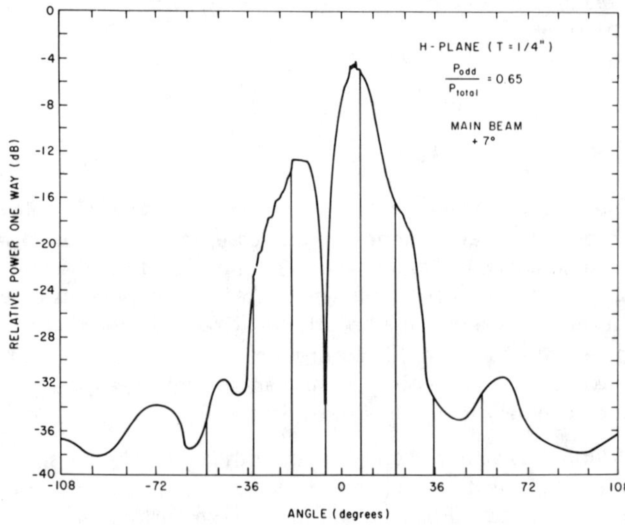

Figure 10.13 - H-Plane End-of-Scan Pattern for Horn with Dielectric Slabs

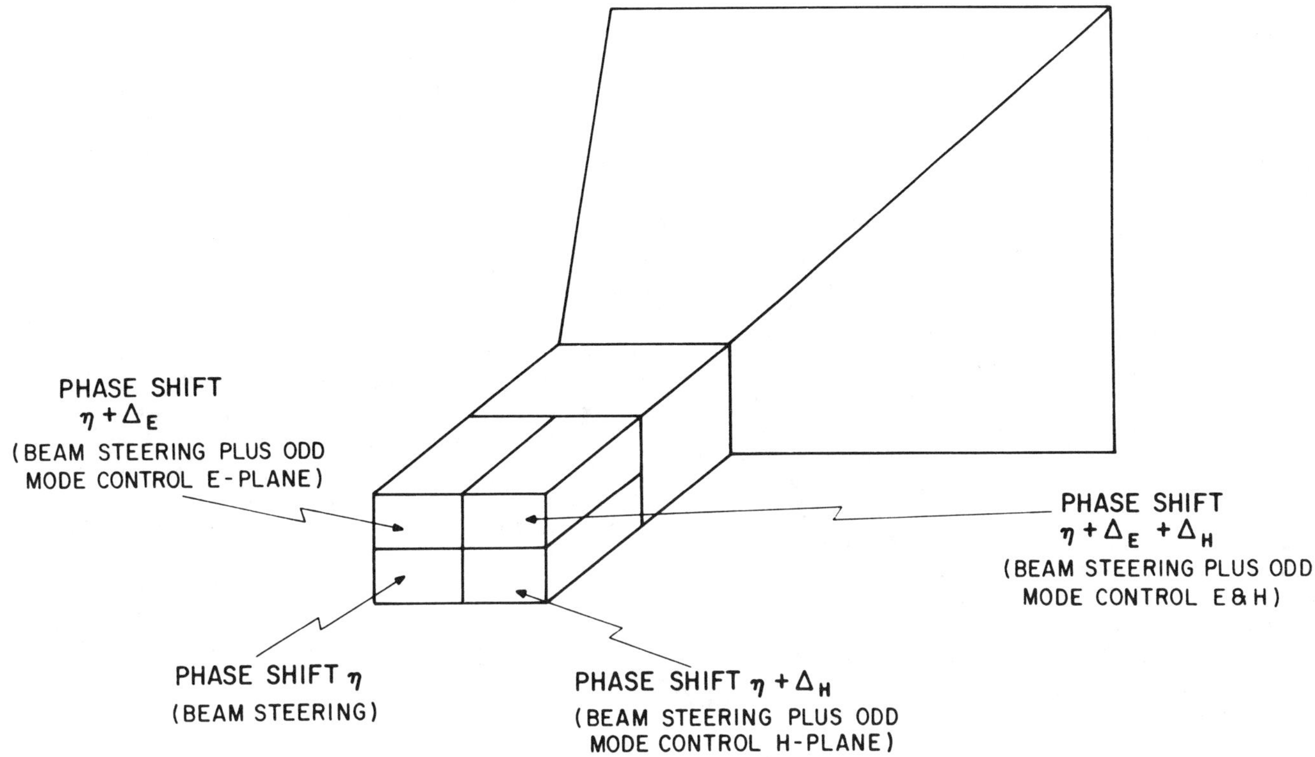

Figure 10.14 - Four Mode Horn and Power Divider Circuit

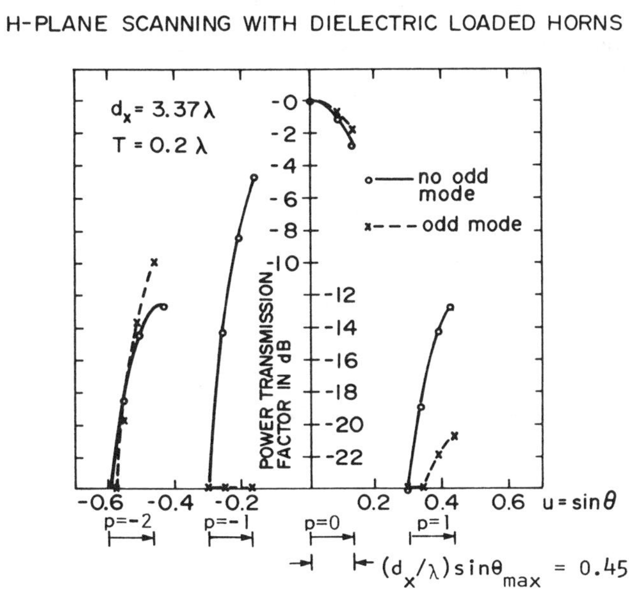

Figure 10.15 - H-Plane Grating Lobe Levels with and without Odd Mode Control

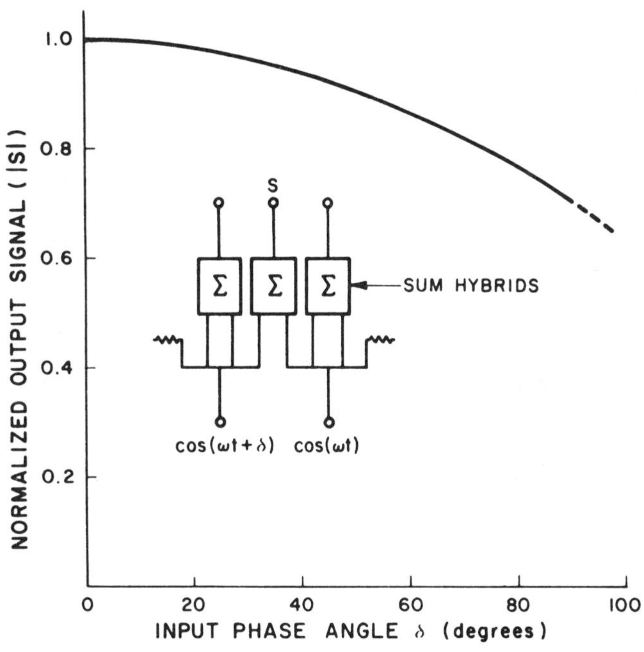

Figure 10.16 - Output of Phase Interpolation Circuit

which is 3/4 of the maximum E-plane scan. Furthermore, this is accompanied by somewhat higher q = -2 grating lobe (at the -10dB level) and a greater gain reduction (nearly 2dB) than the E-plane.

This figure also shows the H-plane scanning results for an element with the dielectric loading but without any odd mode control. As expected, these data show that without the odd mode the p = -1 grating lobe grows very large, but that the case for using odd mode scanning in the H-plane is much weaker than that for the E-plane. In the E-plane, one can scan to $(d/\lambda) \sin\theta = 0.6$ with a -12dB grating lobe using the odd mode, but without the odd mode one can scan only to about $(d/\lambda) \sin\theta = .2$ for a -12dB lobe. Scanning to $(d/\lambda) \sin\theta = 0.3$ results in about a -8.6dB grating lobe. The advantage of the odd mode is therefore a reduction factor of about 1.5 as compared to scanning oversize horns. In the H-plane, Fig. 10.15 shows that at the limit $(d/\lambda) \sin\theta = 0.45$ the main beam is reduced nearly 2dB even with the odd mode, and the grating lobe level (q = -2) is about -10dB. Without the odd mode one can scan to $(d/\lambda) \sin\theta = 0.3$ with a -8.3dB grating lobe (just as in the E-plane) and to $(d/\lambda) \sin\theta = 0.2$ for about a -12dB grating lobe. In this case there is relatively little advantage to the use of the odd mode as compared with using simple dielectric loaded oversize horns.

The arguments made throughout the above paragraphs lead therefore to a system with odd mode control for E-plane scan, with or without odd mode control for H-plane scan. Residual grating lobe levels however are approximately at the -12dB levels, and too high for most applications. The techniques introduced in subsequent sections are addressed to lowering these levels and providing better scan characteristics for arrays of this type.

10.1.4 Overlapped Sub-Array Techniques. Several approaches have been indicated for synthesizing the flat topped sub-array pattern of Fig. 9.5a of Part 1. The orthogonal beam circuit approach of Fig. 9.8 of Part 1 is certainly not a technique that one could use for a very large array and to a lesser extent the HIPSAF antenna suffers from the same problem; very large orthogonal beam matrices are complex, heavy and lossy. Fig. 9.5b of Part 1 illustrates why it is difficult to synthesize lossless circuits that can provide overlapping subarrays with pattern control out to $(d_x/\lambda) \sin\theta = 0.5$. Although the distribution over the central elements are excited predominantly with asymmetrical distributions, and as the element distance from the center increases this becomes even more true. Clearly this excitation cannot be approximated by lowest order distributions in primary apertures. A crude approximation to this distribution is obtained with the phase interpolating circuit described in the next section; a more accurate representation is offered by the multi-mode overlapped subarray.

A. A Phase — Interpolating Circuit. It is possible [9] to devise simple microwave transmission circuits to "interpolate" or derive a signal with phase angle halfway between those of two given signals. This is done by direct addition of the signals when the given phases are less than ±180° apart. The amplitude of the resulting interpolated signal varies as a function of the phase difference between the two given signals. An example of such a circuit arranged to provide excitation for an array is shown in Fig. 10.17.

One can estimate the side lobes associated with a system by considering that two excitations are applied to the array. The first form of excitation has each element driven with signal amplitude unity and progressive phase. The element spacing d/2 is chosen so that no grating lobes exist within the scan region, and so the scan performance of this array can be made highly efficient by conventional scan matching techniques. The second excitation is applied to every other element, and is the difference between the interpolated signal |S| and the uniform progressive phase excitation. This excitation has a lattice of period d, and so has grating lobes within the scan plane. To gain maximum advantage of this technique, the distance d is chosen greater than one wavelength, and so these grating lobes are present at all scan angles. We solve for the amplitude of this signal (1 - |S|) for the case of 20dB grating lobes. This excitation radiates into three points in space, the main beam direction and that of the q = ±1 grating lobes. Consequently, the signal amplitude for each lobe is reduced by multiplication by $1/\sqrt{3}$ (assuming a constant element pattern). In addition, the power gain of the array formed by exciting every other element is ½ of the full array, and so all lobes radiated by it are reduced by multiplication by $1/\sqrt{2}$. The amplitude (1 - |S|) for a 20dB grating lobe signal is found from

$$\frac{(1 - |S|)^2}{2 \cdot 3} = 0.01 \qquad (10.9)$$

or |S| = 0.75. This estimate is conservative because the element pattern will provide further suppression of the grating lobe radiation. It is normalized to a signal of unity amplitudy for the progressive phase distribution.

When two signals of unity amplitude are added in this manner, the interpolated signal amplitude is:

$$S = \tfrac{1}{2} \frac{\sin\delta}{\sin(\delta/2)} \qquad (10.10)$$

where δ is the phase angle between the two given incident signals $[\delta = \frac{2\pi}{\lambda} (d) \sin\theta]$. This amplitude, plotted in Fig. 10.16 is unity for $\delta = 0$, and is 0.707 for $\delta = 90°$. Therefore, if all signals have equal amplitude at $\theta = 0$, this $\delta = 90°$ case is very nearly the 20dB limit. Thus one can scan to the angle

$$(d/\lambda) \sin\theta = 0.20 \qquad (10.11)$$

where, to suppress the q = -1 grating lobe of each element, one must also require that

$$(d/\lambda) \leqslant 1.6$$

Some improvement can be gained by choosing the power division so that the interpolated signal is larger at broadside. Instead of its varying from unity to 0.707, one can choose the excitation so that at broadside its amplitude is 1.25, and its minimum about 0.707. The allowable maximum value of θ corresponds to |s| = 0.565 or $\delta = 110°$, which gives as the approximate maximum scan criterion:

$$(d/\lambda)\sin\theta_m = 0.3 \text{ and } d \leqslant 1.54\lambda \tag{10.12}$$

The $(d/\lambda \sin\theta_m) = 0.25$ limit can be obtained using the hybrid arrangement of Fig. 10.16, with no power loss at broadside, but in order to provide unequal power at broadside requires that the signals which are not interpolated are attenuated by about 1.9dB, corresponding to a loss of 0.85dB in array gain.

The subarray distribution for this technique is the signal applied to each of three consecutive antenna elements with only the center element excited. Using four way power dividers as shown in the figure, the signal at the central antenna is $1/\sqrt{2}$, and the signal applied to the adjacent two elements is $1/2\sqrt{2}$. ahis is only a crude approximation to the distribution of Fig. 9.5 of Part 1, but the circuit is extremely simple as compared with other available techniques, and does provide a factor of two reduction in phase controls over a wide band of frequencies.

B. Multi-Mode Overlapped Subarray. The subarray distribution i(x) can be approximated by higher order mode distributions [10] in horn apertures so that the element spacings can be made equal to the distance d between subarrays. This work is an outgrowth of studies on the active odd-mode control of element radiation patterns, but the developments in overlapped subarrays differ in concept, in means and in results from the multimode scanning technique. The present study describes a passive interconnecting network to synthesize a flat topped, symmetric, sub-array pattern, while the multi-mode scanning technique requires active control of odd mode amplitude and achieves much greater scan per element (although slightly less per phase shifter).

The motivation for choosing to introduce odd modes into the apertures in this case is to better approximate the i(x) pattern at the two elements on either side of center. The technique consists therefore of interconnecting all elements in overlapped sub-arrays of three elements each (nearest neighbor coupling only), so as to produce and control a higher order mode distribution at each element. The basic circuit is much like the phase interpolating circuit of Fig. 10.16 but provides a higher degree of aperture control and allows an element size times scan angle product in the E-plane of approximately

$$(d_y/\lambda)\sin\theta_m = 0.33. \tag{10.13}$$

The array thus uses only 1/3 as many phase shifters for $\pm 7^\circ$ scan and 1/4 as many for $\pm 5^\circ$ scan as a conventional array with 0.9λ element spacing.

Fig. 10.17 shows a schematic of the transmission line circuit for overlapping three elements in an array for E-plane scan. Viewed as an overlapped sub-array with only the central element excited, each coupler with coupling coefficient C and transmission coefficient T is excited by in-phase signals with normalized amplitude $1/\sqrt{2}$. The signals coupled into the circuits at left and right are $C/\sqrt{2}$. For symmetric, lossless directional couplers, T can be taken as real, C as imaginary and $|T|^2 + |C|^2 = 1$. At the sum-hybrid port, the signals at both the left and right of the central antenna are ½C, and the signal at the difference port is +½C for the antenna at left and -½C for that at right. Since C is imaginary, the sum port signals at the left and right antennas are 90° out of phase with that of the central antenna, and these constitute an error in the approximate i(x) distribution for the sub-array. Fortunately this error is of little consequence and has the advantage of providing full aperture efficiency at broadside to outweigh its disadvantage.

In practice, the sum and difference hybrid shown in Fig. 10.17 need only be a junction into a dual mode waveguide, as indicated in the lower figure. Since the modes propagate with different velocities in the horn and dual mode section, its length is chosen to provide an extra 90° delay in the odd mode path as required for the proper aperture field.

An alternative perspective is obtained by assuming that all of the other array input signals are present. In this case the central feed point is still excited by the unity signal and the left end right feed points are excited by signals $e^{j\eta}$ and $e^{-j\eta}$ respectively to form a beam at $\theta = \sin^{-1}[\eta\lambda/2\pi d]$. The sum signal at the central element is $T + C\cos\eta$ and the difference signal is $\pm j\, C\sin\eta$ and is real. The extra $\lambda/4$ delay makes the difference signal imaginary as required.

A similar combination of signals is present at every antenna, and so the net result is that only the even mode contributions are present near broadside, and all the power is distributed to that even mode. Thus, since the two even contributions are 90° out of phase, their power adds to the total input power at broadside. The odd mode term grows as the array is scanned.

Fig. 10.18 compares two sub-array patterns for an infinite array of overlapped waveguides with element spacing equal to the inter-sub array distance 2.78λ, and for square apertures with zero wall thickness. All data can be scaled in $d\sin\theta$ space. The solid and dashed curves represent coupling values of -7.8dB ($|C| = 0.408$) and -5.45dB ($|C| = 0.534$) respectively, while the dotted curve is the single element radiation pattern. The solid and dashed curves coincide throughout the main beam region and so only the dashed curve is shown. The analytical model used for these calculations does not include mutual coupling, but these results were compared with some computed for an array of parallel plane waveguides including all mutual coupling terms and the only pattern differences occurred in sidelobes more than 40° away from broadside. Grating lobe loci are indicated by the shaded area and the horizontal arrow lines below the figure for θ between 0° and 7°. Odd and even mode amplitudes and phases are computed from the array amplitude equation given earlier. The main result achieved by this three element overlap is that the sub-array pattern is squared at the top and has its first null at about 15° for both cases. This null has resulted from the direct cancellation of even and odd mode amplitudes (at about $(d/\lambda)\sin\theta = 0.77$). The second null occuring at 21°, corresponds to the usual even mode element pattern null (at $(d/\lambda)\sin\theta = 1.0$); the odd mode contribution is also null here because its incident signal amplitude is zero. Beyond these points the curve is characterized by several substantial lobes whose peaks do not coincide with grating lobe locations except for relatively wide angles in excess of 60° ($q = 2$).

The main difference between the two curves of Fig. 10.18 is the level of the sidelobe at about 17° which is -20.5dB for the case of -7.8dB coupling, and -14dB for -5.45dB

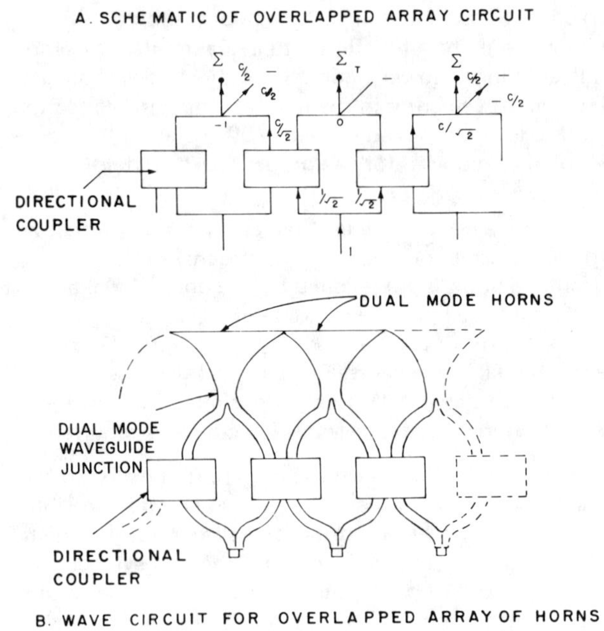

Figure 10.17 - Overlapped Subarray Circuits

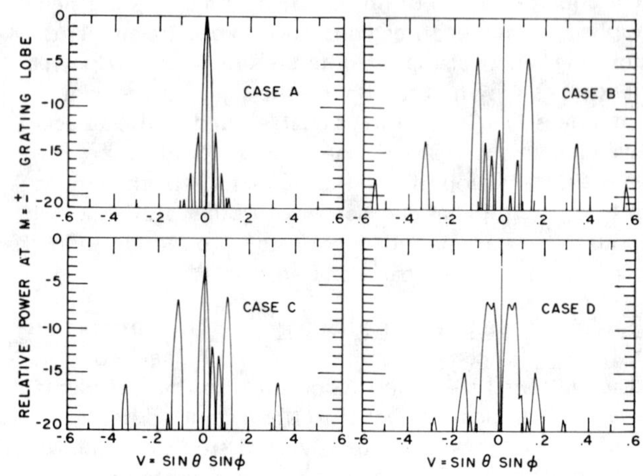

Figure 10.19 - H-Plane (p = ± 1) Grating Lobe Suppression for Eight Element Array with Various Row Displacements

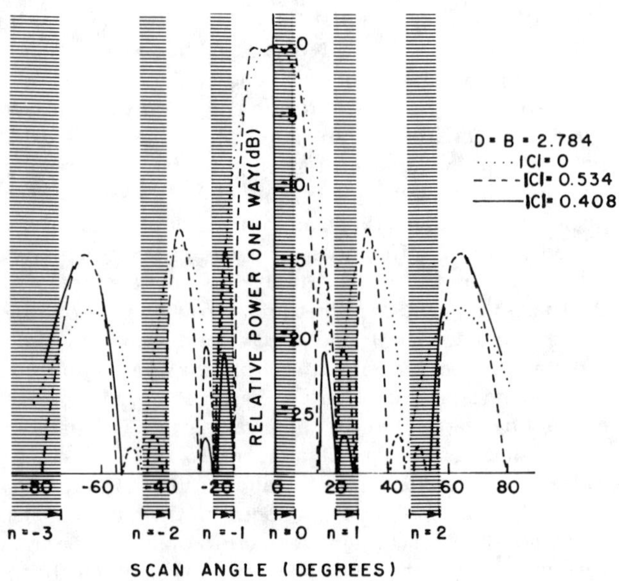

Figure 10.18 - Theoretical Element Patterns for Overlapped Subarray (d = b = 2.784λ)

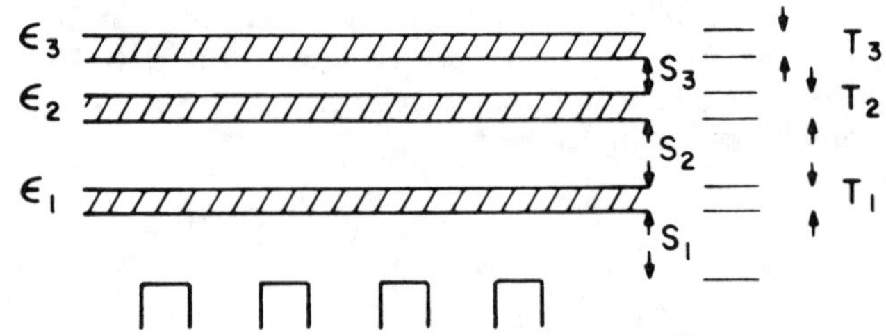

Figure 10.20 - Dielectric Spatial Filter Geometry

coupling. This sidelobe defines the levelof the $q = -1$ grating lobe for positive scan angles. The scan limit is also fixed by this grating lobe position because of the very steep rise in the sub-array pattern just inside of the null at -15°. The $q = -2$ grating lobe sees only a part of the large sidelobe near $\pm 32^\circ$ and the $q = +1$ grating lobe is suppressed until very near the scan limit. The $q = -2$ grating lobe passes through a region with element pattern sidelobe approximately -13dB, but the gain of such an array beam is reduced at least 3dB by the array projection factor. In practice it is also reduced by the finite ground plane effects; as was later confirmed experimentally.

These calculations show that it is possible to synthesize a proper overlapped sub-array pattern with three horn or aperture elements and using even and odd modes. The largest array grating lobes are less than -16dB for beam scan out to $(d/\lambda) \sin\theta = 0.33$ (7° for this case). Circuits have been devised to provide similar overlapped behavior for two planes of scan, but there is not yet sufficient data to compare the relative advantage gained by using the second plane of scan.

10.1.5 Selecting Element Latices for Optimum Scanning. Equation 9.10 in Part 1 shows that by choosing to displace each row of elements by an amount Δ_n with respect to the first row, one obtains an array radiation pattern that is given by the product of the unaltered H-plane pattern $E_x(u)$ and a non-separable E-H plane distribution $E_y(u, v)$, where:

$$E_y(u,v) = \frac{1}{N_y} e_y(v) \sum_{n=1}^{N_y} I_n^E \, e^{j\frac{2\pi}{\lambda}[(v-v_o) n d_y + \Delta_n(u-u_o)]} \qquad (10.14)$$

The maximum value of this pattern is unity and that value is attained at the main beam location $(u,v) = (u_o,v_o)$ for any values $[\Delta_n]$. This pattern also has the same maximum value if $[\Delta_n] = 0$ at the position $v = v_o$ (for all u). Since the total pattern is the product of this distribution and the $E_x(u)$ function, the only major peaks in the combined pattern occur at the grating lobe positions. The v-plane pattern of a uniformally illuminated array ($I_n^E = 1$) with rectangular lattice is shown in Case A of Fig. 10.19 and is the expected -13dB sidelobe pattern characteristic of the uniform array. The form of the sum in equation 10.14 allows one to experiment with the choice of the set $[\Delta_n]$ in order to suppress the peak value of $E_y(u,v)$ as a function of v at the various grating lobe positions $u = u_p$. For example, consider an even number of elements in the E-plane (N_y even). Choosing

$$[\Delta_n] = (o, d_x/2, o, d_x/2, o, d_x/2, \ldots)$$

one obtains at the grating lobe positions $p = \pm 1, \pm 3, \pm 5$, etc., the result: (apart from a complex constant of amplitude unity)

$$E_y(u_p,v) = \frac{1}{N_y} e_y(v) \frac{\sin[N_y\pi(v-v_o)d_y/\lambda]}{\cos[\pi(v-v_o)d_y/\lambda]} \qquad (10.15)$$

This pattern has a zero at $v = v_o$, and an asymmetrical distribution in $(v = v_o)$ with principal maxima of $e_y(v)$ at $(v - v_o) d_y/\lambda = 0.5$. Fig. 10.19 (Case B) shows the function $E_y(u,v)$ with u chosen one half an array beamwidth from the $p = -1$ grating lobe for $N_y = 8$, $d_y = 3.37\lambda$ and $v_o = 0$. This pattern is slightly distorted from the one given in Equation (10.15), which is valid exactly at the grating lobe. Because of the non-symmetrical nature of the array the distortions appear as unsymmetrical sidelobes at and below the -12dB level in comparison to the peak of the beam (at this point half a beamwidth away in the u-plane). The large lobes in this distribution are predicted by Equation (10.15), and have amplitudes about 3.9dB below the peak value of the grating lobe in Case A. In this case the 3.9dB is due entirely to the position of the grating lobes, because the maxima of this function is $e_y(v)$.

At the grating lobes $p = \pm 2, \pm 4, \ldots$, the summation becomes

$$\frac{e_y(v)}{N_y} \frac{\sin[N_y\pi(v-v_o)d_y/\lambda]}{\sin[\pi(v-v_o)d_y/\lambda]} \qquad (10.16)$$

which again is the distribution of Case A and offers no grating lobe suppression.

A third choice is

$$[\Delta_n] = (0, d_x/4, 0, d_x/4, 0, \ldots)$$

For $p = \pm 2, \pm 4 \ldots$ this distribution is given by Equation (10.15), and therefore is shown in Case B. For $p = \pm 1, \pm 3$, etc, the distribution is shown in Case C, and offers about 3dB suppression at $v = 0$, and over 6dB suppression at the other two peaks in v.

The final choice is

$$[\Delta_n] = (0, d_x/4, d_x/2, 3d_x/4, 3d_x/4, d_x/2, d_x/4, 0)$$

As shown in Case D, this distribution offers more than 6dB grating lobe suppression for $p = \pm 1$, independent of the main beam position (u_o, v_o) in direction cosine space.

The examples given above were included to demonstrate independent control of specified grating lobes by lattice selection. The choices have not been optimized, but one can clearly get at least 6dB suppression of the $p = \pm 1$ grating lobes by simply displacing adjacent rows of array elements in the proper manner. The importance of this conclusion to a technique like the multi-mode scanning approach of the last section is that if one chose to use a single odd mode (E-plane) with no H-plane odd mode control and merely scanned out to the $d_x u_{max} = 0.2\lambda$, and chose the $[\Delta_n]$ according to the last example, one would have about -18dB grating lobes in the principal u-plane. Unlike the situation for an array with random element positions, there is practically no penalty to pay for this advantage. The array remains periodic and can be efficiently excited by space feed or corporate feed power dividers, and the phase shift is simply progressive throughout any row with offset phases between the rows.

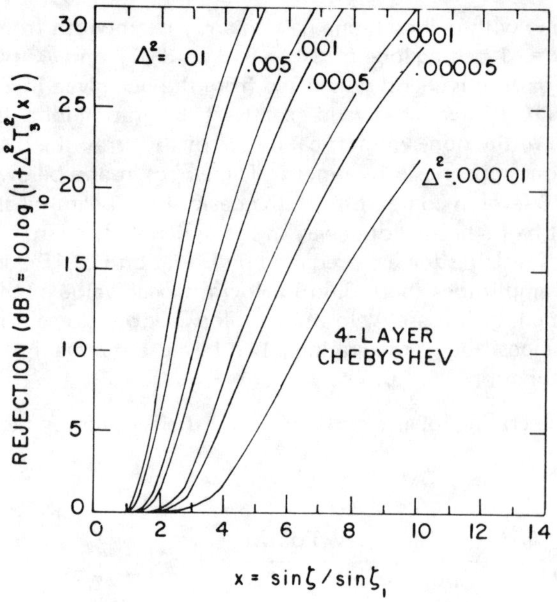

Figure 10.21 - Filter Rejection Characteristics

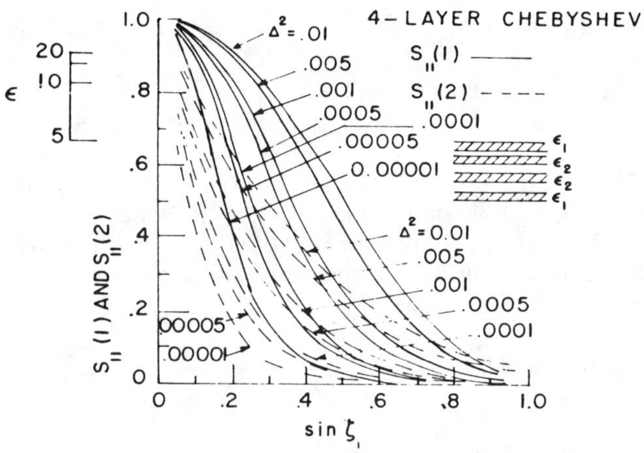

Figure 10.22 - Filter Layer Reflection Coefficients

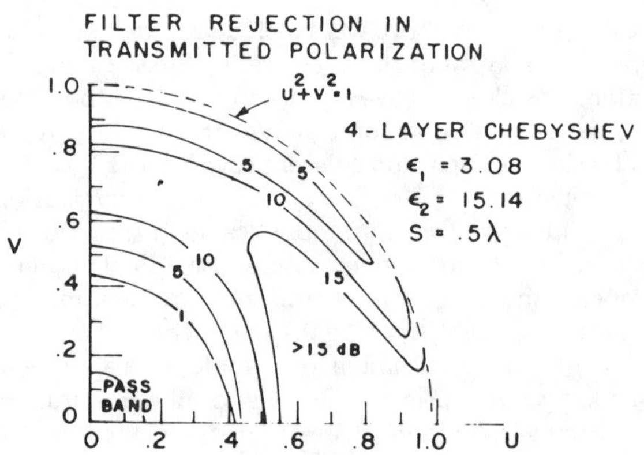

Figure 10.23 - Filter Contour Plot (s = λ/2)

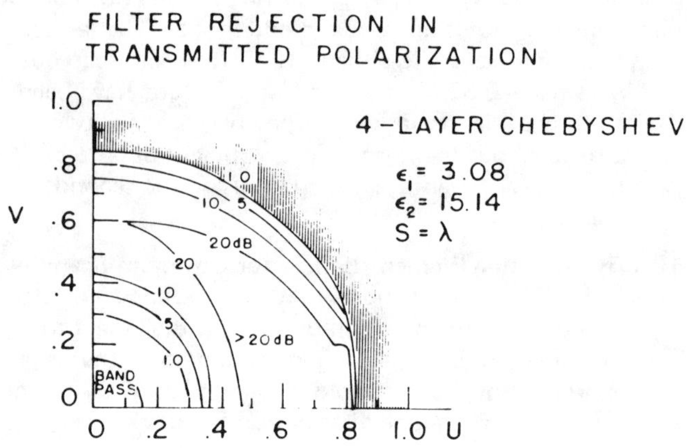

Figure 10.24 - Filter Contour Plot (s = λ)

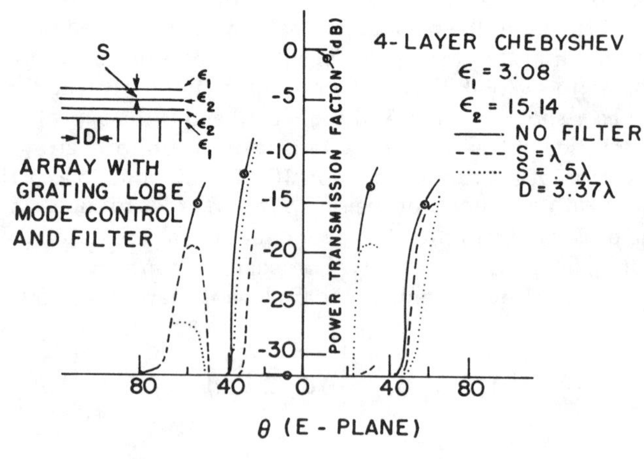

Figure 10.25 - Limited Scan Array with Filter (s = λ)

10.1.6 Dielectric Layers as Spatial Filters: A new technique with application to limited scan antenna design is the synthesis of spatial filters made of dielectric layers. The principles of layered dielectric frequency domain filters and impedance transformers are now well established, and these principles have recently been applied to the design of filters that provide spatial discrimination. [11] The filter geometry, as shown in Fig. 10.20, consists of flat layers of dielectric separated by air, and placed directly over the array face. The examples described here have uniform spacing of 0.5λ and λ, and use quarter wave thick layers of dielectric. The dielectric constant ϵ_j for each layer is determined so as to produce a Chebyshev-type behavior within a given pass band.

Spatial filters can be used to reduce the grating lobes of limited scan array antennas by choosing the spatial pass band to extend from $\sin\theta = 0$ to the maximum scan angle while the stop-band corresponds to the residual grating lobe loci. These characteristics are particularly appropriate for use with the multi-mode scanning designs, for these have good gain characteristics with scan but grating lobes of about -12dB at the scan limit.

Defining

$$\zeta = (2\pi/\lambda)S\cos\theta \tag{10.17}$$

for inter-layer spacings s and u = 0, v = sin θ, and defining the ripple in the inverse of the power transmission factor to be Δ^2, one can write the power loss polynomial as:

$$A_{11}A_{11}^{\ *} = 1 + \Delta^2\ T_m^2\left(\frac{\sin\zeta}{\sin\zeta_1}\right) \tag{10.18}$$

Here the A_{11} is the inverse of the filter transmission coefficient, Δ^2 is a real, positive number and defines a ripple level in the filter passband characteristics, T_m^2 is the square of the m'th order Chebyshev polynomial, ζ_1 defines the end of the stop band, and the T_m polynomial oscillates between ± 1 within the pass band

$$0 \leqslant \sin\zeta \leqslant \sin\zeta_1 \quad .$$

Outside of the passband, the power loss polynomial grows monitonically with sin ζ. Notice that the basic filter design is synthesized for E-plane scan (u = 0).

These characteristics specify the filter. The synthesis process consists of choosing the distance S, transforming the wave matrix description of the power loss ratio into a form that can be identified as the square of a real polynomial, and solving for the dielectric constants ϵ_j of the dielectric layers. Since symmetric filters are chosen there are only two unknowns for a four layer filter.

This synthesis has been carried out and studied in some detail for two, three and four layer filters. Fig. 10.21 shows the rejection possible using a four layer Chebyshev spatial filter for various values of Δ^2. Given the desired maximum scan angle and the location of the grating lobe nearest broadside for the main beam at maximum scan, one can determine the minimum Δ^2 value consistant with the desired rejection level at that grating lobe. Fig. 10.22 shows the reflection coefficients (and the corresponding dielectric constants) for each filter layer as required for a given $\sin\zeta_1$ pass band edge location for the Δ^2 values of Fig. 10.21. A comparison of these two figures indicates that the major tradeoff is the value $\sin\zeta_1$ vs. Δ^2, for to suppress a grating lobe at any given $\sin\zeta$, one either needs a large $\sin\zeta/\sin\zeta_1$ ratio or a large Δ^2. Fig. 10.22 however shows that neither of these parameters can be made too large without requiring excessively high dielectric constants.

The reflection coefficients and the dielectric constants of the filter layers are functions of Δ^2 and $\sin\zeta_1$, and are not direct functions of S/λ. Figs. 10.21 and 10.22 are thus useful for all S, since they can easily show how the required dielectric constants depend on the choice of spacing.

The contour plots in Figs. 10.23 and 10.24 summarize the transmission characteristics (in dB) of several four-element filters for transmitted polarizations in generalized (u,v) space. These show four-layer filters synthesized subject to the parameters $\Delta^2 = 0.0004$ and $\sin\zeta_1 = 0.15$, for S = 0.5λ and S = λ. This choice of parameters leads to filters having identical layers ($\epsilon_1 = 3.08$; $\epsilon_2 = 15.14$), but yielding a different scan performance for each S value.

The circular contours around the origin in u, v space indicate that within the passband at least, the filter properties do not depend strongly on the angle ϕ. For larger θ values, ϕ-dependence becomes more pronounced and filter behavior is noticeably different, with somewhat more rejection in the H-plane (v = 0) than in the E-plane (u = 0). The filter behavior for large scan angles ($\theta > 60^\circ$) is not plotted for the S = λ case because of the complexity of the contour plots and the generally unsatisfactory filter performance at such angles.

Comparison of the two cases does illustrate that for given dielectric layers, much sharper filter skirts are possible with λ spacing than with 0.5λ spacing. This can be an enormous advantage—if the filter properties beyond about $\theta = 50^\circ$ are not important, because the availability, machining properties, and uniformity of materials of lower dielectric constant (less than 15) are far greater than those of materials of higher dielectric constant.

Cross-polarized fields are excited by the layered geometries, but for the cases noted above these contributions are very small within the pass band (-30dB) and they are also small within the filter stop band. The cross polarized contributions rise to a maximum of about -13dB at points between these two regions, but this is still more than adequate suppression for residual grating lobes.

The numerical results of Figs. 10.23 and 10.24 imply incident fields consisting of one or several uncoupled plane waves. The phased array is a more complex distribution. For an infinitely large array, the field is rigorously described in terms of a spectrum of plane waves, some propagating and some cut-off. If a stratified filter is used over the array aperture, then the waves reflected by the filter are cross-coupled by the array face, and the free space properties of the filter may be modified by its proximity to the array. Fig. 10.25 shows the radiation properties of an infinite array of parallel plane waveguides with dimensions chosen to correspond to the array data of Fig. 10.7 and using the multi-mode technique to eliminate the nearest grating lobe. The solid curve is the computed loci of main beam and grating lobe amplitudes for the array with spacings 3.37λ, and the dotted and dashed curves show the rejection provided by the filters of Figs. 10.23 and 10.24.

A comparison of these curves shows that even in the presence of mutual coupling of the elements in the array, the rejection ratios for the filters placed directly on the actual array are almost identical with those of the isolated filter. The rejection properties of the S = λ filter are particularly appropriate to the given array: all of the near grating lboes are reduced by substantial factors. The remaining grating lobe, near 60° at maximum scan, appears approximately at the second filter passband and so is relatively unaltered by the filter. In practice, this lobe is further reduced by the projection factor of the finite array, and would appear at about the -18dB level with or without the filter. Thus, the filter of Fig. 10.24 reduces all grating lobes to approximately -18dB, and reduces all but the 60° grating lobe to about -20dB. This is accomplished with a four-layer Chebyshev design using materials of moderate dielectric constant.

10.2 Design Examples and Conclusion.

Antenna technology has progressed because there is no one antenna that satisfies all requirements. So it was that limited scan antennas were developed; the available techniques were too expensive and when tradeoffs were made between wide angle coverage and limited sector coverage, limited scan antennas found their place. So it is also between the various antenna techniques described in this text. One technique requires fewest phase shifters, another has the most efficient final aperture, still another is simplest and so on. Some of these relative advantages are highlighted in the following examples.

These examples compare various techniques for obtaining about a 1° beam at 10 GHz scanned through a conical sector with 8° half angle with -20dB sidelobes. Designs are compared on this basis, but occasional reference is made to how the selections would change if, instead of a scanning beam, the design called for fixed multiple beam sector coverage.

10.2.1 Single Reflector Designs. The antenna technique studied by Schell [27 – Part 1] and McGahan [28 – Part 1] despite its obvious promise, will be excluded from consideration in these examples because it has not been proposed as a technique for two dimensional scan. With this one exception, single reflector limited scan antennas have element use factors of about 2.5. Since the minimum number of elements for this 1° pencil beam (gain ~ 45dB) is approximately

$$N_0 = 4\left(\frac{\theta_{max}}{\theta_3}\right)^2 = 256 \qquad (10.19)$$

the array would require roughly 640 elements. At 100% aperture efficiency the minimum antenna size would be approximately 57λ diameter (5.6 ft). Single reflector systems hav aperture efficiencies of between 20 and 35%, and if one insists on -20dB sidelobes, Tang [25 – Part 1] has shown that the aperture efficiency begins to suffer. A very optimistic 30% yeilds an antenna diameter of about 10.25 ft. The actual antenna gain is reduces from the 45dB by about 4dB to account for phase shifter losses, blockage effects, spillover and surface matching.

System bandwidth can be about 10%, monopulse nulls can be kept below 20dB over about a 5° or 6° half angle conical sector, but probably not over the 8° sector unless a substantial sacrifice of aperture efficiency can be made. ICR's of 20dB can also be maintained within the 5 or 6° cone, but again there will be substantial difficulty if the full 8° coverage is required.

Finally, there is the ultimate question about sidelobe levels; or to put it another way, "Shall we ask for relief of that spec. now, or wait until we get into the program?" This question can be asked of all of the antennas discussed in this text. It's particularly appropriate here because the answer involves the choice between an on-axis geometry like AGILTRAC, or an off-axis geometry that can be made to give -20dB (but surely not -25dB) sidelobes over most of the 8° sector. The sidelobe question is very important in the selection of antenna type, because if it is neglected the single on-axis array/reflector combination is a very strong choice because of its simplicity, light weight and low cost.

Concerning the alternate design for multiple beams; the phase shifts for this kind of array/reflector combination are not linear and thus one cannot easily devise multiple beam matrices (or lenses) that will provide proper beam formation.

10.2.2 Dual Reflectors or Lenses. Dual reflector or lens geometries are used with arrays because they can be made to scan further and/or have fewer phase shifters than single reflector geometries. One should consider two basically different design concepts for this type of antenna because they lead to markedly different specifications.

The first type of antenna, follows from the Lincoln Laboratory efforts [29 - 30 – Part 1] and has a relatively small subreflector with a diameter about 0.3 times that of the main reflector. The element use factor for a confocal paraboloid geometry is between 2 and 2.5, thus requiring between 512 and 640 phase shifters depending upon whether one uses a true paraboloid or optimizes the reflector contours. In either case the final aperture can be designed to operate with good efficiency and would be about 7.6 ft. diameter. System bandwidths of 10% or more can be maintained. Monopulse null depths of 20dB probably cannot be maintained over the scan sector because of the degree of assymetry required to minimize blockage. Sidelobe levels probably cannot be maintained at the 20dB level without some further compromise in aperture efficiency or array size.

Progressive phases are required for steering, and thus these antennas offer the simplicity of row/column steering and the possibility of use in multiple beam systems. Further reductions in element use factor are possible, at a cost of more complexity, if a limited scan antenna array is used to replace the feed array.

The second option for a dual-reflector or lens system is to use a much larger sub-reflector (one with about 0.7 the diameter of the main reflector.) Properly designed system of this type can combine low sidelobes with element use factors of about 1.5 (requiring about 385 phase shifters). The sidelobe characteristic has not been demonstrated but is predicted theoretically. The final aperture diameter is again about 7.6 ft., but now the sub-reflector is also over five feet in diameter. Bandwidths of 10% are achievable for a reflector geometry and much larger for a lens geometry (but at the expense of a very long structure axially). Non-progressive phase shifts are required, and so adaptation to a multiple beam system requires some effort.

10.2.3 Array Techniques. The antenna designed by Manwarren and Minuti [2] is addressed almost directly to the specifications of this example, and thus should meet the requirements as stated. The advantages of such an antenna are obviously in the relatively few (412) phase shifters, 20dB sidelobe control, about a 6.6 ft. diameter aperture and in principle, good monopulse null formation. Feeding and steering requirements may prove difficult and may begin to erode these obvious advantages. The use of unequal element sizes and pseudorandom grid necessitate a complex and lossy power division scheme using cables, and thus raises system cost. Apart from these several areas of concern, it is clear that this antenna is indeed a viable contender to address the given specifications.

As described, this antenna requires highly non-linear phase progressions and so has no application to the formation of fixed multiple beams within a sector.

Other array techniques include the multi-mode antenna designs and several kinds of achieving spatial overlap. These antennas can be considered separately, but is more illustrative to show how various combinations of approaches can be used to provide alternative solutions.

The multi-mode scanning technique can be used directly to achieve two dimensional scan with the limits

$$(d_x/\lambda) \sin\theta = 0.45 \qquad (10.20)$$

in the H-plane using horn elements that are tailored to have good grating lobe suppression at broadside, and

$$(d_y/\lambda) \sin\theta = 0.6 \qquad (10.21)$$

for the E-plane scan limit.

These numbers combine to give an element use factor of about 3.4 (or 860 phase shifters). This solution is a relatively complex one, implying as it does, the use of three higher order modes, apertures with special H-plane grating lobe control at broadside, and lenses in each element to remove the phase distortion in each horn.

An alternative solution is to use no H-plane correction while still using odd mode control in E and H-planes. This choice obtains H-plane scan out to the usual E-plane scan limit, and the element use factor becomes approximately 2.55 (654 phase shifters). This choice implies H-plane grating lobes at approximately -10dB at broadside, and these can be reduced substantially by stagering the H-plane row positions. Grating lobe levels of less than -20dB can be obtained in this plane for an array of the size described and without scarificing the convenience of progressive phase shifts. Grating lobe control for E-plane is achieved using dielectric layer filters of this type described in section 10.6. Although potentially able to satisfy the specifications of this example, the array described above is still complex, and is basically narrow band band because of the need for controlling four modes in a waveguide geometry. Unwanted higher order modes tend to produce null filling and thus decrease the achievable suppression of the first grating lobe in each plane.

One alternative is to omit the odd mode H-plane control and use row staggering to reduce the H-plane grating lobes. Then, as indicated in Section 10.1.3, one can scan to about

$$(d_x/\lambda) \sin\theta = .25 \qquad (10.22)$$

for a 10dB grating lobe and element use factor of approximately 3.06 (785 elements). This array would still require dielectric slab loaded elements for H-plane pattern control at broadside, and odd mode control in the E-plane. In addition to simplicity, it has several other advantages over the other designs that are well suited to use with the dielectric filter geometries described in section 10.1.6. Using the parameters of this example, one obtains $d_x = 1.8\lambda$, and so $p = -1$ H-plane grating lobe varies from $u_{-1} = -\lambda/d_x, = -0.555$ to approximately -0.42, and this is well within the region of high filter rejection achievable with a dielectric layer structure. Under these circumstances there is no need for H-plane row staggering and one can stagger the columns in the E-plane; this is advantageous because the nearest E-plane grating lobes are nearer to broadside (because $d_y/\lambda = 4.3$) than the H-plane grating lobes, and so are basically more difficult to correct with the filter. The grating lobes of primary concern are the $q = +1$ and $q = -2$ (because the $q = -1$ lobe is suppressed by the odd mode), but the combination of filter and E-plane staggering allow one to eliminate the collimating lenses which maintain good grating lobe suppression in both planes.

Still other possibilities become evident when one considers the use of the several spatial overlap techniques in conjunction with spatial filters and row or column displacement. For example, the phase interpolation scheme of section 10.4.1 was restricted on the basis of maintaining a 20dB grating lobe due to amplitude scalloping and no grating lobe due to element size. Removing these restrictions would allow scan to approximately

$$(d/\lambda) \sin\theta = 0.33 \qquad (10.23)$$

for about -12dB grating lobes. The element grating lobe is not large providing that dielectric slabs are used to move the element pattern null to the $\lambda/2d$ point in $\sin\theta$ space. In this case the grating lobe is at the level corresponding to $(d/2\lambda) \sin\theta = 0.165$, and Fig. 10.15 shows that the amplitude of this lobe is approximately -14dB. The element use factor is about 2.32 for this geometry (595 phase shifters) and -20dB grating lobes should be easily obtainable using the combination of spatial filter and row displacement. The exact choice of parameters would depend upon a detailed tradeoff study, but clearly the satisfaction of design specifications over about 10% frequency range appears within the capabilities of the technique.

A final choice is the use of a multi-mode overlapped phased array element like that described in section 10.1.4 B. This choice would replace the E-plane odd mode control by an overlapped array circuit, and would imply either the use of overlapping or phase interpolation in the H-plane. In either case the resulting element use factor would be about 2.1 (540 phase shifters).

Throughout this listing of options, I have relied upon the spatial filter techniques to justify the claim to 20dB grating lobe suppression. None of the phased array techniques have any difficulty in obtaining 20 or even 30dB near sidelobes, although the latter are nearly impossible with reflector techniques. The array techniques, with the exception of the two aperiodic array schemes, have sidelobes at -10 or -12dB in one plane (after one uses row or column displacement to correct the other plane), and the existance of a realistic spatial filter is imperative if 20dB grating lobe suppression is to be achieved. Aperture efficiencies of 80% are achievable for sidelobes below -20dB, and monopulse null depth is theoretically perfect as long as the choice of (Δ_n) retains symmetry.

The several periodic array techniques require row/column progressive phases for beam steering — although with the addition of a correction term to account for the row or column displacement. Thus, phase control is simplified, as is the use of a multiple beam matrix. The use of odd mode control for multiple beams is best achieved using separate matrices for odd and even modes. A detailed comparison of the number of hybrids for accomplishing the beam formation using active odd mode control with that for a multimode overlapped subarray leads to the conclusion that the passive technique is preferable when multiple beams are required.

10.3 Conclusion.

These examples have served to demonstrate the relative advantages and disadvantages of a number of limited scan approaches. Clearly the single refelctor antennas are the simplest to design and build, although they require more phase shifters and a larger aperture than other approaches, and so may not be least expensive for a particular application. With offset feeds these antennas can achieve -20dB sidelobes over restricted scan sectors. Dual reflector systems have been predicted to yield -20dB sidelobes, but this capability is not yet demonstrated. These systems can use remarkably few phase shifters if the design incorporates an oversize sub-reflector.

Phased array systems that use aperiodic grids now seem likely to provide the -20dB sidelobe control with about as many phase shifters as the dual reflector antennas. Their eventual use in system applications will depend upon cost, complexity and the phase tolerance that can be maintained in a practical system. The collection of array techniques that includes multiple-mode element control and overlapped subarrays leads to moderately small numbers of phase shifters, small apertures and extremely good monopulse null control. They require progressive phase controls and so have greater flexibility for beam shaping and null steering than other techniques. These antennas constitute a new and relatively un-proven technology. They offer the promise of low sidelobes and good polarization characteristics, but these depend critically upon the development of layered spatial filters for the suppression of residual grating lobes.

REFERENCES

[1] W.T. Patton, "Limited Scan Arrays" in *Phased Array Antennas; Proceedings of the 1970 Phased Array Antenna Symposium.* Edited by A.A. Oliner and G.H. Knittel, Artech House, Inc., Dedham, MA, 1972, pp 332-343.

[2] T.A. Manwarren, A.R. Minuti, "Zoom Feed Technique Study", RADC-TR-74-56, Final Technical Report, March 1974.

[3] R. Tang, "Survey of Time-Delay Beam Steering Techniques", in *Phased Array Antennas, Proceedings of the 1970 Phased Array Antenna Symposium,* Artech House Inc., Dedham, MA 1972, pp 254-270.

[4] Whistler, W.T., Hein, C.W. and Kuhn, D.H., "Array Beam Switching Study", General Electric Co., Final Report No. RADC-TR- 67-543.

[5] J. Stangel and J. Ponturieri, "Random Subarray Techniques in Electronic Scan Antenna Design", *IEEE/G-AP Int. Symp.,* Dec. 11, 1972.

[6] Mailloux, R.J. and Forbes, G.R., "An Array Technique with Grating-Lobe Suppression for Limited-Scan Application", *IEEE Tr. AP AP-21,* (No. 5); pp 597-602, Sept. 1973.

[7] R.J. Mailloux, G.R. Forbes, "Experimental Studies of a Multiple Mode Array Technique for Limited Scan Application", AFCRL-TR-73-0685, Physical Sciences Research Papers, No. 575, Air Force Cambridge Research Laboratory, 6 Nov. 1973.

[8] Tsandoulas, G.N. and Fitzgerald, W.D. (1972) "Aperture Efficiency Enhancement in Dielectrically Loaded Horns", *IEEE Trans. on Antennas and Propagation,* AP20, (No. 1) pp 69-74.

[9] R.J. Mailloux, et. al., "Phase Interpolation Circuits for Scanning Phased Arrays", NASA Technical Note, TN D-5865, 1970.

[10] R.J. Mailloux, "An Overlapped Subarray for Limited Scan Application", *IEEE Trans. AP-22,* May 1974.

[11] R.J. Mailloux, "Analysis and Synthesis of Spatial Filters That Have Chebyshev Characteristics", AFCRL-TR-74-0455, Physical Sciences Papers No. 601, Air Force Cambridge Research Laboratories, 13, Sept. 1974.

Computer Control of a Multifunction Radar - Part 1

Dr. Walter Weinstock

RCA, Moorestown, NJ

11.1 Introduction. Today's multifunction radar systems combine phased array radar hardware with modern computer systems. In the radar world, the process of going from systems requirements to hardware parameters is fairly well understood and documented; the companion process in the computer world is not. Furthermore, little exists in the literature about the process of going from overall systems requirements to a balanced allocation of radar and computer requirements. [1, 2] Consequently, hardware and software tradeoffs tend to be driven by problems that are readily understood–the hardware problems. Unbalanced system designs and a disproportionate burden on software development can result if the problem is not approached on a system basis. The purpose of this paper is to disclose the methodology of designing a software system for a multifunction radar. By doing this, it should become clear what the critical concerns are, what the trade-off parameters are, and how the overall systems design process flows into a radar system analysis and a computing system analysis.

This paper starts with the radar systems engineer's point of view — the system is a radar with a computer "box" tied into it. No computer background is assumed on the part of the reader. As we start to probe what this box has to do, we end up going through the process of software architecture — the program systems engineering aspect of the problem.

We will show what functions are handled by the radar and its controller and the sequence of events required to carry out any multifunction operation. Following this, we will examine the controller requirements in some detail and show how the software system is structured to meet these requirements. We will conclude with numerical examples which illustrate the mechanisms by which the radar and its controller saturate.

In the interest of highlighting principles, a generic system has been chosen for the sample design analysis. While its characteristics are consistent with the state-of-the art, its parameters are not intended to reflect any particular system.

11.2 Functional Description of System. The basic jobs of a multifunction radar are the detection and tracking of targets. In the case of detection, this can involve establishing that a target is present in a background of noise, environmental clutter, and interference. Tracking involves the measurement of target parameters on a periodic basis. The parameters which are measured and the sampling rate are determined by the end use of the data. In general, the minimum that is measured are the target coordinates. From periodic measurements of these, target rates can be established and flight path estimated. In addition, future position can be extrapolated. For applications where high accuracy is required, it is generally necessary to have high data rates.

While the complete set of tasks performed by a multifunction radar is application oriented, the set of critical functions is not. An examination of phased arrays for satellite surveillance [3] and ballistic missile defense [4] shows that they require the same basic operations that one encounters in other system applications — such as air defense and air traffic control. Fig. 11.1 presents a top-level block diagram which is representative of these systems.

Note that the radar system has been divided into two parts, the radar and the controller. This is a standard and somewhat arbitrary division which separates the general purpose control computer and its programs from the rest of the radar system. In general, the radar involves all of the hardware associated with the execution of a single dwell. It includes the transmitter, the receiver, the signal processor, and the means for transferring data. The radar controller relates single-dwell information fron the radar with other information. By doing this, it develops track histories and eliminates redundant detections in adjacent beam positions. In essence, the radar is a single-dwell device (even when more than one pulse is required at that beam position to accomplish its purpose - as in the case of MTI or integration) and the controller makes all decisions where data beyond the current dwell is used.

The radar system users are, in general, other data processing systems and operators. The division between the radar and the user functions may not always be clearly definable on a hardware basis. Thus, a single computer may house both the radar control functions and some user processing functions (as in the case of the AN/FPS-85 [3]). Regardless of where the interfaces occur, the radar controller must be able to service a multiplicity of users. It is this requirement, together with the support of a multifunction radar, that demand a "time sharing terminal" type of control computer operation.

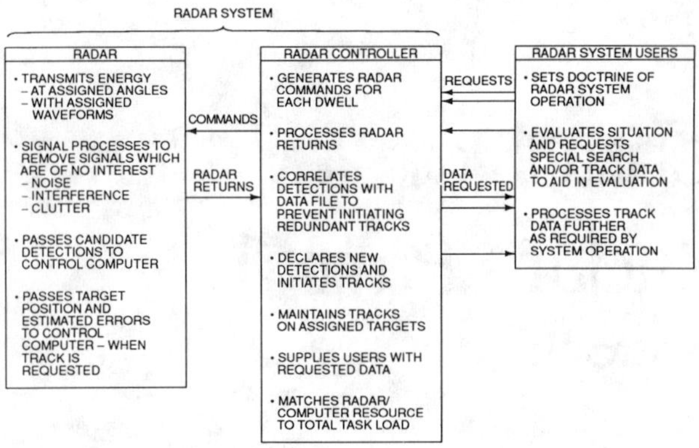

Fig. 11.1 Representative System Block Diagram

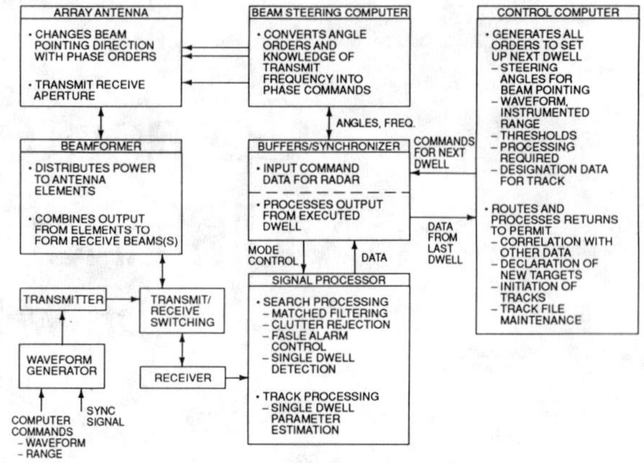

Fig. 11.2 Radar System Block Diagram - Hardware Overview

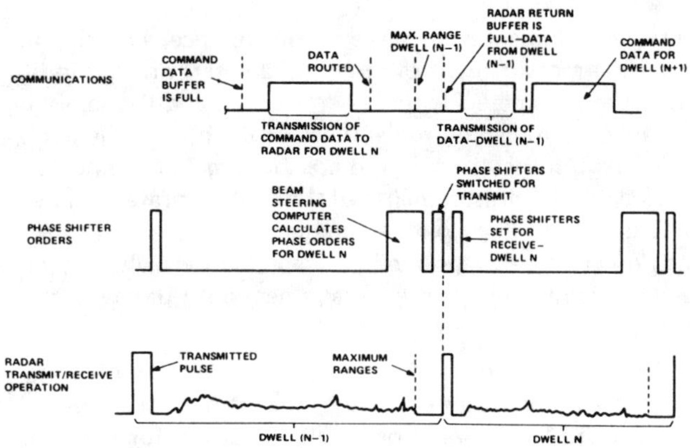

Fig. 11.3 Representative Command/Respond Sequence

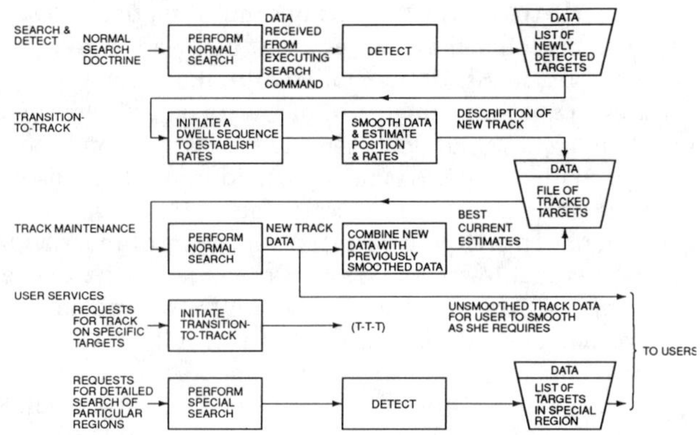

Fig. 11.4 The Basic Radar Processes

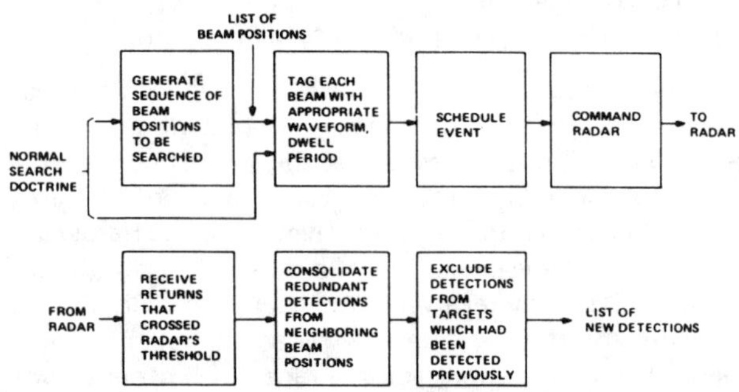

Fig. 11.5 The Sequential Processes Needed for Search and Detect

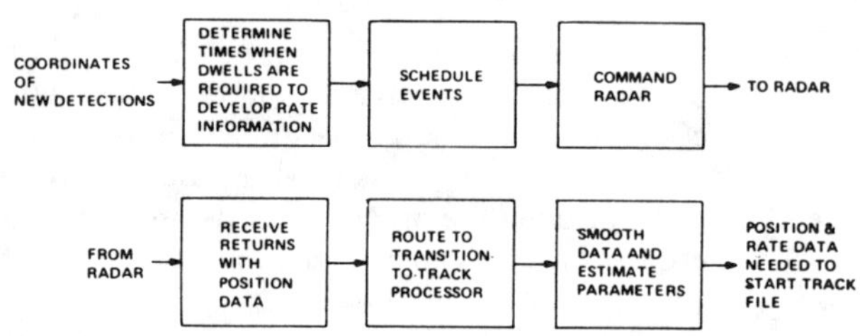

Fig. 11.6 The Sequential Processes Needed for Transition-to-Track

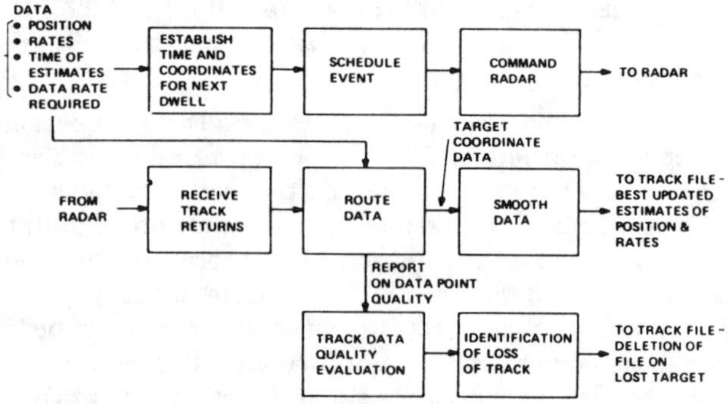

Fig. 11.7 The Sequential Processes for Track Maintenance

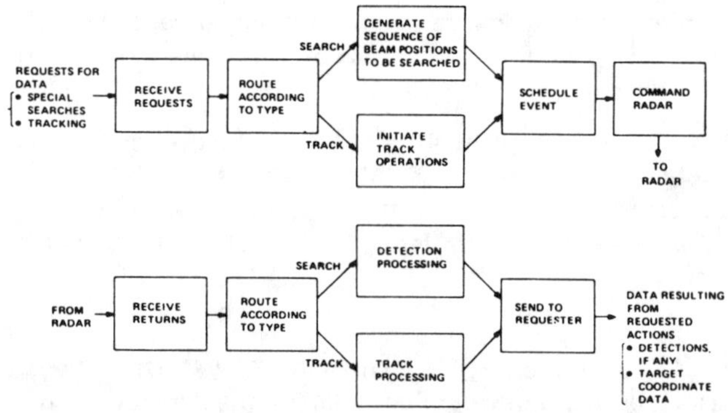

Fig. 11-8 The Sequential Processes Needed for User Services

As we will see, the necessity of being able to handle a number of processing tasks at the same time — by multiplexing computer resources — has a fundamental impact on the computer system requirements. This, coupled with the necessity to do high-priority real-time tasks, drives the computer program design.

11.3 Radar System Operation. Before developing the control computer requirements, let us take a second look at the system block diagram. This time we will do it from the viewpoint of the radar hardware engineer. Since the primary purpose of this paper is to develop the computer control process for a hardware-oriented individual, we will start with a description of the system as he might see it. This is given in Fig. 11.2. A variety of ways exist for feeding power to the multiplicity of elements in the antenna, and generating the phase shift commands to the drivers which set the phase shifters. [5] These are contained within the concept shown in the block diagram, although they can lead to considerably different hardware.

The control computer determines all the conditions for the next radar dwell. It determines the angular position of the beam, the time of transmission, the frequency, the waveform and pulse period to be used, the appropriate detection threshold, and the appropriate support information — such as the dwell's identification number and function. Since the computer operation is asynchronous with the radar, some form of buffering is required. The number of dwell commands accumulated in the buffer is a function of many factors. Among these is efficiency with which the computer can generate batches of sequential commands as compared to generating one command at a time. This topic will be treated in some detail later in this paper.

The accumulating buffers can be in one of three places: they can be part of the control computer's main memory; they can exist as a separate interface entity (as in [3]); or they can be part of the radar hardware. This choice is a subject for sub-system trade-off. The radar synchronizes its transmission to the time commanded by the controller. A common time base is required to effect this operation.

In this way, radar timing synchronization is initiated. The commanded waveform and carrier frequency are generated in the waveform generator and exciter, amplified in the transmitter, and fed to the array antenna.

The phase shifters in the array antenna are set so that the transmitted beam will be in the commanded direction. The controller commands the angular coordinates of the beam and the carrier frequency of the transmission. This information is sufficient to determine the phase shifter settings required at every antenna element. This computation is a natural one for a special purpose computer — the beam steering computer. It is desirable to have this as an integral part of the radar hardware to minimize the problem of communicating the large number of phase-shifter orders involved. Thus, the general-purpose computer gives the three pieces of overall command information — two angles and frequency; the special purpose computer translates these into hundreds, or possibly thousands, of individual phase-shifter commands.

The received signals are processed in the way commanded by the controller. In general, search data is processed differently than track data. In search, we are just trying to establish the presence of a target; tracking requires special processing to allow for the estimation of target parameters. Thus, different paths exist through the signal processing. The routing for any dwell is determined by the purpose of the dwell as stated by the control computer.

The signal processor extracts all the information obtainable from the dwell itself. In the case of search, all of the dwell returns are processed to exclude noise and, in some cases, clutter and interference. Such processing may involve more than one pulse return during the time of the dwell. However, it does not involve information obtained outside of the dwell period. The radar has no memory of past dwells — that is the job of the control computer.

Since search processing can involve hundreds to thousands of resolution elements and since these must be handled during the time of a dwell (typically milliseconds), an extremely high data rate exists. Clearly, this type of problem calls for analog bulk processing or its special-purpose digital equivalent. [6, 7, 8] The signal processor has the task of destroying information which is of no interest. Thus, all of the resolution cells are examined leading — in general — to only a limited number of detections/dwell. The signal processor can reduce the number of candidate target cells by several orders of magnitude. (This is the ratio of all range, angle, doppler cells to those containing target-like returns). The resultant output is something that can be communicated to, and handled by, state-of-the-art general purpose computers.

One of the basic problems of effective signal processing is to minimize the flow of undesired signals, thereby minimizing the possibility of saturating the general-purpose computer. It will become clear in this paper that the control of the saturation problem is one of the most significant factors in determining the computing system design.

To gain some perspective as to how the radar and its controller interface on a timing basis, consider Fig. 11.3. This shows the sequence of events that must occur in parallel with radar transmission/reception in order to effect two-way communications. Basically, a block of command words is communicated to the radar. After these are received, the beam-steering computer calculates the phase-shifter orders for the next transmission. This can be done while the radar is still receiving returns from the last transmission. Once the time of the maximum range return has elapsed, the phase shifters can be switched for the next transmission. At this time, the signal processor is set up to accommodate the next dwell. At the time commanded, the next transmission is executed. When the transmission is completed, reception can begin. Since many systems employ non-reciprocal phase shifters, the array has to have a complementary phase distribution on its aperture to receive at the angle where it transmitted.

While the system is being set up for the next transmission, the data received from the last transmission can be loaded into the buffer for transfer to the control computer. At this point, the information gathered on the last dwell can be transmitted to the computer in the appropriate format. For the scheme just described, which illustrates dwell-by-dwell operation, two-way transfer occurs within the inter-dwell period. The ability to do this puts constraints on the communication rate and the tolerable delays in initiating the communication processes.

11.4 Software System Requirements. We will

now start the process which will yield the design of a software system for multifunction operation.

11.4.1 Functional Requirements. Let us examine Fig. 11.1 again to identify the basic system functions. As we can see, these are:

- Search and detect,
- Transition-to-track after a new detection has been established,
- Continuing track maintenance by acquiring new data and consolidating it with past data, and
- Supporting the various users of the radar by handling their data requests.

There are many ways of categorizing the various radar functions. The preceding list is one way of presenting these as separable functions. Note that functions that make further use of the radar data are considered as part of the user function — rather than the basic radar data. Thus, from the viewpoint of this definition, the processing of tracking data to compute, say, satellite ephemeris or the solution to a fire-control problem, are considered "user" functions.

In general, a data user will process data over a longer span than that required for the target tracking function. The radar's tracking function is optimized for holding the target and extracting data. The radar's output data is smoothed only to the extent required for track maintenance. Subsequent processing to compute quantities like satellite ephemeris will involve a longer memory span and more smoothing. Thus, a hierarchy of memory spans exists within the system. The radar hardware's memory covers only the present dwell. The radar controller's memory covers the number of dwells necessary for track maintenance. The user's memory will, in general, cover a number of these track data intervals, as is exemplified in trajectory calculations and the displays of flight path history.

11.4.2 Basic Processes. Note that each process shown in Fig. 11.4 starts with an input data list or file, and ends producing another. Thus, search and detect has a list of positions in space to investigate, and as a consequence of doing this, generates a file of new detections. This process can continue running indefinitely, with new members being added to the list.

The transition-to-track sequence starts with the list of new detections and generates the data required to describe the new detection in the same way as targets that are undergoing track. This requires that rates be established, given that our initial knowledge of the target is limited to the positional information associated with the detection. Since, at this point of time, there is considerable uncertainty as to where the target will be in the future (we don't know rates yet), data must be acquired quickly. Since quality rate data can be obtained by operating with a large number of samples, over a long time base, the transition-to-track problem is a demanding one. Good rate information must be obtained before the target moves appreciably from its detected position. Thus, suitable rate data is obtained by taking a significant number of samples over a modest time interval. On this basis, the transition-to-track problem is critical because of its demands on systems resources.

Since a single detection dwell can require a sequence of transition dwells — which are critical on a timing basis — care must be exercised to limit transitions to new targets alone. Thus, an extremely important aspect of the detection process involves the correlation of any detection by the radar with previous detections. Such detections can occur for two reasons. First, a previously detected target can be redetected during subsequent scans. Second, if a target is of more than marginal strength, it can be detected in a number of adjacent beam positions. Note that a scan pattern will, in general, not yield these redundant detections in any particular sequence. This is highly dependent on the detailed way that the volume is scanned, and where the target is located. The correlation process is probably one of the most demanding that has to be performed by the controller. Failure to achieve a high degree of success can cause a proliferation of targets and an expanding load on the system.

The track maintenance function is the one that updates the entries in the track file. It continuously generates best estimated position and rates. Its primary job is to guarantee that the quality data will be maintained. The data rate will generally be determined by the end use of the data. To a first approximation, one can consider that the data rate is set by end use requirements and *a priori* knowledge of the target situation. Thus, track maintenance has the job of establishing when the next data point must be obtained, and taking the steps required to obtain it. The minimum rate is that required to hold the target.

Obviously, the track-maintenance problem has some of the time-critical aspects of the transition-to-track problem. The difference is that maintenance only requires the resource commitment on a dwell-by-dwell basis to do its job; transition requires that resource be committed to a number of dwells in sequence to do its present job. The major difference is that if something is wrong during maintenance, a single dwell can be lost; whereas in transition, a sequence of dwells can be lost. As we will see, it is the allocation of critical resources which is the driving force in the computer program design.

The processes described up to this point imply some form of *a priori* tasking of the system. Implicitly, a search doctrine determines what will be searched on a continuing basis, and some form of tactical doctrine will determine data rates on targets in track. Since these processes are in support of an overall mission, with identifiable users, this situation is subject to change. Such change can arise for any one of a number of reasons. One of these is the possibility that the targets seen will exhibit unexpected behavior. Another reason is that the ultimate user can have other sources of data, and might want to task the radar system on this basis. Thus, the users can call for special search or track requirements which are outside of the currently employed operating doctrine. Functionally, this expands the scope of the search-and-detect and transition-and-track maintenance functions. It also leads to a multi-user operation of the system and the need to establish priorities to resolve conflicts over resource allocations.

11.5 Multifunction Processing Requirements.

The four functional processes that we have considered — search and detect, transition-to-track, track maintenance, and user services — have a number of processing requirements in common. To define the data processing system, we must first break down each functional process into its own processing flow. Following this, we will consolidate common requirements. Thus, the task of commanding a dwell execution can be done by one processing module, regardless of the function of the dwell. We shall now consider each process with the view of identifying common processing elements.

11.5.1 Search-and-Detect Processes. The search-and-detect processes are shown in Fig. 11.5. The search part of the process requires that a sequence of radar commands be generated that will allow a specified volume of space to be searched at a specified rate. The extent of this volume and the search rate are, in general doctrinal inputs. Conceptually, there is a list of beam coordinates which must be transferred as radar orders, at the required rate. Practically, it may be more desirable to generate the list as required.

To further define each search dwell, it is necessary to specify the transmitted waveform and the length of the dwell period. In general, the amount of transmitted energy in a dwell depends on the maximum range required. However, it may vary with tactical factors as well. Thus, the transmitted waveform, as well as the maximum range, must be established on a doctrinal basis for the various parts of the radar coverage. Once the angular coordinates of a dwell have been generated, doctrine ties in waveform and dwell interval. Other parameters, such as transmitted frequency, detection thresholds, receiver gain-control settings, etc., can also be tied to the dwell at this point. In effect, a completely described dwell sequence is available — ready for scheduling. It must be recognized that the term "completely described dwell sequence" means that all of the command information necessary to set up and control the radar hardware has been identified.

11.5.2 Radar Dwell Scheduling. The scheduling of the event requires that a time be assigned for the execution of the dwell by the radar. The time assignments should allow the desired volumetric search rate to be realized. Ideally, the system will generate the beam list at the same rate that we wish to scan, and the execution times will be assigned directly. As the search generator adds a new beam to the list, a time would be assigned. Unfortunately, this approach could only be used in a system dedicated to do search. Since other functions, such as track and transit-to-track, must also be handled, scheduling conflicts can arise. Because these other functions are more time critical than search, they would have higher priority in the case of a scheduling conflict.

It is clear that events which want to be executed at approximately the same time must be examined on a priority basis, and execution times must be assigned accordingly. Because of this, the short term average search rate may fluctuate about the desired value. To a first approximation, the scheduling of radar time is done on a task priority basis. There are, however, other resources whose management has a direct impact on the scheduling process. These are computer processing time and the management of radar power. Clearly, one would not schedule an event, regardless of its priority, if the resultant processing requirement or power demand pushes the system beyond its capability. Conceptually, the event scheduling recognizes all these constraints and schedules accordingly.

In practice, there are many ways of achieving these scheduling goals. These are, of course, problem dependent. The type of solution depends on the degree to which the various resource constraints occur simultaneously. In the case where radar time and power dominate, one can assign time and power on a priority basis. [9]

The problem of radar power budgeting is a key one in the allocation of radar resources. On a long time basis, the transmitter duty factor must reflect the average power capability of transmitter system — both its power tubes and its power supply. However, in general, it is possible to exceed the duty factor on a short term basis, as long as the constraints on the average duty factor are met. This means that a dynamic scheduling approach can be employed. Such an approach could adaptively schedule dwells, while satisfying both the short time and long time duty factor constraints.

An alternative to dynamic scheduling is the use of scheduling templates. In this case, the duty factor constraints are met by limiting the average power associated with a batch of dwells. This is accomplished by associating a specific number of dwells of each type with the batch scheduling interval. The slots in the template are filled from queues where dwells of different waveforms are grouped on a priority basis. Template scheduling trades transient response for simplicity in implementation. A high priority group of dwells must be spread over the number of scheduling intervals dictated by the template structure.

In the case where both radar time and processing time are simultaneously constrictive, it may be necessary to allocate the scheduling task to a higher authority — the executive program itself. [10] For the purpose of this paper, we will assume that the radar constraints and processing are handled separately. In this intermediate situation, event scheduling is radar-resource oriented. Processing-resource limitations are handled in an equivalent manner. The data resulting from a priority command may have to be separated on a priority basis, so that processing load can be limited; this is treated in the next section.

Having managed event scheduling to handle the event-priority problem and the radar-power-management problem, a list of events is available for commanding the radar. The process of commanding the radar involves the formatting of the data in a form suitable for communications, and then initiating these communications. Suitable identification must be tied to the commands so that when the results of command execution are sent back, the appropriate actions can be taken. The radar commands are executed at the time designated in the transmitted message. Afterwards, any detections that have resulted are communicated to the controller for subsequent processing.

11.5.3 Controlling False Alarms and Redundant Detections. Since the processes of initiating and maintaining track are very demanding of system resources, it is extremely important to exclude deductions which are of no interest to the system. A large part of this problem is handled in the signal processor where signals due to noise interference, clutter, etc., are excluded by single dwell processing. In addition, areas where spurious signal continuously arise can be blanked out by operator/user actions which inhibit automatic detections in troublesome regions. Thus, detections which are communicated to the computer are likely to be targets of interst. However, even these can present a problem. It is possible to redetect a target which has already been acted on. This can lead to initiating a costly sequence of actions which can result in more than one track file being set up on a single target. Subsequent corelation of tracks can remove the redundancy. By that time, however, a series of high priority resource commitments have already been accomplished, and the system's real capacity has been diminished.

11.5.4 Controlling Multiple Detections. The two basic problems of proliferating detections are multiple detections of a new target and redetecting an old target. The multiple-detection problem arises because of the way volumetric coverage is generated. Search is accomplished by having the radar beam laid down on a predetermined grid in space. Since the beam intensity does not drop to zero at the points where beam width is normally measured, a beam in one position will partly overlap neighboring beam positions. A certain amount of overlap is designed into the system, so that weak targets at the boundary between two beam positions will be detectable. Representative system designs have overlap occurring in the region of the –3dB level on the one-way beam pattern. If a target return is strong, or fluctuates so that it can take on a wide range of values, detection in more than one beam position can occur. The time between these redundant detections depends on the way that the search volume is scanned.

Consider the situation in which a strong return exists where two beams overlap in azimuth. If the search is one in which elevation is scanned and then azimuth is indexed, the time between detections can be an elevation-scan interval. On the other hand, if the target existed at the elevation-overlap point, successive detections can occur. Thus, new radar detections have to be held back until neighboring positions have been scanned. A pool of new detections will develop, which must be correlated against all other new detections developed within the time span of the pool. The processing that develops from this is a function of the number of new detections in this pool, and the degree of similarity of their coordinates.

After this initial correlation has been done, a list of unambiguous detections is available. This must be correlated against existing tracks to insure that we are not dealing with redetecting a tracked target.

11.5.5 Redetection Control. The redetection problem can be handled the same way as one handles a region containing a large number of spurious returns. The radar's detection system can be inhibited in the vicinity of target tracks. This would require preparing a list of all targets that are expected to have the search-beam coordinates, and extrapolating their positions to the time of the search beam's execution. In this way, a set of minimum-extent inhibition gates can be developed. These would have to be transmitted to the radar as part of its command message. Thus, a series of inhibition commands would have to be generated and transmitted to avoid the problem of receiving some or all of the targets back as detections which would have to be correlated. On the other hand, the problems of extrapolating targets and communicating these to the radar can be avoided if one is willing to correlate all detections with the track file.

Thus, two alternate approaches exist for handling the redetection problem. One involves identifying target tracks which are potentially redetectable for each search-beam position, extrapolating, and communicating inhibit commands. The other involves identifying the group of tracks that may correlate with the detections, and then performing the correlation process. However, correlation processing can be expensive when the number of system tracks is large, since the coordinates of each detection must be compared against the entries in the track files. Note that in the case of a new detection, it must be correlated against every track before it is recognized that it is a new target. Thus, the amount of time required for detection processing depends on the number of detections/dwell, the number of possible detections in the vicinity of the dwell, and the number of targets in the track files. While the processing load varies directly with the number of dwells to be handled, it is also a function of a number of other factors in the case of detection.

11.5.6 Transition-to-Track Process. The transition-to-track process is depicted in Fig. 11.6. It is necessary to establish target rates, as well as position, to initiate track with a multifunction radar. This is due to the fact that the tracking data rate can be matched to the problem. Thus, the position of the target, at the time of the next data point, must be predicted. This requires rate information. The transition-to-track process provides the information to start generating the rate data. Since detection produces only position information, the rates must be determined, initially, using during the transition-to-track process. Thus, rate data must be developed before the target moves appreciably. One way to develop rate data, given position only, is to generate a number of samples in a limited period of time. This must be done well enough to predict the first track point.

11.5.7 Track Maintenance. The track maintenance process which follows is depicted in Fig. 11.7. Knowing the data rate required on a target, the time of the next track point is determined. The coordinates for the dwell are extrapolated using past position and rate data. The dwell event is requested for scheduling. It is implicitly assumed that the track waveform and dwell period will be made available as a result of past history on the target.

After the commanded dwell is executed, the track information received must be routed appropriately for further processing. Track information will require special handling so that estimates of radar measurement error can be combined with dwell coordinates to get target position. It should be pointed out that the coordinates in which radar data is gathered is not usually a good set of coordinates for data smoothing. This is due to the fact that a non-maneuvering target flying a straight-line course will produce accelerations in the coordinates of a non-cartesian system. Thus, a straight-line, non-radial flight of a constant-velocity target, shows continuous radial accelerations in spherical coordinates. Radar errors must be handled in radar coordinates so that a positional estimate can be made. This can then be converted into cartesian coordinates for data smoothing and extrapolation.

The received data must also be assessed as to its quality. This must be done on two accounts: the first relates to how it should be blended with past data; the second is to permit an assessment of the track and to determine if any management actions should be taken. Such actions might change the doctrine of energy management associated with the track or, in the extreme case, can result in notifying the user that corrective actions are not useful and the track is effectively lost. If this occurs, the management function must remove the track from the file, or else the processing would attempt to continue updating the track.

11.5.8 User Services. The final process is the user services shown in Fig. 11.8. Basically, these cover special cases of search and detection and transition-to-track/track maintenance, as well as forwarding the data that goes to the users on a doctrinal basis. Fig. 11.8 emphasizes the non-routine operation since additional processing is required to accommodate these. It is clear that even though the user service processing is functionally separated from the processes discussed earlier, the data processing required is not.

The following are the references on signal processing mentioned during the lecture, together with others which may be of general interest.

Allen, J., Computer Architecture for Signal Processing, *Proc. IEEE*, Vol. 63, No. 4 (April 1975), pp. 624-633. This discusses various approaches to programmable signal processors. It shows the types of requirements which such machines are designed to meet, and shows the basis on which such machines can achieve a 100-to-1 speed advantage over general purpose machines, while staying relatively cheaper.

Freeney, S.L., Special-Purpose Hardware for Digital Filtering, *Proc. IEEE*, Vol. 63, No. 4 (April 1975), pp. 633-648. This article is complementary to 6A, giving the non-programmable approach to the signal processor.

Fisher, J.R., Architecture and Applications of the SPS-41 and SPS-81 Programmable Digital Signal Processors, *EASCON '74 Convention Record*, pp. 674-678. Reading this requires some computer background. It is an excellent article by the developer of the only commercially available programmable processors. It elegantly depicts the system concepts and implementations used to develop 100 times the processing speed of a minicomputer with only ten times the number of parts.

Cole, Jr., E.L., and Beaver, E.W., Matrix Controlled Signal Processing, *EASCON '74 Convention Record*, pp. 592-599. This is a clearly written companion to 6C. The implementation approach is different however--an array of microprocessors. The authors start by considering the different architectural concepts that are possible , and logically develop their approach. The article provides a good feel for the major differences between signal processors and general purpose processors.

Computer Control of a Multifunction Radar - Part 2

Dr. Walter Weinstock

RCA, Moorestown, NJ

12.1 Definition of Software System. In the last lecture, the requirements for search and detect, transition-to-track, track maintenance and user services were described. The software system provides the mechanization for fulfilling these requirements.

The nature of this mechanization is dictated by several factors. Perhaps the most significant one of these is the fact that the system is multifunction, and, in general, these functions do not all have the same priority. What this means is that a high priority task can cause a lower priority task to be held in abeyance until it is serviced. An equally demanding requirement is that certain time critical processes - such as those associated with tracking loop closure - must be completed within a prescribed time interval.

The combined requirements of priority handling of multifunction operation and the real time constraint has a profound effect on how the software system can be implemented. As we shall see, this drives us into partitioning the processing functions into pieces or modules. It also requires some type of overall management or executive function to coordinate multifunction operation on a multiplicity of targets.

12.1.1 Priority Processing and Multiprogramming. Systems are rarely designed to handle the worst peak load that will ever be encountered. Rather they are usually designed to handle a nominal peak and have some way of handling all the critical tasks while deferring the low priority ones until peak conditions subside. The same thing is true for the software system which we are considering.

The processing concept must allow the various processes to progress in the machine in parallel and, when peak conditions arise, the less critical processes are held in abeyance until the critical ones are done. Thus, if the processing alternates between search-and-detect and track maintenance during a period of moderate load, the machine may switch over to tracking only if a priority loading arises.

The processing must be capable of switching from doing a number of tasks to doing only high priority ones. It must be capable of responding to these transient peak loads in a timely manner, without throwing away the results of partially-completed lower-priority jobs. Furthermore, it must be capable of handling these problems efficiently relative to the amount of processing effected per unit time. These are the problems that drive the system design of the computer program.

The way that the system handles a mix of functional priorities and target priorities is illustrated in Fig. 12.1. Normally, top priority functions occupy only a limited amount of the processing time. Normally lower priority targets can also be handled. The approach for handling lower priority functions is to do them as time permits. Furthermore, to minimize delays, some of the lower priority processes can be segmented so that they can be stopped at a convenient point and then resumed when the high priority task is completed. For the case of the saturating load, operation on low priority tasks can be suspended until the peak load has tapered.

The preceding discussion has shown that the requirements to handle different processes, with different priority levels, can lead to a processor being multiplexed between different jobs. Multiprogramming is the name given to the technique by which two or more separable computer programs may be allowed to execute within a processor before any one program is completed. In Fig. 12.1, the search-and-detect function was interrupted by a higher priority track function. When this was completed, the search-and-detection process continued to completion. It is clear that some means is required to recognize the presence of a high priority processing requirement, terminate the present effort, initiate the higher priority action, etc. This is done by an executive or supervisory program. It must handle the task of multiplexing control of the machine among various processing elements - on a priority basis. It must be capable of interrupting on-going processing when a higher priority event occurs. In general, it must be able to handle a hierarchy of different priority levels. Modern computing systems have been designed to handle this class of problems. Both the computer hardware and the software reflect this. Some of the basic features that characterize the so-called "third generation" computer systems are: the multiplexing of resources are exemplified by multiprogramming, the requirement to handle multi-level priority interrupts with automatic allocation of resources, and the ability to multiplex efficiently among a number of users.[11] ,[12] ,[13]

12.1.2 Program Architecture. At this point, we are ready to start translating functional requirements into a data processing structure. We will do this in two steps. First we will

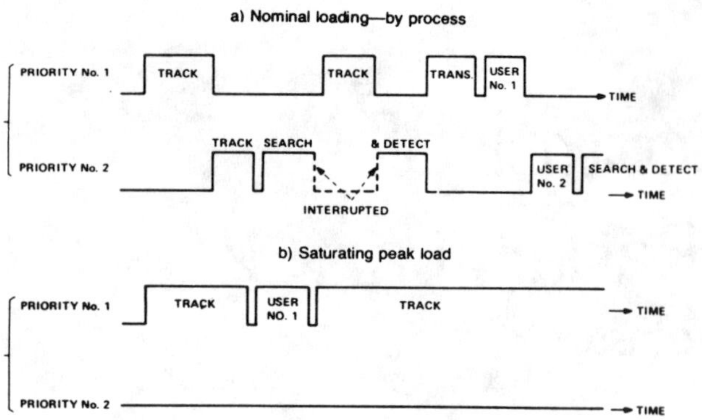

Fig. 12.1 Sample Processing Sequences

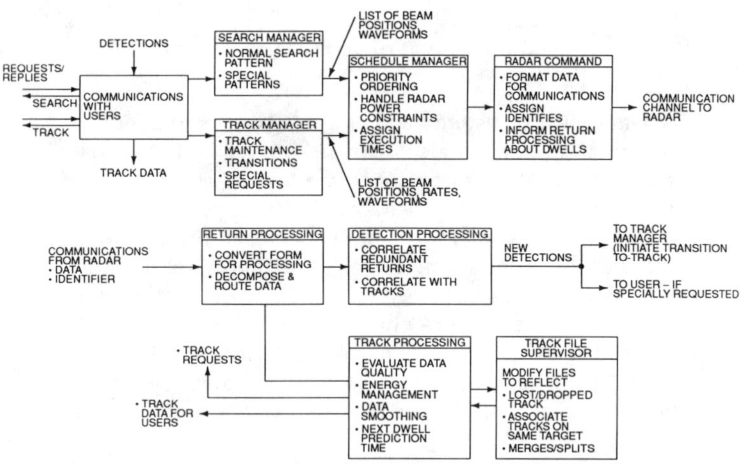

Fig. 12.2 The Allocation of Functions to Processing Modules

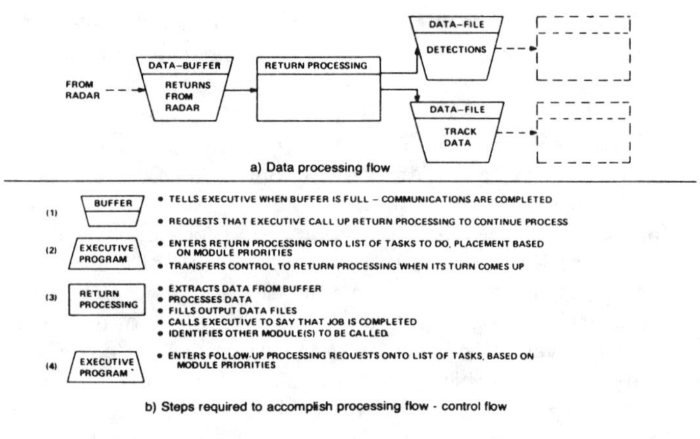

Fig. 12.3 Method of Communicating Between Modules

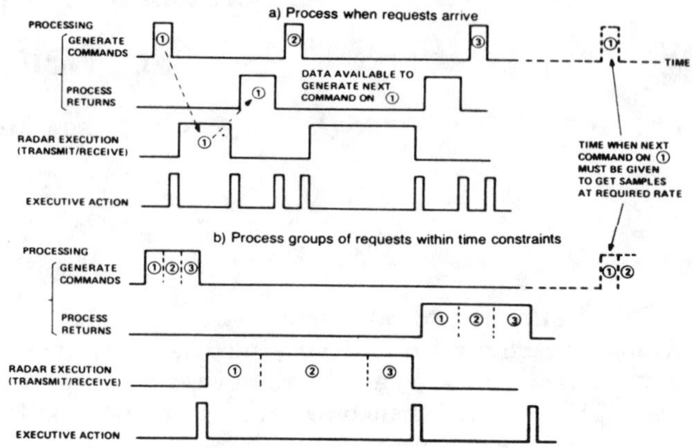

Fig. 12.4 Alternate Ways of Handling High Priority Requests

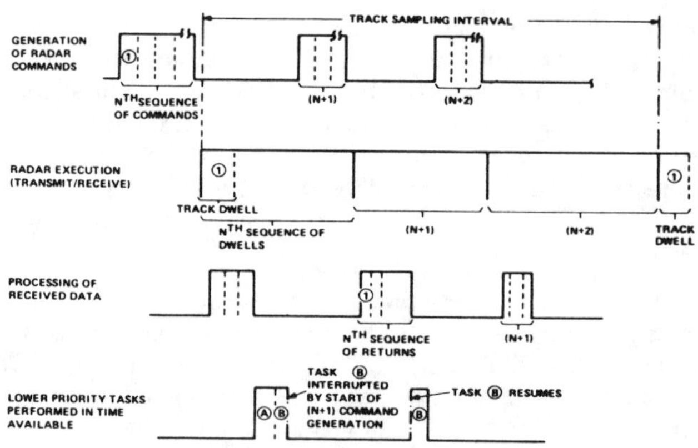

Fig. 12.5 Multiprogramming for Full Processor Utilization and Continuous Radar Operation

consolidate the processing functions defined in Figs. 11.5 through 11.8 of Lecture 11 into a single structure. Then we will reconsider this structure from the viewpoint of handling a large volume of high priority targets. That is, we will optimize the architecture to be responsive to the real time problem under saturating conditions.

We start by noting that each of the four sequential processes calls for processing which serves to interface the computer control system with the radar hardware. In all cases, there is "command radar" and "receive return" processing called for.

The command processing must accumulate all of the information necessary to order the radar so that it can execute the dwells which will be required. The command function must give the particulars of the mode being called for - the waveform to be used, the pulse repetition interval (PRI), the receive control settings such as gain and limit levels, the signal processing configuration, the coordinates of the beam position during the dwell, the transmitted frequency (or frequencies) and the nominal time of dwell execution. All these functions must be performed irrespective of which function the radar is executing.

Furthermore, in the process, it is necessary that this information be in the appropriate format and that a communication channel be opened between the computing system and the radar. Clearly all of these functions should be consolidated into a single processing module - the Radar Command module. This module will be used to interface with the radar regardless of what function the system is performing - search, transition-to-track, track maintenance or user servicing.

Thus, we have a module which is responsible for accumulating the particulars of what is to be done during any dwell, putting the data into the required format and opening the communication channel to the radar. Obviouslv, we will require a companion processing function to handle the problem of introducing the data developed during a dwell into the computing system so that the required processing can be done.

The Return Processing module will handle the companion interface problem. What it does is a function of the mode that the radar was in during the dwell being processed. The data returned during a tracking dwell will be processed differently than data returned during a search dwell. It is Return Processing's job to route the return data appropriately after it has been converted into a suitable form for data processing. In general, this task is not a "black and white" situation. For example, the processing which follows may be dependent on the nature of what is received. Thus, a track dwell which yields more than the expected target in the range interval of interest might initiate detection processing as well as track processing. Likewise, the type of detection processing which the system initiates might be dependent on the state of the external environment, as reported by the radar's signal processing system as part of the radar's output message.

The use of common Radar Command and Return Processing modules was dictated by the fact that the same type of interface processing functions is required in every mode. The need for a common Schedule Manager arises on an entirely different basis.

Each of the various functions such as search, transition-to-track, track maintenance and user servicing will wish to make use of the radar. Each of these processes will, in effect, generate a list of dwells which it would like to have executed in order to accomplish its job. Clearly, there must be some point at which these requests are coordinated so that the use of the radar resource will reflect its responsiveness to time critical problems, priorities by function and target parameters and finally the limitation in radar and processing resource available. This critical consolidation function will be handled by the Schedule Manager module. The detailed way of implementing this function can have a significant effect on system performance. We will discuss this further in the next section, when we treat the problem of optimizing the software system.

The remaining functions in Figs. 11.5 through 11.8 can be grouped into processing modules in any one of a number of ways. Obviously, this significant task involves a number of iterations to handle the real-time constraints. This paper will limit itself to a first-order segmentation in the interest of demonstrating principles. A representative architecture is shown in Fig. 12.2. Subsequent refinement can result in sequential blocks being consolidated, or in individual blocks being broken done into more than one piece.[14] These changes would probably result from the need to handle particular timing restrictions imposed under high load conditions.

As can be seen, the command processing functions of scheduling, generating radar commands, and receiving/routing radar returns have been identified as separate processing modules. Furthermore, certain processes have been grouped in the Search Manager, Track Manager, and the Schedule Manager. This group of modules has the task of generating a complete list of all dwells desired, ranking these on a priority basis, and then arranging for their execution.

Furthermore, note that the critical tracking function has been decomposed into two modules: The first is the Track Processing which must be done to maintain track; the second is the Track File Supervisor which handles the non-time-critical problems of track-file maintenance. This lower priority job can be deferred in times of high loading without a significant penalty in system operation.

The basic operation of the system is as follows: The Search Manager develops a list of beam positions - together with descriptions such as waveforms, dwell period, frequency, etc. - which should be executed in the near future. The Track Manager generates a comparable list. This includes track dwells required to maintain track, transition-to-track dwells, and track points requested by users. The Schedule Manager ranks all these requests on a priority basis, investigates the capability of the radar to support them, and assigns execution times. It is clear that in the case of a track dwell, this processing module must generate the information required to extrapolate the track file to the next track point, since it establishes the time of the this point.

The set of decisions made by the Schedule Manager are translated into a detailed set of orders by Radar Command. Return processing receives the results of the radar executing the commands. Based on the purpose of the dwell (as relayed by Radar Command to Return Processing) and the data that results from the execution, Return Processing segments and routes the received data. Detection Processing or Track Processing follows. Each of these modules can be quite extensive in scope, and may actually represent a series of processes. It is noted that the job of Track File Supervisor has been separated from the basic tracking task, in order to identify the lower priority track operations.

12.1.3 Executive Control of Multifunction Operation. Multifunction operation with a real time system requires that low priority tasks be stopped when time critical processing has to be done. On this basis, all tasks were broken into a sequence of modules. Each module represented a major sub-task in a processing sequence. Conceptually, each module's execution time is reasonably short. This means that - in the event of a priority operation - we can allow the current module to finish its job before we start the higher priority job. This permits tasks to progress in an orderly manner within the system. The responsibility for carrying out this process is the task of the executive or supervisory program. We shall now investigate how it does this and what effect it has on the design and operation of the radar controller.

The Executive program has the task of allocating computer resource to the various processing modules shown in Fig. 12.2. The Executive activates the next highest priority module after the one currently processing is done. A number of things have to exist in order to do this. First, the Executive must have a list of all the modules in the program, together with their priority. Second, the Executive must have some way of knowing which modules are being called for by the different multifunction operations which are in progress within the system. Finally, it must schedule and dispatch modules as the priorities dictate. In order to do all these functions, the Executive itself must use the processing resource of the system. All other tasks must be held in abeyance while these functions are being done. Consequently, there is an overhead cost to support multiprogramming and we will look at this cost in some detail as we develop this subject.

Let us consider a simple example in order to see how the Executive can be made to implement its functions. Let us assume that initially, the system is only doing search. The Search Manager is generating a list of beam positions which are to be searched. The running time allocated to the Search Manager is limited by design. When it has finished its job, it will notify the Executive that it is done, and the machine is available. It will also notify the Executive that it would like the Scheduling Manager to follow it as its successor. The Executive puts the Scheduling Manager on its list of modules awaiting execution, together with its priority. Module priorities are associated with the nature of the multifunction operation.

For the system which we are considering, track maintenance is the highest priority job. As the tracking load increases, we are willing to drop search capability. In fact, if the system is operating at full capability, it can be fully loaded with tracking alone. There is no point in maintaining a generalized search when the system has no capability for initiating tracks on any new detections. On this basis, we might divide the module priorities into three categories: priority 1 - those necessary to maintain track operation, priority 2 - those necessary to support multifunction operation and priority 3 - those which are nice to do when the system has the time.

The priority 1 modules are those necessary to sustain tracking operations: Track Processing, Track Manager, Schedule Manager, Radar Command and Return Processing. Note that the Track File Supervisor is a desirable, but not mandatory tracking function. Priority 2 modules include: Track File Supervisor, Search Manager Detection Processing and Communications with Users. The priority 3 modules would include processing functions which are supportive in nature, such as Background Testing. This priority structure is of course a very simplified version of what one might have in a real system.

Returning to the example, the Scheduling Manager would execute and call for the Radar Command module which in turn does not call for a successor. The process will continue after the radar has executed its commands and transfers its data back to the computing system. The transfer process is recognized by the Executive. The transfer calls for Return Processing. If there were no detections, the process would tend to stop at this point.

In order to insure continuity, the Search Manager must be called again to generate its next list of beam requests. If this module is called up periodically, search will continue when nothing else is happening.

If the radar's signal processor had detected potential targets, then Return Processing would call Detection Processing as its successor. If a new target is detected, then the track Manager is called so that the transition-to-track sequence can be initiated. This in turn will lead to the Schedule Manager being called. At some point, the Search Manager's periodic call will take place. It will be scheduled after the currently queued up priority 1 requests are executed. We will look at some of the detailed timing considerations in the following section.

It is clear that the more modules there are in the program, the more the Executive is involved, and the more the processing resource is used to control the flow of the job rather than its actual execution. On the other hand, if the modules were consolidated into a few long running ones, we would experience difficulties handling time critical tasks which come up while non-critical modules are running. As we shall see in Section 12.2, the reduction of the Executive overhead cost while maintaining real time operation is the key to designing a practical system.

12.1.4 Module Communications. At this point, we will consider how data is communicated and how executive control of computer resources is exercised. After a processing module executes its task, it does two things. First, it puts data into a file which is intended to be used by another processing function. Second, it identifies the processing function(s) that are to use the output of this process. Thus, the way of communicating between processing modules is to establish data to be used subsequently (common data), and notifying the executive program what actions are required to continue the process. Thus, the communication between processing modules is via the common data base and a set of executive program service requests - relative to the actions to follow. The action that actually follows the completion of a particular processing task depends on the priorities in the executive control program.

The executive program must assign priorities (based on a priori rules) to the call-up of the various modules identified in Fig. 12.2. If track maintenance is the most important function to be done, then the second priority tasks will be Search Manager, Detection Processing, Track File Supervisor, and possibly communications with some users. The transfer of control of the computer's resources will depend on the priorities employed in the executive's allocation of resources. It is noted, with the resources just discussed, that under a saturation condition, the job that would have been sustained would have been track maintenance. Search operations would have been reduced in the interest of handling targets that were already entered into the system.

12.1.5 Data and Control Flows. The relationship between data processing flow and control flow is illustrated in Fig. 12.3. Although Fig. 12.3a shows data from a buffer being processed and then sorted into two output tables, the actual events in the machine will probably not flow in this way. As shown in Fig. 12.3b, the assignment of computer resources depends on a priority-ordered list of tasks that the executive has to effect. Actions are performed one at a time — on a task-priority basis. Communications between modules are performed by means of a common data base (one which is accessible to the task modules) and by requests from these modules that the executive initiate appropriate follow-up action.

Thus, three pieces of information are needed to describe how the processing is accomplished: a data-processing flow, a data-base description, and an executive-control flow. The data-processing flow is, more or less, the conventional block diagram of the processing. The data-base description portrays the files that exist in the computer's memory, and which modules can read and/or write in them. This provides the means of interconnection between processing modules. The executive-control flow shows how the control of the machine can pass from one function to the next. Since, in general, the processing elements have multiple interconnections with each other, this is not a simple thing. Thus, the sequence of events that will occur cannot be given in a definite sequence beforehand. They depend on the way that targets and user actions enter the system.

12.2 Concepts for Efficient Real Time Operation Batch Processing for Reduced Overhead.

12.2.1 General Considerations. If the executive program is required to handle calling up different processes, on a priority basis, then the system must pay a price to do this. The price involves the processing required to bring one process to an orderly stop, and then to start the higher priority process. The time required by the executive program to do these operations is overhead. The processing involved relates to computer resource handling, rather than getting any particular task done. Since the time for executive actions is, in general, not insignificant compared to the time required to get a single data point processed, steps must be taken to minimize the overhead.

The problem of efficiently handling data can be seen readily in the saturating peak load case, depicted in Fig. 12.1b. Here we have a sequence of high priority processing requirements. We also have a real-time constraint on the system operation. To illustrate the problem, let us consider two ways of handling these high priority requests. The first will be to process requests as they arise. The other way will be to accumulate requests and batch process. The number of requests that can be accumulated must be limited so that the resultant commands are available at the required time. This is acceptable since the real-time constraints can be satisfied, using lower executive overhead. This is illustrated in Fig. 12.4.

As we can see from Fig. 12.4, the basis for bulk handling of data is that the time to process and execute a track request is less than the time between data samples. Thus, data can be processed in a timely manner with a reduced amount of executive action. The basic reason for this is that there is inherently no difference in processing a list of requests or a single request (a list of one). In both cases, the processing must start by referring to data that must be handled.

The way that the processing requirements are converted into a structure of program segments (or modules) is very much dependent on how we handle the time-critical, peak-load problem just mentioned. The events requiring the most rapid response must be identified. Since, in general, this response time to a request will be longer than the processing time, bulk handling will be possible. The number of targets/requests that can be handled as a group depends on the minimum response time.

Although Fig. 12.4 illustrates the executive overhead problem, it does not show how to make full use of either the radar's execution time or the radar controller's processing time. In the case of the radar, its time should be spent in a continuous series of transmission/reception dwells. In the case of the controller, processing should be occuring continuously in a high load situation. The way to achieve this full utilization of resources is depicted in Fig. 12.5.

A group of commands must be available before the radar gets done executing its present sequence of dwells, if continuous radar operation is to be obtained. Thus, generating future commands must occur in the period that the radar is executing the present set of dwells. In addition, the data from the last sequence of dwells must be processed at this time. As Fig. 12.5 shows, during the interval of the (N+1)st sequence of dwell executions, we must process the returns from the Nth sequence and then generate the commands for the (N+2)nd sequence. If this sequence is not followed, the system will be unable to handle its real-time problem.

The system must process the last batch of received data, the Nth, prior to generating the next set of commands, the (N+2)nd. In the case of tracking, the data from the Nth set is merged with past data so that a best estimate can be made of where to command the beam for dwell (N+2). The length of time that the radar spends executing all the dwells in a batch must be greater than the time it takes to process the returns obtained from the previous batch of dwells, plus the time it takes to generate the commands for the next batch. This situation does not have to be the case for each dwell individually, rather it must be true on a batch basis. Thus, the processing time for commanding the radar and processing its returns cannot exceed the radar execution time. If it does, the commands may be delayed and critical events will be delayed.

Thus, there are three packets of time associated with the process. These are the times for command generation, radar execution, and return processing. Since radar execution times are set, as are the number of executions required per packet, the question that arises is - can the processing for command generation and return processing be done in the remaining time. Implicit in the way this question has been stated is the assumption of a computer hardware configuration. Actually, the basic question is how to trade off between hardware and software in effecting a solution, if there is a processing-overload problem.

The hardware aspect of a solution to the problem, in the limit, could call for another machine with more processing capability. Lesser hardware solutions could call for replacing certain processing, done by the general-purpose computer, with the special hardware. The hardware/software tradeoffs must result in a system that can handle the peak real-time problem. This can be done by suitably taking into account the fact that tasks which are not time critical can be deferred in a peak situation.

The handling of past returns and future commands must be done on a priority basis to avoid delays on critical targets. This is the method for operating when the system tasking exceeds its capability. In general, this type of situation represents an extreme case. In normal operation, one would expect to handle all of the lower priority tasks as well. These can be initiated when they become the highest priority tasks remaining. This is illustrated in Fig. 12.5 for tasks A and B. Such tasks might involve routine search or communicating with a noncritical data user.

Lower priority tasks must be done when time permits. However, they cannot run so long that high priority tasks are unduly delayed. Thus, some method must be provided for segmenting this processing. Two methods are possible. The first involves pre-emption of control by a higher priority task as soon as it arises. This requires that the present task be put aside in such a way that it can be resumed when time permits. An alternate method breaks all lower priority jobs into non-preemptable segments, or modules. Stringent bounds must be put on the execution time of these modules to avoid excessive delays in the transfer of control to the time-critical tasks. The lower priority tasks could be sliced into too many pieces, resulting in short delays, but high executive overhead. The choice of preemption, or time slicing, is a basic problem which will be answered as the software system design is developed.

12.2.2 Fully Utilizing the Radar Resource. We have just discussed how to improve the use of computer processing time by reducing executive overhead. This is one aspect of an even broader problem. That is, how should the system be designed to make the best combined use of both the radar and the processing resources? We shall start our discussion by considering the problem which is companion to the one just discussed. That is, how can the utilization of the radar resource be maximized? There are two aspects to the question. The first is how to use the radar to the limit of its capabilities. The other question is how to make sure that the highest priority tasks are given the proper weight. The answer requires that the requests for radar dwells be priority ordered before the radar dwells are allocated.

The radar constraints that must be contended with are time and power. In general, there are two aspects to the power constraint - short term and long term, since transmitter components and power supplies can handle short term peaks which exceed the long term average capability. Because the pulse energy called for on a transmission can depend on factors other than PRI, the system will be scheduling dwells with different average power levels. Thus, full utilization of the radar means that the high priority dwells are scheduled so that the short term power constraints are not violated while the system is operated to its full long term average power capability.

One generalized method of handling this is shown in Fig. 12.6. The various lists of requests for beam dwells are presented to a priority filtering processor. This processor has available to it the systems priority doctrine. Typically, such a doctrine would classify the relative priority of different types of search and track dwells by functions and possibly by position values. For example, transition-to-track dwells required for rapid acquisition might take priority over updating a current track on a long range target whose coordinates are slowly changing.

Furthermore, the system doctrine may put a limit on how many targets of each priority class will be handling in any scheduling interval - for a batch scheduling system. Such restrictions tend to keep the allocation of resources matched to the expected use of the system.

The output of the priority filter is an ordered list of beam requests which are candidates for scheduling. However, the system must be able to do this without violating the radar's capability. The radar power constraint filter handles this function. If a priority dwell can not be scheduled in the appropriate sequence because of this, it is deferred and rescheduled at a later time.

There are a wide variety of techniques for handling this aspect of the resource management problem. It has been said that there are as many scheduler designs as there are system architects. The basic differences in these design concepts is the way they trade processing burden for better utilization of the system resources. They range in complexity from the simple template scheduler [2] to the completely dynamic scheduler [10].

The simple template handles the combined problems of priority filtering and radar constraints as follows - in the case of a batch system. A series of dwells are allocated for each function (such as transition-to-track, priority short range track, long range track, critical search sector, non-critical search, etc). The track and search queues are entered so that the allocated slots of each type are filled. Implicit in this approach are two assumptions. The first is that the beam request lists will match the template design. For example, let us assume that the template had slots for 2 transition-to-track dwells, 4 track dwells and 4 search dwells — which might represent an average situation. If there were no transitions present at some point in time, these slots would be unoccupied.

A companion consideration is that the slots had to be allocated based on some assumption about average power per dwell. In general, some dwells must call for less than this value. The point is that a fixed scheduling template can not lead to efficient use of the radar's power and time.

This problem can be rectified with varying degrees of sophistication. The price of sophistication is increased processing burden. Instead of having a single fixed template for the scheduling interval, the system can have a small assortment of fixed templates. The size of the various beam request queues can determine which template to use. This will give better time usage, but will not necessarily lead to be power usage.

This problem can be handled by using a partial template — one that has preassigned slots for the critical tasks and an open interval where some of the lower priority tasks can be juggled.

The most general form of the scheduler follows the mechan-

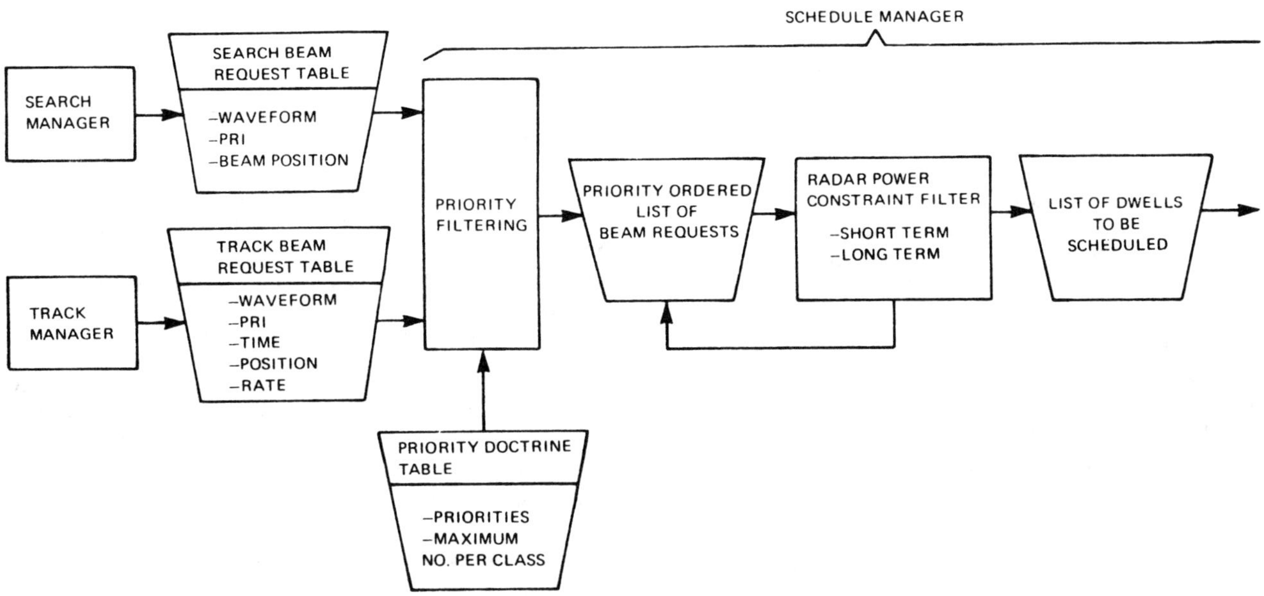

Fig. 12.6 Radar Resource Scheduling

ization depicted in Fig. 12.6. This involves adaptive scheduling. The priority ordered list is presented to the power constraint filter. The filter sees whether any of the power averages will be exceeded if the next request on the list is scheduled. If one of the constraints would be violated, scheduling of that beam is held in abeyance for that particular interval. Clearly, this approach tends to keep the radar operating to the full level of its capability.

The preceeding approaches were based on the assumption that the computer processing resource would be fully utilized and would support the system priority structure. To a first approximation, this is true. The scheduling interval and the number of dwells associated with it were based on having the required processing support to generate commands and process returns. In fact, the size and structure of scheduling templates require that the process burden be assessed in order to determine the structure. Thus, the radar scheduling approaches which have been discussed on consider both radar and computer resources, although the later is considered implicitly. The adaptive approach can be extended to treat both the radar and the processing explicitly.

One way of doing this extends the responsibility of the Executive [10]. In effect, the Executive still continues to allocate the processing resource to task modules but only after the cumulative processing burden is assessed. This is dynamic scheduling with perhaps the highest level of adaptiveness-conceptually. As with any of the other resource management approaches, the choice of the concept has a profound effect on the structure and management of the radar control function.

12.2.3 Further System Considerations. In order to demonstrate principles, certain simplifying assumptions were imbedded in the discussion up to the point. We have considered a single beam radar which stays in a dwell position from energy transmission to the maximum time associated with the preselected PRI. Furthermore, we assumed that even under saturation conditions, conflicts about scheduling critical events could be resolved on a doctrinal basis. We will now comment on these additional considerations.

Some advanced systems have considered the use of multiple receiving beams to effect high resolution search at high data rates. As far as the control system is concerned, the basic problem is the same. There is an amplification of the return processing burden because of the increased magnitude of the problem and because of attendent problems with correlating data. However, the approach called for is an extension of what has been discussed to this point.

Clearly the same thing is true when one considers the application of certain optimization techniques. For example where radar time usage is critical, the dwell interval can be tailored to the maximum range of interest. In the case of tracking, this would be determined by the target itself. While While this does not introduce any significant complication on its own, it has an obvious impact on the processing system itself. Being able to have more efficient use of radar time means that there must be more processing capability. More dwells per unit time means more processing per unit time. Furthermore, this improved use of radar time precludes the use of simple scheduling techniques. If the radar scheduling is going to be tailored for maximum radar usage, the processing must be able to support a varying beam rate also. Otherwise, the system is processor limited.

Another consideration relating to optimization is the optimum scheduling of events to maximize system performance in some measurable way. A good example of this is given by Heffes and Horing [15]. They considered the problem of assigning pulses where the radar is tracking a large number of targets simultaneously. Their problem was one of achieving the highest accuracy by allocating a fixed resource to a group of tracks. Their solution, of course, departs from what has been considered so far, i.e., fixed tracking rates and priority assignments on a doctrinal basis. The impact of the optimization process could be handled by restructuring the Track Manager module. It might also lead to a fundamental change in the scheduling concept. The solution depends on the total system requirements.

Solid state phased arrays present an additional problem. Since their average power and peak power limitations are comparable, it is desirable to operate with almost continuous transmission for efficient operation. This kind of sequence of radar transmissions means that the transmitting beam would be programmed to go from one beam position to the next and be off the air only during signal reception. This particular problem was addressed by Wishner et al [16] who developed a number of algorithms for accomplishing the task. Properly interlacing transmissions and receptions led led to scheduling efficiencies within 20-40% of the optimum (depending on the test problem).

As this section has shown, there are many variations possible in the management of system resources. Even though the same basic principles apply in all cases, there can be significant differences in the implementations. The degree of optimization will dictate the mechanization concept required.

12.3 Saturation Mechanisms.

In order to demonstrate the mechanisms involved, a highly simplified set of requirements and system descriptions will be employed as an example.

12.3.1 Sample System. Let us consider the following case:

System operations : Search and track with priority given to track.

Dwell types: Short range, 1 ms; nominal, 2 ms; extended range 4ms.

Track rates: 1, 2, 4, 8, 16 data points/sec, depending on mission.

Tracking load: Mixed dwell types and data rates. Must be able to handle a nominal tracking load of 120 targets (4-ms dwells at one data point/sec), together with normal search. As a separate requirement, must be able to handle a transient load of 40 short-range, high-data-rate targets (1-ms dwells at 8 data points/sec).

Search load: 200 nominal dwells/sec, and 20 extended range dwells/sec.

Track processing times:

a. Command sequence of *track management, schedule management and radar command:* 0.9 ms/dwell, including executive overhead.

b. Return processing sequence of *return processing* and *track processing:* 2.1 ms/dwell, including executive overhead.

TABLE 1 — SUMMARY COMPARISON OF HIGH LOAD CASES

CASE	SYSTEM REQUIREMENTS	TIME REQUIREMENTS (SEC/SEC) Tracking Radar Execution	Tracking Data Processing	Search Radar Execution	Search Data Processing	Combined Operations Radar Execution	Combined Operations Data Processing
Nominal.	Full search capability: 220 dwells/sec 120 long range tracks @ 1 data points/sec per target	0.480	0.360	0.480	0.396	0.960	0.756
Peak Load.	Maximum search rate possible at full track load 40 short range tracks @ 8 data points/sec per target	0.320	0.960	0.048	0.040	0.386	1.000

Search/detect processing times:

a. Command sequence of *search management, schedule management* and *radar command:* 0.9 ms/dwell, including executive overhead.

b. Return processing sequence of *return processing* and *detection processing:* 0.9 ms/dwell, including executive overhead.

Two types of cases are involved. The nominal case involves a moderate amount of long range tracking. Long range dwells are demanding of radar execution time. However, they also involve a low data rate so that they are not particularly demanding of data processing time. The peak load case, on the other hand, involves high data rate tracking on short range targets. This is not as demanding of radar execution time as the previous case because of reduced target range. However, the data rate is high and the data processing capability becomes the limiting item. A discussion and comparison of these cases is given below. The results are summarized in Table 1.

12.3.2 Peak-Load Case. Let us consider the worst case of the high-data-rate, short-range targets first. Since there are 40 targets and 8 data points/sec per target, there are 320 track dwells/second required. At 1 ms/dwell, this means that 32% of the available radar time must be occupied with tracking. The remaining 68% is available for search and other operations.

The complete processing of a track dwell takes 3.0 ms - 0.9 ms for the command processing and 2.1 ms for the return processing. With 40 targets being tracked at 8 dwells/second, track processing takes 320 x 3 = 960 ms/sec. Thus, 40 targets require 96% of the computer processing time.

The remaining 4% of the processing time, 40 ms/sec, can be devoted to search. If the processor is tied up doing search command and return processing, it spends (0.9 + 0.9) = 1.8 ms/dwell. Thus, 40/1.8 = 22 dwells/sec can be processed. Since the normal search involves (200 + 20) = 220 search dwells/sec, search can be carried out at about 10% of the nominal rate. The radar time required to do this is (20 x 2) + (2 x 4) = 48 ms/sec, or 5% of the time.

The peak load case is processor limited. It can handle the full-track load and search at 10% of the nominal rate. This requires 100% of the processor's time. It requires 37% of the available radar execution time; 32% for track and 5% for search.

12.3.3 Nominal Search. In the nominal case, the radar execution time required for tracking is (120 x 4) = 480 ms/sec. The time required for search is (200 x 2) + (20 x 4) = 480 ms/sec. Thus, 96% of the radar execution time is required.

To process the 120 track points/sec requires (120 x 3.0) = 360 ms/sec. To process the 220 search dwells takes (220 x 1.8) = 396 ms/sec. Thus, 76% of the available processing time is required. Thus, the extended-range dwells called for in the nominal case require radar executions for 96% of the available time, whereas the processing load is at 76% of capacity.

12.3.4 Comparison of Cases. Consequently, the requirement to handle a large number of short-range track points drives the processor utilization to 100%, while requiring that search operations be reduced. Only 37% of the available radar execution time was needed. In general, this type of situation does not stress the radar on a time or a power basis — since we are dealing with a multiplicity of short-range tracks. The companion case, involving long-range targets, forces radar utilization to occur at the 96% level, while the processor operates at the 76% point. This case is demanding of radar time and power, while not stressing the processor unduly. Thus, the mechanism of saturation is sensitive to the target-mix problem imposed on the system.

[1] The author's original work on this subject, Weinstock, W., "Computer Control of a Multifunction Radar", *RCA Reprint RE-18-1-5, RCA Engineer,* Vol, 18, No. 1 (June/July 1972) pp 37-47 seems to be the first paper to address the subject of how a multifunction radar software system is designed.

[2] Baugh, R.A., *Computer Control of Modern Radars,* RCA, 1973. This privately published monograph incorporates reference 1 and extends it to cover a number of software design topics.

[3] Reed, J.E., "The AN/FPS-85 Radar System", *Proc. IEEE,* – Vol. 57, No. 3 (March 1969) pp 324-335. This article is an excellent overview from a system, as well as a subsystem viewpoint. The description of functional operation is an overview of the software system operation. This article appears to be unique in the open literature, because of its handling of the software aspect of the system problem

[4] Johnson, C.M., "Ballistic-Missile Defense Radars", *IEEE Spectrum,* Vol. 7, No. 3 (March 1970) pp. 32-41. The systems description is much more superficial than in Ref. 3. However, the data processing requirements are stated. Furthermore, a direct comparison is made between the PAR radar in this sytem and the AN/FPS-85 in Ref. 3.

[5] See, for example, Robinson, A.S., "Synthesis of Phased Array Radar Systems", *RCA Reprints PE-448, PE-452 and PE-439; RCA Engineer,* Vol. 15, No. 3 (Oct/Nov 1969), p 6-11.

[6] Sobel, H.S., *Introduction to Digital Computer Design*, (Addison-Wesley; Reading, Mass. 1970). Chapter 9 on "A Multimode Radar Application" addresses the general point of using special purpose digital processors in the implementation of a multifunction radar.

[7] Bruner, W.G., Case, W.N., and Leedom, R.C., "Digital Control in Modern Radar", *IEEE EASCON '71 Convention Record,* pp 200-207. This reference is also oriented to the need for digital processing in modern radar applications. The control of multi-mode airborne radar is emphasized.

[8] Cokas, G. and Parsons, J.H., "Data Processing Considerations in Advanced Phased Array Radar Systems", *AGARD Conference Proceedings No. 66,* Advanced Radar Systems, Istanbul, Turkey, 25-29 May 1970. This paper extends the viewpoint of the previous reference. It also projects the increase in computer requirements with time.

[9] Reed, op. cit. p 325.

[10] Scheff, B.H., and Hammel, D.G., "Real Time Computer Control of Phased Array Radars," *Supplement to IEEE Transactions on Aerospace and Electronic Systems,* Vol. AES-3, No. 6 (Nov 1967) pp 198-206. This paper is devoted to examining a solution involving executive control of computer and radar resource allocations, using dynamic scheduling.

[11] An excellent overview of third generation computers, including a comparison with the first two generations, is given by Denning, P.J., "Third Generation Computer Systems", *Computing Surveys,* Vol. 3, No. 4 (Dec. 1971) pp 175-216. This is a systems overview which divorces itself from the specifics of particular computers.

[12] A comparable hardware-oriented overview is given by Flores, I., *Computer Organization* (Prentice-Hall; Englewood Cliffs, N.J.; 1969). The first three chapters of this book summarize the major characteristics of the third generation machines. The rest of the book is devoted to specific computers.

[13] A companion to the two previous references is Wegner, P., *Programming Languages, Information Structures and Machine Organization* (McGraw-Hill; New York; 1968). Specifically, Chapter 1 (pp 1-106) goes into the hardware/software techniques that are associated with the multiprogrammed systems.

[14] Mehling, T., "Computer Program Architectural Design for Weapon System Radar Control", *RCA Reprint 20-1-14, RCA Engineer,* Vol. 20, No. 1 (June/July 1974), pp 58-64.

[15] Heffes, H. and Horing, S., "Optimal Allocation of Tracking Pulses for an Array Radar", *IEEE Trans. On Automatic Control,* Vol. AC-15, No. 1 (Feb. 1970) pp 81-87.

[16] Wishner, R.P. et al; "Radar Pulse Scheduling", *Proceeding Fourth Hawaii International Conference on System Sciences* (S. Levi, Editor), University of Hawaii, Jan 12, 13, 14 1971, pp 386-388.

[17] Salinger, S.N. and Wangsness, D., "Target Handling Capacity of a Phased-Array Tracking Radar", *IEEE Trans. On Aerospace and Electronic Systems,* Vol. AES-8, No. 1 (Jan 1972) pp 43-50. This article explicitly addresses the problem of saturation with a mix of high data rate tracks (acquisition or transition-to-track) and regular tracks.

Glossary of Greek Terms

α	differential phase shift; Lessons 2 and 3, see Fig. 2-1; designated as γ in Fig. 1-3.
α	α of α, β space in Lesson 7, see Fig. 7-6; α, β space designated as u,v space in Lessons 2, 3, 4, 5, 9, and 10.
α	direction cosine angle relative to x-axis in Lesson 8, see Fig. 8-45, and equals complement of this angle in Lessons 2 and 3, see Fig. 3-36a.
α	0.28 dB, range equation constants and dimensional conversion factors; Lesson 7, Sect. 7.2.9; K′ of Eq. (1.2) related to α, specifically: $K' = (4\pi)^2 \alpha$ if same units used in Eq. (1-2) as in range equation of Sect. 7.2.9.
α	position smoothing constant; Lesson 7, Sect. 7.3.2.
α	precomparator phase error; Lesson 7, Fig. 7-27.
α_m	Incident wave amplitude used with Scatter Matrix [S]; Lesson 6, Fig. 6-4.
α_x, α_y	differential phase shift between row and column elements of planar array; Fig. 2-34.
α_1	α at elevation beam position #1; Lesson 7; Sect. 7.2.5.
β	peak radiator element phase error for case of periodic phase error; Fig. 3-26.
β	β of α, β space in Lesson 7, see Fig. 7-6; α, β space designated as u, v space in Lessons 2, 3, 4, 5, 9, and 10.
β	direction cosine relative to y-axis in Lesson 8 (see Fig. 8-45) and equals complement of this angle in Lessons 2 and 3 (see Fig. 3-36a).
β	3 dB beamwidth in degrees in Lesson 1, see Eq. (1-3); θ_3 used instead in Lessons 2 to 6.
β	velocity smoothing constant; Lesson 7, Sect. 7.3.2.
β	Barlow constant for clutter spread; Lesson 7, Sect. 7.4.
β	post comparator phase error; Lesson 7, Fig. 7-27.
β	Parameter describing A.C. Schell's antenna; Lesson 9.
β_m	Scattered wave amplitude used with Scatter Matrix [S]; Lesson 6, Fig. 6-4.
β_1	β at elevation beam position #2; Lesson 7, Sect. 7.2.5.
Γ	active array reflection coefficient for large or infinite array, Lessons 4 and 5.
γ	differential phase shift in Lesson 1, Fig. 1-2; α used instead in Lessons 2 and 3.
γ	post post comparator (limiter) phase error; Fig. 7-27.

Δ	Difference or differential when another symbol; Lesson 8.
Δ	Difference pattern; Lesson 6.
Δ	difference pattern received signal; Sect. 3.1.2.e
Δ_{ATD}	decrease in atmospheric loss between ranges R_I and R_N for search mode; Eq. (1-6) and Sect. 7.5.2.
Δf	bandwidth required for frequency scanning in Hertz; Eq. (1-4).
ΔG_1	precomparator amplitude error; Lesson 7, Fig. 7-27.
Δ_2	post comparator and amplitude error; Fig. 7-27.
Δ_1	fractional amplitude error for the ith radiating element thus if $I_i = 1$ nominally, then with amplitude error $I_i = 1 + \Delta_i$; Lesson 3.
Δ_x (x,y), Δ_y (x,y)	x and y monopulse difference pattern amplitude weighting function at coordinates x, y; Fig. 3-34.
$\Delta\phi$	interelement phase shift in Eq. (1-5); called differential phase shift elsewhere and designated as γ in Fig. 1-2 and as α in Lessons 2 and 3.
Δ^2	Parameter used in dielectric filter study; Lesson 10.
$\overline{\Delta^2}$	mean square value of fractional radiator amplitude error, *i.e.,* average of Δ_1^2; Lesson 3, see Fig. 3-2.
δ	stagger time of MTI canceller; Lesson 7, Sect. 7.4.
δ	Margin with respect to unit circle used in establishing coverage limits unit circle taken at $(1+\delta)$; Lesson 8, Fig. 8-3.
δ	Phase shift between phase-interpolated signals—(not to be confused with regular progressive phases); Lessons 9 and 10.
δ_1	phase error or ith radiator; Lesson 3, see Fig. 3-1.
$\overline{\delta}$	average of δ_1^2; Lesson 3, see Fig. 3-1.
$\overline{\delta^2_{mech}}$	mean square radiator element phase error due to inexact location of elements; Fig. 3-13.
$\overline{\delta\delta^2}$	mean square value of phase shifter randomized quantization error; Lesson 3, see Fig. 3-11.
ϵ	error in predicted position; Lesson 7, Sect. 7.3.2.
ϵ_D	dynamic lag error; Lesson 7, Sect. 7.3.2.
$\epsilon_{R_{MTI}}$	energy reduction at maximum MTI range; Lesson 7, Sect. 7.5.2.
ϵ_{R_N}	search mode energy reduction at N^{th} elevation beam position; Eq. (1-6) and Sect. 7.5.2.
ϵ_{RT}	energy reduction at R_T; Lesson 7, Sect. 7.5.2.
ϵ_{R8}	energy reduction at 8th elevation beam position; Lesson 7, Sect. 7.5.2.
ϵ_s	total energy required for search; Lesson 7, Sect. 7.5.4.
ϵ'_{s1}	total energy requied for search without scan angle management; Lesson 7, Sect. 7.5.4.
ζ, T_m, s, Δ^2	Parameters used in dielectric filter study; Lesson 10.
η	overall antenna efficiency; Fig. 2-19.

η	Phase shift for array pointing; not referenced as a difference; Lessons 9 and 10.
η_t	array taper efficiency; Lessons 2 and 3, see Fig. 2-18.
θ	Direction angle of array beam relative to array normal; Lessons 1, 6, and 8 and Sect. 7.5.2.
θ	one of coordinates of spherical antenna coordinate system; Figs. 2-1 and 2-31.
θ	3 dB beamwidth in Sects. 7.3.2 and 7.4; θ_3 used instead in Lesson 2 and 3; β used in Lesson 1.
θ, ϕ	antenna spherical coordinates.
θ_m	maximum scan angle in Lessons 9 and 10 and Lesson 1, Eq. (1-3) where it is in degrees; $\theta_{o,max}$ used instead in Fig. 2-6.
θ_o	main lobe off broadside angle; Lessons 2 and 3, see Fig. 2-2.
θ_o	3 dB beamwidth of Gaussian pattern in Lesson 7, Fig. 7-5; θ_3 used in Lessons 2 and 3.
$\theta_{o, MAX}$	maximum scan angle θ_o one can have without forming grating lobes; Fig. 2-6.
θ_{ox}, θ_{oy}	complements of θ'_{ox} and θ'_{oy}, respectively; see Fig. 2-32; also designated as α, β in Lessons 2 and 3, see Fig. 3-36; as A, B, in Lesson 7, Fig. 7-6; as 90° – α and 90° – β in Lesson 8.
$\theta'_{ox}, \theta'_{oy}$	the cosines of these angles equal direction cosines of the main lobe axis in Fig. 2-32; designated as α, β in Lesson 8; as A′, B′ in Lesson 7; as 90° – α, 90° – β in Lessons 2 and 3.
θ_1	location of grating lobe; Lessons 2 and 3, see Fig. 2-3.
θ_3	3 dB main lobe width; Lessons 2 to 4, see Fig. 2-3 and 2-33; β used in Lesson 1.
θ_{3x}, θ_{3y}	x-axis and y-axis linear array broadside 3 dB beamwidths of separable planar array; Fig. 2-33; for non-separable array it represents 3 dB broadside beamwidths in φ = 0° and φ = 90° planes, respectively; Fig. 2-35.
θ_3^1, θ_3^2	3 dB beamwidths in planes 1 and 2; Lessons 9 and 10.
θ_{30}	θ_3 at broadside, *i.e.,* for θ_o = 0°; Lessons 2 to 6, see Fig. 2-9.
θ_{300}	θ_{30} for ideal-Taylor weighting; Lesson 2, See Fig. 2-40.
λ	free space signal wavelength; all Lessons; in cm for radar equation of Sect. 7.2.9.
λ_g	waveguide wavelength corresponding to λ; Fig. 1-5.
θg_o	broadside waveguide wavelength; Fig. 1-5.
ξ	scan angle from array normal at which grating lobe first appears; Lesson 7, Sect. 7.2.9.
ξ	polar scan angle from array normal; Lesson 7, Sect. 7.5.2.
ξ_{max}	maximum scan angle from array normal in Lesson 7, Sects. 7.2.5 and 7.5.2; designated as θ_m in Lesson 1 and $\theta_{o, max}$ in Lesson 2.
ξ_o	scan angle from array normal where round trip antenna gain is 3 dB better than at ξ_{max}; Lesson 7, Sect. 7.5.2.

$\prod_{i=1}^{M}$	product of M terms; Lesson 7, Fig. 7-14.
ρ	reflectance of a microwave device; Lesson 6, Fig. 6-7.
ρ	surface reflectivity coefficient; Lesson 7, Sect. 7.3.2.
ρ, R, β	parameters describing A, C, Schell's antenna; Lesson 9.
Σ	receive sum or monopulse reference pattern; Lesson 8.
Σ	sum pattern received signal; Sect. 3.1.2.e.
σ	radar cross section of target; Eq. (1-2), in m^2 in radar equation of Sect. 7.2.9.
σ	main lobe broadening factor for Tapered-Taylor weighting; Lessons 2 and 3, see Fig. 2-40.
σ	standard deviation; Lesson 7, Fig. 7-1 and Sect. 7.5.6.
σ_{AC}	radar cross section of aircraft; Lesson 7, Fig. 7-16.
σ_c	reflected clutter area; Lesson 7, Fig. 7-24.
σ_{MSL}	radar cross section of missile; Lesson 7, Fig. 7-16.
σ_m	multipath error; Lesson 7, Sect. 7.3.2.
σ_o	surface reflectivity of clutter; Lecture 7, Fig. 7-24.
σ_p	propagation and calibration error; Lesson 7, Sect. 7.3.2.
σ_T	target radar cross section in Lesson 7, Fig. 7-24; σ used in Sect. 7.2.9 and Lesson 1.
σ_T	thermal error; Lesson 7, Sect. 7.3.2.
$\sigma_{T_{msi}}$	missile targer radar cross section; Lesson 7, Sect. 7.4.
σ_u^2, σ_v^2	mean square value of errors in u and v due to randomized phase quantization errors; Lesson 3, see Fig. 3-34.
σ_ζ^2	variance of difference pattern phase error of a radiating element (assumed same for all elements); Lesson 3, see Fig. 3-34.
τ	equivalent pulse width; Lesson 7, Fig. 7-3.
τ	RF phase error of monopulse feedhorn in Lesson 7, Fig. 2-27.
τ	Time delay; Sect. 6.10.2.
τ	uncompressed pulse width; Lesson 7, Sects. 7.3.1 and 7.5.6.
τ_c	equivalent pulse length; Lesson 7, Fig. 7-24 and Sect. 7.3.2.
$\tau_{uncompressed}$	uncompressed pulse width; Lesson 7, Sect. 7.2.9.
φ	one of array spherical coordinates; all Lessons, see Fig. 2-31.
φ	Electrical phase angle; Lesson 6, Fig. 6-6.
φ	cylindrical coordinate angle; Lesson 6.
φ	angle of sine squared MTI filter response; Lesson 7, Sect. 7.4.
φ_o	ϕ for peak of mainlobe; Fig. 2-34.
φ_s	phase shift for sth radiating element; Fig. 1-2.
$\ddot{\phi}_1$	normal angular acceleration; Lesson 7, Sect. 7.3.2.

$\ddot{\phi}_2$	tangential angular acceleration; Lesson 7, Sect. 7.3.2.
ϕ_3	3 dB beamwidth of main lobe in ϕ direction; Fig. 2-33.
ω	$2\pi f$; Lessons 4 and 5; ω_o used in Lesson 7.
ϕ_o	2π times transmission frequency; Lesson 7, Sect. 7.4.

Index